OPERATION OF NUTRIENT REMOVAL FACILITIES

WEF Manual of Practice No. 37

2013

Water Environment Federation
601 Wythe Street
Alexandria, VA 22314-1994 USA
http://www.wef.org

Operation of Nutrient Removal Facilities

ISBN 978-1-57278-276-1

About WEF

Founded in 1928, the Water Environment Federation (WEF) is a not-for-profit technical and educational organization of 36,000 individual members and 75 affiliated Member Associations representing water quality professionals around the world. WEF members, Member Associations and staff proudly work to achieve our mission to provide bold leadership, champion innovation, connect water professionals, and leverage knowledge to support clean and safe water worldwide. To learn more, visit www.wef.org.

For information on membership, publications, and conferences, contact

Water Environment Federation
601 Wythe Street
Alexandria, VA 22314-1994 USA
(703) 684-2400
http://www.wef.org

Prepared by **Operation of Nutrient Removal Facilities** Task Force of the **Water Environment Federation**

Nancy Afonso
Archis R. Ambulkar, M.S.
Vincent Apa, P.E., BCEE
Sara Arabi, Ph.D., P.Eng.
Richard G. Atoulikian, PMP, BCEE, P.E.
Vinod Barot, Ph.D.
Somnath Basu, Ph.D., P.E., BCEE
Jason Beck, P.E.
Katherine Bell, Ph.D., P.E., BCEE
Mario Benisch, P.E.
Joshua P. Boltz, Ph.D., P.E., BCEE
Lucas Botero, P.E., BCEE
Bob Bower, B.S., WWTPO 4
Jeanette Brown, P.E., BCEE, D. WRE, F.WEF
Marie S. Burbano, Ph.D., P.E., BCEE
James Clifton
John T. Cookson, Jr., Ph.D.
Rhodes R. Copithorn, P.E., BCEE
John B. Copp, Ph.D.
Catherine Crabb
Allegra K. da Silva, Ph.D.
Bob Dabkowski
Sarah V. Dailey, P.E.
Tania Datta, Ph.D.
Michael John Dempsey, Ph.D., C.Biol.
Timur Deniz, Ph.D., P.E., BCEE
Leon S. Downing, Ph.D., P.E.
Ufuk Erdal, Ph.D., P.E.
Richard Finger
M. Kim Fries, M.A.Sc., P.Eng.
Michael H. Gerardi
Mikel E. Goldblatt, P.E.
Gary M. Grey
Georgine Grissop, P.E., BCEE
John Hahn
Susan Hansler, M.A.Sc., P.Eng.
Ladan Holakoo, Ph.D., P.E., P.Eng.
Sarah Hubbell
Joseph A. Husband, P.E, BCEE
Maria Inneo
Samuel S. Jeyanayagam, Ph.D., P.E., BCEE
Gary R. Johnson, P.E., BCEE
Ishin Kaya, P.Eng.
Michael P. Keleman, MSEV, REHS
Frank Kulick
Robert Lagrange, Ph.D.
Cory Lancaster
Dennis Lindeke
Yanjin Liu, Ph.D., P.E.
Jorj Long
Huijie Lu
Ting Lu, Ph.D.
Nehreen Majed, Ph.D.
Randall B. Marx, Ph.D., P.E.
Neil Massart, P.E.
Samir Mathur, P.E., BCEE
Michael McGhee, P.E.
Jun Meng, M.S., E.I.T
Indra N. Mitra, Ph.D., P.E., MBA, BCEE
George Nakhla, Ph.D., P.E., P.Eng.
Robert C. O'Day, CHMM, QEP, ASP
Annalisa Onnis-Hayden, Ph.D.
Jurek Patoczka, Ph.D., P.E.
Marie-Laure Pellegrin, Ph.D., P.E.
Peter Petersen
Heather M. Phillips, P.E., BCEE
Scott Phipps, P.E.
Alvin Pilobello
Barry Rabinowitz, Ph.D., P.Eng., BCEE
Steven Reusser, P.E.

Contents

List of Figures

List of Tables

Preface

An increasing number of water resource recovery facilities are being required to remove nutrients, and many facilities that already accomplish nutrient removal are facing increasingly stringent performance requirements. At the same time, operators are under pressure to meet these treatment requirements while simultaneously minimizing operations and maintenance costs. Successfully operating a stable, reliable, and efficient facility to remove nitrogen and/or phosphorus requires a high level of operator attention, skill, and knowledge. This manual provides up-to-date guidance regarding the successful, efficient operation of nutrient removal facilities.

The manual is written primarily for plant managers and operators but will also be useful for design engineers and regulators. Early chapters of the manual provide technical information regarding fundamental biological and chemical processes that are in use at nutrient removal facilities. Later chapters illustrate configurations with which these processes are used at water resource recovery facilities and the ways that operators may use, monitor, and control these processes to meet their facilities'y treatment goals.

Additional review was provided by Armen S. Casparian, M.Sc., RPIH, NRCC-CHO, and Abedin Zainul.

Authors' and reviewers' efforts were supported by the following organizations:

AECOM, Kitchener, Ontario, and Markham, Ontario
Advanced Bioprocess Development Ltd., Manchester, United Kingdom
American Water, Voorhees, New Jersey
Black & Veatch Corporation, Kansas City, Missouri, and Markham, Ontario
Brentwood Industries, Reading, Pennsylvania
Brinjac Engineering Inc., Harrisburg, Pennsylvania
Carollo Engineers, Salt Lake City, Utah
CDM Smith, Cambridge, Massachusetts
CH2M HILL, Inc., Burnaby, British Columbia; Calgary, Alberta; Chantilly, Virginia; Santa Ana, California; and Tampa, Florida
City of Olathe, Kansas
Conestoga-Rovers & Associates (CRA), Waterloo, Ontario
Donohue and Associates, Sheboygan, Wisconsin
Ekologix Earth Friendly Solutions Inc., Waterloo, Ontario
Entex Technologies Inc., Chapel Hill, North Carolina
Environmental Operating Solutions, Inc., Windsor, Connecticut

Florence & Hutcheson, Inc./ICA, Nashville, Tennessee
GHD, Bowie, Maryland
Hach Company, Loveland, Colorado
Hatch Mott MacDonald, Millburn, New Jersey
Hazen and Sawyer, Columbus, Ohio, and New York, New York
HDR Engineering, Inc., Cleveland, Ohio; Edmonds, Washington; and Omaha, Nebraska
HDR | HydroQual, Mahwah, New Jersey
HRM Environmental, Cedar Park, Texas
InSinkErator, Racine, Wisconsin
Lagrange Consulting, Snellville, Georgia
Madison Metropolitan Sewerage District, Madison, Wisconsin
Malcolm Pirnie, Inc., White Plains, New York
Manchester Metropolitan University, Manchester, United Kingdom
Manhattan College, Bronx, New York
Metro Council Environmental Services, St. Paul, Minnesota
Metropolitan Sewer District of Greater Cincinnati, Cincinnati, Ohio
MWH Americas, Inc., Denver, Colorado
Northeastern University, Boston, Massachusetts
Praxair, Inc., Burr Ridge, Illinois
PG Environmental, LLC, Herndon, Virginia
Primodal Inc., Hamilton, Ontario
Shell Technology Center, Houston, Texas
Simsbury Water Pollution Control, Simsbury, Connecticut
Stantec Consulting Services Inc., Phoenix, Arizona
Tetra Tech, Inc., Denver, Colorado, and Greenville, South Carolina
The Water Planet Company, New London, Connecticut
Thomas E. Wilson Environmental Engineers LLC, Barrington, Illinois
Union County Public Works–Union County, Monroe, North Carolina
University of Western Ontario, London, Ontario, Canada
URS Corporation, Columbus, Ohio
Veolia Water North America, Indianapolis, Indiana
Water Authority, Grand Cayman, Cayman Islands
Water Pollution Biology, Williamsport, Pennsylvania
Water/Wastewater Treatment Greater Philadelphia Area, Pennsylvania
Wentworth Institute of Technology, Boston, Massachusetts
World Water Works, Inc., Oklahoma City, Oklahoma
YSI Incorporated, Yellow Springs, Ohio

Chapter 1

Introduction

Gary J. ReVoir II, P.E.

1.0 INTRODUCTION

A *nutrient* is defined as a substance used in an organism's metabolism that must be taken in from its environment. Nutrients are essential to all forms of life, including those found within any natural waterbody. Natural levels of nutrients in the environment help in the growth of plants and animals. Aquatic life as a whole is supported and maintained through a healthy balance of the right amount and form of nutrients.

However, elevated levels of nutrients, particularly nitrogen and phosphorus, above critical levels specific to each waterbody can result in excess growth of aquatic plant life, especially algae. The largest contributors of nitrogen and phosphorus to freshwater and marine ecosystems are local runoff (nonpoint sources), discharges from a local industrial facility, and municipal water resource recovery facilities (WRRFs) (point sources). A point-source pollutant can be traced to the specific location where the contribution enters the waterbody, whereas non-point-source pollutants enter a body of water from many diffuse sources.

There are numerous WRRFs worldwide that are required by local regulations to reduce nitrogen and phosphorus concentrations within their effluent. Effective nitrogen removal from a wastewater stream relies heavily on biological processes, while phosphorus removal may require a biological process, chemical precipitation, or a combination of both. Successful operation of a biological nutrient removal (BNR) facility requires both operator knowledge and involvement in the daily operations of the facility. To successfully achieve the required level of nutrient reduction within the treatment process, the operator must perform additional testing, sampling, analysis, decision making, and monitoring beyond that required for a standard secondary process, which is primarily focused on biochemical oxygen demand and suspended solids removal. In addition, successful operation of a BNR facility requires a thorough understanding of process control and troubleshooting.

In 1972, the U.S. Environmental Protection Agency (U.S. EPA) enacted the Clean Water Act (CWA), creating a federal program designed to achieve the goals of protecting and restoring waters of the United States (U.S. EPA, 1972). Since that time, many additional regulations and technological advances have been made to further protect the nation's natural waterbodies. *The Water Quality Act of 1987* (U.S. EPA, 1987) authorized individual states to establish their own water quality standards. A provision of CWA mandated that each state prepare a water quality inventory that lists the state's impaired waters. The total maximum daily load (TMDL) process was established based on a combination of these rules to specifically meet the overall goals on a local level. A TMDL is the maximum daily amount of any pollutant that can enter a specific waterbody before the quality of the water is deemed unfit for its designated uses. Thus, the TMDL of a pollutant is the threshold or upper limit loading for each specific body of water. Total maximum daily loads must be established for both point-source and non-point-source pollutants.

As part of the TMDL program, each state is responsible for allocating loads among point and nonpoint sources identified under each TMDL. The U.S. EPA encourages states to consider a range of options that are both technically feasible and demonstrate consistency and that are based on cost effectiveness and equity. Final allocation determinations are policy decisions and should reflect public perceptions about acceptable tradeoffs between these measures.

2.0 OBJECTIVE

With the global objective of reducing point-source contributions of nutrient loadings to local waterbodies, states have placed regulatory limits on numerous WRRFs specifically related to nitrogen and phosphorus. The purpose of this manual is to provide facility managers and operators with an understanding of the theory and typical design requirements for processes currently used for nitrogen and phosphorus removal to optimize their facilities' ability to consistently meet the state's and U.S. EPA's goals, and to provide general guidance on process control and troubleshooting methodologies. This manual benefits design engineers and other technical professionals by broadening their understanding of BNR processes and enhancing their ability to relate to the facility's daily operations following construction and implementation of their design.

3.0 FORMAT

The chapters of this manual are organized such that the reader develops an understanding of the theory before reading about process control parameters and requirements and troubleshooting methodologies. Chapter 2 explains pertinent characteristics of raw wastewater and its overall effect on nutrient removal processes within a WRRF. In particular, this chapter provides the distinction between typical influent characteristics of a raw wastewater flow stream and typical effluent permit limits often required of a BNR facility. The differential between influent and effluent concentrations sets the stage for the remainder of the manual to describe successful configurations of both fixed-film and suspended growth processes for removing nitrogen and phosphorus.

The theory and basics of nitrification processes are introduced in Chapter 3. Nitrification is the first step in the two-step biological process by which ammonia-nitrogen is converted to nitrite-nitrogen and, subsequently, nitrate-nitrogen using two specific groups of autotrophic bacteria. The text in this chapter provides a stepwise process for both operators and design engineers to best optimize nitrification. A brief discussion on emerging processes that bypass the conversion of nitrite to nitrate and convert nitrite directly to nitrogen gas (denitrification) is also included.

Although use of biofilm reactors in municipal wastewater treatment has been reduced significantly over the past few decades, fixed-film technologies are regaining popularity. Chapter 4 presents a brief history of this classic treatment method and also presents theory and examples of the nitrifying process in biofilm reactors. Biofilm processes have long been used in both rotating biological contactors and trickling filters. Use of biofilm reactors has been reduced greatly over the years, resulting in a large increase in the use of suspended growth in the activated sludge process. Nutrient removal requirements are causing fixed-film processes to be reconsidered in new configurations, particularly in combination with the suspended growth process to enhance current facilities.

The overall process of nitrogen removal involves both nitrification and denitrification. Chapter 5 begins the discussion on denitrification and presents stoichiometry, optimal conditions, and their overall application in variations to the standard activated sludge process.

The discussion in Chapter 6 begins to integrate concepts of previous chapters and describes design details and benefits of combined nitrification and denitrification treatment systems. Such systems are used to achieve carbonaceous biochemical oxygen demand removal, nitrification, and denitrification in a single sludge configuration. Building on previous discussions, Chapter 6 also provides an in-depth discussion of design criteria and the expected performance of various process configurations for nitrogen removal.

Chapter 7 includes an overview of enhanced biological phosphorus removal (EBPR). This chapter focuses on process fundamentals, the microbiology of the process, and the biochemistry of biological phosphorus removal. Additionally, the chapter provides facility operators with key operational considerations and effects on sludge handling with biological phosphorus removal processes in addition to a summary of ongoing research related to EBPR.

The discussion in Chapter 8 continues with phosphorus removal and presents concepts and processes related to the chemical precipitation of phosphorus in a WRRF. Removal of phosphorus using chemical precipitation is common and has been in practice for many decades, either as a primary removal process in conjunction with a biological removal process or as a supplemental, polishing, or backup process.

Chapter 9 provides a narrative that builds on the fundamentals of EBPR presented in Chapter 7 and describes the components of a functional EBPR system. Alternate configurations of EBPR process and operating strategies are presented.

Chapter 10 describes the interaction between nitrogen and phosphorus removal in combined systems. Removal of phosphorus in a combined nitrogen and phosphorus removal facility can be achieved both chemically and biologically; however, the biological alternative has a number of significant advantages. This chapter focuses on combined nitrogen and phosphorus removal systems and their operational conditions, process control strategies, and troubleshooting.

With tightening budgets and increasingly stringent regulations, operators are under constant pressure to optimize the performance of nutrient removal processes in terms of effluent quality and/or reducing ongoing operations costs associated with power and chemical use. As such, Chapter 11 provides a summary of cost-effective operational considerations for various nutrient removal processes.

Chapter 12 addresses concerns and management strategies of recycling nutrients from solids management processes. These recycle streams can represent a significant nutrient load on the nutrient removal process and must be managed to ensure compliance with effluent quality goals and to improve treatment efficiency.

Chapter 13 provides a summary of process control using oxidation–reduction potential (ORP) and dissolved oxygen levels as monitoring parameters. Operating a BNR process is more complex than conventional processes and requires a high level of process control, operator involvement, and knowledge. Using online process control parameters such as ORP and dissolved oxygen is significantly beneficial to the operation of BNR systems. Experience has also shown that consistent use of online monitoring of specific parameters better optimizes the BNR process, reduces costs, and improves reliability.

Chapter 14 provides a summary of process control, instrumentation, and automation related to operation of BNR WRRFs. Nutrient removal facilities can be complex, requiring significant monitoring and control of processes to meet effluent requirements. This chapter provides a description of the details of BNR processes as related to the various parameters for operators to monitor and control within their WRRFs.

Chapter 15 describes in detail laboratory analysis performed as part of the operation and monitoring of BNR processes. This chapter outlines sample collection methods related to frequency, sample type, and sample locations within the process and their use for analysis and process optimization.

Chapter 16 presents several case studies covering full-scale facilities that are designed and operated for BNR to meet effluent total nitrogen limits. Each facility covered in this chapter has a unique process modification to the activated sludge process specifically designed to reduce nitrogen levels in the effluent. The chapter includes a case study of four-stage Bardenpho, modified Ludzack–Ettinger, a sequencing batch reactor, and integrated fixed-film activated sludge, among others.

Similar to Chapter 16, Chapter 17 includes several case studies, but covers only full-scale facilities that are specifically designed and operated to meet effluent total phosphorus limits between 0.05 and 1.0 mg/L for optimal EBPR. All the facilities presented in this chapter have been in operation for a minimum of 2 years. The case studies provide a unique summary of the challenges encountered by facility operators during the startup and commissioning phases of the project.

Finally, a BNR troubleshooting guide is provided as an appendix to the manual. Operators can use this guide to quickly identify potential problem sources that may be encountered during the operation of a BNR facility and possible solutions.

4.0 REFERENCES

U.S. Environmental Protection Agency (1972) *Federal Water Pollution Control Amendments of 1972;* U.S. Environmental Protection Agency: Washington, D.C.

U.S. Environmental Protection Agency (1987) *Water Quality Act of 1987;* U.S. Environmental Protection Agency: Washington, D.C.

Chapter 2

Wastewater Constituents that Affect Nutrient Removal

Daniel L. Theobald; Archis R. Ambulkar, M.S.; and John Hahn

1.0 INTRODUCTION

Wastewater characteristics, influent loadings, and effluent limits dictate treatment needs and processes involving the removal of organics, solids, and nutrients in a water resource recovery facility (WRRF). This chapter focuses on wastewater characteristics that affect nitrogen and phosphorus removal.

1.1 Wastewater Characteristics and Nutrients

Wastewater characteristics and influent nutrient loadings to WRRFs depend on factors such as generation sources, collection system operation, and inflow/infiltration. These characteristics are typically measured as biochemical oxygen demand (BOD), total suspended solids (TSS), pH, and macronutrients (e.g., nitrogen and phosphorus). Table 2.1 summarizes these parameters and Chapter 15 discusses procedures for their measurement.

Excessive nitrogen and phosphorus discharges from WRRFs to receiving waters have environmental consequences, including eutrophication processes that result in excessive growth of plant and algae, ammonia toxicity, and nitrate contamination in groundwater. Ammonia toxicity causes fish mortality and health issues in aquatic reproduction. Nitrate contamination in groundwater can result in public health issues such as blood disorders affecting infants (e.g., methemoglobinemia, which is commonly referred to as "blue baby" syndrome).

1.2 Nutrient Sources for Wastewater

1.2.1 Nitrogen Sources

Ammonia (NH_3), ammonium ion (NH_4^+), nitrite (NO_2^-), nitrate (NO_3^-), and organic nitrogen are common forms of nitrogen found in wastewater. Decaying

TABLE 2.1 Typical raw wastewater characteristics.

		Concentration[b]		
Contaminants	**Units[2]**	**Low-strength**	**Medium-strength**	**High-strength**
Solids, total	mg/L	390	720	1230
Dissolved, total	mg/L	270	500	860
Fixed	mg/L	160	300	520
Volatile	mg/L	110	300	340
Suspended solids, total	mg/L	120	210	400
Fixed	mg/L	25	50	85
Volatile	mg/L	95	160	315
Settleable solids	mg/L	5	10	20
Biochemical oxygen demand, 5-day, 20 °C (BOD_5, 20 °C)	mg/L	110	190	350
Total organic carbon	mg/L	80	140	260
Chemical oxygen demand	mg/L	250	430	800
Nitrogen (total as N)	mg/L	20	40	70
Organic	mg/L	8	15	25
Free ammonia	mg/L	12	25	45
Nitrites	mg/L	0	0	0
Nitrates	mg/L	0	0	0
Phosphorus (total as P)	mg/L	4	7	12
Organic	mg/L	1	2	4
Inorganic	mg/L	3	5	10
Chlorides	mg/L	30	50	90
Sulfate	mg/L	20	30	50
Oil and grease	mg/L	50	90	100
Volatile organic compounds	mg/L	<100	100 to 400	>400
Total coliform	No./100 mL	10^6 to 10^8	10^7 to 10^9	10^7 to 10^{10}
Fecal coliform	No./100 mL	10^3 to 10^5	10^4 to 10^6	10^4 to 10^8
Cryptosporidum oocysts	No./100 mL	10^{-1} to 10^0	10^{-1} to 10^1	10^{-1} to 10^2
Giardia lamblia cysts	No./100 mL	10^{-1} to 10^1	10^{-1} to 10^2	10^{-1} to 10^3

[a]mg/L = g/m^3.

[b]Low-strength is based on an approximate wastewater flowrate of 750 L/cap·d (200 gpd/cap), medium-strength is based on an approximate wastewater flowrate of 460 L/cap·d (120 gpd/cap), and high-strength is based on an approximate wastewater flowrate of 240 L/cap·d (60 gpd/cap).

[c]Values should be increased by the amount of constituent present in the domestic water supply.

plants, animals, and their wastes; human wastes; and fertilizers are significant sources of nitrogen.

1.2.2 *Phosphorus Sources*

Fertilizers, manure, detergents, household cleaning products, and human and animal waste are typical sources of phosphorus. Common forms of phosphorus are orthophosphate, polyphosphates (condensed phosphates), and organic phosphates (phospholipids and nucleotides). Phosphorus can be present in wastewater in both soluble and particulate forms.

2.0 WASTEWATER CHARACTERISTICS THAT AFFECT NUTRIENT REMOVAL

Raw wastewater characteristics are typically characterized based on physical, chemical, and biological properties, as shown in Table 2.1.

2.1 Physical Characteristics

2.1.1 *Temperature*

Wastewater temperature affects the solubility of gases and reaction rate kinetics for biological and chemical processes. Microorganisms are sensitive to temperature and exhibit slower growth in colder temperatures and faster growth in warmer temperatures (optimum bioactivity between 25 and 35 °C). Microbial activity particularly affects dissolved oxygen concentrations and nitrification and denitrification processes (optimum bioactivity of 25 to 35 °C).

2.1.2 *Total Solids*

Wastewater solids consist of total suspended solids (TSS) and total dissolved solids (TDS), with each portion further categorized as volatile solids and fixed solids. Volatile solids are an indicator of biodegradability. The biodegradable fraction of volatile suspended solids (VSS) contributes to BOD, nitrogen, and phosphorus load, and is typically 70 to 80% of TSS for domestic wastewater. Higher VSS content is typical of domestic wastes; however, combined sewers typically lower this ratio because of the contribution of inert solids. A current issue is wastewater generated from hydro fracturing in the petroleum industries, which can contain high dissolved solids (up to 219 000 mg/L) and chloride (up to 169 000 mg/L) (Society of Petroleum Engineers, 2009).

2.1.3 *Dissolved Oxygen*

Sufficient dissolved oxygen is required for nitrification processes. Because nitrification involves the oxidation of ammonia and organic nitrogen to nitrite and nitrate, the presence of dissolved oxygen in wastewater processes is necessary.

Because the denitrification process requires anoxic conditions (i.e., no dissolved oxygen), lower dissolved oxygen is required for efficient conversion of nitrite and nitrate to nitrogen gas. Aeration equipment at a WRRF provides the oxygen needed for the nitrification process.

2.2 Chemical Characteristics

This section presents common chemical characteristics of wastewater (Table 2.1).

2.2.1 *pH*

pH is defined as the negative log of hydrogen ion concentration represented by a numerical scale from 1 to 14 standard units, with lower than pH 7 (neutral condition) indicating acidic conditions and higher than pH 7 indicating basic conditions. Biological processes are sensitive to the pH of wastewater, with the best results observed close to neutral values (6 to 9 pH range). Domestic wastewater typically falls within this range, whereas industrial wastewater pH may vary significantly.

2.2.2 *Alkalinity*

Alkalinity is the buffering capacity of a solution to resist change in pH. The ions present in wastewater that generally contribute most of the alkalinity are carbonate (CO_3^{2-}), bicarbonate (HCO_3^-), and hydroxyl (OH^-) ions. Total alkalinity is the sum of all these alkalinities. Alkalinity is required for nitrification processes and is partly gained back through the denitrification process. Chapter 11 presents an in-depth discussion of alkalinity.

2.2.3 *Ultimate Biochemical Oxygen Demand*

Wastewater BOD is a measure of oxygen consumed during biochemical oxidation of organic matter by microorganisms. Products of this oxidation process include carbon dioxide (CO_2) (nitrite-nitrogen and nitrate-nitrogen, water, new cells, and debris), which is used to create new cell mass and to maintain cells. Oxygen required to complete these reactions is referred to as *BOD*. Oxidation of organic matter and nitrogen (as ammonia) contributes to this ultimate BOD.

2.2.4 *Carbonaceous Oxygen Demand*

Carbonaceous oxygen demand (COD) represents oxygen consumed during a laboratory procedure that chemically oxidizes organic matter in wastewater. The COD value is always higher than the BOD value because it also takes into account reducing chemicals that are not available for biological uptake. The raw wastewater COD to BOD ratio indicates ranges between 2.0 and 2.2 to 1. A higher ratio indicates the presence of a significant amount of non-biodegradable or refractory materials, which is an indication of industrial wastewater or inhibitory substances. Chapter 11 provides details on COD fraction quantification.

2.2.5 *Total Nitrogen and Related Fractions*

The typical nitrogen concentration range in domestic wastewater includes organic and inorganic compounds (ammonia, nitrite, and nitrate). Nitrogen comprises 50% of untreated domestic wastewater in organic form, whereas industrial wastewater could have higher nitrite and nitrate concentrations. Chapter 11 contains detailed information on nitrogen, related fractions, and operational characteristics. Figure 2.1 shows transformations of nitrogen illustrated by nitrogen cycle.

2.2.6 *Total Phosphorus and Related Fractions*

As Table 2.1 indicates, domestic wastewater total phosphorus concentrations typically range between 4 and 7 mg/L. Sidestreams within an enhanced biological nutrient removal (EBNR) facility can return significant phosphorus concentrations back from sidestreams to the process (potentially >200 mg/L). The presence of industrial discharges (such as detergents, etc.) and the nature of

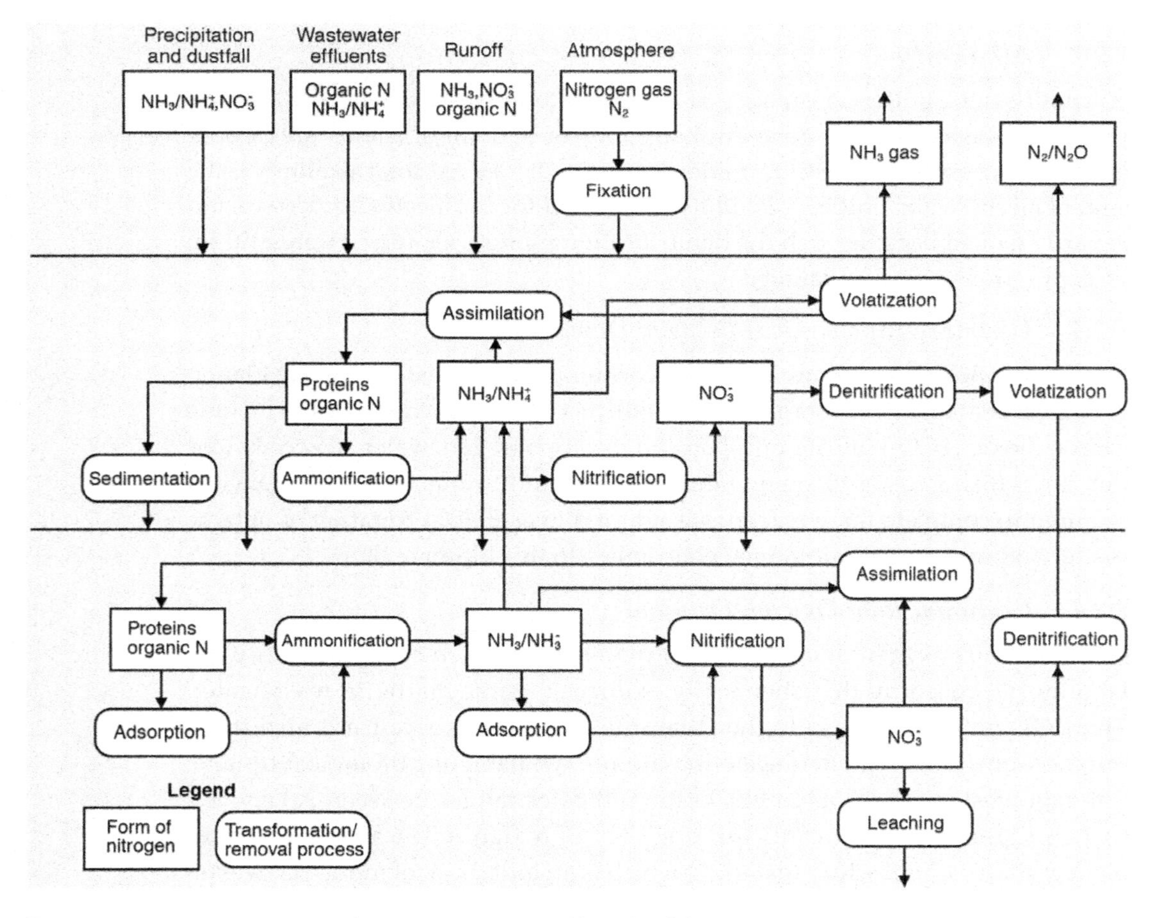

FIGURE 2.1 Nitrogen transformations (Metcalf and Eddy, 2003; WEF, 2005).

drinking water treatment processes that may contain corrosion inhibitors rich in phosphorus also indirectly affect wastewater total phosphorus concentrations. Phosphorus in wastewater is comprised of both inorganic and organic forms. The inorganic forms, which are soluble, include orthophosphate and polyphosphates. The orthophosphate form (PO_4^{3-}) is the simpler form of phosphorus and typically accounts for 70 to 90% of the total phosphorus. It is also the form that is available for biological metabolism without further breakdown and is precipitated by metal salts during chemical phosphorus removal. Related fractions of orthophosphate are associated with pH. Between pH 6.8 and 6.5, an inversion of proportion occurs between hydrogen phosphate ions and dihydrogen phosphate ions (61 to 39% to 38 to 62%, respectively) (Campbell and Reece, 2005). Therefore, chemical addition to further lower phosphorus would be most effectively introduced to react in an environment having pH near 6.8 (or slightly below) to achieve the greatest effect.

Polyphosphates are complex in nature and are broken down to orthophosphate (or derivations thereof) during the treatment process. Organically bound phosphorus includes a variety of more complex forms of phosphorus that derive from proteins, amino acids, and nucleic acids that, to some extent, are degraded and are present as waste products.

2.2.7 *Trace Metals*

Trace amounts of metals help support microbial growth and metabolism during biological treatment processes. Trace metals such as cadmium (Cd), chromium (Cr), copper (Cu), iron (Fe), lead (Pb), manganese (Mn), mercury (Hg), nickel (Ni), and zinc (Zn) are important constituents in water (Metcalf and Eddy, 2003). Most of these metals help with biological growth, and requirements for the presence of trace metals have been recognized. However, if trace metals are present in excess quantity, they may be toxic to microorganisms and eventually inhibit biological treatment processes. Sources of these metals include residential dwellings, groundwater infiltration, and commercial and industrial dischargers. Analyzing trace metals in treated effluent helps assess the suitability of water for reuse.

2.2.8 *Volatile Fatty Acids—Carbon*

Volatile fatty acids (VFAs) are an integral component of wastewater, especially within an EBNR facility. Chapters 7 and 11 provide detailed information on VFAs.

2.3 Biological Characteristics

Biological characteristics discussed in this section include specific microorganisms and toxicity.

2.3.1 *Specific Microorganisms*

Biological processes for wastewater treatment consist of mixed communities containing a wide variety of microorganisms including bacteria, protozoa, fungi, and rotifers. Typical bacterial types include aerobic, anaerobic, and facultative bacteria. Aerobic bacteria need an oxygen-rich environment, whereas anaerobic bacteria require no oxygen. Facultative bacteria are capable of living in the presence and absence of oxygen. During the biological treatment process, the microorganism mass (mixed liquor suspended solids) initiates various biochemical reactions that convert complex compounds into simpler forms.

Populations of polyphosphate-accumulating organisms (PAOs) and glycogen-accumulating organisms (GAOs) are found in EBPR facilities. Glycogen-accumulating organisms directly compete with PAOs, particularly at lower pH levels (i.e., below 7.0 standard units) (see Chapter 7 for more detailed information on GAOs). Table 2.2 classifies common types of bacteria by electron donor, electron acceptor, sources of cell carbon, and end products.

2.3.2 *Toxicity*

Toxicity tests are typically conducted to assess the suitability of treated effluent parameters on receiving water and environmental conditions. An awareness of toxic levels for microorganisms provided on a material safety data sheet should provide guidance to prevent harmful "overfeeds" to both process and side-stream operations. Two typical types of toxicities analyzed are acute toxicity and chronic toxicity. Acute toxicity focuses on exposure that would result in significant response shortly after exposure (typically observed within 48 to 96 hours). Chronic toxicity describes exposure resulting in sublethal response over a long period of time, often one-tenth of the life span or longer (Metcalf and Eddy, 2003).

3.0 EFFECT OF WATER RESOURCE RECOVERY FACILITY FLOW AND LOADING

Water resource recovery facility design and operation should address hydraulic and mass loading conditions to achieve effluent limits. Depending on the time of day, month, season, and so on, hydraulic and mass loads can vary significantly. To dampen the effect of peak flows and loads on treatment processes, some WRRFs provide equalization units for temporary storage of excess wastewater. In addition, appropriate design criteria for various pieces of process equipment may differ based on average, maximum, or peak flow conditions. It is important to understand these different flow and mass loading conditions for WRRF design and process optimization. Figure 2.2 illustrates a typical diurnal flow and loading profile. Flowrates and corresponding loadings significantly affect system performance. Various common flow-related terminologies are described in the following sections.

3.1 Average Daily Flows

- Average daily flow and monthly average flow—total wastewater inflow volume during the specific time period, divided by the number of days in

TABLE 2.2 Classification of microorganisms (Metcalf and Eddy, 2003).

Type of bacteria	Common reaction name	Carbon source	Electron donor substrate oxidized	Electron acceptor	Products
Aerobic heterotrophic	Aerobic oxidation	Organic compounds	Organic compounds	O_2	CO_2, H_2O
Aerobic autotrophic	Nitrification	CO_2	NH_3^-, NO_2^-	O_2	NO_2^-, NO_3^-
	Iron oxidation	CO_2	Fe (II)	O_2	Ferric iron Fe (III)
	Sulfur oxidation	CO_2	H_2S,S° $S_2O_3^{2-}$	O_2	SO_4^{2-}
Facultative heterotrophic	Denitrification anoxic reaction	Organic compounds	Organic compounds	Organic compounds	Volatile fatty acids (acetate, propionate, butyrate)
	Iron reduction	Organic compounds	Organic compounds	Fe (III)	Fe (II), CO_2, H_2O
	Sulfate reduction	Organic compounds	Organic compounds	SO4	H_2S, CO_2, H_2O
	Methanogenesis	Organic compounds	Volatile fatty acids	CO_2	Methane

that period, and total wastewater inflow volume during a 1-month period, divided by the number of days in that month, respectively;

- Annual average daily flow—total wastewater inflow volume during any consecutive 365 days, divided by 365; and
- Three-month average daily flow—total wastewater inflow volume during 3 consecutive months, divided by the number of days in this period.

3.2 Maximum Flows

- Maximum daily flow—largest volume of wastewater inflow during any consecutive 24-hour period and
- Maximum monthly average flow—the highest monthly average flow during any one calendar year.

3.3 Peak Flows

- Peak hourly flow—maximum inflow averaged during 1 hour and
- Peak instantaneous flow—maximum instantaneous inflow at any given time.

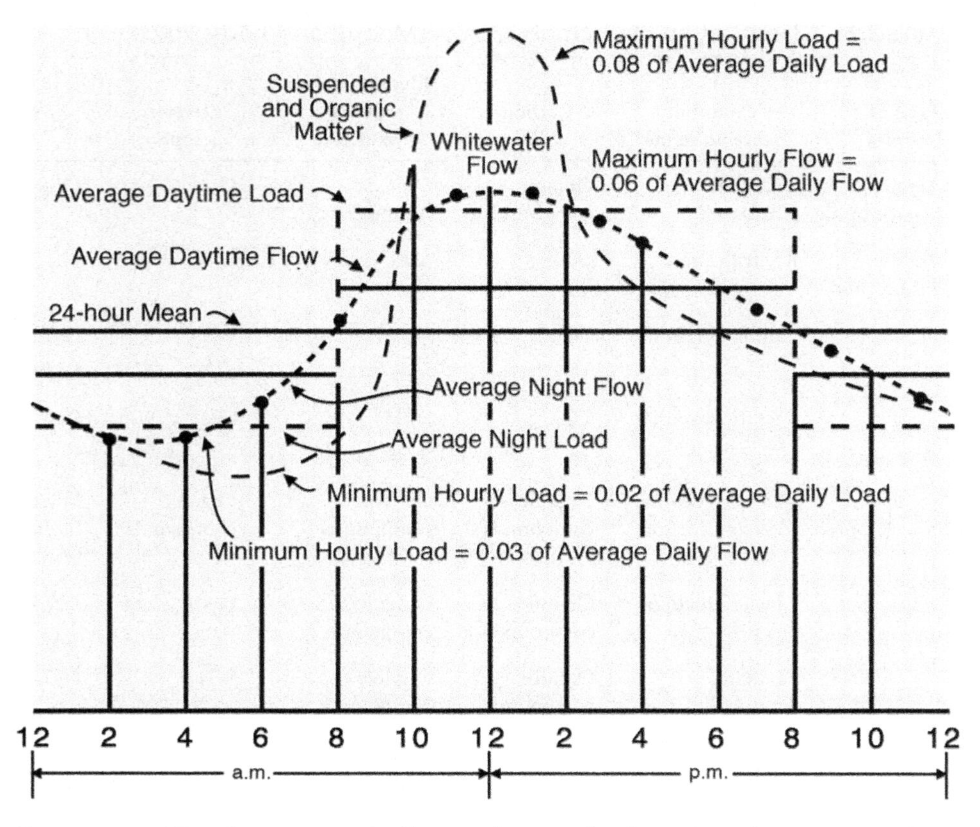

FIGURE 2.2 Hourly variation in flow and strength of municipal wastewater (Fair et al., 1971).

3.4 Internal Sidestream Flows

Sidestreams represent flows generated within a WRRF that are recycled back to be treated. These streams are typically sent back to the head of the WRRF. In general, they are low in volumes (about 5 to 10% compared to inflows), and may be rich in soluble organic, solids, and nutrients. Sidestreams can affect aeration demands and BNR significantly if they are intermittent (e.g., if dewatering only takes place 8 hours per day, 2 or 3 days per week).

4.0 PRETREATMENT REQUIREMENTS FOR INDUSTRIAL DISCHARGERS

The U.S. Environmental Protection Agency's (U.S. EPA's) National Pretreatment Program (http://cfpub.epa.gov/npdes/pretreatment/pstandards.cfm) identifies specific requirements that apply to all industrial users or nondomestic sources of wastewater before entering the sewer system. These objectives are achieved by enforcing the discharge standards outlined in the following sections.

4.1 Prohibited Discharge Standards

Prohibited discharge standards forbid the discharge of any pollutant(s) to a WRRF that causes pass through or interference. These national standards apply to all industrial users discharging to a WRRF, regardless of whether or not the WRRF has an approved pretreatment program or whether the industrial user has been issued a control mechanism or permit. These standards are intended to provide protection for publicly owned treatment works. The prohibited discharge standards are listed in 40 CFR 403.5 (http://www.gpo.gov/fdsys/pkg/CFR-2010-title40-vol28/xml/CFR-2010-title40-vol28-part403.xml).

4.2 Categorical Pretreatment Standards

Categorical pretreatment standards limit pollutant discharges to WRRFs from specific process wastewaters of particular industrial categories. These national technology-based standards apply regardless of whether or not the WRRF has an approved pretreatment program or the industrial user has been issued a control mechanism or permit. Such industries are called *categorical industrial users*. The standards are promulgated by U.S. EPA in accordance with Section 307 of the Clean Water Act and are designated in "Effluent Guidelines & Limitations" (Parts 405–471) as "Pretreatment Standards for Existing Sources" and "Pretreatment Standards for New Sources".

4.3 Local Limits

Local limits reflect specific needs and capabilities at individual WRRFs and are designed to protect the WRRF and its receiving waters and sludge disposal practices. Regulations of 40 CFR 403.8(f)(4) state that WRRF pretreatment programs must develop local limits or demonstrate that they are not necessary; in addition, 40 CFR 403.5(c) states that local limits are needed when pollutants are received that could result in pass through or interference at the WRRF. Essentially, local limits translate the General Prohibited Discharge Standards of 40 CFR 403.5 to site-specific needs.

5.0 NUTRIENT EFFLUENT PERMIT REQUIREMENTS

To preserve and improve the quality of receiving waters, WRRFs are required to comply with effluent discharge permit requirements. Types of permits or concepts used for developing discharge limits are discussed in the following sections.

5.1 Technology-Based Permit Limits

For technology-based permits, effluent requirements are based on specific treatment levels achievable by the available technology, that is, considering the "best available technology" or "limit of technology". This treatment level might not be related to the assimilative capacity of receiving streams or water quality standards.

5.1.1 *Monthly Average-Based Permit Limits*

Monthly average-based permit limits are based on 30-day average effluent limits. This reporting period requires the WRRF to operate at optimum efficiency at all times. If few samples indicate high levels of total nitrogen, for example, then the facility must operate at a higher level of performance for the rest of the month to maintain compliance with the monthly average. Typically, there are allowances for higher daily or weekly results, providing flexibility for peak daily or weekly influent loads.

5.1.2 *Annual Average-Based Permit Limits*

An effluent permit that is based on an annual average during a 1-year period allows more flexibility to operate at or near the permit level. If there is an excursion above the annual average limit, then there is more time to operate at a higher level of treatment to maintain compliance with the annual average limit.

5.1.3 *Seasonal-Based Permit Limits*

Seasonal permits require a different permit level in the summer vs the winter; this may allow the facility to be operated in a "less aggressive" manner during the "off season". For example, a facility with a seasonal nitrification permit may be able to operate at a lower solids retention time in the winter than otherwise would be required for a year-round permit.

5.2 Water-Quality-Based Permit Limits

Water-quality-based permits are based on assimilative capacity of the receiving stream and, in general, are based on an annual average load allocation from the facility. As flows increase, the level of treatment required to meet the permit increases. At lower flows below design capacity, the allowable effluent concentration is greater. An annual load allocation allows some flexibility in seasonal operation if the treatment facility is capable of operating below load allocation on an average monthly basis or if it is not yet at design capacity. In some states, water-quality-based limits are enforced in terms of a concentration limit, which governs how the treatment facility must be operated.

6.0 REFERENCES

Campbell, N. A.; Reece, J. B. (2005) *Biology,* 7th ed.; Benjamin Cummings: San Francisco, California.

Fair, G. M.; Geyer, J. C.; Okun, D. A. (1971) *Elements of Water Supply & Wastewater Disposal,* 2nd ed.; Wiley & Sons: New York.

Metcalf and Eddy, Inc. (2003) *Wastewater Engineering Treatment and Reuse Textbook,* 4th ed.; Metcalf and Eddy, Inc.: Wakefield, Massachusetts.

Society of Petroleum Engineers (2009) *Marcellus Shale Post-Frac Flowback Waters—Where Is All the Salt Coming from and What are the Implications?* Blauch, M.E., Myers, R.R.; Moore, T.R.; Lipinski, B.A., Houston, N.A., Eds.; Society of Petroleum Engineers: Richardson, Texas.

Chapter 3

Nitrification

Jeanette Brown, P.E., BCEE, D. WRE, and
Robert R. Sharp, Ph.D., P.E.

1.0 NITRIFICATION

Nitrification is the process of biological oxidation of ammonia (which exists mostly as ammonium nitrogen [NH_4-N] in typical wastewater) to nitrite-nitrogen (NO_2-N) and further oxidation to nitrate-nitrogen (NO_3-N). Nitrification is important because it converts toxic ammonia to nontoxic forms of nitrogen and is the first step in biological nitrogen removal (denitrification), which will be discussed in Chapter 5. Conversions to nitrite and nitrate involve two specific groups of autotrophic bacteria: ammonia-oxidizing bacteria (AOB) and nitrate-oxidizing bacteria (NOB). Ammonia-oxidizing bacteria carry out the oxidation of ammonia to nitrite and NOB carry out nitrite conversion to nitrate. Both these reactions should operate at optimal rates for the production of nitrate. Ammonia-oxidizing bacteria and NOB are often referred to as *nitrifiers* and, although classified together, they are not related phylogenically (Block et al., 1991). Autotrophic bacteria, specifically chemoautotrophic bacteria, differ from heterotrophic bacteria that consume organic material (biochemical oxygen demand [BOD]) in that chemoautotrophic bacteria synthesize cellular material from inorganic carbon (HCO_3^-) under typical operating conditions. Oxidation of ammonia or nitrite provides the energy needed for cell synthesis. These bacteria are obligate aerobes because they grow only when dissolved oxygen is available; however, the lack of oxygen is not lethal because nitrifiers periodically withstand both anaerobic and anoxic conditions. There are emerging processes that bypass the conversion of nitrite to nitrate and convert nitrite directly to nitrogen gas (denitritation), thus completing the nitrogen removal process. These processes will be discussed in Chapter 12.

1.1 Stoichiometry and Kinetics

The stoichiometry of biochemical reactions associated with nitrification defines the proportion of reactants and products involved in this process. Understanding stoichiometry is important because it defines the basic inputs and outputs for each of the steps in the process and can determine which of these inputs will limit the reaction. The stoichiometric equation that defines the molar ratios for the oxidation of ammonia to nitrite by AOB is

$$NH_4^+ + 1.5\ O_2 \Rightarrow NO_2^- + 2\ H_2O + 2H \qquad (3.1)$$

Similarly, the stoichiometric equation that describes the oxidation of nitrite to nitrate by NOB is

$$NO_2^- + 0.5\ O_2 \Rightarrow NO_3^- \qquad (3.2)$$

These reactions also generate biomass associated with the growth of AOB and NOB. Unlike heterotrophs, AOB and NOB obtain carbon for cell growth from an inorganic source, that is, carbon dioxide. The total yield of these nitrifying organisms for the combination of both steps in the process is substantially lower than for heterotrophs, generally ranging from 0.06 to 0.20 g volatile suspended solids (VSS)/g NH_4-N oxidized. The stoichiometric equations do not reflect that a portion of nitrogen is used in assimilation (bacterial growth). However, because the quantity of nitrogen removed through this assimilation process into nitrifier biomass is typically less than 2% of the ammonia nitrified, it can generally be ignored.

The overall energy reaction for the two-step nitrification process combines eqs 3.1 and 3.2, as follows:

$$NH_4^+ + 2\ O_2 \Rightarrow NO_3^- + 2\ H + H_2O \quad (3.3)$$

The overall stoichiometric equation that accommodates both ammonia oxidation and cell synthesis reactions is given by

$$NH_4^+ + 1.863\ O_2 + 0.098\ CO_2 \Rightarrow 0.0196\ C_5H_7O_2N + 0.98\ NO_3^- + 0.094\ H_2O + 1.88\ H_2CO_3 \quad (3.4)$$

Based on the stoichiometry of the overall energy reaction, 2 mol of oxygen is required to oxidize 1 mol of ammonia to nitrate, which is equivalent to 4.57 g O_2/g NH_4^+-N oxidized.

During nitrification, 2 mol of hydrogen ion are also formed. The formation of 1 mol of hydrogen is equivalent to the consumption of 1 mol of alkalinity or 0.5 mol of $CaCO_3$ (50 g $CaCO_3$). The consumption of alkalinity during nitrification may decrease the pH of the wastewater if there is insufficient alkalinity available and can adversely affect the nitrification process, as discussed later in this chapter. It is important to understand the differences between heterotrophic bacteria, which synthesize cellular material from organic carbon (BOD), and autotrophic bacteria (AOB and NOB), which synthesize cellular material from carbon dioxide. Table 3.1 presents typical oxygen and alkalinity relationships of the nitrification process.

1.2 Nitrification Kinetics

1.2.1 Biomass Growth and Ammonia Utilization

The first step in nitrification (i.e., conversion of ammonia to nitrite) is the rate-limiting step because AOB typically have lower specific growth rates than NOB. Growth of AOB and NOB is the result of the oxidation of ammonia and nitrite, respectively. The Monod specific growth rate equation is used to describe the

TABLE 3.1 Key nitrification relationships.

Parameter	Coefficient units
Oxygen use	4.57 g O_2/g NH_4-N
Alkalinity consumption	7.14 g alkalinity (as $CaCO_3$)/g NH_4-N

effect of limiting substrates on microbial growth (Monod, 1949). Both ammonia and dissolved oxygen are substrates for AOB growth. The concentrations of either or both of these substances could be low enough to limit the specific AOB growth rate in wastewater treatment systems.

It is important to understand that nitrifiers cannot store food (ammonia). Therefore, they must metabolize it before the mixed liquor discharges to the clarifier and an anaerobic environment. Thus, for a given solids retention time (SRT), the process must provide a hydraulic detention time large enough for the slowly metabolizing nitrifiers to complete nitrification of all of the ammonia.

Assuming no alkalinity limitation, the growth rate of AOB can be expressed as

$$\mu_{AOB} = \mu_{n\text{-}max,AOB}\,[S_{nh}/(S_{nh} + K_{ns,AOB})][S_o/(S_o + K_{o,AOB})] \quad (3.5)$$

Where

μ_{AOB} = specific growth rate of AOB biomass (g biomass formed/g biomass present per day in system), d^{-1};
$\mu_{n\text{-}max,AOB}$ = maximum specific growth rate of AOB (g biomass formed/g biomass present per day in system), d^{-1};
$K_{ns,AOB}$ = half-saturation coefficient for AOB (mg N/L);
S_{nh} = growth-limiting substrate (NH_4-N) concentration (mg N/L);
S_o = dissolved oxygen concentration of bulk mixed liquor or wastewater (mg O_2/L); and
$K_{o,AOB}$ = oxygen half-saturation coefficient for AOB (mg O_2/L).

Equation 3.5 is a simplification because it does not explicitly consider free ammonia, the substrate for AOB. Figure 3.1 shows specific growth as a function of the NH_4-N concentration when dissolved oxygen is not limiting. The specific growth rate increases almost linearly (approximating first-order kinetics) with respect to NH_4-N concentration near the origin of the plot where the NH_4-N concentration is low. At higher NH_4-N concentrations, the specific growth rate approaches an asymptotic value (the maximum specific growth rate), thus exhibiting zero-order behavior. The NH_4-N half-saturation coefficient is the NH_4-N concentration at which the specific growth rate of the AOBs is one-half of its maximum value. A similar plot could be drawn to depict the relationship between the specific growth rate and dissolved oxygen concentration when NH_4-N is not limiting. Under this condition, the oxygen half-saturation coefficient would be the dissolved oxygen concentration, where the specific growth rate is one-half of its maximum value.

Nitrate-oxidizing bacteria can use both ammonia and nitrite (nitrous acid) as nitrogen sources. Considering NH_4-N and NO_2-N as nitrogen sources, the specific growth rate equation can be written as

$$\mu_{NOB} = \mu_{n\text{-}max,NOB}\,[S_{nh}/(S_{nh} + K_{ns,NOB})]\,[S_o/(S_o + K_{o,NOB})]\,[S_{NO2}/(S_{NO2} + K_{NO2,NOB})] \quad (3.6)$$

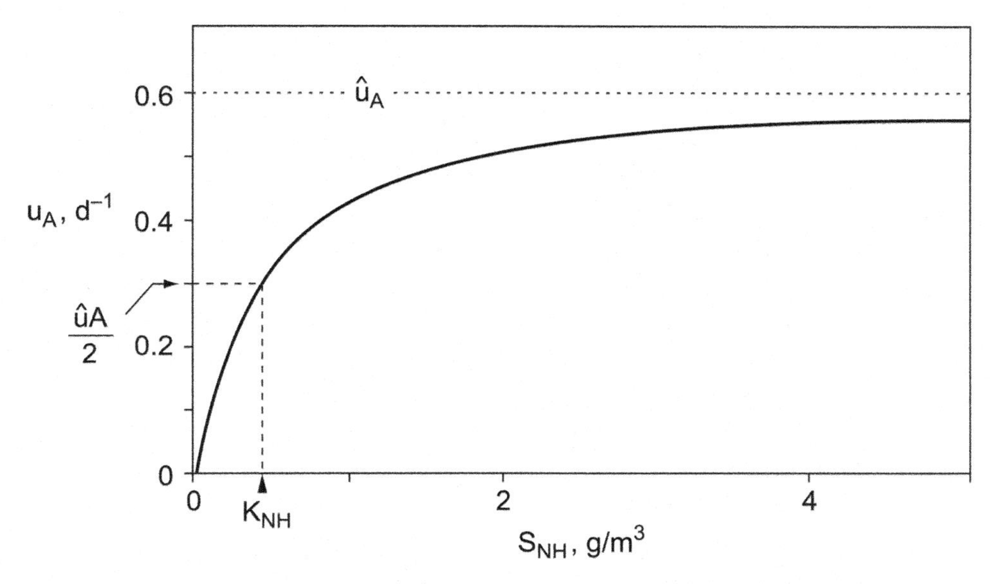

FIGURE 3.1 Relationship between specific growth rate of AOB and ammonia-nitrogen concentration as predicted by the Monod equation (dissolved oxygen is assumed to be nonlimiting).

Where

μ_{NOB} = specific growth rate of NOB biomass (g biomass formed/g biomass present per day in system), d^{-1};

$\mu_{n\text{-}max,NOB}$ = maximum specific growth rate of NOB (g biomass formed/g biomass present per day in system), d^{-1};

S_{nh} = growth-limiting substrate (NH_4-N) concentration (mg N/L);

$K_{ns,NOB}$ = half-saturation coefficient for NOB (mg N/L);

S_o = dissolved oxygen concentration of bulk mixed liquor or wastewater (mg O_2/L);

$K_{o,NOB}$ = oxygen half-saturation coefficient for NOB (mg O_2/L);

S_{NO2} = growth limiting substrate (NO_2-N) concentration (mg N/L); and

$K_{NO2,NOB}$ = NO_2-N half-saturation coefficient for NOB (mg N/L).

Where nitrous acid (NO_2-N) is the only source of nitrogen, the expression can be written as

$$\mu_{NOB} = \mu_{n\text{-}max,NOB}\,[S_o/S_o + K_{o,NOB}][S_{NO2}/S_{NO2} + K_{NO2,NOB}] \qquad (3.7)$$

Nitrous acid typically does not accumulate in large concentrations in biological treatment systems under stable operation because the maximum growth rate of NOB is significantly higher than that of AOB. As a result, the growth rate of AOB generally controls the overall rate of nitrification (i.e., as soon as nitrate is produced by AOB, it is used by NOB). However, facilities that transition into and out of nitrification or have dissolved oxygen limitations or toxic substances entering the facility may experience conditions where the NOB growth rate is unable to keep up with the AOB growth rate, resulting in elevated effluent nitrite concentrations. This condition is known as *nitrite lock* and results in a significant increase in effluent chlorine demand caused by high levels of nitrite in the effluent (5.1 mg of chlorine is consumed per milligram of nitrite). If the effluent has a high nitrate concentration but low ammonia concentration (typical of treatment facilities that are nitrifying), chlorine reacts as free chlorine and is removed by the reaction with nitrite, thus increasing the amount of chlorine needed to meet the required effluent coliform concentration by 5 : 1. When significant ammonia is present, chlorine reacts preferentially with ammonia (as opposed to nitrite) to produce mostly mono-chloramine, which is an effective disinfectant.

In addition, accumulation of nitrite in a nitrification process can indicate the presence of an inhibitory or toxic substance in the wastewater, which may require further evaluation, as discussed later in this chapter. In situations where effluent permits contain both seasonal nitrification and seasonal disinfection requirements (using chlorine), it is advisable to establish complete nitrification before the start of chlorination.

Bacterial growth rate of AOB, as defined by the time required to double their population, is 10 to 20 times slower than that of aerobic heterotrophic bacteria. Because the SRT of an activated sludge nitrification system is inversely proportional to the net growth rate of the slowest growing nitrifiers (AOB), the aerobic SRT needed to effectively nitrify must be significantly longer than the typical SRT needed to maintain a stable heterotrophic bacterial population (i.e., secondary treatment). Typically, the SRT is between 4 and 6 days for a nitrifying facility, which correlates to a food-to-microorganism ratio of approximately 0.15. The SRT in a biological system is typically defined as

$$\text{SRT} = \Theta = (\text{Mass of biological solids in the reactor} / \text{Mass of biological solids leaving the system/day}). \quad (3.8)$$

Because determination of the mass of active biological solids can be difficult, the overall solids maintained in, and wasted from, the reactor can be used to determine SRT. At steady state, the solids leaving the system will equal the solids produced within the system plus any inert solids that enter the system and pass through the process. Therefore, if the unit solids production rate of the biological treatment system (i.e., the net yield) is known, the required mass of reactor solids can be determined to achieve the target SRT. Typically, net yield values are reported as the mass of solids produced per each unit mass of carbonaceous oxygen demand or 5-day BOD (BOD_5) removed.

The growth rate of nitrifying microorganisms in the system is related to SRT, as follows:

$$\Theta = 1/(\mu_n - b_n) \tag{3.9}$$

Where

μ_n = specific growth rate of nitrifiers (g VSS added·d/g VSS in system) and
b_n = endogenous decay coefficient for nitrifiers (g VSS destroyed·d/g VSS in the system).

A number of environmental factors significantly affect the nitrifier growth rate, which, in turn, influences the minimum SRT required to allow for accumulation and maintenance of a sufficient population of nitrifying organisms. The effects of significant environmental factors that influence nitrification are presented in the following sections.

1.2.2 *Environmental Conditions*

1.2.2.1 *Wastewater Temperature*

Growth rates of AOB and NOB are particularly sensitive to the liquid temperature in which they live. Nitrification occurs in wastewater temperatures from 4 to 45 °C (39 to 113 °F), with an optimum rate occurring at about 30 °C (86 °F) (U.S. EPA, 1993a). However, most water resource recovery facilities (WRRFs) operate with a liquid temperature between 10 and 25 °C (50 and 77 °F). It is generally recognized that the nitrification rate doubles for every 8 to 10 °C rise in temperature. A number of relationships between maximum nitrifier growth rate (μn-max) and wastewater temperature have been developed. The most commonly accepted expression for this relationship for wastewater over a temperature range from 5 to 30 °C (41 to 86 °F) is the following (U.S. EPA, 1993a):

$$\mu_{n\text{-}max} = 0.47\ e^{0.098(T-15)} \tag{3.10}$$

Where

$\mu_{n\text{-}max}$ = maximum specific growth rate of microorganisms (g nitrifiers/g nitrifiers in system·d) and
T = wastewater temperature (°C).

This relationship between maximum nitrifier growth rate and wastewater temperature is illustrated graphically in Figure 3.2. It is important to note that the endogenous decay rate (b_n) for nitrifying organisms is also temperature dependent and that the coefficients must be corrected for temperature.

1.2.2.2 *Dissolved Oxygen Concentration*

Nitrifiers are obligate aerobes, meaning they are only able to grow under aerobic conditions although they can survive in anaerobic or anoxic conditions. Consequently, the dissolved oxygen concentration in the bulk liquid can have a

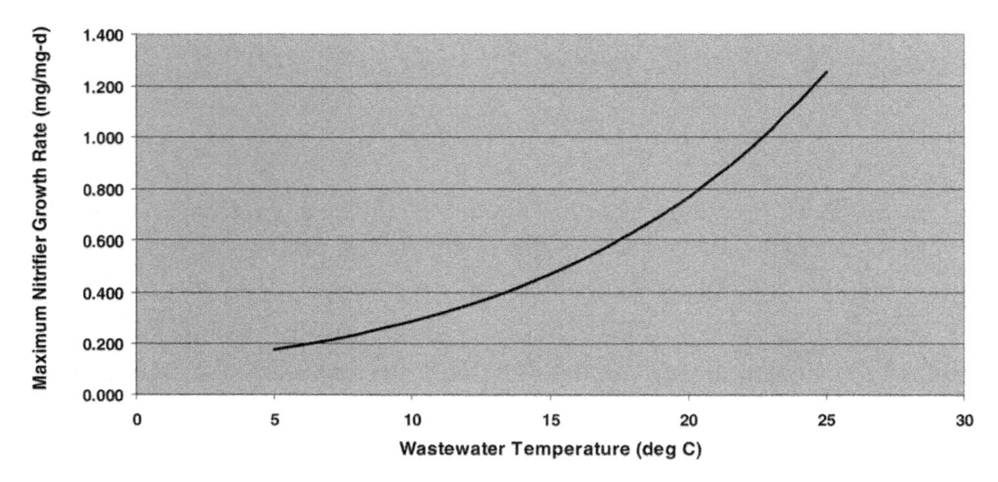

FIGURE 3.2 Influence of temperature on the nitrification rate.

significant effect on the nitrifier growth rate. The value at which dissolved oxygen concentration reduces the rate of nitrification varies on a site-specific basis, depending on temperature, organic loading rate, SRT, and diffusional limitations (i.e., floc size and density). It is generally accepted that nitrification is not limited at dissolved oxygen concentrations greater than 2.0 mg/L. However, this may not be true for systems using high biomass concentrations or biofilm systems or in cyclical systems or transition zones between unaerated and aerated reactors, where dissolved oxygen concentrations less than 2.0 mg/L may affect nitrification. In addition, some activated sludge systems may have large dense floc (possibly indicated by very low sludge volume indexes [<65]), which will require higher dissolved oxygen levels to achieve consistent and effective nitrification because of poor oxygen mass-transfer efficiencies within the dense floc. Operators should check the diurnal variation of dissolved oxygen concentration, if possible, because there may be a significant difference between peak and low loading periods.

1.2.2.3 pH and Alkalinity

Equations 3.3 and 3.4 describe the effect that nitrification has on alkalinity levels within the treatment reactors. These equations illustrate that, in situations where low-alkalinity wastewater is being treated, the nitrification process can have a substantial effect on alkalinity levels and, ultimately, pH. Reactor pH levels have been shown to have a significant effect on the rate of nitrification (U.S. EPA, 1993a). An important factor to consider with regard to the effects of pH on the nitrification process is the degree of acclimation that the particular system has achieved. Wide swings in pH have been demonstrated to be detrimental to nitrification performance, although acclimation generally allows satisfactorily performance with consistent pH control within the range of 6.5 to 8.0 standard pH units. It is generally recommended that sufficient alkalinity be present throughout the nitrifying reactors by maintaining a minimum effluent alkalinity of at least 50 and preferably 100 mg/L (as $CaCO_3$). Low pH shifts the ammonia and ammonium equilibrium more toward ammonium, reducing the actual concentration of ammonia, which is the organisms' food. Low pH slows the nitrifier metabolic rate because the food concentration is reduced and requires higher detention time to compensate. Because the acid dissociation constant (pK_a) of ammonia is 9.25, a pH of 7.25 allows only 1% of the total ammonia to be in ammonia form; pH 6.75 reduces the concentration to less than 0.5%.

1.2.2.4 *Inhibition*

Nitrifiers are particularly susceptible to inhibition from a variety of organic and inorganic substances. In addition, nitrifiers are particularly susceptible to wide fluctuations in the concentration of inhibitory substances, but may exhibit only minor effects if these substances are in low concentrations and are consistently applied to the system.

Nitrifier performance can also be significantly affected by heavy metals, including nickel (at or above 0.25 mg/L), chromium (at or above 0.25 mg/L), and copper (at or above 0.10 mg/L) (Metcalf and Eddy, 2003). Un-ionized or free ammonia can be inhibitory to AOB and NOB, depending on temperature and pH conditions in the reactor. At a pH of 7.0 and a temperature of 20 °C, inhibition is expected to begin at an ammonia-plus-ammonia (NH_3-N + NH_4^+-N) concentration of 1000 and 20 mg/L for AOB and NOB, respectively (U.S. EPA, 1993a).

1.2.2.5 *Flow and Load Variations*

Provided that environmental conditions do not limit growth of nitrifying organisms, the quantity or mass of AOB and NOB that grow in the system will be a function of the applied ammonia load. As such, variations in flow and nitrogen load to the system that result in either a significantly reduced hydraulic retention time (HRT) or increased ammonia load may result in an increase in effluent ammonia. Short HRT systems, such as many biofilm processes, are more likely to experience this reduction in process efficiency, even during normal diurnal variations. Longer HRT systems, such as extended aeration, sequencing batch reactor, and oxidation ditch activated sludge systems, are less likely to experience increased effluent ammonia levels because of variations in flow and ammonia load.

2.0 SUSPENDED GROWTH SYSTEMS

2.1 General

Suspended growth biological treatment systems operate in a fashion that allows control of the amount of biomass in the process and, therefore, the net growth rate of the biomass. The operating range available is limited by the physical capacities of the system, most notably, aeration tank volume, aeration capacity, clarifier surface area, and return activated sludge pump capacity. The ability to control the overall biomass growth rate within the system provides the opportunity to manipulate the system to achieve the concurrent growth of heterotrophic (carbonaceous BOD [CBOD]-consuming) and autotrophic (nitrifier) biomass populations.

Suspended growth nitrification can be achieved in various reactor configurations; however, these configurations must be designed and operated to meet the following two overriding criteria:

1. The biomass inventory must be retained in the system for a sufficient period of time to allow a stable population of nitrifiers to develop and be maintained in the process (i.e., there must be sufficient SRT) and
2. The HRT of the system must be long enough to allow the biomass ample time to remove the pollutant load (BOD and ammonia) entering the system to the extent necessary to maintain permit compliance.

The $SRT_{aerobic}$ is the average period of time that any solid or particle is retained in the aerated portion of a suspended growth process reactor. The $SRT_{aerobic}$, also referred to as *aerobic mean cell residence time* or *aerobic sludge age*, is most commonly defined as the mass of solids in the aerated portion of the reactor tanks divided by the mass of solids removed from the system (wasted) per day, as shown in the following equation:

$$SRT_{aerobic} = \text{Mass of MLSS in aerobic portion of reactor} / \text{Mass of MLSS wasted per day} \quad (3.11)$$

This equation has been simplified based on the following assumptions:

- Only the mixed liquor suspended solids (MLSS) inventory in the aerated portion of the system is considered to contribute to nitrification because the nitrifier growth rate is assumed to be negligible at low or zero dissolved oxygen conditions (i.e., anoxic zones) and
- The mass of effluent total suspended solids (TSS) is ignored because it is significantly lower than the mass of MLSS in the waste sludge, and typically makes up a small fraction of the total sludge wasted. (If it is necessary to calculate SRT during process upsets or wet weather conditions, then effluent TSS should be included in the calculation.)

These two assumptions are generally true for most full-scale applications. However, it is important to note that the mass of effluent TSS may be significant in situations where the effluent contains high concentrations of effluent TSS, such as during a secondary clarifier washout event or in activated sludge systems where the quantity of waste sludge generated is small as a result of the low organic loading (e.g., separate sludge nitrification systems). Separate activated sludge systems are discussed in Section 2.3 of this chapter.

The overall approach for operating a suspended growth nitrification system presented here is to determine the target $SRT_{aerobic}$ using the equations presented and adjust sludge wasting rates to maintain the actual $SRT_{aerobic}$ of the system at or above the target $SRT_{aerobic}$. Because nitrification is sensitive to wastewater temperature, dissolved oxygen concentration, pH, and ammonia concentrations, each of these factors may have a significant effect on the target $SRT_{aerobic}$ determination and each often requires a different target $SRT_{aerobic}$ during specific operating conditions (i.e., colder temperatures will often result in a higher target $SRT_{aerobic}$).

2.1.1 Determining the Target Aerobic Solids Retention Time

The following steps can be used to determine the target $SRT_{aerobic}$:

1. Calculate the maximum nitrifier growth rate (μ_n-max) based on wastewater temperature (eq 3.10);
2. Calculate the specific nitrifier growth rate (μn) based on the reactor ammonia concentration. Assuming pH values are stable and approximately neutral ($6.5 < pH < 8.0$) and dissolved oxygen concentrations of greater than 2.0 mg/L are maintained, proceed to the next step;
3. Calculate the minimum aerobic SRT, which equates to the SRT when nitrifiers are reproducing as fast as they are being wasted out of the system;
4. Select the process design factor (PDF) to account for the variability in influent wastewater characteristics and operation and the safety factor. The PDF is the product of a peaking factor and a safety factor and typically ranges between 1.5 and 3.0. The peaking factor can be determined to account for the variability in process influent wastewater characteristics; internal facility recycles; facility operation (return activated sludge [RAS], waste activated sludge, and dissolved oxygen control); effluent permit requirements; and the physical process configuration. Highly variable influent wastewater characteristics, intermittent or poorly controlled internal facility recycles, stringent permit limits, intermittent or manually controlled operation, and smaller process tanks warrant larger peaking factor values. Consistent influent and internal facility recycles; less frequent and stringent permit requirements (effluent quality and/or permit compliance period); well-automated processes; and facilities with larger, more forgiving process tanks or equalization allow smaller peaking factor values. The safety factor is generally a means of reflecting the level of uncertainty in design or process performance. Use of a safety factor may be warranted in situations with newer, less proven technologies, and may be omitted if

all variations to the process influent and operation are well defined and accounted for in the design (Grady and Lim, 1980); and

5. Calculate the target $SRT_{aerobic}$ by multiplying the minimum $SRT_{aerobic}$ times the PDF.

Use of daily values for wastewater temperature and $SRT_{aerobic}$ should be avoided. Instead, moving average values are recommended to dampen intermittent operations and sampling variability. Generally, a 7-day moving average is satisfactory to minimize the effect of variability without creating a significant "lag" effect on calculated values. Figure 3.3 illustrates daily vs 7-day moving average $SRT_{aerobic}$ for an operating facility. Note how SRT increases during colder weather months and decreases during warmer months. The graph in Figure 3.4 shows the calculated minimum and target $SRT_{aerobic}$ values vs temperature with an assumed PDF of 2.5.

2.1.2 *Example 3.1—Single Sludge Suspended Growth Nitrification*

The following is an example procedure for determining the target $SRT_{aerobic}$ for a single sludge suspended growth nitrification system.

The following secondary influent conditions and secondary effluent requirements are given:

- Average daily flow = 4000 m^3/d (1.06 mgd);
- Secondary influent BOD_5 = 150 mg/L;

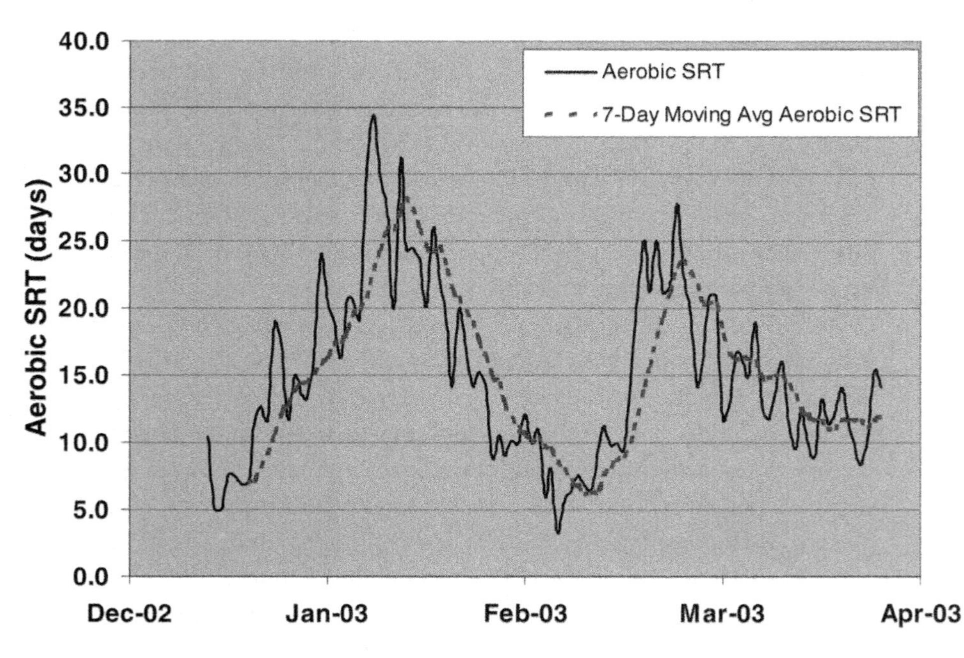

FIGURE 3.3 Daily vs 7-day moving average SRT.

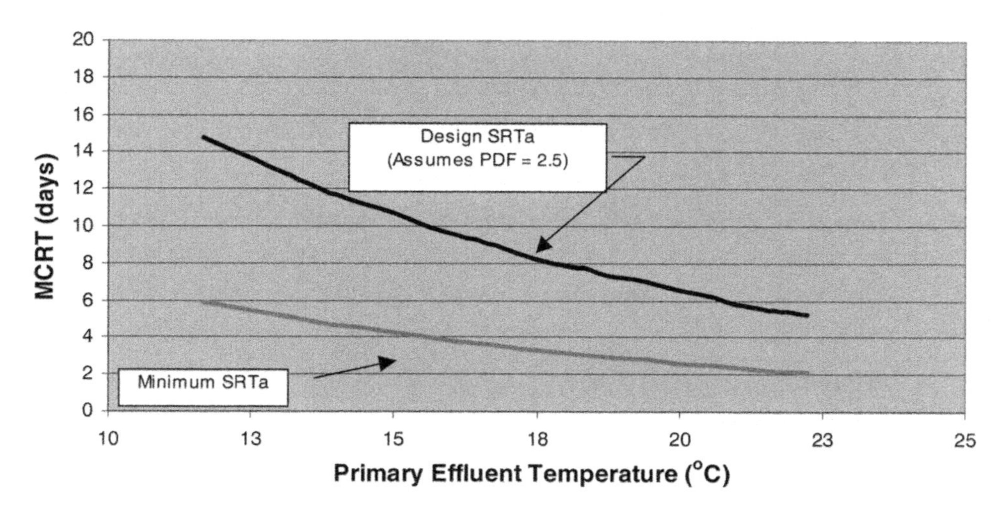

FIGURE 3.4 Nitrification aerobic SRT requirements.

- Secondary influent total Kjeldahl nitrogen (TKN) = 35 mg/L;
- Secondary influent ammonia-nitrogen (NH_4-N) = 21 mg/L;
- Secondary influent wastewater temperature = 15 °C;
- Maximum monthly and average flow ratio = 1.3;
- Diurnal peak and average flow ratio = 1.2; and
- Secondary effluent NH_4-N = 1.0 mg/L (note that secondary influent flow, BOD, and TKN loading should be based on the most stringent permit criteria [monthly, weekly, etc.] anticipated for the time period).

The target $SRT_{aerobic}$ for a single sludge suspended growth nitrification system can be determined using the following steps:

1. Calculate the nitrifier maximum growth rate as follows:

$$\mu n\text{-max} = 0.47\ e^{0.098(15-15)} = 0.47/d$$

2. Calculate the specific growth rate as follows:

$$\mu n = 0.47\ (1\ mg/L/[1\ mg/L + 1\ mg/L]) = 0.235/d$$

3. Calculate the minimum aerobic SRT as follows:

$$\text{Minimum } SRT_{aerobic} = 1/\mu n = 1/0.235/d = 4.3 \text{ days}$$

Note that minimum $SRT_{aerobic}$ is the SRT when nitrifiers are growing just as fast as they are wasted from the system and is an inherently unstable operating condition.

4. Determine the PDF as follows:

   PDF = Peaking factor × safety factor

   Peaking factor = Maximum month flow factor × diurnal flow factor
   Peaking factor = 1.3 × 1.2 = 1.56

   The safety factor is intended to account for variability in secondary influent wastewater characteristics and process operation (dissolved oxygen control, sludge wasting, etc.) and process performance uncertainty. For example,

   Safety factor = 1.5 (selected)
   PDF = 1.56 × 1.5 = 2.34

5. Calculate the target $SRT_{aerobic}$ as follows:

   Target $SRT_{aerobic}$ = Minimum $SRT_{aerobic}$ × PDF

   Target $SRT_{aerobic}$ = 4.3 days × 2.34 = 10.1 days

2.2 Single Sludge Systems

Single sludge suspended growth nitrification systems are defined as a process configuration including only one set of clarifiers per train of reactors for oxidation of both CBOD and nitrogenous (TKN and ammonia) pollutants. Because of this configuration, these systems generally operate at higher CBOD-to-TKN ratios than separate sludge systems. Because of the common practice of incorporating denitrification and/or biological phosphorus removal with nitrifying systems, single sludge biological nutrient removal (BNR) systems have become the predominate approach when nitrification is required. Figure 3.5 presents a schematic of a single sludge nitrification system.

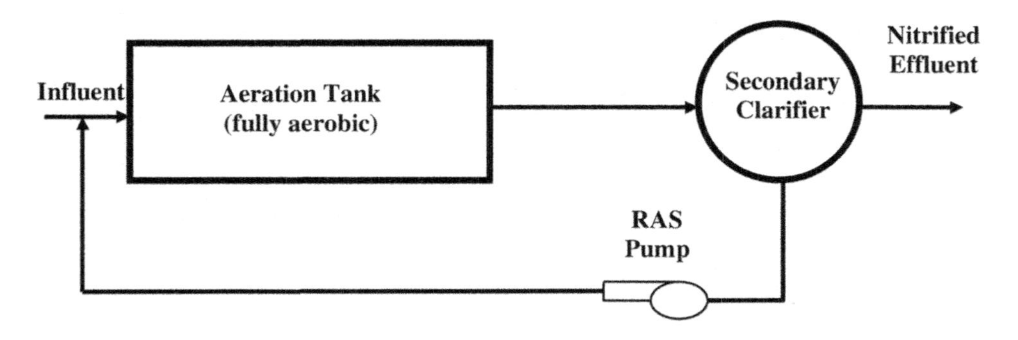

FIGURE 3.5 Single sludge nitrification system.

2.3 Separate Sludge Systems

Separate sludge suspended growth systems, as shown in Figure 3.6, are defined as a process configuration that includes a separate set of reactors and clarifiers for each step of the process, typically for CBOD removal and nitrification. Separate sludge nitrification is characterized by low CBOD-to-TKN ratios (typically less than 5 : 1 and often less than 2 : 1) in the second stage. This influent loading to the nitrification system typically results in particularly low waste activated sludge generation rates in the nitrification stage. In some situations, a portion of the primary effluent is bypassed to the separate sludge nitrification system to provide sufficient loading to generate more biomass than is lost in the effluent TSS.

Separate sludge nitrification systems have been implemented for a variety of reasons, including the following:

- To provide tertiary treatment following an upstream biofilm (trickling filter) process;
- To protect the nitrification system from an influent wastewater that contains materials toxic to nitrifiers (i.e., metals); and
- To allow optimized operation of the separate carbonaceous and nitrifying activated sludge systems to meet specific effluent requirements or to provide a fully nitrified wastewater to a downstream denitrification process.

In general, separate sludge suspended growth nitrifying systems have become less common recently, primarily because of increased costs associated with the construction of separate clarifiers for each stage and the tendency to integrate BNR (nitrification and denitrification) systems together. Separate stage systems have

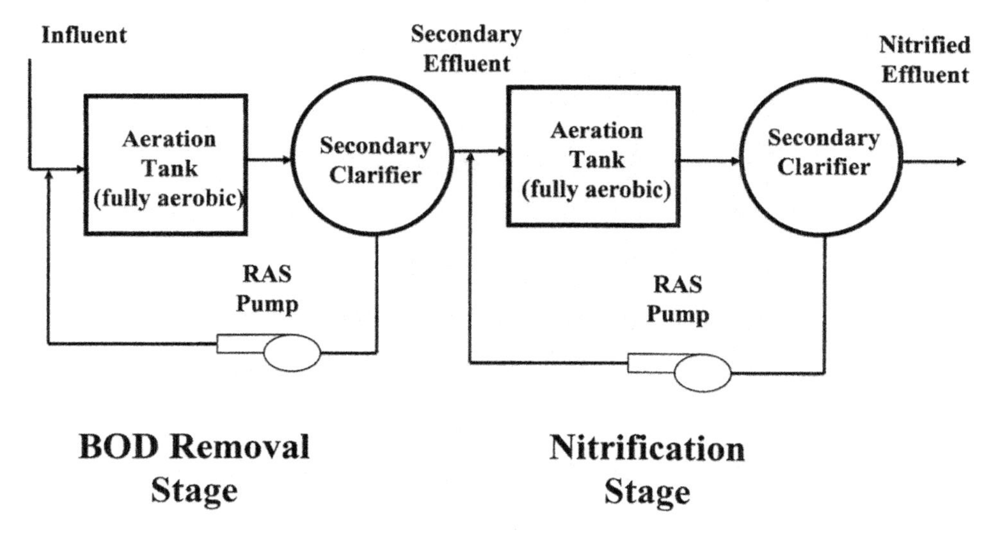

FIGURE 3.6 Separate sludge nitrification system.

experienced a number of operational challenges mainly related to low biomass growth rates in the nitrification stage and "weak" floc structure that is susceptible to shearing, which can result in effluent solids. However, there are some large separate sludge WRRFs (e.g., Blue Plains Wastewater Treatment Plant in Washington, D.C.) that have a long history of successful operation and performance.

3.0 BIOFILM SYSTEMS

3.1 General

Biological wastewater treatment processes with biomass attached to some type of inert media are termed *biofilm reactors* (a.k.a., fixed film, attached growth, or fixed growth). Figure 3.7 depicts a general biofilm diagram. True biofilm systems are differentiated from coupled treatment systems that incorporate a separate biofilm reactor followed by an activated sludge process without intermediate clarification and hybrid systems that include suspended and biofilm processes within the same reactor. Biofilm processes and technologies used for nitrification are briefly described in this section. A more complete description of biofilm processes used in wastewater treatment is presented in Chapter 4.

In general, biofilm treatment processes are easy to operate and are resilient to shock loads, but also have significantly less flexibility than suspended growth treatment systems. Unlike suspended growth nitrification systems, the onset and accumulation of autotrophic bacteria that accomplish nitrification are limited by

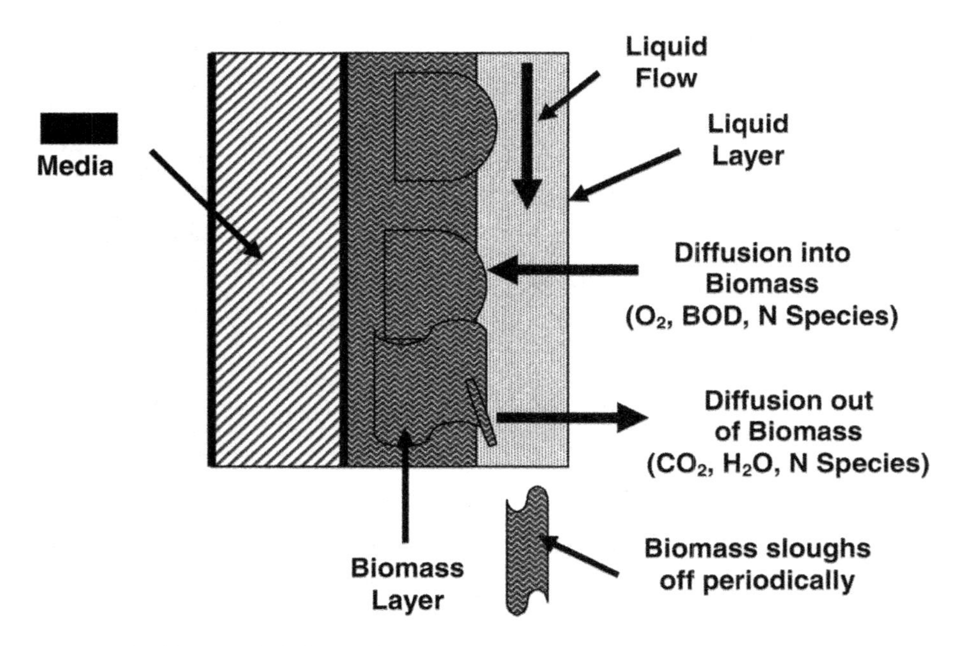

FIGURE 3.7 General biofilm schematic.

the growth of heterotrophic bacteria that predominate until the substrate (soluble BOD_5 or CBOD) for those microorganisms is mostly exhausted. In general, AOB and NOB will not become a significant portion of the biomass growth on the media surface until soluble BOD_5 is less than 15 mg/L or CBOD is less than 20 mg/L. Therefore, CBOD removal and nitrification are generally considered sequential processes in biofilm systems. In other words, if nitrification is to occur in a biofilm reactor, competition from the heterotrophs for oxygen and space on the media surface must be reduced. This can be accomplished via upstream CBOD removal processes or within the biofilm system by providing sufficient surface area to first reduce the CBOD and then nitrify the ammonia load. Specifically,

- An organic loading rate of 7.3 to 9.8 g CBOD/m^2·d (1.5 to 2.0 lb CBOD/d/1000 sq ft) (media surface area) will reduce CBOD to less than 20 mg/L so nitrification can begin and
- An ammonia–nitrogen loading rate of 1.0 to 2.0 g TKN/m^2·d (0.2 to 0.4 lb TKN/d/1000 sq ft) (media surface area) will reduce the ammonia-nitrogen rate to less than 2.0 mg/L NH_3-N.

These values are simply approximations for a multitude of biofilm reactors (U.S. EPA, 1993a). The actual performance of any particular system will depend on the type and quantity of media in use, wastewater characteristics, and other important factors including temperature, dissolved oxygen concentration, and pH.

Another considerable difference between suspended and biofilm systems is the influence that diffusion mass transfer has on performance. Soluble matter (soluble BOD, ammonia, and oxygen) must diffuse into the bacterial film to be used. In suspended growth systems, biomass is suspended within liquid and surrounded by substrate (BOD) and an electron acceptor (oxygen or nitrate) on all sides. Diffusion of these constituents and other necessary nutrients can enter from all around the floc particle. Biofilm systems, by their very nature, have much less biomass surface area in contact with the passing liquid or gas to allow diffusion into the biomass, making the transfer of substrate and electron acceptor into the biofilm much more likely to influence or limit pollutant removal rates.

3.2 Trickling Filters

Trickling filters are aerobic biofilm reactors consisting of open-air media that support biomass growth. Settled primary effluent or fine-screened wastewater is distributed at the top of the filter and allowed to "trickle" down through the media as a thin film over the accumulated biofilm. Air flows through the media either concurrently or counter-currently with the liquid, driven by either convection and/or by low-pressure fans. There should be no standing or stagnant liquid within the trickling filter reactor that would hinder the flow of air throughout the entire height of the media. The media support a mixed aerobic biomass that removes CBOD and, if conditions allow, ammonia by sorption and assimilation into the biomass. Excess biomass periodically "sloughs" off the trickling filter media and is typically removed via downstream clarifiers.

Historically, trickling filters have been classified based on hydraulic and organic loading rates, as listed in Table 3.2. Trickling filter media typically consist of one of a number of types of plastic media. Older trickling filters, mainly roughing filters, may use rock for media. Generally, rock trickling filters are limited to a depth of less than 3 m (10 ft) because of weight. Many rock trickling filters have been replaced with plastic media filters or other treatment processes. Plastic media trickling filters can be more than 7.6-m (25-ft) deep (U.S. EPA, 1993a). Plastic media offer much greater available surface area per unit volume and less weight than rock trickling filters, allowing taller and more efficient filters. Table 3.3

TABLE 3.2 Historical classification of trickling filters.

Design characteristics	Low or standard rate[a]	Intermediate rate[a]	High rate[a]	Super rate[a]	Roughing
Media	Stone	Stone	Stone	Plastic	Stone/plastic
Hydraulic loading					
mgd/ac[b]	1 to 4	4 to 10	10 to 40	15 to 90	60 to 180
gpd/sq ft[c]	25 to 90	90 to 230	230 to 900	350 to 2100	1400 to 4200
Organic loading					
lb BOD_5/ac ft[d]	200 to 600	700 to 1400	1000 to 1500	—	—
lb BOD_5/d/1000 cu ft[e,f]	5 to 15	15 to 30	30 to 150	≤300	>100
Recirculation	Minimum	Typically	Always	Typically	Not typically required
Filter flies	Many	Varies	Few	Few	Few
Sloughing	Intermittent	Intermittent	Continuous[g]	Continuous[g]	Continuous[g]
Depth, ft[h]	6 to 8	6 to 8	3 to 8	≤40	3 to 20
BOD removal, %[i]	80 to 85	50 to 70	40 to 80	65 to 85	40 to 85
Effluent quality	Well nitrified	Some nitrification	No nitrification	Limited nitrification	No nitrification

[a]Historical/obsolete terminology.
[b]mgd/ac × 9353 = m3/ha·d.
[c]gpd/sq ft ÷ 2121 = (L/m²·s).
[d]lb/ac ft ÷ 2725 = kg/(m3·d).
[e]Excluding recirculation.
[f]lb/1000 cu ft ÷ 62.42 = kg/(m3·d).
[g]May be intermittent up to a total hydraulic rate of between 0.7 and 1.0 gpd/sq ft.
[h]ft × 0.3048 = m.
[i]Including subsequent setting.

TABLE 3.3 Comparative physical properties of trickling filter media.

Media type	Nominal size, in. × in.[a]	Unit weight, lb/cu ft[b]	Specific surface area, sq ft/cu ft[c]	Void space, %	Application[d]
Plastic (bundle)	24 × 24 × 48	2 to 5	27 to 32	>95	C, CN, N
	24 × 24 × 48	4 to 6	42 to 45	>94	N
Rock	1 to 3	90	19	50	CN, N
Rock	2 to 4	100	14	60	C, CN, N
Plastic (random)	Varies	2 to 4	25 to 35	>95	C, CN, N
	Varies	3 to 5	42 to 50	>94	N
Wood	48 × 48 × 1.875	10.3	14		C, CN

[a]in. × 25.4 = mm.
[b]lb/cu ft × 16.02 = kg/m^3.
[c]sq ft/cu ft × 3.281 = m^2/m^3.
[d]C = $CBOD_5R$, CN = $CBOD_5R$ and NODR, and N = tertiary NODR.

presents a comparison of the physical properties of various types of trickling filter media.

Because the biomass growing on trickling filters is not submerged, the media must be continuously wetted. Therefore, recirculation is commonly used to maintain consistent flow across the media and to provide for continuous operation of hydraulically driven distribution mechanisms. Recirculation also improves performance by reseeding media with sloughed biomass, dampening variations of influent wastewater concentration, and improving consistent flow distribution across the filter. Clarification of filter recycle before recirculation has been shown to improve trickling filter performance as a result of TSS removal. However, improved trickling filter performance can easily be negated if the influent plus recirculation hydraulic loads negatively affect clarifier operation. Therefore, recirculation through clarifiers should only be practiced if clarifiers have been sized to handle recycle flows or during periods when influent plus recycle flows do not exceed clarifier capacity or negatively affect performance.

Figure 3.8 illustrates four trickling filter process flow diagrams. During trickling filter retrofit projects, consideration should be given to replacing hydraulically driven flow distribution with mechanically driven, variable-speed flow distribution and adding forced ventilation to improve operation flexibility and overall trickling filter performance. The addition of covers to trickling filters can reduce temperature effects and improve cold weather performance.

Ammonium–nitrogen and dissolved oxygen concentrations and temperature variations can act individually or in combination to reduce the observed

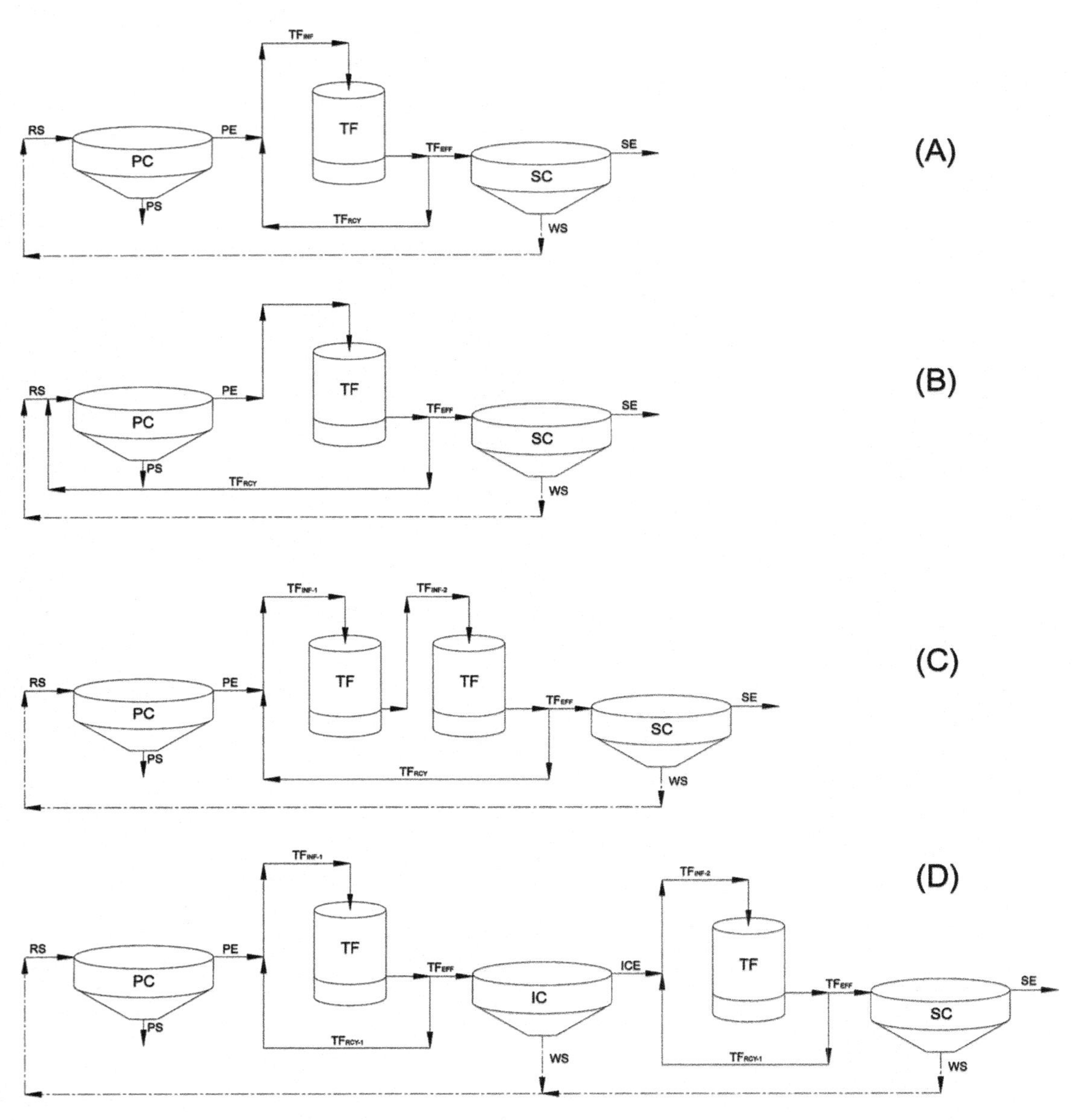

FIGURE 3.8 Trickling filter recirculation layouts.

nitrification rate in trickling filters. It is generally believed that nitrification rates are reduced as a result of substrate limitations starting at ammonium–nitrogen concentrations somewhere between 2.0 and 4.0 mg/L (WEF and ASCE, 1998). Figure 3.9 provides the approximate upper and lower limits of ammonium removal rates measured at five nitrifying trickling filter facilities (Okey and Albertson, 1989). It should be noted that the values used to develop Figure 3.9 have not

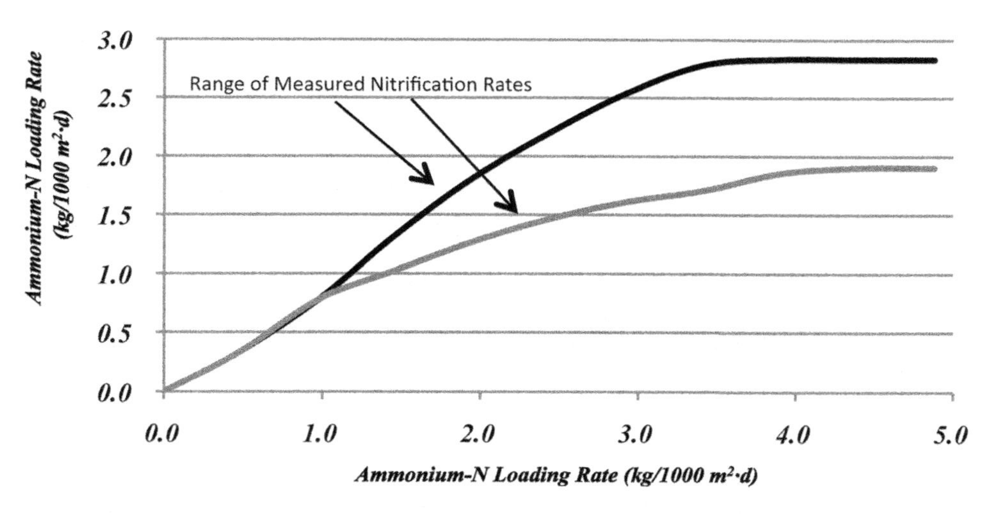

FIGURE 3.9 Trickling filter nitrification (Okey and Albertson, 1989).

been corrected for temperature or dissolved oxygen concentration. It has been hypothesized that ammonium or dissolved oxygen limitations can mask the effect of temperature on the nitrification process (Okey and Albertson, 1989). Because nitrification produces hydrogen ions, the pH of the trickling filter effluent can be depressed. This may result in a reduction in nitrification rates in the portions of the filter affected. The effects of lower pH on process performance are typically limited to situations in which recirculation rates are low. The pH levels can be kept in check by alkalinity addition, as discussed earlier in this chapter.

3.3 Rotating Biological Contactors

Rotating biological contactors (RBCs) consist of a series of circular plastic disks mounted on a horizontal shaft. The most common configuration consists of the shaft mounted above a tank and the disks approximately 40% submerged in the wastewater. In some instances, RBCs are almost fully submerged, and oxygen transfer is accomplished fully by diffused aeration. The shaft is rotated at 1 to 2 rpm, alternately exposing the plastic disks to wastewater and air (WEF and ASCE, 2009). Bacteria and microorganisms attach themselves to the disks and form a biofilm covering the media surface. The microorganisms respond to the environmental conditions present, and the types and numbers of organisms present on the media will vary from stage to stage. A stage may consist of a portion of one shaft, one complete shaft, or even multiple shafts. Excess biomass periodically "sloughs" off the RBC media and is typically removed via downstream clarifiers. Figure 3.10 shows a schematic of a typical RBC system.

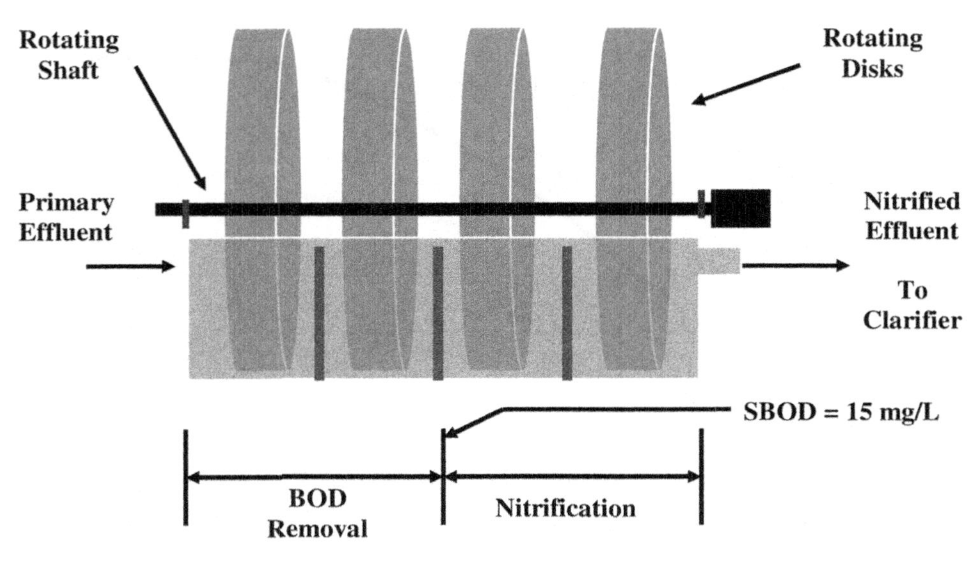

FIGURE 3.10 Rotating biological contactor schematic.

Unlike trickling filters, RBCs cannot readily be used for heavy organic loading or "roughing" applications. Early in their history, RBCs experienced a number of structural shaft failures associated with excessive and unequal biomass buildup on the disks. Shaft designs were modified to alleviate this problem, and RBCs are well suited to provide a high level of secondary treatment, nitrification, and, in fully submerged applications, denitrification. Rotating biological contactors are typically provided in standard 8.2-m- (27-ft-) long shafts with disks that are 3.7 m (12 ft) in diameter. Media densities range from 9290 m^2 (100 000 sq ft) per shaft to 16 722 m^2 (180 000 sq ft) per shaft. Rotating biological contactors that are used for CBOD removal are typically configured with a lower media density to avoid media clogging, and use of higher density media is typically reserved for CBOD polishing or nitrification.

As with other biofilm systems, nitrification using RBCs is subject to the wastewater characteristics entering the reactor. Nitrification will not commence until CBOD is reduced sufficiently to allow the nitrifiers to compete successfully on the media surface against heterotrophic bacteria. Removal of CBOD in a RBC is typically limited by a maximum first-stage BOD_5 loading of 24 to 29 $g/m^2{\cdot}d$ (5 to 6 lb/d/1000 sq ft), a soluble BOD loading of 12 $g/m^2{\cdot}d$ (2.5 lb/d/1000 sq ft) on any RBC stage, and an overall BOD_5 loading of approximately 10 $g/m^2{\cdot}d$ (2 lb/d/1000 sq ft) to reduce effluent concentrations sufficiently to commence nitrification. The surface area requirements used for nitrifier growth cannot be met until the organic (CBOD) loading has been reduced. Therefore, in determining the overall media requirements needed to achieve nitrification, the amount of media needed for ammonium removal must be added to the amount needed for CBOD removal.

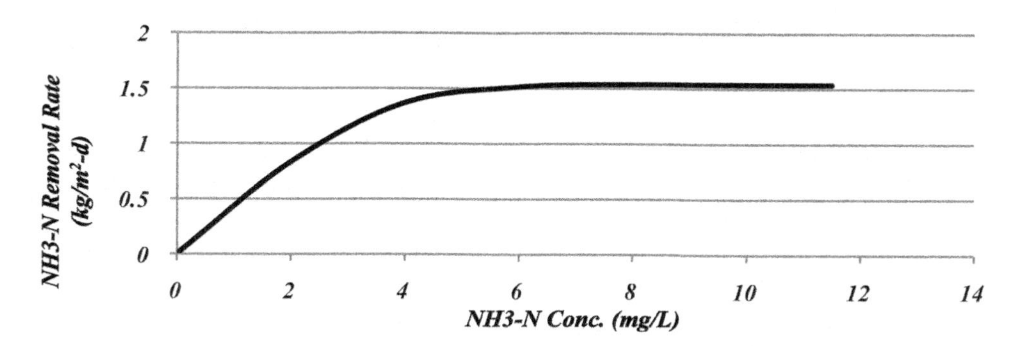

FIGURE 3.11 Rotating biological contactor nitrification rates at 20 °C.

Once CBOD is reduced sufficiently to start nitrification, the following step-by-step procedure can be used to determine surface area requirements and the number of RBC shafts needed to reduce the effluent ammonia concentration:

1. Determine the ammonium–nitrogen removal rate for the effluent ammonia concentration required from Figure 3.11;
2. If the wastewater temperature is less than 13 °C (55 °F), calculate the temperature correction factor that needs to be applied to the RBC surface area required from Figure 3.12; and
3. Apply the temperature correction factor to the ammonium removal rate to determine the surface area required for nitrification.

In the event that the effluent ammonia concentration required is less than 5 mg/L, the surface area requirement should be determined in two steps: (1) from

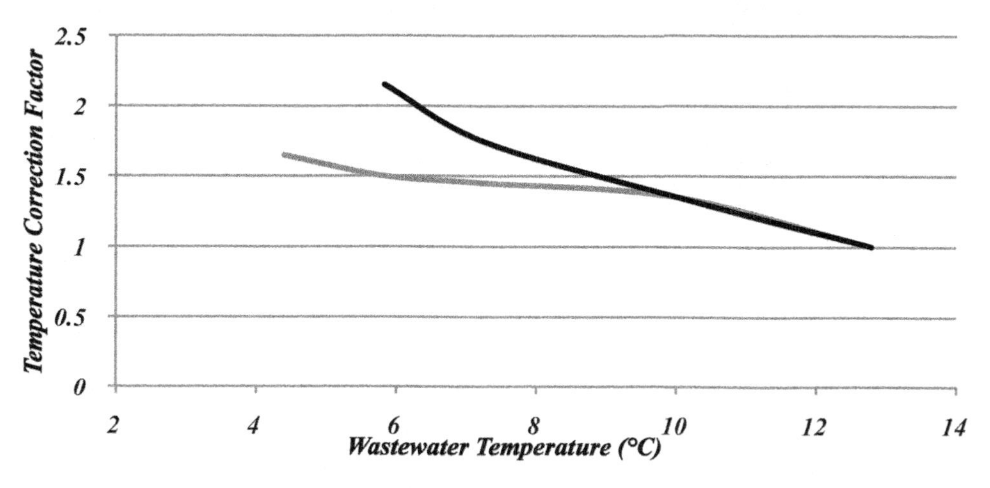

FIGURE 3.12 Rotating biological contactor temperature correction.

the influent concentration down to 5 mg/L and then (2) from 5 mg/L to the effluent concentration required (U.S. EPA, 1993a).

Rotating biological contactor systems are typically configured with multiple stages, particularly when low effluent BOD_5 or ammonia limits are required. Table 3.4 presents two examples of one RBC manufacturer's staging recommendations.

3.4 Biological Aerated Filter

A biological aerated filter (BAF) is a biofilm treatment process that combines aerobic biological treatment with filtration, often eliminating the need for a separate solids removal process. Biological aerated filters can be configured either as upflow or downflow units, with either a fixed or floating bed of media. The BAFs provide a physical configuration for biomass to either attach to or be trapped between the filter media. Air is sparged into the bottom of the filter to provide an aerobic environment, allowing the biomass to oxidize CBOD or ammonia as it passes through the filter. As the biomass within the filter builds up, liquid flow is restricted, creating an increase in head loss through the filter (WEF, 2000). The filter is periodically backwashed to remove excess solids. However, the backwash is intentionally not so vigorous to remove all the biological solids present; most of these biological solids are retained to provide for continued treatment of the wastewater. Figure 3.13 presents an upflow fixed-bed BAF configuration.

4.0 COUPLED SYSTEMS

A *coupled process* can be defined as a process configuration that combines two different treatment processes. Most coupled processes consist of a biofilm reactor

TABLE 3.4 Rotating biological contactor manufacturer-recommended staging (U.S. EPA, 1993b).

	Carbon oxidation		Nitrification	
	Effluent BOD_5	Number stages	Effluent ammonia–N	Number stages
Envirex	>25 mg/L	1	5 mg/L	1
	15 to 25 mg/L	1 to 2	<5 mg/L	Based on first-order kinetics
	10 to 15 mg/L	2 to 3		
	<10 mg/L	3 to 4		
Lyco	<40% removal	1	<40% removal	1
	35 to 65% removal	2	35 to 65% removal	3
	60 to 85% removal	3	60 to 85% removal	3
	80 to 90% removal	4	80 to 90% removal	4

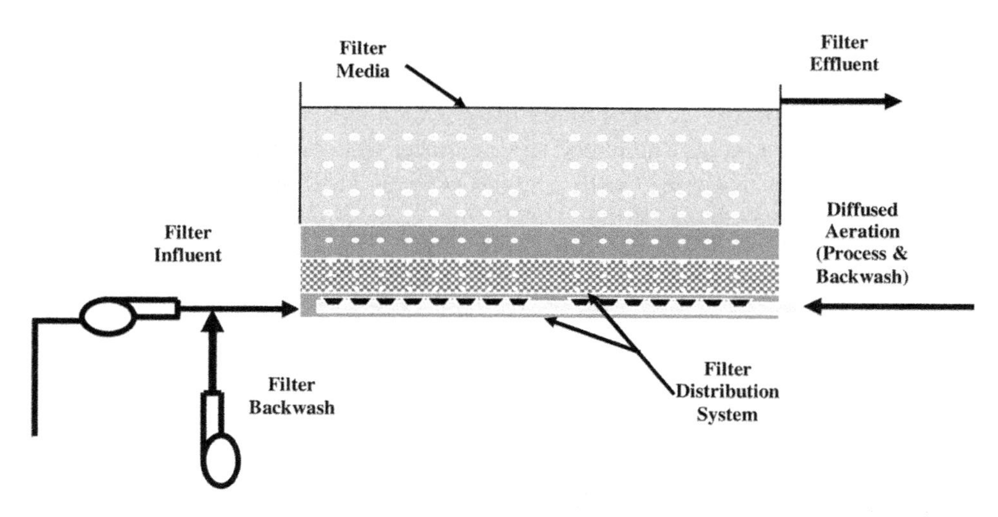

FIGURE 3.13 Biological aerated upflow filter.

followed by a suspended growth reactor. The most commonly used coupled processes currently in use are the trickling filter and solids contact (TF/SC) and the roughing filter and activated sludge (RF/AS) processes. Figure 3.14 shows one configuration of the RF/AS process. Combining the two processes is often a means of cost-effectively upgrading an existing facility and may also be used for new designs of secondary treatment facilities. Using dual processes offers the opportunity to use the best attributes of each process while limiting some of their individual weaknesses. Because most coupled process configurations include a trickling or roughing filter as the initial process, they are not necessarily

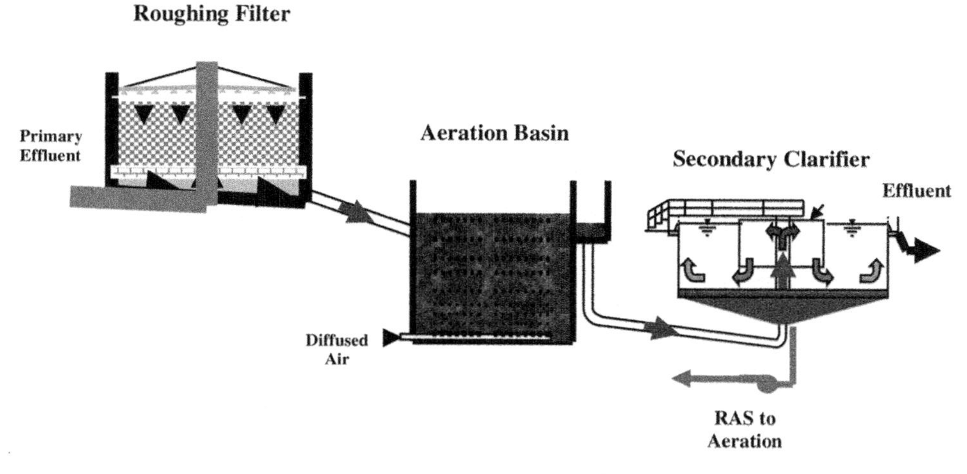

FIGURE 3.14 The RF/AS process.

an optimum configuration for total nitrogen removal. However, upgrading a facility using a coupled process is often a cost-effective alternative to implementing nitrification or a modest level of total nitrogen removal, especially if the trickling filters already exist and the activated sludge process represents the upgrade. Some operational issues associated with coupled systems include the accumulation of filter snails in the activated sludge basin, which can require additional operation and maintenance attention, and the presence of nuisance filter flies.

Integrated fixed-film activated sludge (IFAS) and moving bed bioreactor (MBBR) processes have become increasingly popular technologies for WRRFs trying to increase nitrification and total nitrogen removal capacity. Both processes use biofilms, with IFAS incorporating the biofilm process to an aeration tank, and MBBRs relying solely on biofilm processes in a flow-through tank configuration. The IFAS process uses RAS, as shown in Figure 3.15a, while the MBBR does not use RAS recycle (Figure 3.15b). Both processes can be used to increase nitrification capacity and, although they can be designed and built as new processes, they are often implemented as upgrades to existing activated sludge facilities.

The IFAS process uses either fixed or free-floating media that can be made of plastic, foam, or fabric. The fixed media are anchored within the reactor and typically resemble a structured plastic filter media (net or mesh) or bundles of ropes. Media are installed in an activated sludge basin to encourage the establishment of biofilm in the tank, thus increasing the solids inventory in the basin. Table 3.5 presents a brief description of different types of fixed and free-floating media that are currently in use in IFAS systems. Integrated fixed-film activated sludge systems are often implemented to improve nitrification capacity at existing facilities without increasing the solids load rate on secondary clarifiers. The IFAS systems have been shown to achieve 100% nitrification with 50% or less of aeration volume compared to conventional activated sludge systems.

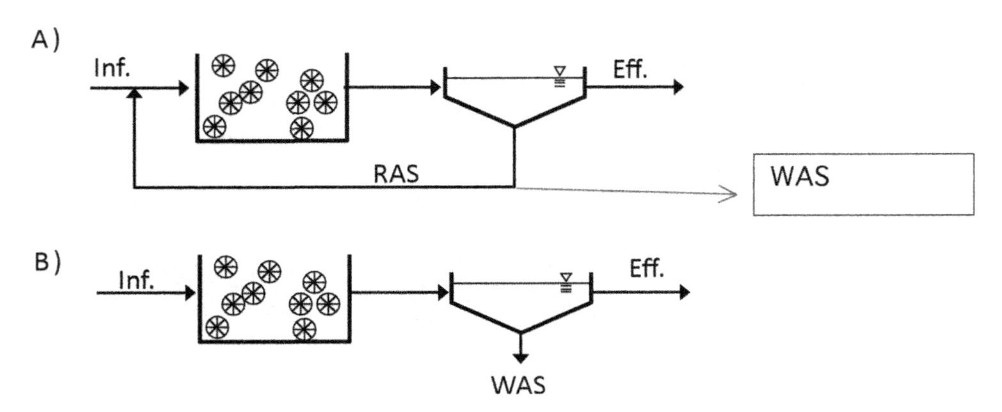

FIGURE 3.15 (a) Schematic of a simple IFAS application showing RAS recycle from a secondary clarifier and (b) schematic of a simple MBBR application.

TABLE 3.5 Types of media used in IFAS systems.

Media type	Advantages	Disadvantages
	Fixed media	
Fabric web type	Simple and quick to install, low initial cost, low maintenance, no risk of media loss	Prone to fouling (rags and red worm infestations), accumulation and trapping of debris, may require primary treatment upgrade
Rope type	Simple and quick to install, low initial cost, low maintenance, no risk of media loss	Prone to fouling (rags and red worm infestations), media breakage and entanglement, field assembly required
Poly vinyl chloride sheet (trickling filter media)	Simple and quick to install, low initial cost, low maintenance, no risk of media loss	Structural media may impede mixing and air distribution, excess biomass accumulation may cause clogging, rags can cause clogging
	Free-floating media	
Polypropolyene finned cylinders	Excellent mixing, staged implementation, may convert to MBBR	Potential for media loss (abrasion and washout), fouling of aeration system and screens, difficult to maintain the aeration system, need to account for media distribution

The key to the IFAS process is the increased total biomass inventory that can be achieved by combining the attached biomass with the suspended MLSS. The amount of biomass that can be retained on the different types of media depends on multiple factors, including loading rate, dissolved oxygen concentration, media type, temperature, mixing, suspended activated sludge biomass concentration, and overall SRT. Because the attached biomass is retained in the activated sludge basin and not sent to the clarifiers, use of IFAS technology can increase the capacity of the activated sludge system in the same tank volume without significantly affecting the clarifiers.

To implement IFAS technology, the following design and operational issues must be considered:

- Mixing requirements—enough mixing energy is needed to ensure that free-floating media remain well mixed and uniformly distributed in the tank. The mixing requirement is also critical for sloughing of biomass and maintenance of a thin biofilm to reduce mass-transfer limitations;
- Aeration requirements for nitrification systems—a higher dissolved oxygen concentration (greater than 3.0 to 4.0 mg/L) is typically required in the suspended phase to overcome the biofilm mass-transfer limitation and ensure that the biofilm is completely aerobic. In addition, many IFAS systems require the use of course-bubble diffusers as opposed to fine-bubble diffusers, which can result in additional blower capacity;

- Effluent screens—when free-floating media are used, effluent screens or media retention sieves (typically 6-mm screens) must be installed in the bioreactors to retain the media within the reactors. The requirement that all tank effluent pass through the screens can result in a foam trap that can increase foam accumulation in the IFAS tanks, which may require special foam control tools to control (i.e., defoamant surface spray and selective surface wasting). In addition, the screens may also result in the accumulation of floatables and other debris, which can block or clog the effluent screens. For this reason, upgrades to preliminary and primary treatment may need to be considered when implementing an IFAS system; and
- Media retention and distribution—when floating media are used, aeration diffuser placement and power need to be carefully considered to ensure that the media are well distributed throughout the tank and that they do not accumulate at the end of the tank because of excessive bulk flow velocities. Recirculation of the media to the head of the tank and/or implementation of a hydraulic knife at the effluent screens may be needed to avoid media buildup at the screens. Incidents of effluent screen failures resulting in total or partial loss of floating media have been reported.

Facility operators considering installing an IFAS system as an upgrade may also need to consider upgrading influent screens (e.g., 6-mm screens for facilities with primary clarifiers and 3-mm screens for facilities without primary clarifiers). Additionally, the effluent screen or sieve approach velocity should be monitored and maintained at a level (<35 mg/h) that reduces media stacking toward the effluent end of the tank. Finally, operators can control the approach velocity by reducing RAS rates. Tools to control froth and scum accumulation may also be needed for some installations.

One significant operational issue associated with IFAS systems is the need to remove media for tanks and aeration system maintenance. During such an event, the facility must also have plans for transport and storage of the media when removed. Fixed-media IFAS systems avoid many of the media retention and washout issues, but will still require equipment and facilities for removal during aeration tank maintenance.

Integrated fixed-film activated sludge systems can be used to increase both nitrification and denitrification capacity, but are most commonly used to increase nitrification capacity. Figure 3.16 depicts a schematic of IFAS for a 4-stage BNR process.

Moving bed bioreactors are similar to free-floating media IFAS systems, except there is no RAS recycle and the process relies almost entirely on biofilm activity. Many of the same issues associated with free-floating IFAS media are also a concern in MBBR systems (i.e., media loss, foaming, etc.). The key advantage of MBBR processes is their ability to retain slow-growing organisms and to maintain

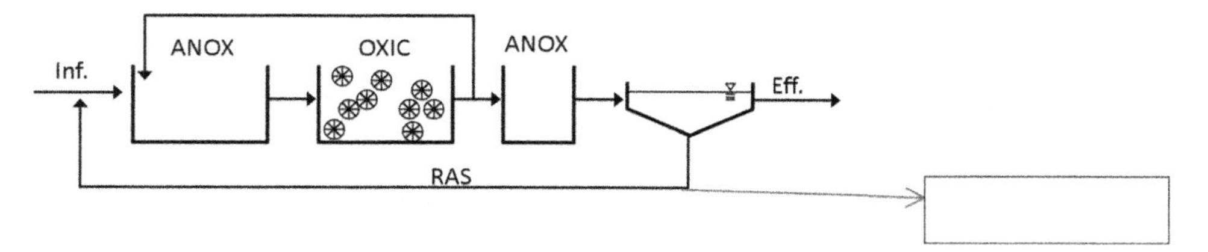

FIGURE 3.16 Schematic of IFAS implementation in a BNR process with media placed in the oxic zone to increase nitrification capacity.

a high equivalent SRT. For this reason, MBBRs are currently being implemented as separate centrate treatment technologies that use anaerobic ammonia oxidation (Anammox) biomass and deammonification processes. These types of systems require specialized biomass to be retained in the reactor for long SRTs (>30 days). One such process is the MBBR nitritation and Anammox process, which allows for high levels of specialized biomass to be retained for long SRTs while providing the specific biofilm niche needed to carry out this two-path BNR removal process. Additional information on Anammox, IFAS, and MBBR processes can be found in Water Environment Federation's (2010) *Biofilm Reactors* and in Chapters 4 and 6 of this manual.

5.0 OPERATIONAL CONSIDERATIONS

Attention to all operating considerations is important to ensure an efficient process and one that consistently meets required permits. Operational considerations include

- Dissolved oxygen,
- Alkalinity,
- Inhibition,
- Hydraulic retention time, and
- Nitrite formation.

For more information on these topics, see Section 1.2.2.2.

6.0 REFERENCES

Block, E.; Koopd, H. P.; Harms, H.; Alers, A. (1991) The Biochemistry of Nitrifying Organisms; In *Variations of Autotrophic Life;* Shively, J. M, Barton, L. L., Eds.; Academic Press: New York.

Grady, C. P. L.; Lim, H. C. (1980) *Biological Wastewater Treatment: Theory and Applications;* Marcel Dekker: New York.

Metcalf and Eddy, Inc. (2003) *Wastewater Engineering: Treatment and Reuse*, 4th ed.; Tchobanoglous, G., Burton, F. L., Stensel, H. D., Eds.; McGraw-Hill: New York.

Okey, R. W.; Albertson, O. E. (1989) Diffusion's Role in Regulating Rate and Masking Temperature Effects in Fixed Film Nitrification. *J.—Water Pollut. Control Fed.*, **61**, 510.

U.S. Environmental Protection Agency (1993a) *Nitrogen Control Manual;* EPA-625/R-93-010; U.S. Environmental Protection Agency: Cincinnati, Ohio.

U.S. Environmental Protection Agency (1993b) *RBC Nitrification Design Using Zero-Order Kinetics;* EPA-600/J-94-168; U.S. Environmental Protection Agency: Cincinnati, Ohio.

Water Environment Federation (2010) *Biofilm Reactors*; Manual of Practice No. 35; Water Environment Federation: Alexandria, Virginia.

Water Environment Federation; American Society of Civil Engineers; Environmental and Water Resources Institute (2009) *Design of Municipal Wastewater Treatment Plants*, 5th ed., WEF Manual of Practice No. 8; ASCE Manuals and Reports on Engineering Practice No. 76; McGraw-Hill: New York.

7.0 SUGGESTED READINGS

McCarty, P. L. (1970) Phosphorus and Nitrogen Removal in Biological Systems. *Proceedings of the Wastewater Reclamation and Reuse Workshop*; Lake Tahoe, California, June 25–27; p 226.

Richardson, M. (1985) *Nitrification Inhibition in the Treatment of Sewage;* The Royal Society of Chemistry, Burlington House: Thames Water, Reading, Pennsylvania.

Sharma, B.; Ahlert, R. C. (1977) Nitrification and Nitrogen Removal. *Water Res.* (G.B.), **11**, 897.

Chapter 4

Nitrification in Biofilm Reactors

Leon S. Downing, Ph.D., P.E.;
Joshua P. Boltz, Ph.D., P.E., BCEE;
Michael John Dempsey, Ph.D., C.Biol.;
Sarah Hubbell; and Frank Kulick

1.0 INTRODUCTION

Nitrification is the oxidation of ammonium to nitrate using specialized bacteria that occur naturally in soil and water. While Chapter 3 discussed nitrification in general, this chapter will focus on nitrification control in biofilm systems. The term, *biofilm,* is typically used for a community of microorganisms growing on a surface. Historically, bacteria growing on an attached surface in a wastewater system were referred to as *fixed-film reactors*. Recently, there has been a shift in the industry to referring to these systems as *biofilm reactors,* as evidenced by Water Environment Federation's (WEF's) publication, *Biofilm Reactors* (WEF, 2011). Such biofilms can be problematic when they occur inside pipes, valves, and so on, but advantageous when they grow on media in a biological treatment process. Often, the same microorganisms occur in both situations; the only difference is where they occur and how they are described (e.g., *fouling* when microorganisms are not wanted and *biofilms* when they are wanted).

Biofilm reactors are often implemented for increased nitrification in wastewater applications because the fixed nature of bacteria in biofilms allows for retention and proliferation of the "slow"-growing nitrifying bacteria. Given the increasing importance of nutrient removal using smaller footprint systems, nitrifying biofilm reactors constitute an important technology for water resource recovery operators. This chapter focuses on nitrifying biofilm reactors, although biofilm reactors can be implemented for other treatment goals as well. Understanding some of the basic fundamentals of reactions in biofilms and how to control these reactions in full-scale reactors are critical pieces of knowledge for effectively operating nitrifying biofilm reactors.

2.0 BASIC BIOFILM STRUCTURE AND FUNCTION

Understanding the operation of a biofilm reactor is dependent on understanding how biofilms grow. Physical transport of nutrients and oxygen into the biofilm layer must occur for biological activity to take place. The rate at which materials from wastewater are transported to the biofilm is dominated by diffusion. Diffusion is directly related to liquid velocity in the reactor. Liquid velocity is increased by increasing the flow of the liquid, either through pumping or mixing. As mixing intensity is increased, the rate of diffusion into the biofilm increases. This increased diffusion rate results in a higher rate of transport of "food" to bacteria in the biofilm and higher biological activity.

In addition to understanding the rate of food transport into the biofilm, the ecology of the biofilm is also important to understand. Each layer of biofilm consumes substrates, nutrients, and oxygen, resulting in a lower concentration of each of these compounds in every subsequent layer. The different concentrations of substrate, oxygen, and nutrients in each layer result in varying selective pressures on the ecology, which then results in a stratification of bacteria. Typically, heterotrophs are found on the outer edge of an aerobic biofilm, followed by nitrifying bacteria, and then anoxic and anaerobic metabolism at the base of the biofilm (Okabe et al., 1996). Oxidation of organic (carbonaceous) matter and ammonia (nitrification) can occur in the outer aerobic zone, while denitrification can occur in the anoxic base of the biofilm.

A simplified schematic of the stratification of a biofilm, with each subsequent layer being exposed to different compound concentrations, is shown in Figure 4.1. Figure 4.1 also shows idealized concentration profiles of substrates being consumed (dissolved oxygen, carbonaceous biochemical oxygen demand [$CBOD_5$], and ammonia [NH_4^+-N]) or produced (nitrate-nitrogen [NO_3^--N]) by the organisms in a biofilm. Substrate concentrations in the wastewater (liquid) are scaled to represent a secondary effluent concentration of 20, 30, and 10 mg/L of BOD, suspended solids, and NH_4^+-N, respectively, which diffuses from the bulk water through the mass-transfer boundary layer (MTBL) and into the biofilm. The reduced concentration of NO_3^--N compared to NH_4^+-N represents 10% of the nitrogen being assimilated into biomass and the change in concentration across the boundary layer is proportional to the individual substrate diffusion coefficient, with carbonaceous BOD (CBOD) idealized as glucose. The outer biofilm

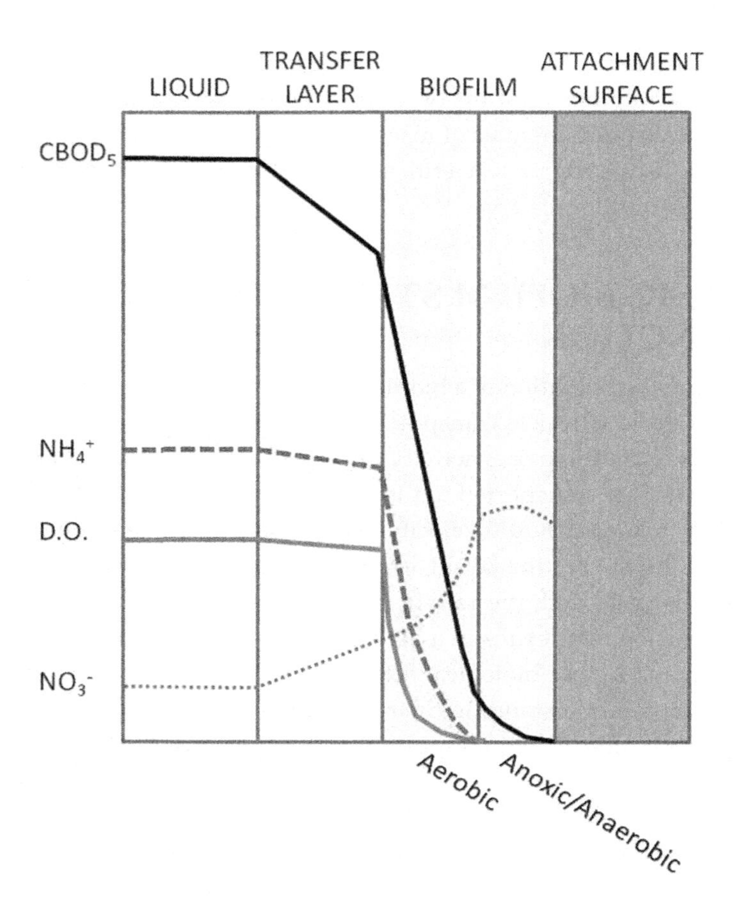

FIGURE 4.1 Idealized concentration profiles of substrates being consumed (dissolved oxygen, CBOD, and NH_4^+-N) or produced (NO_3^--N) by organisms in a biofilm. Substrate concentrations in the wastewater (liquid) represent a 20 : 30 : 10 (mg/L) secondary effluent of BOD, suspended solids, and NH_4^+-N, respectively, and a bulk liquid dissolved oxygen concentration of 7 mg/L.

is aerobic, but the inner biofilm is anoxic or anaerobic depending on the nitrate concentrations. Biofilm grows on a biomass support material.

Within the biofilm, as the concentrations of substrates (i.e., BOD and ammonia) change with penetration depth, the selective pressure for bacterial selection also changes. For each layer deeper in the biofilm shown in Figure 4.1, a lower BOD, ammonia, and dissolved oxygen concentration is observed. In contrast, the concentration of materials being produced (e.g., nitrate [NO_3^-]) rises as it diffuses out of the biofilm because it is being formed as the end product of ammonia oxidation (i.e., nitrification) by the nitrifiers.

The depth of oxygen penetration into a biofilm has been examined experimentally. Microelectrodes, which are essentially small-tipped dissolved oxygen sensors (i.e., the tip diameter is less than 10 μm), have been used to examine oxygen penetration into biofilms. In a biofilm of a trickling filter treating municipal wastewater, the dissolved oxygen concentration was found to reach zero at a depth of only 400 μm (0.016 in.) under steady-state conditions with raw wastewater. When the activity of the biofilm was increased by feeding the system an artificial nutrient broth, the dissolved oxygen concentration reached zero at a depth of only 100 μm (0.004 in.) (Kuenen et al., 1986). Using biofilm from an eel aquaculture trickling filter, Schramm et al. (1996) demonstrated that oxygen consumption and nitrate production only occurred in the outer 50 μm (0.002 in.) of the biofilm. These results highlight the rapid consumption of oxygen within a biofilm and how efficient transfer of oxygen into a biofilm is critical to optimize the amount of biofilm exhibiting aerobic activity.

Gradients develop within a biofilm as a result of different rates of diffusion from the wastewater and different rates of consumption by biofilm microorganisms. For a nitrifying biofilm, a partially penetrated biofilm exists when either ammonia or oxygen is depleted before reaching the bottom of the biofilm. Biofilms grown in full-scale reactors might be completely penetrated immediately after a hydraulically triggered detachment event or during an organic shock load, but it is likely that most biofilm reactors generally operate with partially penetrated biofilms. Once a nitrifying biofilm is partially penetrated, increasing the biofilm thickness has no nitrification benefit because the ammonia, oxygen, and/or macronutrients cannot penetrate to the full depth of the biofilm and, therefore, cells that are deeper than the penetration depth will be starved of these nutrients. However, the inner portions of a partially penetrated biofilm can become anoxic, leading to some levels of denitrification in the biofilm even if the biofilm is in an aerobic basin (Points et al., 2010). If biofilm thickness becomes excessive, biofilm zones existing near the biofilm support may be deprived of oxygen and/or macronutrients, resulting in anaerobic conditions in this ecological niche and potentially uncontrolled biofilm detachment (or sloughing) and/or odors.

3.0 NITRIFYING BIOFILM OPERATION—OVERVIEW

The primary difference between a suspended growth activated sludge reactor and a biofilm reactor is that suspended growth reactors are biomass-limited, whereas biofilm reactors are controlled by resistance to mass transfer (i.e., diffusion). Mass transfer is the driving force to convey pollutants from the liquid to the biofilm, where biological and chemical reactions reduce pollutant concentrations. While a wide range of nitrifying biofilm reactor technologies are currently available and implemented in practice, the overriding principles for operating biofilm reactors for improved nitrification can be summarized by the following two areas: mass-transfer control and biofilm thickness management.

Although nitrification rate control through mass-transfer control and biofilm management are discussed in separate sections in this chapter, they are intimately related, and modifications to either of the parameters will have a direct effect on the other parameter.

3.1 Nitrification Rate Control—Mass Transfer

The nitrification rate of the biofilm is tied to the thickness of the mass-transfer boundary layer (MTBL) (also known as *transfer layer* or *diffusion distance*) and concentrations in the bulk liquid. This mass transfer is described by Fick's Law. The thickness of the boundary layer can be thought of as the distance pollutants must travel from the well-mixed liquid to reach the biofilm. The nitrification rate is inversely proportional to diffusion distance, meaning that smaller diffusion distances result in higher nitrification rates. The smaller this diffusion distance, the "faster" pollutants can reach the biofilm, resulting in a higher amount of "food" (or substrate) in the biofilm for bacteria to consume.

The distance pollutants have to travel to reach the biofilm is one aspect of how much food reaches the bacteria. The other factor is the concentration of substrate in the liquid. Higher liquid concentrations lead to a larger mass of pollutants reaching the biofilm, resulting in increased bacterial activity. Pollutant concentrations in the liquid are controlled by influent loadings and effluent requirements; however, the oxygen concentration can be manipulated for most biofilm reactor technologies by increasing the oxygen concentration in the wastewater, thereby increasing reaction rates in the biofilm.

3.1.1 Decreasing Diffusion Distance

Penetration of food into the biofilm can be directly tied to the thickness of the mass-transfer boundary layer, or diffusion distance, and to reactions within the biofilm. *Transfer layer* is a term related to mass-transfer dynamics and can be thought of as the diffusion distance for substrate into the biofilm. The faster the liquid is moving, the shorter the diffusion distance and the more efficient the transfer of compounds into the biofilm.

While a high liquid velocity would theoretically lead to a small diffusion distance and, hence, high reaction rates, a balance must be achieved for the biofilm to remain attached to the support media. Achieving sufficient liquid velocity to achieve high levels of mass transfer into the biofilm and maintain adequate biofilm thickness is dependent on wastewater characteristics and the biofilm reactor technology implemented. Some operational control measures to decrease the diffusion distance into the biofilm include increased mechanical mixing, increased aeration flowrate, and increased flushing rates. For each of these techniques, more intense mixing or flow results in higher liquid velocity, thereby decreasing the diffusion distance. Details of these specific means for mass-transfer control for each technology will be discussed later in this chapter.

3.1.2 *Liquid Concentration Management*

For a nitrifying biofilm reactor, it is important to address bulk liquid BOD, ammonia, and oxygen concentrations. The higher the liquid concentration of a nutrient, the further it will penetrate a biofilm. As the concentration of BOD transferred into the biofilm increases, increased heterotrophic competition occurs for space and oxygen, resulting in decreased nitrification rates. To help limit the effect of BOD on nitrification rates, a BOD loading rate to the biofilm reactor should be maintained that minimizes diffusion of BOD in the biofilms. This loading varies for different reactor technologies. If the BOD loading rate exceeds the recommended loading for a given technology, it is critical to either (1) increase BOD removal upstream of the biofilm reactor or (2) increase the surface area within the reactor for biofilm attachment. Depending on the type of biofilm reactor, adding media to increase the biofilm attachment surface area can be difficult and/or costly. Operational control to limit the effect of excessive BOD in the reactor can be accomplished by increasing BOD removal upstream of the nitrifying biofilm reactor through methods such as chemically enhanced primary treatment (CEPT) or upstream biological processes.

Increasing the dissolved oxygen concentration in the liquid will also result in increased oxygen penetration into the biofilm and, potentially, much thicker biofilms. This results in a thicker aerobic layer, which can allow for both the higher heterotrophic population and the nitrifying population to coexist in the same biofilm. Although this can be an effective adjustment, it can also lead to excessive biofilm thickness. Excessive thickness, in turn, can lead to sloughing events and elevated effluent total suspended solids (TSS) concentrations.

Lower bulk liquid ammonia concentrations will result in slower nitrification rates in the biofilm because of the decreased transfer of ammonia into the biofilm. When stringent effluent ammonia limitations are in place (e.g., lower than 1 mg/L), a multistaged nitrifying biofilm reactor can be implemented, with the first stage having a higher bulk liquid ammonia concentration (and thus a higher nitrification rate) and the later stage having a lower bulk liquid ammonia concentration (and thus selecting for a biofilm ecology that thrives in an environment with a low ammonia concentration).

For both bulk liquid BOD and ammonia concentrations, increasing the attachment surface area can decrease the bulk liquid concentration of these pollutants because more surface is available for the growth of active biofilm. Thus, for the same biofilm thickness, more surface area results in more biomass. The resulting increase in active biofilm volume can increase reaction rates, thereby decreasing the bulk liquid concentrations. However, for several biofilm reactor technologies, it is not possible to increase the attachment surface area without significant infrastructure changes.

Controlling the dissolved oxygen concentration in bulk liquid can be the most direct operational control parameter. Whether the dissolved oxygen concentration is controlled via mechanical mixing, diffused aeration, or water recirculation and flushing, it is critical to ensure that sufficient oxygen is present in the liquid so that oxygen can diffuse throughout the entire biofilm thickness. As discussed previously, the dissolved oxygen may only penetrate a biofilm to a depth of 400 μm (0.016 in.) under steady-state conditions in a wastewater system (Kuenen et al., 1986). In addition, increasing the bulk liquid dissolved oxygen typically results in increased fluid velocity and, hence, a shorter diffusion distance and higher transfer rates into the biofilm. A higher dissolved oxygen concentration in the bulk liquid or a smaller diffusion distance are directly tied to a higher biofilm oxygen concentration, resulting in a higher level of activity within the biofilm and, therefore, higher nitrification rates.

3.1.3 *Relationship of Diffusion Thickness and Liquid Concentration*

For nitrifying biofilm reactors, controlling the penetration of dissolved oxygen into the biofilm is directly tied to mechanisms for mixing, whether they are mechanical mixing, flushing water rates, diffused aeration, or other means of fluid movement. As the liquid velocity at the biofilm surface is increased, so, too, is the dissolved-oxygen-transfer rate into the biofilm. The higher liquid velocity reduces the diffusion distance, which also increases the nitrification rate. However, the liquid velocity must be monitored to ensure that velocities are not getting so high that they scour the biofilm off the attachment surface.

The point of excessive fluid velocity is highly dependent on the reactor type and specific field conditions (temperature, BOD loadings, alkalinity, etc.). From an operations standpoint, increasing the fluid velocity to decrease diffusion distance and increase the dissolved oxygen concentration should be made in a stepwise manner. These stepwise increases well help ensure that excessive biofilm detachment is not being induced from an attempt to increase the nitrification rate. Slowly increasing aeration rates, flushing rates, or backwash cycles can help prevent excessive biofilm loss. Losing the biofilm will obviously have a detrimental effect regardless of the bulk liquid concentrations because, without sufficient biofilm, nitrification will not occur at a sufficient rate.

3.2 Nitrification Rate Control—Biofilm Thickness Management

Mass-transfer control affects the transfer of substrates into the biofilm, but biofilm thickness and ecology affect the chemical and biological reactions that occur within the biofilm. A biofilm system will respond to varying environmental conditions (e.g., temperature and alkalinity) and electron donor, electron acceptor, and/or macronutrient availability. The physical response will be a change in biofilm mass, which is mainly observed as a change in biofilm thickness. This change in biofilm thickness results in a change to the substrate penetration depth into the biofilm, which affects the ecology, reaction rates, and overall health of the biofilm.

In reactors with removable media (e.g., moving bed biofilm reactors [MBBRs] and biologically aerated filters [BAFs]), this biofilm thickness information can be measured and used quantitatively in process models and qualitatively as an indicator of nitrification rate and the presence of anoxic and anaerobic layers within the biofilm.

3.2.1 *Substrate Penetration*

Substrate concentration gradients within a biofilm result from either partial or complete penetration by any soluble substrate diffusing into the biofilm. For a nitrifying biofilm, partial penetration exists when the ammonia, oxygen, and/or macronutrient is depleted before reaching the biofilm support. Biofilms grown in full-scale reactors might be completely penetrated immediately after a hydraulically triggered detachment event, during an organic shock load, or during certain points of the diurnal flow pattern. However, it is likely that most biofilm reactors generally operate with partially penetrated biofilms.

For facilities with low-level ammonia and/or phosphorus effluent limitations, care should be taken in the final stage of a nitrifying biofilm system to ensure that ammonia and phosphorus, in addition to oxygen, penetrate the biofilm fully. The concentration of ammonia and phosphorus affects this penetration, as does maintaining a shorter diffusion distance.

3.2.2 *Biofilm Ecology*

Biofilm ecology is controlled by substrate concentrations within the biofilm, which are directly tied to the mass-transfer controls discussed previously. For a nitrifying biofilm reactor, ecology concerns are mainly associated with the competition for oxygen between heterotrophic bacteria consuming BOD and autotrophic, nitrifying bacteria oxidizing ammonia. Substantial work has been conducted concerning the competition between these two groups of bacteria in biofilms for wastewater treatment (Fernandez-Polanco et al., 2000; Nogueira et al., 2005; Okabe et al., 1996). As the influent BOD concentration increases, heterotrophs and nitrifiers form a distinct stratification, with nitrifiers in the interior and heterotrophs at the exterior (Satoh et al., 2000). This stratification at high bulk liquid BOD has been verified in multiple reactor types (Cole et al., 2004; Downing and Nerenberg, 2008a, 2008b; Fernandez-Polanco et al., 2000; Hibiya et al., 2003; Okabe et al., 1996; Satoh et al., 2004; Terada et al., 2003).

For nitrifying biofilm reactors, the key to selecting for a biofilm dominated by nitrifying bacteria is to maintain a bulk liquid BOD concentration low enough to prevent excessive heterotrophic growth and to achieve sufficient oxygen penetration to maintain a high level of nitrification activity. The mass-transfer factors are controlled by the operational tools discussed previously, but the management of mass-transfer factors also has a positive effect on biofilm ecology (e.g., increased oxygen transfer increases aerobic biofilm volume and the volume of nitrifying bacteria), resulting in further optimization of nitrification rates.

3.2.3 *Biofilm Detachment*

Detachment by physical removal or grazing is a natural event that is necessary for successful operation of a biofilm reactor. Detachment is the only process by which biomass is wasted in a biofilm reactor (in a hybrid reactor, where suspended solids are present, wasting would occur with removal of the suspended growth). Bryers (1984) described four biofilm detachment processes: abrasion, erosion, sloughing, and predator grazing. *Abrasion* (initiated by particle collision) and *erosion* (initiated by fluid-induced shear in the vicinity of the biofilm surface) refer to the removal of small groups of cells and represent a relatively constant removal of biofilm biomass. *Sloughing* refers to the loss of large segments of biofilm approximately equal to the biofilm thickness (Morgenroth, 2003). Predators such as macrofauna and microfauna graze biofilms and consume biomass (Boltz et al., 2008). Depending on the reactor type, the detachment mechanism may be one or more of the aforementioned processes. Whichever method is implemented in a biofilm reactor, controlled and consistent detachment of biomass is critical to a healthy, stable, nitrifying biofilm reactor. Just as an activated sludge system needs to have consistent wasting of biomass, so does a biofilm system. If the only wasting of biofilm is via large, uncontrolled sloughing events, not only will the facility likely experience spikes in effluent TSS, but the remaining biofilm will likely experience periods of decreased nitrification capacity.

Excess biofilm accumulation can result in increased hydraulic head loss, odors, reduced media specific surface area, increased biofilm weight (which can exceed design structural loads), and media clogging. All biofilm reactor types rely on biofilm thickness control, which is commonly achieved by increasing the liquid velocity. This control occurs either at a constant rate or during periodic increases in fluid velocity to shear excess biofilm. If excess biofilms develop, substrates and/or nutrients do not penetrate to the base of the biofilm. As noted previously, these nutrient-deficient zones at the base of the biofilm can result in large biomass inactivity at the base of the biofilm, where attachment mechanisms are important. This can result in large biomass loss events (sloughing), reducing the overall biomass in the reactor and detrimentally affecting the observed nitrification rate.

3.3 General Operation and Control Parameters for Nitrifying Biofilm Reactors

Effective biofilm reactor operation consists of actively controlling the reactor's liquid velocity and managing biofilm thickness. Active control allows for management of mass transfer into the biofilm, management of the diffusion distance, and management of the biofilm thickness. For many technologies, the same operational tool is used to achieve both of these control objectives. As will be discussed for each technology later in this chapter, adjusting fluid velocity is one of the main operational tools available to a biofilm reactor operator. Whether adjusting flushing rates, aeration rates, backwash cycles, or scour periods, changes in biofilm reactor operation are generally geared toward management of the fluid velocity. This requires operational observation of biofilm thickness, biofilm consistency, nitrification rates, and, potentially, oxygen concentrations within the reactor. For each technology, methods for monitoring these parameters and operational responses to changes in parameters will be discussed.

For common nitrifying biofilm reactor technologies, the main operational decisions will involve increasing or decreasing the liquid velocity. A summary of operational goals and general operation and control parameters is shown in Table 4.1; the table also includes an indication of whether the liquid velocity should be increased or decreased in response to the operational change. Like all wastewater processes, these control parameters are not independent of each other. However, an understanding of their effect on biofilm reactor performance and how to effectively control each area is important to successful operation of a biofilm reactor system.

4.0 BIOFILM REACTOR ANCILLARY COMPONENTS

While operation of the nitrifying biofilm reactor itself is the main focus of this chapter, the effect of operations associated with several ancillary processes not directly tied to the biofilm is important for overall permit compliance and maintenance procedures.

4.1 Influent Screening

Several biofilm processes require ultrafine screening of influent wastewater to limit the accumulation of rags and debris within the biofilm reactor. Several biofilm reactor manufacturers recommend the use of 3-mm screens for facilities without primary clarifiers and 6-mm screens for facilities with primary clarifiers.

4.2 Phosphorus Removal

Unless an activated sludge system is used in conjunction with a biofilm reactor, phosphorus removal is typically achieved through biological assimilation and chemical precipitation. Enhanced biological phosphorus removal is only achieved when biomass is exposed to cyclic anaerobic and aerobic conditions, which are only achievable in a system where the bacteria are exposed to both anaerobic and aerobic conditions (see Chapter 7). This cycling between anaerobic and aerobic conditions is typically not achieved in biofilm reactor systems.

When chemical precipitation for phosphorus removal is used in conjunction with a biofilm process, it is critical that enough soluble phosphorus enters the

TABLE 4.1 General operation and control parameters for nitrifying biofilm reactors.

Operational goal	Operation modification	Liquid velocity change	Operation and control
Improve nitrification rate—mass transfer limitation	Increase transfer into the biofilm	Increase	Increase oxygen transfer rate to increase dissolved oxygen concentration in biofilm
			Increase upstream BOD removal to reduce heterotrophic competition in the biofilm
			Increase fluid velocity to decrease MTBL thickness
Improve nitrification rate—biofilm thickness limitation	Increase biofilm thickness	Decrease	Decrease fluid velocity/mixing intensity
			Increase dissolved oxygen concentration
			Increase biofilm attachment area by adding additional media if sufficient reactor capacity exists (% fill < maximum fill achievable)
Prevent biofilm septicity—biofilm thickness limitation	Decrease biofilm thickness	Increase	Increase fluid velocity and mixing intensity to increase shear force
			Implement cyclical backwash cycles
			Increase biofilm attachment area by adding additional media
Prevent biofilm septicity—mass transfer limitation	Increase oxygen transfer	Increase	Increase oxygen transfer rate to increase dissolved oxygen concentration in biofilm
Prevent excessive biomass loss (sloughing)—mass transfer limitation	Increase transfer into the biofilm	Increase	Ensure sufficient phosphorus and nitrogen entering the biofilm reactor
			Increase oxygen transfer rate to increase dissolved oxygen concentration in biofilm
			Ensure sufficient alkalinity to buffer pH
			Prevent oxygen and nutrient starvation
Prevent excessive biomass loss (sloughing)—biofilm thickness limitation	Decrease shear force on the biofilm	Decrease	Decrease fluid velocity and mixing intensity

biofilm reactor to ensure diffusion of phosphorus to the base of the biofilm. The soluble phosphorus concentration in the biofilm reactor is a critical component of phosphorus removal optimization in a biofilm system. Frequently, a multipoint addition system is used for chemical phosphorus removal in a biofilm system. Depending on site-specific hydraulic conditions and biofilm thickness, different phosphorus concentrations may be required to enter the biofilm reactor to prevent nutrient deficiency at the base of the biofilm and associated excessive biofilm sloughing events. If excessive sloughing events occur, it may be beneficial to examine the phosphorus concentration entering the biofilm process to determine if the biofilm is receiving sufficient phosphorus.

4.3 Solids Separation

Differences in solids generated in a biofilm system greatly affect settling performance. The ability of aerobic biofilm reactors to achieve effluent performance is dependent on the operation of the downstream settling process. Suspended solids remaining in the effluent contribute to BOD and organic nitrogen and phosphorus and have the potential to consume residual dissolved oxygen, which may result in nitrogen gas generation because of denitrification in the clarifier sludge blanket, causing sludge to float. Solids separation performance is affected by many factors, which are characterized in *Design of Municipal Wastewater Treatment Plants* (WEF et al., 2009). Effluent solids particles from a biofilm reactor are denser than activated sludge flocs, but poor bioflocculation results in a large range of biofilm aggregate sizes. This, in turn, results in a wide distribution of particle sizes and a fraction of biofilm reactor effluent solids that will almost never settle in a conventional clarifier (Ødegaard et al., 2010).

Because of the production of effluent solids aggregates with poor settling characteristics, several operational modifications have been developed to improve flocculation before solids separation. Solids contact zones, where settled solids from final clarification are recycled to an aerobic tank downstream of the biofilm process, have been implemented for several biofilm reactor technologies. This has resulted in production of increased solids settleability because of bioflocculation in the aerobic contact zone. Chemical flocculation has also been implemented for improved flocculation, in which a small flocculation basin is added between the biofilm reactor and clarifier. Chemical flocculants are also effective at precipitating phosphorus and chemical-based flocculation downstream of a biofilm process is an option for nitrifying biofilm reactor facilities where an effluent phosphorus permit is in place.

Detailed operational considerations for the solids separation process itself are the same for biofilm reactors as they are for other treatment systems. Dissolved air flotation, conventional clarification, high-rate clarification, and filtration have all been successfully applied to various biofilm reactor technologies (Ødegaard et al., 2010). Detailed discussions on solids separation technologies can be found in *Design of Municipal Wastewater Treatment Plants* (WEF et al., 2009). It should be noted that despite the lower effluent suspended solids concentrations and lower yields observed from biofilm reactors, it is typically not recommended to forgo the solids separation step because of occasional sloughing events that cause a high level of variability in effluent TSS concentrations.

4.4 Microfauna Management

Growth of microfauna has been a significant concern for many biofilm reactor installations. Microfauna vary depending on whether a biofilm reactor is a submerged biofilm or a non-submerged biofilm. For non-submerged biofilm reactors (e.g., trickling filters and rotating biological contactors [RBCs]), filter flies proliferate when the biofilm reactors have semidry pockets of old biomass available for grazing by fly larvae. Increasing wetting rates of the reactor will assist in reducing the suitability of the environment for propagation of filter flies.

Snails can become a significant problem, particularly in nitrifying trickling filters. Snails consume biofilm as a source of food and grow in dark, moist environments. In some instances, snails consume nitrifying bacteria faster than they are produced, which can result in loss of ammonia removal efficiencies. Snails also excrete ammonia wastes. There are several approaches to reduce snail populations. One approach is ammonia washing, which is a process in which the biofilm reactor is isolated during off-peak hours and the influent and recirculation flow stream is charged with water containing approximately 100 mg/L of ammonia (Daigger and Boltz, 2011). Sludge centrate, digester supernatant, or other ammonia-rich water may be used for this purpose. The addition of sufficient caustic to raise the pH to 9.5 or 10 is necessary to ensure a high concentration of toxic, un-ionized ammonia. The ammonia solution should be recirculated for several hours or a day if possible. The high strength of dissolved gaseous ammonia will discourage snail growth and the high recirculation rate will purge the snail shells and eggs from the tower. The elevated pH should be maintained by adding more caustic, if necessary. This procedure should be repeated, as necessary, to mitigate the snail population.

Ponding is a problem typically associated with rock trickling filters that occurs when the buildup of biomass, plant life, or inert solids does not permit proper flow of wastewater through the rock matrix. Increasing Spülkraft flushing intensity (the "SK rate") (see section 5.1.1) or taking the filter out of service to "dry out" the biomass or plant life may improve flow. If excessive grit or inert material is the cause, then media replacement is likely necessary.

For submerged biofilms, the main microfauna challenge has been associated with redworm proliferation. Several biofilm reactor systems have experienced a bloom of a type of redworm population that feeds on the biomass on the media. The worms are obligate aerobes. Thus, high levels of dissolved oxygen favor their growth. Occasional, un-aerated operation (in the range of 1 to 3 hours) can help to limit the worm growth potential. Maintaining a healthy thin biomass through shearing by dedicated aeration can help keep worms at bay, although this results in increased oxygen concentration, which can select for more worm growth. Controlling the worms consists of creating an anoxic condition in the area of the worm bloom by turning the air off for several hours. Nitrifying bacteria will survive this treatment because they can tolerate periods of oxygen depletion of at least 24 hours.

4.5 Peak Flow Management

Unlike suspended growth reactors, biofilm reactors are not at risk of biomass "washout" during peak flow events. Fluid velocity must be maintained below the point that would create excess biofilm detachment during peak flow events, but the attached biofilm will not be carried with the fluid flow as a suspended biomass would be. In addition, because of the lower effluent suspended solids concentration from a biofilm reactor, downstream clarifiers are typically not controlled by solids loading rates as an activated sludge clarifier would be.

5.0 NITRIFYING BIOFILM REACTOR TECHNOLOGIES

A wide range of biofilm reactor technologies are currently available and being implemented for nitrification. Figure 4.2 is a schematic of some of the most common nitrifying biofilm reactor technologies. From an application standpoint, the trickling filter, MBBR, and integrated fixed-film activated sludge (IFAS) are the most commonly installed reactor technologies, although full-scale examples of all of the reactor types shown in Figure 4.2 do exist. A detailed discussion of these three most common reactor technologies and a brief description of each of the additional technologies are included in this section. It is important to remember that the fundamentals of biofilm reactor operation are the same for each technology; only the mechanisms for control vary from reactor technology to reactor technology. For further information pertaining to the functions, designs, and details of each reactor technology, refer to *Biofilm Reactors* (WEF, 2011).

5.1 Trickling Filter

Trickling filters are aerobic biofilm reactors that receive either primary-treated or fine-screened wastewater that is distributed or trickled over the media. Nitrifying trickling filters receive secondary-treated wastewater typically from carbon-oxidizing trickling filters. As the wastewater travels down through the trickling filter contacting the biofilm, air flows either up or down through the trickling filter by natural convection or low-pressure fans. The direction of the airflow for natural convection depends on the density difference between the air within the trickling filter and exterior ambient air densities, which can vary daily with air

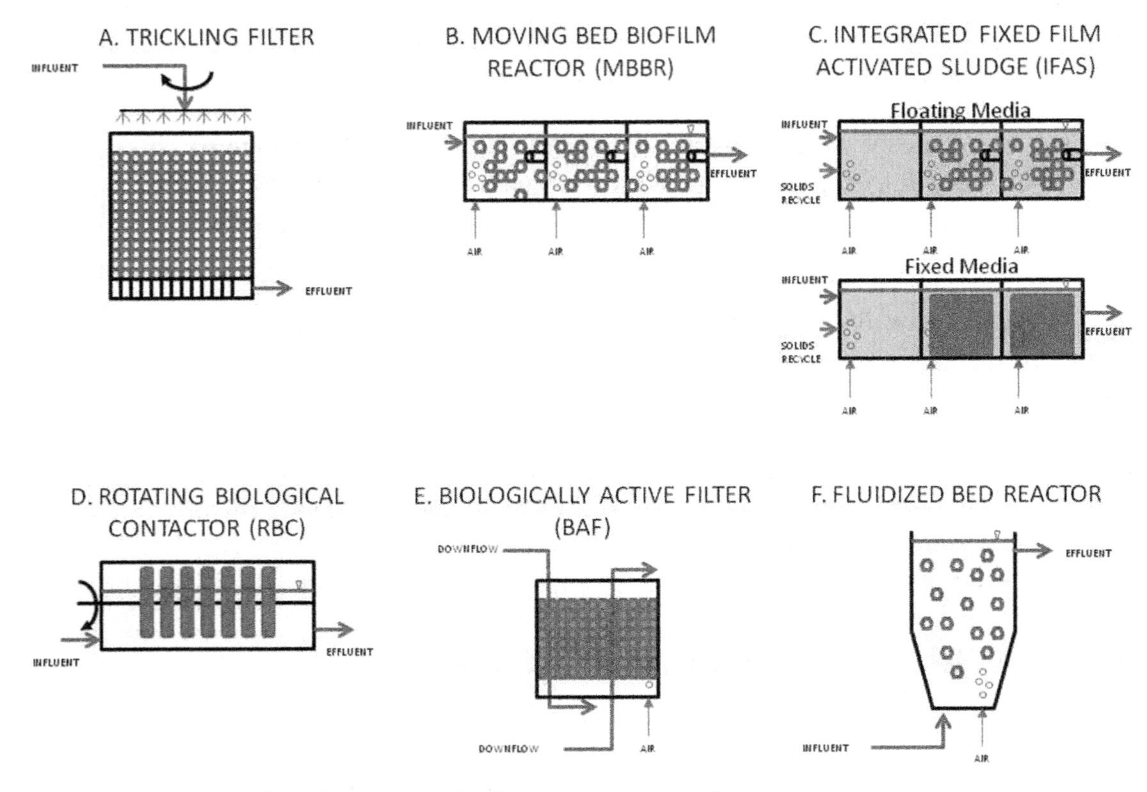

FIGURE 4.2 Examples of nitrifying biofilm reactor technologies.

temperature. *Diurnal stagnation* is the term used to describe the condition where air densities are equal and airflow is limited. During diurnal stagnation conditions, odors proliferate from the trickling filter. Treated wastewater leaves the media and enters the underdrains where it is collected and either recirculated to the top of the trickling filter or exited from the trickling filter process.

Trickling filters are classified by hydraulic and organic loading rates. As defined in *Design of Municipal Wastewater Treatment Plants* (WEF et al., 2009), filters are classified as either roughing, carbon-oxidizing, carbon-oxidizing and nitrifying, or nitrifying filters. The classifications are based on the intended service for BOD, BOD and ammonia, and ammonia removal, and give loading and expected removal efficiencies.

The trickling filter consists of media, a containment structure, underdrains, and a distributor. If forced ventilation is used, low-pressure fans and a cover are also required. The media serve to provide surface area for biofilm growth and can be structured sheet, rock, or random dump-type media. Structured sheet media can be cross flow, vertical flow, or a combination of the two. Cross-flow media act to continually redistribute flow to ensure utilization of the entire media surface.

Low-density surface area media (100 m^2/m^3)(31 sq ft/cu ft) are generally used for BOD removal, medium-density surface area media (160 m^2/m^3) (48 sq ft/cu ft) are used for nitrification and low-load BOD removal, and high-density surface area media (225 m^2/m^3) (68 sq ft/cu ft) are used for tertiary nitrification applications. Vertical-flow media are used when an unusually high solids concentration is applied to the trickling filter or when high-rate roughing is desired where sloughed biomass is more easily flushed from the media. Vertical-flow media are typically used in conjunction with two layers of cross-flow media above for improved distribution. Rock and random dump media are not self-supporting and require structural support to contain the media. *Design of Municipal Wastewater Treatment Plants* (WEF et al., 2009) presents a comparison of the physical properties of various types of trickling filter media. Underdrains collect treated wastewater and convey it for recirculation or to downstream unit processes, creating a plenum for the transfer of air throughout the media (Grady et al., 1999).

5.1.1 *Mass-Transfer Control*

Flow distribution, application rates, and recirculation rates are the main control mechanisms for trickling filter reactors. Distribution of flow to the top of the trickling filter is accomplished through fixed nozzles or a rotating distributor. Harrison and Timpany (1988) reported fixed nozzles as giving reduced performance because of incomplete and noncyclic wetting. The rotary distributor can be operated hydraulically or motorized; however, motorized distributors offer more operational control. Reversing jets can be used to reduce the speed of hydraulic rotary distributors; however, damage to structured sheet media can occur at the radial location of the reversing jets on the distribution arm. Periodic pan testing of the trickling filter distributor is necessary to ensure proper media wetting and distribution of wastewater. Periodic cleaning and adjustment of rotary distributor orifice plates are typically required to correct poor distribution.

Wetting rate refers to the average flow per unit plan view surface area of the tricking filter, typically given in liters per square meters per minute (gallons per minute per square feet), and is a function of the applied flow (i.e., influent plus recirculation) and tower diameter. Spülkraft flushing intensity, or, the SK rate (millimeters per pass), is the instantaneous rate of wastewater applied to the trickling filter. The SK rate is calculated using the following equation:

$$SK = THL/(N_a{}^*\omega_d) \quad (4.1)$$

Where

SK = Spülkraft flushing intensity (mm/pass);
THL = total hydraulic loading (influent plus recirculation), m^3/d;
N_a = number of distribution arms (typically two or four); and
ω_d = rotational speed, rev/min.

Correct trickling filter hydraulics are critical for proper operation. Improper flushing at inadequate SK rates retains excess biomass and macrofauna (e.g., filter flies and snails). Both the wetting rate and SK rate depend on the flowrate to the distributor, which is influenced greatly by recirculation. During low-flow periods, recirculation is necessary to maintain a minimum flow for the rotation of hydraulic rotary distributors and thus keep the biomass wetted. Similarly, for the SK rate, the distributor speed changes with flowrate because of the jet effect. Therefore, the SK rate must be assessed for operation at various flowrates to determine optimum performance. Recommended operating and flushing dosing rates for rotary distributors range from 25 to 230 mm/pass (1 to 9 in./pass) (WEF et al., 2009).

Recirculation typically occurs from underdrains and is controlled through the placement of a weir. Clarification of the filter effluent before recirculation improves trickling filter performance as a result of TSS removal. However, improved trickling filter performance can easily be negated if the facility influent plus recirculation hydraulic loads negatively affect clarifier operation. Additionally, recirculation returned to primary clarification can produce anaerobic conditions in the feed and, subsequently, in the top of the trickling filter, generating nuisance *Beggiatoa* populations. *Beggiatoa* use sulfur compounds from anaerobic decay, which is indicated by a white "plating" or "matting" biomass. Proper recirculation returns dissolved oxygen to the top of the trickling filter, which improves performance in instances where limited dissolved oxygen exists in the trickling filter influent, especially in shallow filters.

Trickling filters that provide both BOD and ammonia removal can benefit from increased recirculation by reducing BOD concentrations applied to the media, thereby limiting competition between heterotrophs and autotrophs. The lower liquid BOD concentration limits BOD diffusion into the biofilm. Utilization of the entire media surface area for ammonia removal can be achieved from the dilution effects. Empirical models, such as the modified Velz equation, show the effects of recirculation on BOD removal (WEF et al., 2009). However, excessive recirculation can bypass biomass on the media and cause excessive loss of biomass, resulting in poor performance.

Covers are used to control odors and in instances where ambient air temperatures are relatively low to eliminate freezing of the trickling filter. When covers are used, low-pressure fans are required for air circulation. Forced ventilation provides both warmer air and increased oxygen transfer to the biofilm.

5.1.2 *Solids Separation*

Aerobic solids contact chambers and chemical flocculation have both been shown to improve solids settling. In facilities without aerobic solids contact basins between the clarifiers and trickling filter, addition of chemical coagulants for phosphorus removal downstream of the trickling filter can significantly improve solids settleability.

Macrofauna, mainly in the form of snails, are typically removed during the clarification process. Clarification before recirculation can minimize potential plugging of media from snail shells, which affects both air and water flow. Periodic removal of snail shells from underdrains may be necessary to maintain trickling filter performance and extend the life of media in configurations that recirculate directly from the trickling filter. Proper design of the media support system can also minimize the potential for plugging; in extreme instances, vertical-flow media replacement is necessary based on the improved flushing it provides.

5.1.3 *Nitrification Rate*

For nitrification to take place, the CBOD concentration of wastewater must be sufficiently low to limit competition between heterotrophic and autotrophic populations because a bulk liquid BOD concentration in excess of 30 mg/L essentially causes nitrification to cease (WEF, 2000). This is increasingly important in combined carbon- and ammonia-oxidizing trickling filters. Carbon-oxidizing trickling filters remove the more easily biodegradable soluble BOD, while the particulate BOD remains relatively unchanged. Diurnal flow variations may cause conditions in the trickling filter to cycle across this threshold value, thereby eliminating nitrification. Increased recirculation can improve nitrification performance by increasing oxygen transfer to the biofilm and decreasing the MTBL.

Ammonium-nitrogen concentration, dissolved oxygen concentration, alkalinity deficiency, and temperature factors affect the nitrification rate in trickling filters. It is generally understood that nitrification rates are reduced as a result of ammonia limitations starting at ammonium-nitrogen concentrations between 1 and 4 mg/L (WEF et al., 2009). Figure 4.3 provides the approximate upper and lower limits of ammonium removal rates measured at five nitrifying trickling filter facilities (Okey and Albertson, 1989). It is important to note that the values used to develop Figure 4.3 have not been corrected for temperature or dissolved oxygen concentration. For combined carbon and ammonia oxidation, trickling filter effluent ammonia concentration is generally accepted to be limited to 2 mg/L (see Chapter 13, Section 7.6, of *Design of Municipal Wastewater Treatment Plants* [WEF et al., 2009]); however, in lightly loaded trickling filters operated under proper conditions, lower ammonia concentrations may be achieved because a selection for a microbial ecology well suited for low ammonia concentrations will develop.

5.1.4 *Trickling Filter Startup and Shutdown*

After all mechanical equipment have been checked out according to the manufacturer's recommendations, the flow of wastewater to the filter should be begun at the lowest rate possible while maintaining the highest possible recirculation rate. In addition, a pan test should be performed to ensure proper distribution and wetting of the media. The initial effluent may not be suitable for discharge,

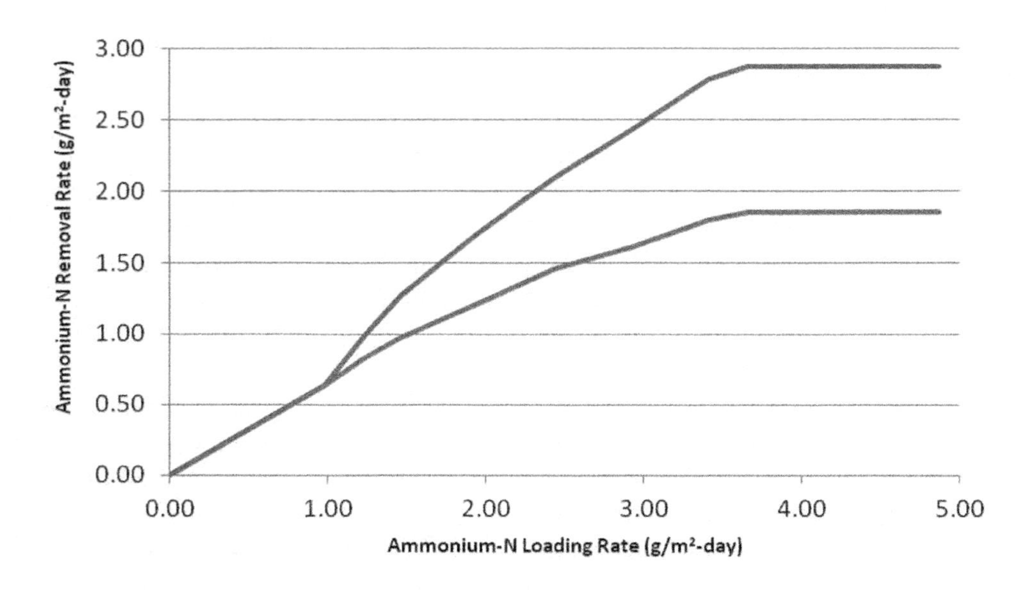

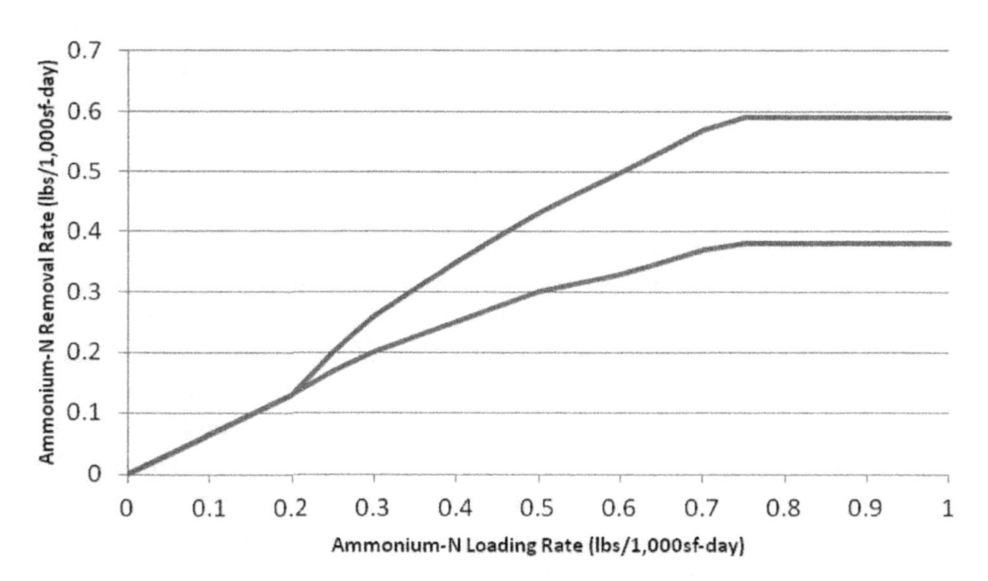

FIGURE 4.3 Trickling filter nitrification (Okey and Albertson, 1989).

but should improve rapidly after the first week. Recirculating effluent to other operating units may be advisable to avoid discharge violations.

A shutdown procedure is necessary to prevent the possible generation of offensive odors, accumulation of dried biomass that may clog the filter when restarted, and damage to the media from extreme heat or cold. To flush biomass from the media, the mechanical distributor should be used at the slowest speed recommended by the manufacturer at the highest possible flowrate for an extended period of time. The flushing water may be chlorinated to clean the media and kill biomass. The chlorinated water should be recirculated until the chlorine residual is safe to discharge. For domed filters, the condition of the media and the temperature in the dome should be monitored at least once per shift to avoid buildup of heat or ammonia gas from the decomposition of any remaining biomass. It is important to continue to operate ventilating and odor control fans to cool the domed space and remove accumulated gases.

5.2 Moving Bed Biofilm Reactors

Moving bed biofilm reactors can be designed for aerobic or anoxic biofilm conditions. This discussion will focus on the operation of nitrifying MBBR systems. The MBBR is strictly a biofilm reactor; it does not incorporate suspended growth (although some may be present because of detachment) and has no solids recycling from the final clarification step. The MBBR relies on the addition of floating media in a submerged reactor compartment. Moving bed biofilm reactors contain a plastic biofilm carrier volume ranging from 25 to 67% of the liquid volume. This parameter is referred to as the *carrier fill*. Retention sieves typically are installed on one MBBR wall and allow treated effluent to flow through to the next treatment step while retaining free-moving plastic biofilm carriers. An MBBR may be a single reactor or several reactors in series. Reactors with a length-to-width ratio greater than 1.5 to 1 can result in nonuniform distribution of free-moving plastic biofilm carriers. Carbon-oxidation, nitrification, or combined carbon-oxidation and nitrification MBBRs use a diffused aeration system to uniformly distribute plastic biofilm carriers and meet process oxygen requirements. Plastic biofilm carriers must be removed before draining and servicing or repairing air diffusers. The MBBR is capable of meeting similar carbon-oxidation and nitrogen removal as activated sludge (Andreottola et al., 2000).

5.2.1 *Mass-Transfer Control*

The main control mechanism for mass transfer in a nitrifying MBBR is the airflow rate. Air is introduced to an MBBR system through coarse-bubble diffusers mounted on the floor of the reactor. The diffusers are designed to produce a rolling pattern in the tank, effectively achieving oxygen transfer and mixing with the aeration system. Because of the presence of biofilm carriers in the reactor, coarse-bubble diffusers have been reported to have a higher oxygen-transfer efficiency than coarse-bubble diffusers in an activated sludge system, with the MBBR oxygen-transfer efficiency in the range of 2.5 to 3.5% per meter of submergence (0.8 to 1.2% per foot of submergence) (McQuarrie and Boltz, 2010).

Increasing the airflow increases both the fluid velocity over the biofilm and the dissolved oxygen concentration in the bulk liquid. For mixing requirements, an airflow rate of 5 to 10 $m^3/(m^2{\cdot}h)$ (0.27 to 0.52 scfm/sq ft) is typical. These airflows ensure sufficient mixing of the floating media and sufficient fluid velocity across the biofilm growing in the carrier. To ensure sufficient oxygen supply, operating with a dissolved oxygen concentration in the range of 4 to 6 mg/L is typically recommended. For effective operations, the balance between increasing the airflow to increase shear force and increasing the dissolved oxygen concentration to produce increased biofilm thickness must be achieved. This will be a site-specific condition.

5.2.2 *Solids Separation*

Retention of floating media in an MBBR is accomplished via several retention sieves in the reactor itself. Retention sieves are required at the effluent side of each of the media zones in the aeration basin to prevent media loss or migration from the aeration basin. The screen design has evolved from flat screens mounted on the face of the cell wall to cylindrical screens that are mounted horizontally facing into the flow (see examples in Figure 4.4). This type of design is effective in preventing clogging. The media itself tend to scrub the screens clean. Flat screens, where used, should be equipped with air knives to encourage media scouring and to avoid excessive head loss resulting from accumulations on the screen surface. Preventive maintenance schedules should include screen inspections and cleaning.

Moving bed biofilm reactors are low hydraulic retention time (HRT) biofilm reactors that have suspended solids remaining in the effluent stream as a result of biological transformation processes. Therefore, MBBR process performance is dependent on a downstream liquid–solid separation unit. Generally, the suspended solids concentration in the MBBR effluent stream has been reported to be 150 to 250 g suspended solids/m^3 when treating medium- to high-strength municipal wastewater (Ødegaard et al., 2010). Solids separation after the retention sieves has been achieved using a wide range of technologies, including dissolved air flotation, conventional clarification, high-rate clarification, and filtration (Ødegaard et al., 2010). For increased bioflocculation of MBBR effluent solids, aerobic solids contact and chemical coagulation have both been implemented with success. For chemical coagulation, it is important to achieve sufficient mixing and contact time to ensure proper flocculation before the final clarifiers. For systems required to meet an effluent phosphorus limitation, the addition of a chemical

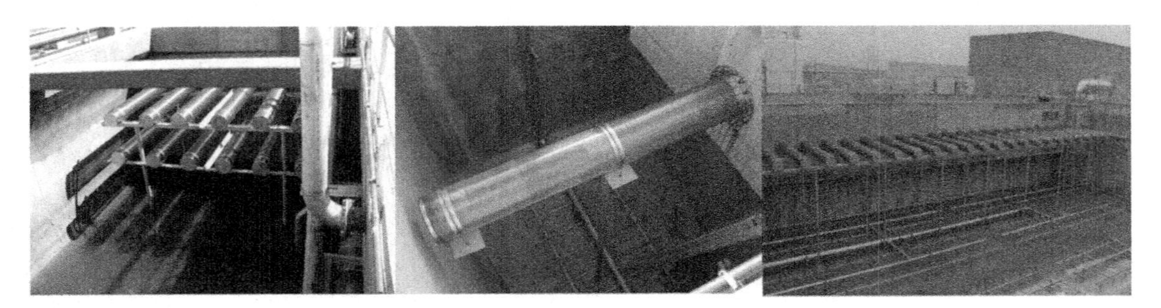

FIGURE 4.4 Examples of MBBR retention sieves (photographs of Kruger, Inc., sieves).

coagulant after the MBBR process can increase the flocculation of MBBR effluent TSS, reducing the final effluent TSS concentration.

5.2.3 *Nitrification Rate*

In applications where sufficient alkalinity and ammonia are present (at least initially), the nitrification rate will increase as organic loading is reduced until the dissolved oxygen concentration becomes rate-limiting. In well-established nitrifying MBBR biofilms, the availability of oxygen limits the rate of nitrification on the carriers as long as the ratio of oxygen to ammonium-nitrogen is below 2.0. Thus, unlike suspended growth systems, the reaction rate in a moving bed reactor exhibits a linear or near-linear dependence on the bulk dissolved oxygen concentration under oxygen-limited conditions. The increased mixing energy under higher aeration velocities also helps to improve transfer from the bulk liquid to the biofilm. If organic loading is held constant (i.e., biofilm thickness and composition), the nitrification rate can be expected to increase linearly with an increasing dissolved oxygen concentration. Until fairly low bulk liquid ammonia concentrations (1 to 2 mg/L) are observed in the reactor, an increase in the bulk dissolved oxygen level helps to increase the nitrification rate.

Although reduced nitrification rates are observed at lower temperatures, at the governing "winter-time" design temperature, this effect can be offset by the combined effect of higher attached biofilm concentrations on the carriers typically observed at lower temperatures (less than 10 °C) (Bjornberg et al., 2009) and by maintaining a higher bulk dissolved oxygen concentration in the reactor. At lower wastewater temperatures, a higher biomass (gram per square meter) is commonly observed. In addition, a higher bulk dissolved oxygen concentration can be attained without an increase in aeration velocity as a result of its higher solubility at lower liquid temperatures. The result of this balance of oxygen concentration and biomass density is that the nitrification activity per unit surface area of carrier can be maintained effectively despite a reduction in specific biofilm activity.

5.2.4 *Biomass Growth*

The operator should monitor the accumulation of growth inside the protected surface area of plastic media. If growth appears to be excessive, the roll pattern may be too gentle, resulting in too weak an aeration-induced shear force. If the roll pattern is sufficient, but there is excessive biofilm, organic loading may be excessive. To rectify this, the organic loading may have to be reduced or additional media may have to be installed. Reducing organic loading can be challenging because implementation of CEPT in primary clarifiers can lead to phosphorus limitation in the biofilm. Adding media is often a last resort because this is analogous to a capacity design change. Airflow control is typically the main control mechanism in the MBBR, with increased airflow rates used to increase biofilm shearing. Biomass growth can range from 5 to 30 mg TSS/m^2, depending on loading, temperature, and operating conditions (McQuarrie and Boltz, 2010).

An MBBR is typically a multizone system, with each zone representing a separate, completely mixed zone. Media located in upstream zones of a multizone MBBR typically will have a thicker biofilm, with a greater portion of the biofilm composed of heterotrophs (BOD oxidizers). Media in downstream zones will have a thinner biofilm with a higher proportion of autotrophs (nitrifiers).

Because the media are free-floating and completely mixed within a zone of the MBBR, the biofilm will adjust itself to conditions in the bulk liquid. During extended periods of high load, the biofilm will get thicker; conversely, during extended periods of low load, the biofilm will get thinner. Typically, the entire aeration basin will contain the media, even if they are divided into several cells. Therefore, there is no need to relocate media to other sections of the basin in response to changing load patterns.

5.2.5 *Moving Bed Biofilm Reactor Startup Considerations*

Although establishment of a heterotrophic biofilm occurs quickly in an MBBR, the nitrifying biofilm can take longer to establish. If excessive heterotrophic biofilm develops during startup in the nitrifying portions of the reactors, the shift from this heterotrophic biofilm to a nitrifying biofilm can be prolonged. To avoid excessive heterotrophic growth during startup, it is important to maintain a low bulk liquid BOD concentration in the nitrifying portion of the MBBR by either lightly loading the system from a hydraulic loading rate at startup or by increasing BOD removal in the primary clarifiers.

Foaming episodes may occur occasionally in MBBRs, especially during startup or process upset. The foam will become trapped in the MBBR reactor because the partition wall between two successive reactors extends above the water surface. A nonsilicone defoaming agent should be used when foam must be controlled in a reactor. Use of a silicone-based defoamer will coat the carrier media, impeding diffusion to the biofilm, which may affect performance. It is extremely important to use defoamers that do not use silicon-compound agents because these types of defoamers are incompatible with plastic-carrier media.

5.2.6 *Media Transfer and Inventory Management*

Although there is little to no maintenance required inside the tank of an MBBR, it is still prudent to consider how one might manage transferring and storing media on-site should a reactor ever need to be taken out of service for in-basin maintenance. The contents of a reactor, including carriers, can be transferred using a 10-cm (4-in.) recessed impeller pump. Depending on the design fill fraction, it may be possible to transfer the media to another reactor. The drawback of this is that it is difficult to reapportion the media once two reactors are combined. After pumping the media back to the reactor, the only reasonably accurate method for measuring the media fill fraction is to dewater the reactor and measure the height of the media in both reactors to check the fill fraction. Preferably, another tank or idle unit process could be used for temporary storage of the media. This way, it is easier to ensure that the proper fill fraction has been restored to the reactor.

5.3 Integrated Fixed-Film Activated Sludge

Integrated fixed-film activated sludge facilities can be designed by adding biofilm support media (biofilm carrier particles) to activated sludge basins. The IFAS process can use three categories of biofilm: fixed bed, plastic carrier moving media, and sponge-type media. During the past decade, plastic carrier media have become the most common carriers for IFAS systems. These carriers are similar, if not identical, to carriers used for an MBBR system.

The principle of the IFAS system is to enhance BOD removal and nitrification above the removal that could have been achieved using mixed liquor suspended solids in a suspended growth-only reactor of like volume. The interaction and exchange between the biofilm and the mixed liquor increases the operational complexity relative to activated sludge and MBBRs as the operator is now managing both a suspended growth and biofilm reactor. Therefore, for an IFAS system, the operator must understand both the nitrification control discussed in the previous section and the nitrification control discussed in Chapter 3 of this manual. Several of the key concepts for operation are summarized in the following subsections. For a floating-media-type IFAS system, the main operational control is the same as that for an MBBR.

In some existing IFAS systems, the bulk liquid solids retention time (SRT) is still maintained in the range for nitrification. The HRT would limit nitrification in the system without the addition of a biofilm process, which is the main advantage of the addition of IFAS attachment. In other systems, the bulk liquid SRT is not maintained at a high enough value for nitrification, potentially because of the limited solids capacity of the final clarifiers.

5.3.1 Nitrification Rate

Nitrification occurs in both the suspended growth and biofilm in an IFAS system. For nitrification in the biofilm portion of IFAS, the nitrification rate is driven by the reactions occurring in the bulk liquid (nitrification and BOD oxidation) and dissolved oxygen concentration. The IFAS media are typically placed after the first one-fourth to one-third of the aeration basin to limit exposure of the biofilm to high BOD concentrations. The following steps can be taken to increase nitrification in an IFAS system:

- Decrease the mixed liquor wasting rate to increase the mixed liquor suspended solids concentration and increase the nitrifying bacteria population in the suspended growth,
- Increase the dissolved oxygen concentration in the aeration basin zones with IFAS up to 4 to 6 mg/L,

- Increase the biofilm attachment area by increasing media in the system, and
- Increase BOD removal upstream of the aeration basins.

Increasing the dissolved oxygen concentration in the aeration basin zones with IFAS media is the most readily controlled parameter. Increasing the airflow rate increases the dissolved oxygen concentration and the fluid velocity over the biofilm, resulting in increased mass transfer. Longer-term adjustments to the typical operating mixed liquor concentration can also be addressed.

5.3.2 *Integrated Fixed-Film Activated Sludge Startup Considerations*

Development of a heterotrophic biofilm on IFAS carriers can limit nitrifying biofilm development and lead to extended startup periods. Because biofilm zones in the IFAS system are typically designed for the lower BOD regions of the aeration basin, it is important to limit the BOD concentration in these zones during startup to limit the formation of excess heterotrophic biofilm. The attachment media can also be added in stages to allow for proper wetting of the media, which, in turn, allows for staged biofilm development.

5.3.3 *Interaction of Biofilm and Suspended Bacteria*

By accomplishing a portion of treatment in a biofilm, IFAS helps reduce the overall operating mixed liquor concentration. This reduces the solids loading to the final clarifier, resulting in increased capacity for clarifiers that are not hydraulically limited. Biofilm erosion from the carriers also has a seeding effect on suspended growth because the biofilm that is sheared from the carrier can remain in the bulk liquid.

5.3.4 *Biofilm Thickness Control*

For floating media, biofilm thickness control is similar to MBBR operation. Several approaches have been developed for biofilm thickness management for fixed-media IFAS systems. Many fixed IFAS systems are installed with fine-bubble aeration, which does not provide the liquid velocity to dislodge biomass. One of the more common strategies is to use periodic, high-intensity air sparging to increase fluid velocity around the biofilm structure. This sparging can also produce large air bubbles that dislodge biomass. If an active biofilm management strategy such as air sparging is not implemented in fixed-media IFAS systems, the biofilm modules risk excess biofilm accumulation and decreased nitrification rates because of lack of oxygen transfer into the inner portions of the biofilm.

5.4 Additional Biofilm Reactor Technologies

A brief summary of additional biofilm reactor technologies is presented in the following subsections. Each of these reactors would be governed by the same principles as previously discussed, with specific control mechanisms differing for each technology.

5.4.1 *Rotating Biological Contactors*

As a secondary treatment process, RBCs have been applied in instances where average effluent water-quality standards are less than or equal to 30-mg/L BOD and TSS. When the RBC is used in conjunction with effluent filtration, the process is capable of meeting more stringent effluent water-quality limits of 10-mg/L BOD and TSS. Nitrification RBCs can produce effluent having less than 1 mg/L of ammonia-nitrogen remaining in the effluent stream, although this is not typical. The RBC uses a cylindrical, synthetic media bundle that is mounted on a horizontal shaft. The media are partially submerged (typically 40%) and slowly rotate (1 to 1.6 rpm) to expose the biofilm to substrates in the bulk of the liquid (when submerged) and to air (when not submerged).

The RBC process has the following advantages: operational simplicity, low energy costs, and rapid recovery from shock loadings. However, the literature has documented several examples of RBC malfunctions resulting from shaft, media, or media support system structural failure; poor treatment performance; accumulation of nuisance macrofauna; poor biofilm thickness control; and inadequate performance of air-drive systems for shaft rotation.

5.4.2 *Biological Filters*

Biological wastewater treatment and suspended solids removal are carried out in BAFs under either aerobic or anoxic conditions. In a BAF, the media act simultaneously to support the growth of biomass and as a filtration medium to retain filtered solids. Accumulated solids are removed from the BAF through backwashing. There is a direct interaction between media characteristics and the process because the configuration (i.e., sunken media or floating media) and flow and backwash regimes depend on media density. Media may be natural mineral, structured plastic, or random plastic. Backwashing the filters maximizes capture and run times and guarantees proper effluent quality. Proper backwashing requires filter bed expansion and rigorous scouring, followed by efficient rinsing. Poor filter cleaning will result in shortened filter runs, accumulation of solids, and deteriorating performance. Accumulation of solids and media ("mudballing") produces short-circuiting of water flow and can result in excessive media loss.

5.4.3 *Expanded and Fluidized Bed Reactors*

Expanded and fluidized bed biofilm reactors are similar in that they contain a bed of small particles (approximately 1 mm) of a suitable biomass support material on which biofilm grows and through which wastewater flows upward. Because the upflow velocity is fixed but flow through the works is not, there is automatically a variable recirculation rate. This type of reactor can hold a high concentration of active biofilm (up to 40 g m^{-3}) with a large surface area (900 to 2800 m^2 m^{-3}). Because the bed is expanded, it does not trap solids and, therefore, does not need backwashing. Being pseudo plug flow, the concentration of NH_3-N in the final effluent can be reduced to low levels (i.e., below 1 mg/L). Originally, dissolved oxygen was supplied by introducing air through a sparger or diffuser (Figure 4.2); however, recent improvements to design use countercurrent aeration in a separate column, which results in improved oxygen transfer. Management of the system is through control of the organic loading rate, to minimize competition from heterotrophic bacteria (which depend on the effectiveness of the upstream process), and control of the dissolved oxygen concentration at the top of the bed to ensure adequate oxygen supply. Occasional blooms of red oligochaete worms can be controlled by turning off aeration for a few hours, which kills the worms but not the nitrifying bacteria.

5.4.4 *Submerged Fixed-Media Reactors*

Structured sheet media have been installed for use in both aerobic biofilm and IFAS applications. Low- and medium-density cross-flow trickling filter media have historically been used for these installations. The structured sheet media modules are fixed in place using a retention system, with diffusers placed beneath the media towers. Similar to the mixed-media configuration of trickling filters, where cross-flow and vertical-flow media are used in combination to provide improved distribution of wastewater and the ability to flush solids, the submerged structured sheet media system uses cross-flow distribution media for improved air distribution below vertical-flow media for greater surface area and air-lift pumping. The diffused air provides air-lift pumping of wastewater through the vertical-flow media and downward flow occurs through "downcomer" regions where no media are installed. This allows for repeated recirculation within the tower, which results in contact of the wastewater with the fixed biomass and provides air scouring simply from process air requirements.

Biofilm control occurs in the vertical-flow media as air bubbles pass the attached biofilm. The high mixing rate developed from air-lift pumping minimizes the thickness of the mass-transfer boundary layer, enhancing nitrification performance. The media surface areas range from 157 m^2/m^3 (48 sq ft/cu ft) to 315 m^2/m^3 (96 sq ft/cu ft), with surface nitrification rates as high as 1.66 $g/(m^2 \cdot d)$ at 20 °C (Ye et al., 2010). The bulk liquid dissolved oxygen concentration is typically maintained between 3 and 5 mg/L. The structured sheet has a limited effect on the oxygen-transfer efficiency of fine-bubble diffusers that are used to improve oxygen-transfer efficiency for the system (Zhu et al., 2011).

5.5 Biofilm Reactor Technologies—General Operations Table

Operation of each of these biofilm reactor technologies and emerging and yet-to-be-developed technologies will generally follow the same principles outlined in Table 4.1. The specific mechanisms for adjustment vary by technology, and technology-specific operational tools to achieve the main operational goals are summarized in Table 4.2.

TABLE 4.2 Biofilm reactor technology operational table.

Operational goal	Biofilm modification	Liquid velocity change	Operation and control—general	Operation and control—trickling filter	Operation and control—MBBR	Operation and control—IFAS
Improve nitrification rate—mass transfer limitation	Increase mass transfer to the biofilm	Increase	Increase oxygen transfer rate to increase dissolved oxygen concentration in biofilm Increase upstream BOD removal to reduce heterotrophic competition in the biofilm Increase fluid velocity to decrease MTBL thickness	Increase NTF recirculation rate to transfer more dissolved oxygen to bulk phase (i.e., mix with influent) Increase primary trickling filter (i.e., BOD removal) recirculation rate to improve BOD removal rate (this will also raise inlet dissolved oxygen to NTF)	Increase aeration rate to increase oxygen in bulk liquid Increase aeration rate to decrease MTBL thickness Increase BOD removal in upstream MBBR zones	Increase aeration rate to increase oxygen in bulk liquid Increase aeration rate to decrease MTBL thickness Increase mixed liquor suspended solids to decrease bulk liquid BOD concentration
Improve nitrification rate—biofilm thickness limitation	Increase biofilm thickness	Decrease	Decrease fluid velocity and mixing intensity	Reduce flushing intensity (SK rate)	Decrease aeration rate to decrease shear force on biofilm and decrease biofilm erosion	Decrease mixed liquor suspended solids concentration/decrease bulk liquid SRT to increase bulk liquid concentrations
			Increase biofilm attachment area by adding additional media if sufficient reactor capacity exists (% fill < maximum fill achievable)	Increase primary trickling filter (i.e., BOD removal) recirculation rate to improve BOD removal rate		Decrease aeration rate to decrease shear force on biofilm and decrease biofilm erosion

(continued)

Table 4.2 Biofilm reactor technology operational table (*Continued*).

Operational goal	Biofilm modification	Liquid velocity change	Operation and control—general	Operation and control—trickling filter	Operation and control—MBBR	Operation and control—IFAS
				If operated for combined BOD and ammonia removal, increased recirculation may also improve performance (applied soluble BOD concentration < ~30 mg/L) Operate trickling filters in series to reduce heterotrophic and autotrophic competition (if currently operated in parallel)		
Prevent biofilm septicity—Biofilm thickness limitation	Decrease biofilm thickness	Increase	Increase fluid velocity and mixing intensity to increase shear force Implement cyclical backwash cycles Increase biofilm attachment area by adding additional media	Increase flushing intensity (SK rate) Increase recirculation rate to increase bulk dissolved oxygen concentration (increased recycle rate may potentially increase SK rate; verify through calculation because distributor speed may increase for a hydraulically propelled distributor)	Increase aeration intensity to increase shear force and resulting biofilm ersosion Adjust aeration roll pattern to increase shear force	Increase mixed liquor suspended solids/increase bulk liquid SRT to decrease bulk liquid BOD and ammonia concentrations Increase aeration rate to increase shear force and biofilm erosion Adjust aeration roll pattern to increase shear force

Prevent biofilm septicity—Mass trasnfer limiation	Increase oxygen mass transfer	Increase	Increase oxygen transfer rate to increase dissolved oxygen concentration in biofilm	Increase recirculation rate to increase bulk dissolved oxygen concentration	Increase aeration rate to increase oxygen in bulk liquid Increase aeration rate to decrease MTBL thickness and increase mass transfer rate	Increase aeration rate to increase oxygen in bulk liquid Increase aeration rate to decrease MTBL thickness and increase mass transfer rate Increase mixed liquor suspended solids to decrease bulk liquid BOD concentration
Prevent excessive biomass loss (sloughing)—mass transfer limitation	Increase mass transfer to the biofilm	Increase	Ensure sufficient phosphorus and nitrogen entering biofilm reactor Increase oxygen transfer rate to increase dissolved oxygen concentration in biofilm Ensure sufficient alkalinity to buffer pH Prevent oxygen and nutrient starvation	Ensure sufficient phosphorus and nitrogen entering biofilm reactor Increase oxygen transfer rate by increasing recirculation rate to increase dissolved oxygen concentration in biofilm Ensure sufficient alkalinity to buffer pH	Ensure sufficient phosphorus and nitrogen entering biofilm reactor Increase aeration rate to decrease MTBL thickness and increase mass transfer rate Ensure sufficient alkalinity to buffer pH	Ensure sufficient phosphorus and nitrogen entering biofilm reactor Increase aeration rate to decrease MTBL thickness and increase mass transfer rate Increase mixed liquor suspended solids to decrease bulk liquid BOD concentration Ensure sufficient alkalinity to buffer pH
Prevent excessive biomass loss (sloughing)—biofilm thickness limitation	Decrease shear force on biofilm	Decrease	Decrease fluid velocity and mixing intensity	Reduce flushing intensity (SK rate) while still maintaining adequete wetting	Reduce aeration rate to reduce shear force and biofilm erosion	Reduce aeration rate to reduce shear force and biofilm erosion

6.0 BIOFILM REACTOR DESIGN

An in-depth discussion of the principles and practices of biofilm reactor design is presented in *Biofilm Reactors* (WEF, 2011). This publication covers details of biofilm rector design and should be referenced for questions pertaining to specific reactor selection and evaluation. Recommended loading rates can be used as a preliminary evaluation of overall loading to an operating biofilm reactor technology and can help determine if the system is significantly overloaded or underloaded. These loadings only represent a cursory evaluation of the system; given site-specific environmental conditions, these applicable loading rates can vary significantly.

7.0 REFERENCES

Andreottola, G.; Foladori, P.; Ragazzi, M.; Tatano, F. (2000) Experimental Comparison Between MBBR and Activated Sludge System for the Treatment of Municipal Wastewater. *Water Sci. Technol.*, **41** (4–5), 375–382.

Bjornberg, C; Lin, W; Zimmerman, R. (2009) Effect of Temperature on Biofilm Growth Dynamics and Nitrification Kinetics in a Full-Scale MBBR System. *Proceedings of the 82nd Annual Water Environment Federation Annual Technical Exhibition and Conference* [CD-ROM]; Orlando, Florida, Oct 10–14; Water Environment Federation: Alexandria, Virginia; pp 4407–4426.

Boltz, J. P.; Goodwin, S. J.; Rippon, D.; Daigger, G. T. (2008) A Review of Operational Control Strategies for Snails and other Macro-Fauna Infestations in Trickling Filters. *Water Pract.*, **2** (4).

Bryers, J. D. (1984) Biofilm Formation and Chemostat Dynamics: Pure and Mixed Culture Considerations. *Biotechnol. Bioeng.*, **26**, 948–958.

Cole, A. C.; Semmens, M. J.; LaPara, T. M. (2004) Stratification of Activity and Bacterial Community Structure in Biofilms Grown on Membranes Transferring Oxygen. *Appl. Environ. Microbiol.*, **70** (4), 1982–1989.

Daigger, G. T.; Boltz, J. P. (2011) Trickling Filter and Trickling Filter-Suspended Growth Process Design and Operation: A State-of-the-Art Review. *Water Environ. Res.*, **83** (5), 388–404.

Downing, L.; Nerenberg, R. (2008a) Effect of Bulk Liquid BOD Concentration on Activity and Microbial Community Structure of a Nitrifying, Membrane-Aerated Biofilm. *Appl. Microbiol. Biotechnol.*, **81**, 153–162.

Downing, L.; Nerenberg, R. (2008b) Effect of Oxygen Gradients on Performance and Microbial Community Structure of a Nitrifying, Membrane-Aerated Biofilm. *Biotechnol. Bioeng.*, **101** (6), 1193–1204.

Fernandez-Polanco, F.; Mendez, E.; Uruena, M. A.; Villaverde, S.; Garcia, P. A. (2000) Spatial Distribution of Heterotrophs and Nitrifiers in a Submerged Biofilter for Nitrification. *Water Res.*, **34** (16), 4081–4089.

Grady, L. E.; Daigger, G. T.; Lim, H. (1999) *Biological Wastewater Treatment*, 2nd ed.; Marcel Dekker: New York.

Harrison, J. R.; Timpany, P. L. (1988) Design Considerations with the Trickling Filter Solids Contact Process. *Proceedings of the Joint Canadian Society of Civil Engineers, ASCE National Conference on Environmental Engineering;* Vancouver, British Columbia; pp 753–762.

Hibiya, K.; Terada, A.; Tsuneda, S.; Hirata, A. (2003) Simultaneous Nitrification and Denitrification by Controlling Vertical and Horizontal Microenvironment in a Membrane-Aerated Biofilm Reactor. *J. Biotechnol.*, **100** (1), 23–32.

Kuenen, J. G.; Jorgensen, B. B.; Revsbech, N. P. (1986) Oxygen Microprofiles of Trickling Filter Biofilms. *Water Res.*, **20** (12), 1589–1598.

McQuarrie, J. P.; Boltz, J. P. (2010) Moving Bed Biofilm Reactor Technology: Process Applications, Design, and Performance. *Water Environ. Res.*, **83** (6), 560–575.

Morgenroth, E. (2003) Detachment: An Often Overlooked Phenomenon in Biofilm Research. In *Biofilm in Wastewater Treatment*, Wuertz, S., Bishop, P., Wilderer, P., Eds.; IWA Publishing: London.

Nogueira, R.; Elenter, D.; Brito, A.; Melo, L. F.; Wagner, M.; Morgenroth, E. (2005) Evaluating Heterotrophic Growth in a Nitrifying Biofilm Reactor Using Fluorescence In Situ Hybridization and Mathematical Modeling. *Water Sci. Technol.*, **52** (7), 135–141.

Oakey, R. W.; Albertson, O. E. (1989) Diffusion's Role in Regulating Rate and Masking Temperature Effects in Fixed Film Nitrification. *J.—Water Pollut. Control Fed.*, **61**, 510.

Ødegaard, H.; Cimbritz, M.; Christensson, M.; Dahl, C. P. (2010) Separation of Biomass from Moving Bed Biofilm Reactors (MBBRs). *Proceedings of the Water Environment Federation and International Water Association Biofilm Reactor Technology Conference;* Portland, Oregon, Aug 14–18; Water Environment Federation: Alexandria, Virginia.

Okabe, S.; Hiratia, K.; Ozawa, Y.; Watanabe, Y. (1996) Spatial Microbial Distributions of Nitrifiers and Heterotrophs in Mixed-Population Biofilms. *Biotechnol. Bioeng.*, **50** (1), 24–35.

Points, A.; Downing, L.; Yu, J. (2010) Nitrification and Denitrification in an Aerobic Integrated Fixed-Film Activated Sludge System. *Proceedings of the 83rd Annual*

Water Environment Federation Technical Exhibition and Conference [CD-ROM]; New Orleans, Louisiana, Oct 2–6; Water Environment Federation: Alexandria, Virginia; pp 556–559.

Satoh, H.; Okabe, S.; Norimatsu, N.; Watanabe, Y. (2000) Significance of Substrate C/N Ratio on the Structurea and Activity of Nitrifying Biofilms Determined by In Situ Hybridization and the Use of Microelectrodes. *Water Sci. Technol.*, **41** (5), 317–321.

Satoh, H.; Ono, H.; Rulin, B.; Kamo, J.; Okabe, S.; Fukushi, K.-I. (2004) Macroscale and Microscale Analyses of Nitrification and Denitrification in Biofilms Attached on Membrane Aerated Biofilm Reactors. *Water Res.*, **38** (6), 1633–1641.

Schramm, A.; Larsen, L. H.; Revsbech, N. P.; Ramsing, N. B.; Amann, R.; Schleifer, K.-H. (1996) Structure and Function of a Nitrifying Biofilm as Determined by In Situ Hybridization and the Use of microelectrodes. *Appl. Environ. Microbiol.*, **62**, 4641–4647.

Terada, A.; Hibiya, K.; Nagai, J.; Tsuneda, S.; Hirata, A. (2003) Nitrogen Removal Characteristics and Biofilm Analysis of a Membrane-Aerated Biofilm Reactor Applicable to High-Strength Nitrogenous Wastewater Treatment. *J. Biosci. Bioeng.*, **95** (2), 170–178.

Water Environment Federation (2000) *Aerobic Fixed-Growth Reactors*; Special Publication; Water Environment Federation: Alexandria, Virginia.

Water Environment Federation (2011) *Biofilm Reactors*; WEF Manual of Practice No. 35; McGraw-Hill: New York.

Water Environment Federation; American Society of Civil Engineers; Environmental and Water Resources Institute (2009) *Design of Municipal Wastewater Treatment Plants*, 5th ed.; WEF Manual of Practice No. 8; ASCE Manual and Report on Engineering Practice No. 76; McGraw-Hill: New York.

Ye, J.; Kulick, F. M., III; McDowell, C. S. (2010) Biofilm Performance of High Surface Area Density Vertical-Flow Structured Sheet Media for IFAS and Fixed Bed Biofilm Reactor (FBBR) Applications. *Proceedings of the Water Environment Federation, Biofilms 2010.*

Zhu, J.; Kulick, F. M., III; Koch, K. (2011) Impact of Structured Sheet Media on the Oxygen Transfer Efficiency of Fine and Coarse Bubble Diffusers. *Proceedings of the 84th Annual Water Environment Federation Technical Exhibition and Conference,* [CD-ROM]; Los Angeles, California, Oct 15–19; Water Environment Federation: Alexandria, Virginia.

Chapter 5

Denitrification

Jeanette Brown, P.E., BCEE, D. WRE;
Gary R. Johnson, P.E., BCEE; Huijie Lu; and
Robert R. Sharp, Ph.D., P.E.

1.0 PROCESS FUNDAMENTALS

1.1 Chemical Thermodynamics

The chemical transformation or degradation of a compound by microbial action is based on chemical thermodynamics. The reaction that results is that which yields the highest positive energy. All microbial species obtain energy as a result of this favorable energy flow. If thermodynamics are not favorable, the transformation of chemicals will not occur. This energy flow is obtained from the coupling of oxidation–reduction reactions, which are frequently referred to as *redox reactions*. The amount of free energy depends on the Gibbs free energies of the substrates and the products. A useful concept for solution-based redox reactions is electron activity (pE). Electron activity is a measure of the available electrons in much the same manner that pH measures the availability of protons. The amount of free energy that can be obtained by microorganisms from redox reactions is directly proportional to the electron activity. Understanding if an energy yield is positive or negative is a key component of knowing how compounds can be biologically degraded or transformed to desired chemical species.

Understanding microbial metabolism modes is important to successful operation of denitrification processes and troubleshooting poor denitrification performance. Microorganisms have developed wide varieties of metabolism systems. These are characterized by the nature of the reductant and the oxidant. For aerobic respiration, the electron acceptor is molecular oxygen. Anaerobic respiration uses an oxidized inorganic or organic compound other than oxygen as the electron acceptor. It is the electron acceptor that provides the name for the metabolism mode.

Microorganisms take advantage of this by using the transfer of electrons to produce energy-rich bonds within the microorganism's cell. Microorganisms in the presence of a potential energy source produce enzymes that catalyze the natural thermodynamic reaction to yield an increase in the rate of energy flow

(Cookson, 1995). For denitrification, the electron acceptor must be nitrate (NO_3^-). The oxidation reaction is nitrate oxidizing the organic compound or compounds measured as biochemical oxygen demand (BOD). The organic compound loses electrons, which are transferred to nitrate. This overall reaction consists of two half reactions: (1) the oxidation of organic compounds and (2) the reduction of nitrate. The end products are carbon dioxide and nitrogen gas. Other chemicals in water can serve as electron acceptors (e.g., oxygen, sulfate, ferric iron, another organic compound, and carbon dioxide). The electron acceptor that enters this reaction with the organic compound establishes the metabolism mode.

If thermodynamics are not favorable for nitrate to be the electron acceptor, denitrification will not occur. Certain changes in water chemistry can result in unfavorable conditions for denitrification. Any compound and, in some instances, gases, that can serve as an electron acceptor and have a higher pE than nitrate will block denitrification. One example is the presence of oxygen. Understanding these principles can help solve potential problems resulting from changes in water chemistry that can change chemical thermodynamics.

1.2 Stoichiometry and Kinetics

Heterotrophic denitrification is the dissimilatory reduction of ionic nitrogen oxides such as nitrate (NO_3^-) and nitrite (NO_2^-) to gaseous oxides such as nitric oxide (NO) and nitrous oxide (N_2O) and, eventually, to nitrogen gas (N_2) under anoxic conditions using organic electron donors and assimilative carbon sources (Knowles, 1982). Nitrogen oxides act as terminal electron acceptors in the absence of oxygen. Various organic carbons, such as methanol, acetate, glycerol, or glucose, can serve as electron donors and carbon sources for growth. The overall reaction when using methanol as a carbon source for denitrification is as follows (note that cell growth is not taken into account):

$$5CH_3OH + 6NO_3^- \rightarrow 3N_2 + 5CO_2 + 7H_2O + 6OH^- \qquad (5.1)$$

The amount of nitrate that can be removed from a given wastewater is a function of organic carbon availability and environmental and operating conditions, and is primarily determined as the stoichiometric ratio between the electron donor and acceptor (i.e., carbonaceous oxygen demand [COD]-to-nitrate—as nitrogen—ratio [C/N]). Many studies have reported a wide range of COD/N required for satisfactory denitrification, that is, between 3 and 7 g COD/g N (Chiu and Chung 2003; Li et al., 2008; Sobieszuk and Szewczyk, 2006). However, this ratio largely depends on the nature of the influent concentrations of biodegradable COD and nitrogen species, solids retention time (SRT), and the microbial ecology of the denitrification process itself. For instance, if the influent wastewater COD/N is not sufficient, external carbon source is required to achieve complete nitrate reduction to N_2. Furthermore, denitrifying bacteria with a higher growth yield coefficient require a higher dosage of an external electron donor, as follows (Grady et al., 1999):

$$\text{Carbon dose} = \frac{2.86}{1-Y} \times (NO_3^-{}_{in}\text{-}NO_3^-{}_{eff}) \tag{5.2}$$

Where

$(NO_3^-{}_{in}\text{-}NO_3^-{}_{eff})$ = the degree of denitrification required (mg N/d, calculated by multiplying the difference between the influent and effluent concentrations of nitrate [mg N/L] by the flowrate [L/d]);

Y = the anoxic growth yield of denitrifying bacteria on specific carbon sources (mg biomass COD produced/mg COD removed); and

carbon dose = in mg COD/d.

On the other hand, the kinetics of denitrification directly influence anoxic reactor zone sizing in biological nutrient removal (BNR) reactors. Microbial kinetics are generally described by the Monod model (eq 5.3 and Figure 5.1). Biokinetic parameters that are essential to full-scale denitrification process design and operation, such as maximum specific growth rate (μ_{max}) and half-saturation coefficients (Ks), can be obtained by nonlinear regressions of the following equations:

$$\mu = \frac{\mu_{max} S_{COD}}{1 + k_{s,COD} S_{COD}} \tag{5.3}$$

$$\frac{dX_H}{dt} = \mu X_H = \frac{\mu_{max} S_{COD}}{1 + k_{s,COD} S_{COD}} X_H \tag{5.4}$$

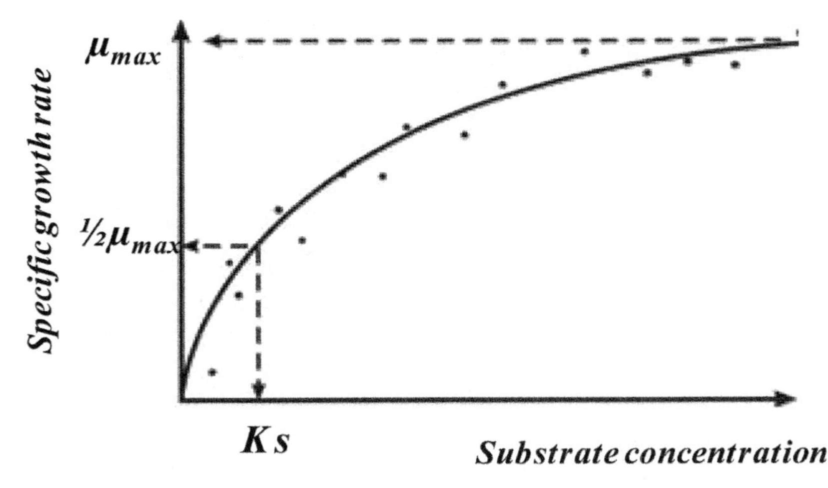

FIGURE 5.1 Monod model that defines the relationship between the growth rate (μ) and the concentration of the limiting nutrient (S) (Monod, 1949).

$$\frac{dS_{NO}}{dt} = \frac{1-Y_H}{2.86Y_H}\frac{dX_H}{dt} \tag{5.5}$$

$$\frac{dS_{COD}}{dt} = \frac{1}{Y_H}\frac{dX_H}{dt} \tag{5.6}$$

Where

μ = the specific growth rate (t^{-1});
μ_{max} = the maximum specific growth rate (t^{-1});
$K_{s,COD}$ = the half-saturation coefficient of organic carbon (mg/L));
Y_H = the growth yield of heterotrophs (mg COD biomass formed/mg COD removed);
X_H = the biomass concentration (mg/L); and
S = the substrate concentration (mg/L, S_{COD}: organic carbon; S_{NO}: nitrate).

Using methanol as the carbon source, values for μ_{max} range from 0.2 to 2.0 d^{-1} according to Gaudy and Gaudy (1980). The μ_{max} is also a function of temperature, as reported by Stensel et al. (1973). The specific substrate use rate (q) is related to μ by the growth yield, Y (eq 5.7). The specific denitrification rate (SDNR) is a direct measure of denitrification activity (eq 5.8) and can be obtained by denitrification rate batch assays, as follows:

$$q = \frac{\mu}{Y} \tag{5.7}$$

$$\frac{dS_{NO}}{dt} = \text{sDNR} \times X_H \tag{5.8}$$

A double-substrate Monod model can be used to describe the effects of both carbon and nitrate concentrations on the specific denitrification rate (eq 5.9). However, the half-saturation coefficient of nitrate ($K_{S,NO}$) is less than 1 mg/N and has little influence on measured SDNR values (Marazioti et al., 2003; Wild et al., 1995). Figure 5.2 shows SNDR values as a function of the food-to-microorganism (F/M) ratio in an anoxic reactor and the content of readily biodegradable BOD in the feed wastewater to the anoxic reactor. If the F/M is 0.5 g BOD/g mixed liquor suspended solids (MLSS)/d in the anoxic reactor and readily biodegradable BOD comprises 30% of the total BOD, SNDR becomes approximately 0.125 g NO_3-N/g MLSS/d, as follows:

$$\text{sDNR} = \text{sDNR}_{max} \times \frac{S_{COD}}{K_{S,COD} + S_{COD}} \times \frac{S_{NO}}{K_{S,NO} + S_{NO}} \tag{5.9}$$

For treatment processes with a single anoxic zone, the denitrifying reactor can be sized based on the *SDNR* values, the concentration of volatile suspended

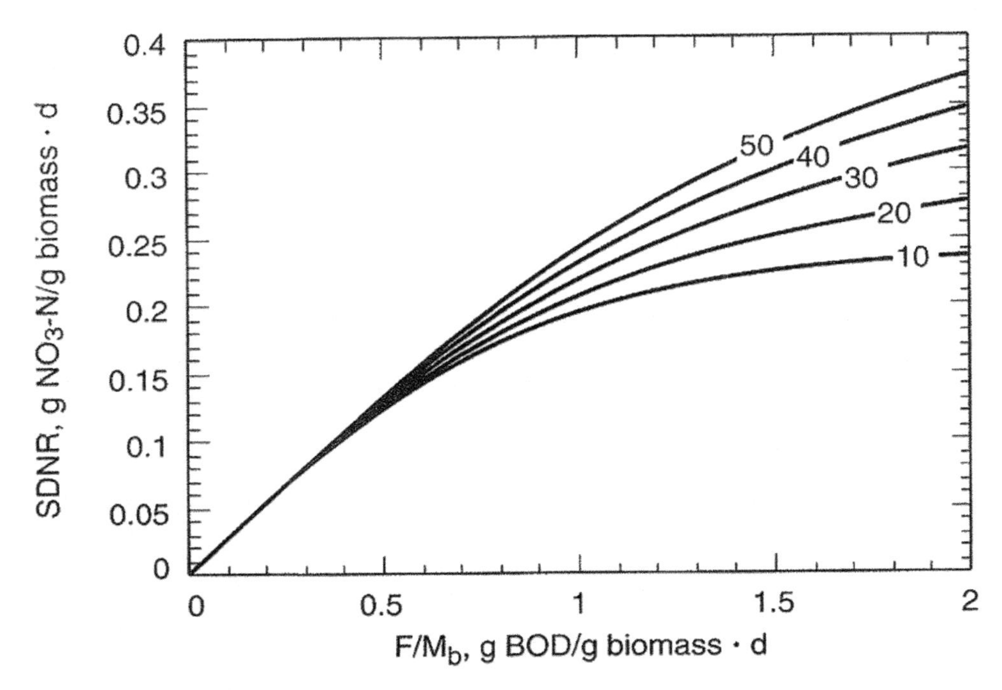

FIGURE 5.2 Specific denitrification rate as a function of F/M and the ratio of readily biodegradable BOD to total BOD (Metcalf and Eddy, 2003).

solids (VSS) in the mixed liquor, and the degree of denitrification required, as follows:

$$V_{ANX} = \frac{S_{NO,in} - S_{NO,eff}}{\mathrm{sDNR} \cdot X_{VSS}} \tag{5.10}$$

Where

V_{ANX} = the calculated volume of the anoxic zone (L),
$S_{NO,in}$-$S_{NO,eff}$ = the degree of denitrification required (mg N/d),
SDNR = the specific denitrification rate (mg N/mg VSS/d), and
X_{VSS} = the concentration of VSS in the anoxic zone (mg VSS/L).

1.3 Denitrifying Organisms

Unlike nitrifying bacteria, which are phylogenetically closely related (Purkhold et al., 2000), denitrifying bacteria are distributed in a large variety of physiological and taxonomic groups belonging to various subclasses of *Proteobacteria* (Table 5.1). Because methanol is the most commonly used carbon source supporting wastewater denitrification, methylotrophic denitrifiers have been widely studied (Anthony, 2011; Labbe et al., 2007; Nurse, 1980; Sperl and Hoare, 1971). Methanol denitrification selects special groups of bacteria (Christensson et al., 1994;

TABLE 5.1 Genera of denitrifying bacteria detected in activated sludge (Gerardi, 2006).

Achromobacter	*Escherichia*	*Neisseria*
Acinetobacter	*Flavobacterium*	*Paracoccus*
Agrobacterium	*Gluconobacer*	*Propionibacterium*
Alcaligens	*Holobacterium*	*Pseudomonas*
Bacillus	*Hyphomicrobium*	*Rhizobium*
Chromobacterium	*Kingella*	*Rhodopseudomonas*
Corynebacterium	*Methanonas*	*Spirillum*
Denitrobacillus	*Moraxella*	*Thiobacillus*
Enterobacter	*Xanthomonas*	

Ginige et al., 2004) that can be further divided into three subgroups: obligate, restricted facultative, and typical facultative methylotrophs (de Vries et al., 1990). The obligate methylotroph can only use C1 compounds (e.g., *Hyphomicrobium methylovorum* KM146 and *Methylococcus capsulatus*) (de Vries et al., 1990). The C1 compounds are those that contain only one carbon atom (e.g., methanol). The restricted methylotroph can grow on a limited range of complex organic compounds (e.g., *Methylophilus glucosoxydans* and *Methylophilus rhizosphaerae*) (de Vries et al., 1990). Typical facultative methylotrophs, such as *Methylobacterium extorquens* AM1, can grow on a wide range of polycarbon compounds (Chistoserdova et al., 2003; O'Keeffe and Anthony, 1980).

1.4 Favorable Operating Conditions

1.4.1 Anoxic

Oxygen inhibits denitrification by providing a better electron acceptor for denitrifying species to generate energy. The Gibbs standard free energy of water–oxygen is −78.73 kJ/eeq (the definition is as follows: 1 eeq = amount of substrate releasing 1 mol e^- during total carbonaceous oxidation [e.g., substrate ≥ CO_2, NH_4^+]; note that 1 eeq of an organic electron donor = 8 g COD) and that for nitrate-nitrogen is −72.20 kJ/eeq (Rittmann and McCarty, 2001). The threshold oxygen inhibition concentration is approximately 0.2 mg O_2/L (Knowles, 1982). In practice, oxidation–reduction potential (ORP), which is a measure of the activity or strength of oxidizers and reducers in relation to their concentration in wastewater, has been used to indicate aerobic, anoxic, and anaerobic conditions. The ORP generally shows a strong response to dissolved oxygen, especially at low concentrations, and is a better monitoring and controlling parameter under anoxic and anaerobic conditions than dissolved oxygen concentrations. In general, ORP values lower than −200 mV indicate anaerobic conditions, values between −200 and +200 mV indicate anoxic conditions, and values higher than +200 mV are for aerobic conditions. The ORP in an anoxic reactor must be below +50 mV to trigger a denitrification reaction. Typically, denitrification occurs between −50 and +50 mV. It is important to note that ORP values would be expected using an ORP probe with a silver-to-silver chloride ratio reference electrode. However, if another reference electrode system is used, such as a hydrogen or a calomel electrode, the ORP values will be different. For instance, a hydrogen electrode registers an ORP approximately 200 mV more positive than the silver-to-silver chloride ratio reference. The inhibition of oxygen to denitrification is reversible, and the response time of the enzymatic system to the electron acceptor switch is generally in the magnitude of minutes (Kornaros and Lyberatos, 1998). An expended Monod expression of

$$\frac{dS_{NO}}{dt} = \left(\frac{1-Y}{2.86Y}\right)\mu_{max}\left(\frac{S_{COD}}{K_{S,COD} + S_{COD}}\right)\left(\frac{S_{NO}}{K_{S,NO} + S_{NO}}\right)\left(\frac{K_{I,O}}{K_{I,O} + S_O}\right)\eta X_{bH} \quad (5.11)$$

Where

S_S, S_{NO}, and S_O = concentrations of the organic carbon source, nitrate-nitrogen, and oxygen (in COD mg/L, mg N/L, and mg O_2/L, respectively);

$K_{S,COD}$ and $K_{S,NO}$ = the half-saturation coefficients of the organic carbon source (mg COD/L) and nitrate (mg N/L), respectively;

$K_{I,O}$ = the inhibition coefficient of oxygen on nitrate reduction, which was found to be around 0.1 mg O_2/L using both mixed and pure cultures (Grady et al., 1999; Kornaros and Lyberatos, 1998); and

η = the correlation factor for growth under anoxic conditions (dimensionless, with the typical value of 0.8).

1.4.2 Carbonaceous Oxygen Demand-to-Nitrate Ratio

The performance of single sludge predenitrification processes depends on the composition of influent wastewater, and low ratios of COD/N can lead to poor denitrification, whereas relatively high COD/N can detrimentally affect nitrification (Sharma and Gupta, 2004). The most critical form of COD is readily biodegradable, soluble COD that can be metabolized quickly and efficiently by organisms.

For postdenitrification systems with external carbon augmentation, the optimal C/N will vary depending on the forms of carbon source. Tam et al. (1994) reported an optimal 5-day BOD (BOD_5)/NOx-N (NO_x is either nitrate-nitrogen [NO_3-N] or nitrite-nitrogen [NO_2-N]) of 2.48 for a denitrification process with methanol and acetic acid as the carbon source. A theoretical optimal carbon-to-nitrogen ratio (C/N) may be calculated using the following stoichiometric relationship (using methanol as an example):

$$6\,NO_3^- + 5\,CH_3OH + H_2CO_3 \rightarrow 3\,N_2 + 8\,H_2O + 6\,HCO_3^- \quad (5.12)$$

Converting this to a mass basis reduces the equation (in grams) to the following:

$$84g\,NO_3^-\text{-}N + 160g\,CH_3OH + 62g\,H_2CO_3 \rightarrow 84g\,N_2 + 144g\,H_2O + 366g\,HCO_3^- \quad (5.13)$$

One gram of methanol has the theoretical oxygen equivalent of 1.5 g of oxygen. Therefore, 1.91 g methanol/g NO_3^--N is required to reduce nitrate completely. In addition to providing energy for the reduction of nitrate, the methanol or carbonaceous BOD (CBOD) consumed also creates new biomass. As a result, more methanol or CBOD are required than are presented in eqs 5.12 and 5.13 to reduce each unit of nitrate. The amount of new biomass generated and the portion used for denitrification are specific to each compound. Equations for calculating methanol consumption using nitrate, nitrite, and oxygen are as follows (U.S. EPA, 1975):

$$NO_3^- + 1.08\,CH_3OH + 0.24\,H_2CO_3 \rightarrow 0.04\,C_5H_7NO_2 + 0.48\,N_2 + 1.23\,H_2O + HCO_3^- \quad (5.14)$$

$$NO_2^- + 0.67\,CH_3OH + 0.53\,H_2CO_3 \rightarrow 0.056\,C_5H_7NO_2 + 0.47\,N_2 + 1.68\,H_2O + HCO_3^- \quad (5.15)$$

$$O_2^- + 0.93\,CH_3OH + 0.056\,NO_3^- \rightarrow 0.056\,C_5H_7NO_2 + 1.04\,N_2 + 0.59\,H_2CO_3 + 0.056\,HCO_3^- \quad (5.16)$$

These three equations can be converted into a methanol dose based on nitrate, nitrite, and dissolved oxygen concentrations, as follows:

$$\text{Methanol dose (mg/L)} = 2.47\,NO_3^-\text{-}N + 1.53\,NO_2^-\text{-}N + 0.87\,\text{Dissolved oxygen} \quad (5.17)$$

Provided that dissolved oxygen discharged or recycled to a denitrification reactor is minimal, the general rule that 3 mg/L BOD_5 is removed for every 1 mg/L NO_3^--N denitrified is a reasonable approximation.

1.4.3 *Temperature*

Similar to other heterotrophs, the effect of temperature on the kinetics of denitrifying bacteria is mathematically approximated based on the following Arrhenius equation:

$$k = Ae^{-Ea/RT}$$
$$k_1 = k_2\theta^{T_1 - T_2} \quad (5.18)$$

Where

k = specific denitrification rate at temperature T (mg NO_3-N/mg VSS/d),
A = the prefactor,
Ea = activation energy in J/mol,
R = ideal gas constant (8.314 J/mol/K),
T = absolute temperature, and
θ = temperature coefficient (dimensionless).

Denitrification rate vs temperature is, therefore, a bell-shaped curve, and the maximum denitrification rate is obtained at the optimal temperature (ranging from 20 to 60 °C [Dawson and Murphy, 1972; Knowles, 1982]), and then declines rapidly above this temperature. Various θ values have been reported based on a number of site-specific studies, yielding a range from 1.03 to 1.20 (Dawson and Murphy, 1972).

1.4.4 *External Carbon Source*

Nearly all separate-stage denitrification reactors and even some denitrification reactors within single sludge systems use some form of supplemental carbon or substrate augmentation. Most separate-stage denitrification processes follow CBOD and nitrification processes. Therefore, there is a limited quantity of biodegradable organic carbon available to serve as the substrate for denitrification. The form of substrate and, in particular, the ease or difficulty of biodegradation of that substance is a key parameter related to the rate and efficiency of the denitrification process. Several organic carbon sources have historically been used to augment the rate of the denitrification process, including methanol, ethanol, acetate, molasses, and brewery wastes. Several industrial wastes were also tested as organic carbon sources. Some organic wastes, such as formaldehyde and dextrose waste, were less efficiently degraded than distillery oils or methanol, but the majority of food industry wastes exhibit high denitrification rates and carbon-to-nitrogen ratios from 2 to 6. The addition of residue from food industrial processes also shows improved nitrogen removal performance at a full-scale municipal water resource recovery facility (WRRF) in Finland. The use of anaerobic fermentation products from the organic fraction of municipal solid waste has been suggested for enhanced denitrification (Bolzonella et al., 2001) in an integrated BNR and solid waste management system, which has already been applied in full-scale size at the Treviso (Italy) WRRF.

Soluble and readily degradable substrates support the highest rates of denitrification. Although methanol is the most widely used exogenous carbon source, it is not as efficient as other carbon substrates (e.g., ethanol and acetate on the kinetic basis) (Onnis-Hayden and Gu, 2008). Denitrifying bacteria enriched by ethanol and acetate also grow at higher yields. The stoichiometric and kinetic coefficients of denitrification with various carbon sources are summarized in Table 5.2.

TABLE 5.2 Representative stoichiometric and kinetic parameters at 20 °C for denitrifying biomass grown on different carbon sources.

Carbon source	COD/N	Y (g VSS/g COD)	μ_{max} (d^{-1})	SDNR (mg N/g VSS·h)	Reference
Methanol	4.1–4.5	0.23–0.25	0.77–2	32–91	Christensson et al. (1994)
	4.7		0.4–1.0	6.07	Mokhayeri et al. (2006)
		0.4		3.2	Peng et al. (2007)
Ethanol		0.42		9.6	Peng et al. (2007)
		0.22		27	Akunna et al. (1993)
	5.877	0.36			McCarty et al. (1969)
Acetate		0.65		12	Peng et al. (2007)
	3.5		1.2–3.5		Mokhayeri et al. (2006)
		0.3		3.6	Kujawa and Klapwijk, 1999
MicroC1000™	7.87	0.44	1.99	3.31	Akunna et al. (1993)
MicroC2000™	6.46	0.39	2.05	9.34	Du et al. (2011)
Glucose				2.7	Du et al. (2011)
					Akunna et al. (1993)
		0.38			Muller et al. (2003)
Biodiesel	4.8			1.8	Ramalingam et al. (2007)

Adaptation of denitrifying communities to change external carbon sources may involve both enrichment of new populations and the regulation of enzyme expression in existing populations. Methanol addition often requires an adaptation period of up to several months before denitrification rates significantly increase (Nyberg et al., 1992). This may be attributed to necessary population shifts in the microbial community (Ginige et al., 2004) to enrich methylotrophs. In contrast, activated sludge responds immediately to acetate and ethanol (Hallin et al., 1996; Hallin and Pell, 1998; Isaacs et al., 1994) because these two carbons select *Azoarcus*, *Dechloromonas*, *Thauera*, and *Acidovorax*-like denitrifiers, which have broad substrate specificities (Hallin and Pell, 1998; Hwang et al., 2006). Denitrifying species identified in activated sludge processes with different carbon sources are summarized in Table 5.3.

1.5 Benefits of Denitrification

1.5.1 Permit Compliance

Denitrification reduces nitrite and nitrate ions that are discharged, thereby protecting the quality of the receiving water and preventing eutrophication. Activated sludge processes that have a total nitrogen discharge limit must denitrify.

Table 5.3 Denitrifying populations grown on different types of carbon sources.

Carbon source	Dominant functional groups	Techniques	References
Methanol	*Methylophilales, Methyloversatilis Hyphomicrobium, Paracoccus*	Culture isolation, SIP[a], 16S rRNA gene sequencing, FISH[b]	Ginige et al. (2004), Blaszcytk et al. (1985), Lemmer et al. (1997), Bayshtok et al. (2008), Bayshtok et al. (2009)
Ethanol	*Methyloversatilis, Azoarcus Dechloromonas, Pseudomonas Hydrogenophaga*	Stable isotope probing, 16S rRNA sequencing	Hwang et al. (2006), Baysthtok et al.(2008)
Acetate	*Comamonas, Acidovorax, Thauera, Dechloromonas, Paracoccus, Rhodobacter*	Stable isotope probing, 16S rRNA gene sequencing, FISH[c]	Osaka et al. (2006), Ginige et al. (2005)
Glycerol	*Comamonas, Bradyrhizobium, Diaphorobacter, Tessaracoccus*	Stable isotope probing	Lu and Chandran (2010)
Glucose	*Arthrobacter, Phenylobacterium, Pseudomonas, Proteobacterium*	Denaturing gradient gel electrophoresis	Nakatsu et al. (2005)
Landfill leachate	*Thauera, Acidovorax, Alcaligenes*	16S rRNA gene sequencing	Etchebehere et al. (2001)

[a]SIP = stable isotope probing.
[b,c]FISH = fluorescent in situ hybridization.

1.5.2 *Control of Undesired Filamentous Organism Growth*

Denitrification promotes the formation of firm and dense floc particles. Firm and dense floc particles are resistant to shearing action and exhibit desired settling characteristics. An anoxic environment produced through denitrification favors the growth of facultative anaerobic, floc-forming bacteria and discourages the growth of strict aerobic, filamentous organisms and weak facultative anaerobic, filamentous organisms. Strict aerobic, filamentous organisms can only use free molecular oxygen. Therefore, their growth is stopped or the organisms die in an anoxic environment. Although weak facultative, anaerobic, filamentous organisms can use free molecular oxygen and nitrite ions or nitrate ions, these filamentous organisms cannot compete successfully for nitrite ions or nitrate ions with facultative anaerobic, floc-forming bacteria.

1.5.3 *Return of Alkalinity*

Approximately 3.57 mg of alkalinity as $CaCO_3$ is produced for each milligram of nitrate converted to molecular nitrogen. Nearly all activated sludge processes that denitrify must nitrify, which results in a consumption of alkalinity for about

7.14 mg as $CaCO_3$/mg NH_4^+ oxidized. Therefore, the alkalinity that is returned during denitrification is approximately one-half the amount of alkalinity that is lost during nitrification.

1.5.4 Energy Savings

Each gram of nitrate consumed in respiration saves 2.86 g of oxygen. If an activated sludge facility is configured so that facultative bacteria can use the nitrate produced during nitrification while consuming CBOD, the facility can realize substantial energy cost savings as a result of the decrease in aeration.

2.0 DENITRIFICATION SYSTEMS

2.1 Suspended Growth Systems

Single-sludge, suspended growth denitrification processes accomplish COD removal, nitrification, and denitrification with the same biomass. In these systems, biomass is exposed to different environmental conditions within various zones and shares the same set of secondary clarifiers. These systems achieve COD and nitrogen removal using linked reactors in series and a common set of secondary clarifiers. This consolidation of processes offers the potential of reduced capital and operating costs and, at the same time, a greater operational challenge related to proper balancing that is sometimes required between the different process environments.

Combined nitrification and denitrification processes can be classified in various ways such as biofilm vs suspended growth, flow regime, staging of process sequences, or method of aeration. The common aspect consistent throughout all of these processes is alternating environments that are used sequentially to nitrify and denitrify. Nitrification must be completed to a large extent before denitrification can be achieved. Various factors such as dissolved oxygen concentration (high or low) and toxic compounds can influence the efficiency of these systems. In addition, the following three parameters can limit denitrification and, ultimately, total nitrogen removal:

- Nitrate—incomplete nitrification will result in low concentrations of nitrate, reducing nitrogen removal through denitrification and resulting in high-effluent total nitrogen concentrations;
- Substrate (CBOD)—insufficient carbon in the preanoxic zone reduces denitrification efficiency; and
- Denitrification capacity (detention time)—insufficient reactor volume reduces the detention time necessary for denitrification.

Another limiting factor is the capacity of secondary clarifiers. Secondary clarifiers for BNR facilities must be designed conservatively to avoid washout (WEF et al., 2009).

2.2 Preanoxic Processes

Preanoxic processes use a single anoxic (nonaerated) zone followed by a single aerated zone for nitrogen removal with internal nitrate recycle to the unaerated zone. These processes will typically produce effluent total nitrogen concentrations between 6 and 10 mg N/L. These processes were developed to take advantage of the carbonaceous substrate available in the influent wastewater by placing the anoxic zone upstream of the nitrification reactor as shown in Figure 5.3 (Ludzack and Ettinger, 1962). The most common of these is the modified Ludzack–Ettinger (MLE) process, which will be discussed in more detail in Chapter 6. The MLE process uses COD present in primary effluent to reduce the nitrate in the internal recycle entering the preanoxic zone to nitrogen gas.

2.3 Preanoxic and Postanoxic Processes

To achieve low-effluent total nitrogen concentrations (<6 mg total nitrogen/L), a postanoxic zone is added (Figure 5.4) in which additional nitrate is denitrified. At this point in the process, most of the COD has been depleted and an external carbon source may be necessary. Depending on the permit limits and climate, carbon produced during endogenous decay of the MLSS may be sufficient to meet effluent total nitrogen limits. However, in colder climates or low-effluent total nitrogen concentrations, an external carbon source is added. An example of this process is the four-stage Bardenpho process, which will be discussed in Chapter 6. The four-stage Bardenpho as well as the enhanced MLE processes will achieve effluent total nitrogen concentrations lower than 6 mg/L, assuming adequate secondary clarification capacity and an external carbon source.

2.4 Step-Feed Alternate Aerobic–Anoxic Processes

Step-feed systems can carry a higher solids inventory and thus a higher SRT in the same volume as plug flow systems while keeping the solids load to the final

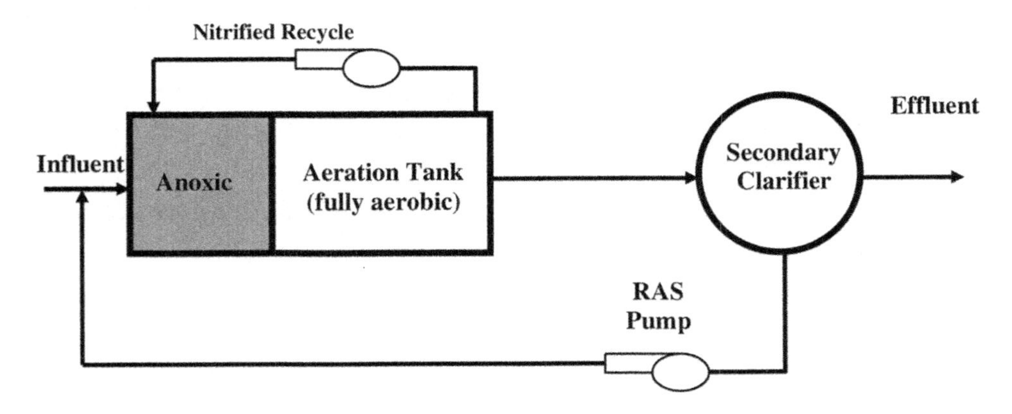

FIGURE 5.3 Modified Ludzack–Ettinger process.

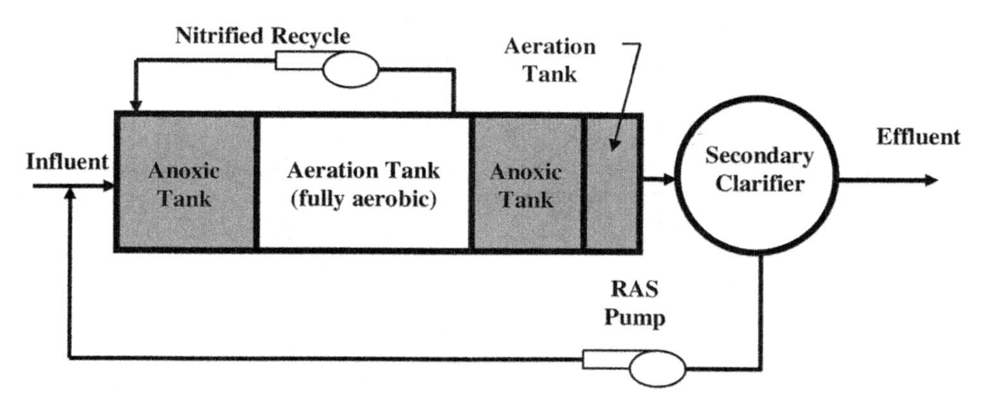

FIGURE 5.4 Four-stage Bardenpho process.

clarifiers at the same concentration. The process train typically consists of three to four sets of anoxic and aerobic reactors placed in series. The anoxic and aerobic stages are equal in number and alternate, starting with an anoxic stage. The feed is introduced in steps in each anoxic stage such that the last anoxic stage functions as a postanoxic reactor, followed by a last aerobic stage for removal of residual BOD following the principle explained previously. Because of the presence of multiple stages of alternating anoxic and aerobic stages, there is no need for an internal recycle stream over and above the return activated sludge (RAS) stream.

However, not all step-feed processes can achieve an effluent total nitrogen concentration lower than 6 mg/L unless their influent total nitrogen concentration is low. Essentially, any nitrogen added to the last pass will either be discharged as ammonia (if incomplete nitrification) or as nitrate-nitrogen or nitrite-nitrogen because there is no downstream anoxic zone for denitrifying the influent ammonia that was converted to nitrate. For more information on this topic, refer to Chapter 6.

2.5 Oxidation Ditch

Typical oxidation-ditch treatment systems consist of a single or multichannel configuration within a ring-, oval-, or horseshoe-shaped basin (Figure 5.5). Horizontally or vertically mounted aerators provide circulation, oxygen transfer, and aeration in the ditch. Flow to the oxidation ditch is aerated and mixed with RAS from a secondary clarifier. Surface aerators, such as brush rotors, disc aerators, draft-tube aerators, or fine-bubble diffusers are used to circulate the mixed liquor. The mixing process entrains oxygen into the mixed liquor to foster microbial growth and the motive velocity ensures contact of microorganisms with incoming wastewater. For nitrogen removal, an anoxic zone is added upstream of the ditch along with mixed liquor recirculation from the aerobic zone to the tank to achieve higher levels of denitrification. Because of the flow pattern and "infinite

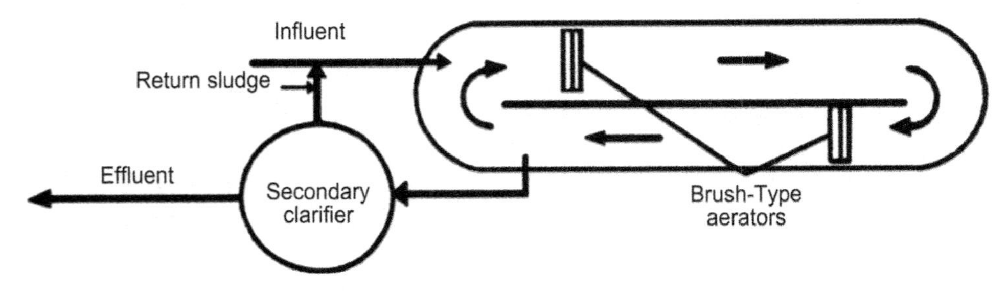

FIGURE 5.5 Oxidation ditch reactor.

recycle", oxidation ditches can produce low-effluent nitrogen concentrations. For more information on this topic, refer to Chapter 6.

2.6 Sequencing Batch Reactor

A sequencing batch reactor (SBR) operates in semibatch mode in identical, repeating cycles (Figure 5.6). This process will be described in detail in Chapter 6. Each cycle treats a new batch of influent in multiple, sequential steps. Unlike in a continuous activated sludge facility in which the unit processes are spatially separated, in an SBR the events (or steps) of filling, biochemical reaction, solids settling, and decanting of treated and clarified effluent occur in the same tank. This sequence also represents a two-stage reactor system with the preanoxic stage preceding the aeration stage and a clarifier in series. An anaerobic stage can be included after the preanoxic stage to accomplish enhanced biological phosphorus removal, if desired. For more information on this topic, refer to Chapter 6.

2.7 Membrane Bioreactor

In membrane bioreactor (MBR) technology, mixed liquor from the activated sludge process is subjected to solids–liquid separation by microfiltration membranes instead of by gravity separation in a secondary clarifier. The membranes are placed in situ in the aeration tanks or ex situ in a separate structure. When

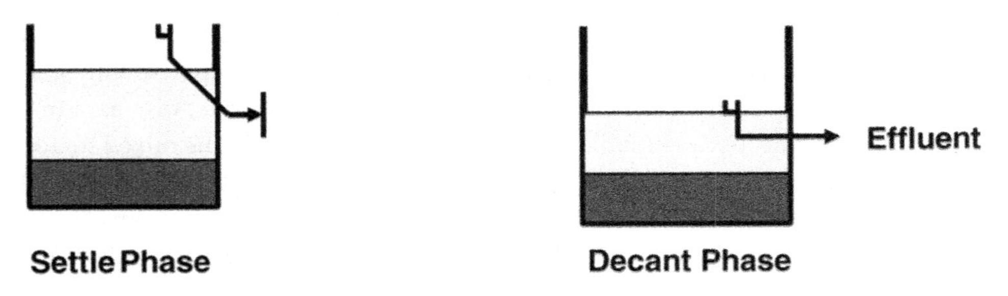

FIGURE 5.6 Sequencing batch reactor process cycle.

placed in situ, the membrane replaces the secondary clarifier, and the MLSS concentration can be increased to 10 000 mg/L or more. When used ex situ, the primary purpose is to provide high-quality water for reuse. The membrane composition, type, and placement in the overall treatment train are vendor specific.

Membranes are sensitive to coarse and abrasive solids and, to protect them from such solids, the influent wastewater stream from the primary clarifier is passed through a set of fine-gauge (<2 mm) screens to remove small inorganic and abrasive materials. The screened primary effluent flows into the reactor stages for BOD oxidation and nitrogen removal. Fine-bubble diffusers are used for aeration as in the case of a conventional activated sludge process. In a typical ex situ process, the mixed liquor flows into the MBR tank under the influence of a negative pressure created across the membranes by a set of permeate pumps. The MBR tanks are equipped with cassettes of membranes that carry out high-efficiency solids–liquid separation from the mixed liquor. They concentrate the incoming mixed liquor into a thicker liquid as the retentate (sludge) phase. The permeate phase consists of clear liquid that is typically less than 1 mg/L of TSS and BOD. Effluent total nitrogen concentrations of 5 mg/L or less can be achieved for processes incorporating nitrification and denitrification. However, the final effluent total nitrogen concentration depends on wastewater temperature and characteristics and process design. For more information on this topic, refer to Chapter 6.

3.0 CARBON AUGMENTATION

A carbon source is necessary for denitrification for nitrogen removal to serve as an electron donor. The fundamentals of denitrification (stoichiometry and kinetics) are covered in Section 1.1 of this chapter. In the denitrification process, a readily biodegradable carbon source must be available. These electron donor or carbon requirements can be met by influent wastewater soluble BOD, from natural or enhanced cell decay and lysis, or by the addition of an external supplemental carbon source (Onnis-Hayden and Gu, 2008). This section will discuss when an external or internal carbon source is needed and will provide information on various types of carbon sources, their applications, and handling and dosage requirements.

3.1 The Need for Carbon Use to Meet Effluent Nitrogen Limits

When total nitrogen, total inorganic nitrogen, and nitrate-nitrogen effluent limits are in place, it is necessary to have a sufficient supply of available carbon for denitrification. This is especially the case when effluent total nitrogen limits are 10 mg/L or less. If effluent total nitrogen limits are 5 mg/L or less, having a source of external supplemental carbon available to consistently meet the limit is typically required.

3.1.1 *Internal Carbon Sources*

Internal carbon sources are present in most WRRFs. Carbon is present in raw influent wastewater and endogenous carbon is present in the postanoxic zone after aerobic treatment. Internal carbon is also available from volatile fatty acids (VFAs) from sludge fermentation or from a separate septage stream received at the WRRF.

3.1.1.1 Raw Wastewater and Primary Effluent

Raw influent wastewater and primary effluent contain BOD and COD that are necessary for denitrification. The readily biodegradable fraction of COD is used for denitrification and typically represents 15 to 25% of the total COD in the influent to the anoxic reactor (Metcalf and Eddy, 2003). Water resource recovery facilities without primary clarifiers generally have higher amounts of carbon present in the unsettled wastewater than facilities with primary clarification. The influent wastewater characteristics of each individual treatment facility need to be evaluated to determine the availability of degradable carbon available for denitrification in the wastewater.

Influent or primary effluent BOD or COD will typically need to be 4 to 5 times the total Kjeldahl nitrogen (TKN) for effective denitrification in the primary anoxic zone. When available COD is less than this, denitrification will still take place until the available COD is consumed.

3.1.1.2 Endogenous Carbon

Endogenous carbon results from cell mass decay and is typically used as a carbon source in postanoxic processes in which all of the available influent carbon has been consumed in the aerobic treatment process. Specific denitrification rates (SDNRs) are low, that is, from 0.01 to 0.04 g NO_3-N/g MLVSS under endogenous respiration (WEF, 2011). Slow SDNRs require a large postanoxic zone to accomplish denitrification.

3.1.1.3 Sludge Fermentation

Volatile fatty acids resulting from sludge fermentation are typically used as a carbon source with biological phosphorus removal to drive biological release and subsequent uptake of soluble phosphorus. Use of VFAs for biological phosphorus is described in Chapter 7.

For nitrogen removal, VFAs can serve as an electron donor in the form of acetic and propionic acids. The amount of soluble COD in fermentate is estimated to be between 400 and 800 mg/L. This is not a high concentration for supplemental carbon addition and, therefore, will require a considerable volume of primary sludge fermentate available for denitrification. Mixed liquor or RAS fermentation is also feasible at facilities that do not have primary clarifiers or facilities that use primary sludge to enhance digester biogas production vs fermentation.

3.1.1.4 *Cell Lysis*

Cell lysis is a process in which cell walls break down and the cell contents, which contain carbon, are released to the mixed liquor. This process occurs naturally in the endogenous respiration process discussed in Section 3.1.1.2. Cell lysis can also result from an external force (electrical or mechanical) to burst cells and release carbon. Proprietary processes such as pulse electric field, ultrasound disintegration, and hydro dynamic cavitation provide a carbon source for denitrification from cell lysis (Metcalf and Eddy, 2003).

3.1.1.5 *Septage*

Septage can potentially be a source of carbon for denitrification; however, careful analysis of the septage is required. Along with COD, septage contains high levels of TKN, which can negate the benefit as a carbon source in the septage (WEF et al., 2009).

3.1.2 *External Carbon Sources*

Sections 3.1.2.1 through 3.1.2.7 describe the most commonly available external carbon sources used for denitrification. Additional information on evaluating external carbon sources is available in Water Environment Research Foundation's *Protocol to Evaluate Alternative External Carbon Sources for Denitrification at Full-Scale Wastewater Treatment Plants* (Gu and Onnis-Hayden, 2010).

3.1.2.1 *Methanol*

Historically, methanol (CH_3OH) has been the most commonly used external carbon source for denitrification. Methanol use in both suspended growth and biofilm denitrification systems has been well documented in the literature. Advantages of methanol include a low COD/N requirement and a low sludge or solids yield. Methanol requires a period of acclimation in a biological process that can vary from a week to several weeks depending on the wastewater temperature. Methanol is used by a specific group of facultative methylotroph organisms, as discussed earlier in this chapter. These organisms are slow growing and a population must be established for denitrification to occur.

On a unit-cost basis, methanol is often the least costly of the external carbon sources. However, methanol has been subject to historical pricing fluctuations as it is manufactured from natural gas and almost exclusively imported from outside of the United States. Methanol is a flammable liquid and requires specialized pumping, storage, and handling facilities that are regulated under Sections 30, Flammable and Combustible Liquids Code, and 820, Fire Protection in Wastewater Treatment and Collection Facilities, of the National Fire Protection Association codes (NFPA, 2012a; 2012b) (see Section 3.2).

3.1.2.2 Ethanol

Ethanol (CH_3CH_2OH) is classified as a flammable liquid and requires the same specialized handling requirements as methanol. Despite the high COD concentration, ethanol is typically not cost effective compared to methanol because of a higher unit cost and higher biomass yield (COD/N). As a result, ethanol use for biological denitrification in wastewater applications is sparse at best. Pure ethanol is regulated by the Alcohol and Tobacco Tax and Trade Bureau; therefore, denatured ethanol is the most suitable option for biological denitrification.

3.1.2.3 Sugar Solutions

Sugar solutions are categorized as carbohydrate-based solutions that may include compounds such as glucose, sucrose, tri and polysaccharides, molasses, and corn sweeteners, in addition to other commodity carbohydrate compounds. Sugar solutions are widely available and nonhazardous, making them suitable options for smaller facilities that are sensitive to safety. Denitrifying organisms grown on carbohydrate solutions result in a higher biomass yield that can create operational challenges, particularly with biofilm denitrification systems. Sugar solutions have a higher freezing point than methanol and are subject to biological degradation over time. To allow for a long shelf life for stability products such as MicroC 1000™, a proprietary carbohydrate solution contains a preservative for long-term stability.

Molasses has been used as a supplemental carbon source; however, it may contain nitrogen and phosphorus that must be evaluated in the product before use to avoid adding additional nutrients to the wastewater that will need to be removed. In some applications where molasses has been used with UV disinfection, there have been issues where coating of the surface of UV lamps has been observed, resulting in additional maintenance for cleaning the UV lamps.

3.1.2.4 Acetic Acid Solutions

Acetic acid (CH_3COOH) has been used in wastewater treatment applications for both denitrification and biological phosphorous removal. Acetic acid has favorable kinetics and does not require an acclimation period. Commonly available acetic acid solution strengths are 20, 56, 80, and 100%. More dilute solutions are used when concerns about handling hazardous materials need to be minimized. Higher-strength solutions require special handling and solutions greater than 80% may be categorized as flammable liquids. Acetic acid has a high freezing point and heated storage may be necessary in colder climates. Acetic acid is typically more expensive than other available supplemental carbon sources.

3.1.2.5 Sodium and Potassium Acetate

Acetate salts such as sodium and potassium acetate have the same functional group (CH_3COO^-) as acetic acid, but are available in a crystalline or powder form that requires being dissolved in water before use. The solubility of these acetate salts in water results in handling and storage challenges, even at moderate temperatures. Sodium and potassium acetate are nonhazardous; however, they are not widely used because of their high cost relative to other carbon sources.

3.1.2.6 *Glycerol*

Glycerol, or glycerin ($C_3H_8O_3$), has gained popularity as a supplemental carbon source at WRRFs in the last several years for both suspended growth and biofilm processes. Crude glycerin is a byproduct of the biodiesel production process. Unrefined crude glycerin may contain undesirable fats, soaps, biodiesel, and other intermediate and end products of the biodiesel process. Unrefined crude glycerin can also contain high concentrations of methanol, which results in a flammable classification. A number of refined glycerin products are available that have a fixed glycerin concentration (typically 70%) and COD greater than 1 kg/L, and that have had methanol and other undesirable products removed from the crude glycerin.

Glycerin provides favorable kinetics, in addition to a low freezing point, and can be used in most climates without freeze protection. With the exception of methanol, glycerin-based products are often the least expensive option when evaluated on a dollar(s)-per-kilogram-of-nitrogen-removed basis. Glycerin does not require an acclimation period or specialized storage and feed equipment.

3.1.2.7 *Waste Products*

A number of waste products such a soda bottling wastes, acetic acid solutions, brewery wastes, and cheese whey have been used as supplemental carbon sources. These waste streams can provide a local source of carbon that can be used for denitrification. Concerns with any waste product include consistency and type of COD, nitrogen and phosphorus content, availability on a continuous basis, cost, and the potential for toxic compounds in the waste product. A careful analysis should be performed on the waste product from an analytical, pilot-testing, and financial perspective before a product is used full scale in a treatment process.

3.2 Supplemental Carbon Handling and Storage Facilities

Supplemental carbon handling and storage facilities are designed for either flammable or nonflammable systems. Nonflammable carbon sources can potentially be used in systems originally designed for flammable liquids. Flammable liquids fall under the requirements of the National Fire Protection Association (NFPA) and cannot be used in nonflammable-rated storage and handling facilities.

3.2.1 *Regulations*

Use of certain external supplemental carbon sources may be regulated by NFPA, the Occupational Safety and Health Administration, and the U.S. Environmental Protection Agency. The regulations and their enforcement are site specific and often fall under the jurisdiction of the state and local fire marshal. In the aftermath of the 2006 methanol fire and explosion at the Bethune Point Wastewater Treatment Plant in Daytona Beach, Florida, the U.S. Chemical Safety and Hazard Investigation Board issued a report that spelled out specific design changes and training requirements for the use of flammable liquids at WRRFs (NFPA, 2012a; 2012b).

Therefore, the local fire marshal office should be consulted for specific local requirements for bulk liquid storage and handling requirements because they vary by jurisdiction.

3.2.2 *Storage Facilities*

Bulk storage tanks can be made of carbon steel, stainless steel, fiberglass, and reinforced polyethylene. The type of external carbon or waste product will require analysis by the tank manufacturer to ensure compatibility. Carbon products are also available in intermediate bulk carrier totes and 208-L (55-gal) drums to be used for small use or temporary installations. Applications should include either a double-wall tank or a suitable containment structure in case of a spill or failure of the bulk tank.

3.2.3 *Pumping and Conveyance Facilities*

Pumping and conveyance systems, like bulk storage tanks, are specific to the carbon source to be used. Generally, nonflammable products can be used in all systems, and flammable products require specialized NFPA-approved systems. Piping systems can range from polyvinyl chloride to stainless steel or carbon steel. Some hazardous flammable products require specialized double-wall containment piping (NFPA, 2012a; 2012b). Chemical metering pumps are used in most applications for product dosing. Common pump types include hose or peristaltic pumps, diaphragm pumps, and gear pumps.

3.3 Carbon Dosage Points and Mixing Requirements

Poor mixing and high dissolved oxygen in the anoxic zone or process can result in high carbon use and poor denitrification performance. For any carbon source to be effective, proper dosing and feed point specific to the process are needed. In suspended growth systems in general, the external carbon source needs to feed into a well-mixed point at the head of the process tank. Dissolved oxygen will need to be limited to less than 0.5 mg/L entering the anoxic process to not consume excess supplemental carbon before achieving anoxic conditions for denitrification.

Anoxic tank mixing can be accomplished with a number of different mixing technologies. With all mixing technologies, it is important that the entire anoxic process tank be completely mixed with no dead zones. Mixers should use the minimum mixing energy to completely mix the process tank and not add additional dissolved oxygen to the process.

With biofilm processes, it is necessary to limit influent dissolved oxygen entering the filter process or excessive supplement carbon will need to be fed to the process to lower dissolved oxygen to anoxic conditions before denitrification can be accomplished. Minimizing dissolved oxygen can be accomplished by limiting turbulent influent pumping conditions to the filter and minimizing the drop into the filter from influent feed weirs.

5.0 REFERENCES

Akunna, J. C.; Bizeau, C.; Moletta, R. (1993) Nitrate and Nitrite Reductions with Anaerobic Sludge Using Various Carbon Sources: Glucose, Glycerol, Acetic Acid, Lactic Acid and Methanol. *Water Res.*, **27** (8), 1303–1312.

Anthony, C. (2011) How Half a Century of Research Was Required to Understand Bacterial Growth on C1 and C2 Compounds; The Story of the Serine Cycle and the Ethylmalonyl-CoA Pathway. *Sci. Prog.*, **94** (2), 109–137.

Baytshtok, V.; Kim, S.; Yu, R.; Park, H.; Chandran, K. (2008) Molecular and Biokinetic Characterization of Methylotrophic Denitrification Using Nitrate and Nitrite as Terminal Electron Acceptors. *Water Sci. Technol.*, **58** (2), 359–365.

Baytshtok, V.; Lu, H. J.; Park, H.; Kim, S.; Yu, R.; Chandran, K. (2009) Impact of Varying Electron Donors on the Molecular Microbial Ecology and Biokinetics of Methylotrophic Denitrifying Bacteria. *Biotechnol. Bioeng.*, **102** (6), 1527–1536.

Blaszczyk, M.; Galka, E.; Sakowicz, E.; Mycielski, R. (1985) Denitrification of High Concentrations of Nitrites and Nitrates in Synthetic Medium with Different Sources of Organic Carbon. III. Methanol. *Acta Microbiol. Pol.*, **34** (2), 195–205.

Chistoserdova, L.; Chen, S.-W.; Lapidus, A.; Lidstrom, M. E. (2003) Methylotrophy in *Methylobacterium extorquens* AM1 from a Genomic Point of View. *J. Bacteriol.*, **185** (10), 2980–2987.

Chiu, Y. C.; Chung, M. S. (2003) Determination of Optimal COD/Nitrate Ratio for Biological Denitrification. *Int. Biodeterioration Biodegrad.*, **51**, 43–49.

Christensson, M.; Lie, E.; Welander, T. (1994) A Comparison Between Ethanol and Methanol as Carbon Sources for Denitrification. *Water Sci. Technol.*, **30** (6), 83–90.

Dawson, R. N.; Murphy, K. L. (1972) The Temperature Dependency of Biological Denitrification. *Water Res.*, **6** (1), 71–83.

de Vries, G. E; Kües, U.; Stahl, U. (1990) Physiology and Genetics of Methylotrophic Bacteria. *FEMS Microbiol. Lett.*, **75** (1), 57–101.

Du, Y. Q.; Liu, S.; Onnis-Hayden, A.; Gu, A. Z. (2010) *Investigation of MicroCG™ and Microcglycerin™ a Alternative Electron Donor/Carbon Sources for Enhancing Biological Nutrient Removal in Wastewater Treatment.* Technical Report prepared for Environmental Operation Solutions, Inc.

Etchebehere, C.; Errazquin, I.; Barrandeguy, E.; Dabert, P.; Moletta, R.; Muxi, L. (2001) Evaluation of the Denitrifying Microbiota of Anoxic Reactors. *FEMS Microbiol. Ecol.,* **35** (3), 259–265.

Gaudy, A. F. J.; Gaudy, E. T. (1980) *Microbiology for Environmental Scientists and Engineers;* McGraw-Hill: New York.

Gerardi, M. H. (2006) *Wastewater Bacteria;* Wiley-Interscience: Hoboken, New Jersey; pp 92–93.

Ginige, M. P.; Hugenholtz, P.; Damis, H.; Wagner, M.; Keller, J.; Blackall, L. (2004) Use of Stable-Isotope Probing, Full-Cycle rRNA Analysis, and Fluorescence In Situ Hybridization-Microautoradiography to Study a Methanol-Fed Denitrifying Microbial Community. *Appl. Environ. Microbiol.,* **70** (1), 588–596.

Ginige, M. P.; Keller, J.; Blackall, L. L. (2005) Investigation of an Acetate-Fed Denitrifying Microbial Community by Stable Isotope Probing, Full-Cycle rRNA Analysis, and Fluorescent In Situ Hybridization-Microautoradiography. *Appl. Environ. Microbiol.,* **71** (12), 8683–8691.

Grady, C. P. L. Jr.; Daigger, G. T.; Lim, H. C. (1999) *Biological Wastewater Treatment,* 2nd ed.; Marcel Dekker: New York.

Gu, A.; Onnis-Hayden, A. (2010) *Protocol to Evaluate Alternative External Carbon Sources for Denitrification at Full-Scale Wastewater Treatment Plants;* WERF Report NUTR1R067b; Water Environment Research Foundation: Alexandria, Virginia.

Hallin, S.; Pell, M. (1998) Metabolic Properties of Denitrifying Bacteria Adapting to Methanol and Ethanol in Activated Sludge. *Water Res.,* **32** (1), 13–18.

Hallin, S.; Rothman, M. ; Pell, M. (1996) Adaptation of Denitrifying Bacteria to Acetate and Methanol in Activated Sludge. *Water Res.,* **30** (6), 1445–1450.

Hwang C.; Wu W. M.; Gentry, T.; Carley, J.; Carroll, S.; Schadt, C.; Watson, D.; Jardine, P.; Zhou, J.; Hickey, R.; Criddle, C.; Fields, M. (2006) Changes in Bacterial Community Structure Correlate with Initial Operating Conditions of a Field-Scale Denitrifying Fluidized Bed Reactor. *Appl. Microbiol. Biotechnol.,* **71** (5), 748–760.

Isaacs, S. H.; Henze, M.; Søeberg, H.; Kümmel, M. (1994) External Carbon Source Addition as a Means to Control an Activated Sludge Nutrient Removal Process. *Water Res.,* **28** (3), 511–520.

Knowles, R. (1982) Denitrification, *Microbiol. Rev.,* **46** (1), 43–70.

Kornaros, M.; Lyberatos, G. (1998) Kinetic Modelling of *Pseudomonas denitrificans* Growth and Denitrification under Aerobic, Anoxic and Transient Operating Conditions. *Water Res.,* **32** (6), 1912–1922.

Kujawa, K.; Klapwijk, B. (1999) A Method to Estimate Denitrification Potential for Predenitrification Systems Using NUR Batch Test. *Water Res.,* **33** (10), 2291–2300.

Labbe, N.; Laurin, V.; Juteau, P.; Parent, S.; Villemur, R. (2007) Microbiological Community Structure of the Biofilm of a Methanol-Fed, Marine Denitrification System, and Identification of the Methanol-Utilizing Microorganisms. *Microb. Ecol.*, **53** (4), 621–630.

Lemmer, H.; Zaglauer, A.; Neef, A.; Meier, H.; Amann, R. (1997) Denitrification in a Methanol-Fed Fixed-Bed Reactor. Part 2: Composition and Ecology of the Bacterial Community in the Biofilms. *Water Res.*, **31** (8), 1903–1908.

Li, Y.-M.; Li, J.; Zheng, G.-H.; Luan, J.-F.; Fu, Q.-S.; Gu, G.-W. (2008) Effects of the COD/NO_3-N Ratio and pH on the Accumulation of Denitrification Intermediates with Available Pyridine as a Sole Electron Donor and Carbon Source. *Environ. Technol.*, **29** (12), 1297–1306.

Lu, H.; Chandran, K. (2010) Diagnosis and Quantification of Glycerol Assimilating Denitrifying Bacteria in an Integrated Fixed-Film Activated Sludge Reactor via ^{13}C DNA Stable-Isotope Probing. *Environ. Sci. Technol.*, **44** (23), 8943–8949.

Ludzack, F. J.; Ettinger, M. B. (1962) Controlling Operation to Minimize Activated Sludge Effluent Nitrogen. *J.—Water Pollut. Control Fed.*, **34,** 920–931.

Marazioti, C.; Kornaros, M.; Lyberatos, G. (2003) Kinetic Modeling of a Mixed Culture of *Pseudomonas denitrificans* and *Bacillus subtilis* under Aerobic and Anoxic Operating Conditions. *Water Res.*, **37** (6), 1239–1251.

McCarty, P. L.; Beck, L.; Amant, P. S. (1969) Biological Denitrification of Wastewaters by Addition of Organic Materials. *Proceedings of the 24th Industrial Waste Conference (Engineering Extension Series)*; Lafayette, Indiana, May 6–8; Purdue University: Lafayette, Indiana; pp 1271–1286.

Metcalf and Eddy, Inc. (2003) *Wastewater Engineering: Treatment and Reuse*; McGraw-Hill: New York.

Mokhayeri, Y.; Nichols, A.; Murtry, S. (2006) Examining the Influence of Substrates and Temperature on Maximum Specific Growth Rate of Denitrifiers. *Water Sci. Technol.*, **54** (8), 155–162.

Monod, J. (1949) The Growth of Bacterial Cultures. *Annu. Rev. Microbiol.*, **3**, 371–394.

Muller, A.; Wentzel, M. C.; Loewenthal, R. E.; Ekama, G. A. (2003) Heterotroph Anoxic Yield in Anoxic Aerobic Activated Sludge Systems Treating Municipal Wastewater. *Water Res.*, **37** (10), 2435–2441.

Nakatsu, C. H.; Carmosini, N.; Baldwin, B.; Beasley, F.; Kourtev, P.; Konopka, A. (2005) Soil Microbial Community Responses to Additions of Organic Carbon Substrates and Heavy Metals (Pb and Cr). *Appl. Environ. Microbiol.*, **71** (12), 7679–7689.

National Fire Protection Association (2012a) *NFPA 30: Flammable and Combustible Liquids Code;* National Fire Protection Association: Quincy, Massachusetts.

National Fire Protection Association (2012b) *NFPA 820: Standard for Fire Protection in Wastewater Treatment and Collection Facilities;* National Fire Protection Association: Quincy, Massachusetts.

Nurse, G. R. (1980) Denitrification with Methanol: Microbiology and Biochemistry. *Water Res.,* **14** (5), 531–537.

Nyberg, U.; Aspegren, H.; Andersson, B.; la Cour Jansen, J.; Villadsen, I. S. (1992) Full-Scale Application of Nitrogen Removal with Methanol as Carbon Source. *Water Sci. Technol.,* **26** (5-6), 1077–1086.

O'Keeffe, D. T.; Anthony, C. (1980) The Two Cytochromes in the Facultative Methylotroph *Pseudomonas* AM1. *J. Biochem.,* **192** (2), 411–419.

Onnis-Hayden, A.; Gu, A. Z. (2008) Comparisons of Organic Sources for Denitrification: Biodegradability, Denitrification Rates, Kinetic Constants and Practical Implication for their Application in WWTPs. *Proceedings of the 81st Annual Water Environment Federation Technical Exhibition and Conference* [CD-ROM]; Chicago, Illinois, Oct 18–22; Water Environment Federation: Alexandria, Virginia.

Osaka, T.; Yoshie, S.; Tsuneda, S.; Hirata, A.; Iwami, N.; Inamori, Y. (2006) Identification of Acetate- or Methanol-Assimilating Bacteria under Nitrate-Reducing Conditions by Stable-Isotope Probing. *Microb. Ecol.,* **52** (2), 253–266.

Peng, Y.-z.; Ma, Y.; Wang, S.-y. (2007) Denitrification Potential Enhancement by Addition of External Carbon Sources in a Pre-Denitrification Process. *J. Environ. Sci.,* **19** (3), 284–289.

Purkhold, U.; Pommerening-Röser, A.; Juretschko, S.; Schmid, M. C.; Koops, H.-P.; Wagner, M. (2000) Phylogeny of All Recognized Species of Ammonia Oxidizers Based on Comparative 16S rRNA and *amoA* Sequence Analysis: Implications for Molecular Diversity Surveys. *Appl. Environ. Microbiol.,* **66** (12), 5368–5382.

Ramalingam, K., Fillos, J.; Deur, A.; Beckmann, K. (2007) Specific Denitrification Rates with Alternate External Sources of Organic Carbon. *Proceedings of the 2nd External Carbon Source WERF Workshop;* Washington, D.C., Dec 12–13; Water Environment Research Foundation: Alexandria, Virginia.

Rittmann, B. E.; McCarty, P. L. (2001) *Environmental Biotechnology: Principles and Applications;* McGraw-Hill: New York.

Sharma, R.; Gupta, S. K. (2004) Influence of Chemical Oxygen Demand/Total Kjeldahl Nitrogen Ratio and Sludge Age on Nitrification of Nitrogenous Wastewater. *Water Environ. Res.,* **76** (2), 155–161.

Sobieszuk, P.; Szewczyk, K. W. (2006) Estimation of (C/N) Ratio for Microbial Denitrification. *Environ. Technol.,* **27** (1), 103–108.

Sperl, G. T.; Hoare, D. S. (1971) Denitrification with Methanol: A Selective Enrichment for *Hyphomicrobium* Species. *J. Bacteriol.,* **108** (2), 733–736.

Stensel, H. D.; Loehr, R. C.; Lawrence, A. W. (1973) Biological Kinetics of Suspended-Growth Denitrification. *J.—Water Pollut. Control Fed.,* **45** (2), 249–261.

Tam, N. F. Y.; Leung, G. L. W.; Wong, Y. S. (1994) The Effects of External Carbon Loading on Nitrogen Removal in Sequencing Batch Reactors. *Water Sci. Technol.,* **30** (6), 73–81.

U.S. Chemical Safety and Hazard Investigation Board (2007) Methanol Tank Explosion and Fire, Bethune Point Wastewater Treatment Plant, City of Daytona Beach, Florida; Report No. 2006-03-I-FL, March.

U.S. Environmental Protection Agency (1975) *Process Design Manual for Nitrogen Control;* U.S. Environmental Protection Agency: National Environmental Research Center: Washington, D.C.

Water Environment Federation; American Society of Civil Engineers; Environmental and Water Resources Institute (2009) *Design of Municipal Wastewater Treatment Plants,* 5th ed.; WEF Manual of Practice No. 8; ASCE Manual and Report on Engineering Practice No. 76; McGraw-Hill: New York.

Water Environment Federation (2011) *Nutrient Removal;* WEF Manual of Practice No. 34; McGraw-Hill: New York.

Wild, D.; von Schulthess, R.; Gujer, W. (1995) Structured Modelling of Denitrification Intermediates. *Water Sci. Technol.,* **31** (2), 45–54.

Chapter 6

Combined Nitrifying and Denitrifying Systems

William C. McConnell, P.E., BCEE; Vincent Apa, P.E., BCEE; Jason Beck, P.E.; and Joel C. Rife, P.E., BCEE

1.0 INTRODUCTION

Combined nitrification and denitrification (NDN) treatment systems achieve 5-day biochemical oxygen demand (BOD) removal and NDN in a single-sludge configuration using linked reactors or zones in series and a common process for solids separation (clarifiers or membranes). Combining these processes in one system provides the potential for initial economic capital and ongoing operating costs. It also presents a significant operational challenge related to proper balancing that is sometimes required between the different process environments.

Combined NDN systems are classified in various ways, including fixed vs suspended growth, continuous-flow vs batch treatment, staging of process sequences, and method of aeration. A common thread that is consistent throughout all of these processes is the alternating environment that is used sequentially to nitrify and denitrify.

2.0 BASIC CONSIDERATIONS

Many process options exist to accomplish NDN in a combined system. The most commonly applied configurations of these are described in Section 3.0. The right process configuration for any particular facility is influenced by several factors, the most significant of which are described in the following subsections.

2.1 Effluent Requirements

Fundamental to the selection of the appropriate process configuration is the level of treatment that must be provided and the process effluent quality required. Effluent total nitrogen requirements have tended to be implemented in the following tiers, based on a general and common understanding of the relative difficulty and costs of compliance: (1) a relatively easily attainable concentration in the range of 8 to 10 mg/L, (2) a more stringent concentration in the range of 3 to 8 mg/L, and (3) a very stringent concentration of less than 3 mg/L. As described in Section 3.0, different process configurations are capable of meeting different effluent total nitrogen results; achieving lower effluent total nitrogen requires the use of more complex and costly process configurations.

2.2 Economics

Selecting which process configuration to use in both the design and operation phases must consider economic effects. Achieving sufficient nitrification (typically complete nitrification in total nitrogen removal systems) requires having enough of the bioreactor volume under aeration to achieve this throughout the diurnal cycle of wastewater flow and loading and with a sufficient safety factor to protect the sensitive nitrifying organism population from toxic loads, low temperature, and other factors that can cause loss of nitrifiers. The amount of bioreactor volume under aeration increases capital costs and is the biggest factor in operating costs. Because of the inability to fully characterize the variability of most wastewater influents and the erratic sensitivity of nitrifiers, there is a tendency to have an excessive volume under aeration. In addition to higher costs, overaeration of mixed liquor can interfere with efficient denitrification. Therefore, it is important to accurately define the operational parameters affecting nitrification for each facility as much as possible before deciding on aerated volumes. Future operational parameters are frequently difficult to accurately define and predict; as such, many facilities are equipped with zones capable of being operated in either aerated or anoxic mode. These zones are referred to as *swing zones,* and can provide operators with a great deal of flexibility to fine-tune the anoxic and oxic volume ratio in response to loading variation, thereby potentially reducing operating costs. Swing zones can increase capital costs because the swing zone must be equipped with both mixers and aeration diffusion equipment. Denitrification volumes are easier to define because of the stability of denitrifying organisms; however, the efficiency of the denitrification process is dependent on a number of complex and interrelated factors that can only be evaluated through the use of mechanistic models. For this reason, models must be used to arrive at the bioreactor configuration with the lowest capital and operating cost. (Refer to *An Introduction to Process Modeling for Designers* [WEF, 2009] for a detailed discussion of process models.)

2.3 Operational Considerations

Operational control of the NDN process is important to achieve permit compliance and prevent operational issues with other aspects of the facility, such as chlorine dosage control and denitrification in the final clarifiers. The most important operational considerations include the following:

- Establishing a reliable and repeatable sampling program with the most important factors being carbonaceous oxygen demand (COD) and total Kjeldahl nitrogen (TKN) concentrations entering the bioreactor,
- Providing sufficient aeration for complete nitrification while avoiding overaeration of the process,
- Finding the optimum mixed liquor and return activated sludge (RAS) recycle rates and avoiding excessive dissolved oxygen in the mixed liquor recycle,
- Operating the process at a high enough aerobic solids retention time (SRT) to ensure complete nitrification while avoiding a buildup of nuisance filamentous or foaming organisms,
- Selecting a swing-zone operating mode (aerobic or anoxic) to provide sufficient nitrification or denitrification capacity based on loading,

- Maintaining sufficient alkalinity in the bioreactor and monitoring pH, and
- Providing supplemental carbon to the process if required.

2.4 Sustainability

Sustainability, that is, using approaches that balance current environmental, economic, and social approaches without compromising the ability of future generations to meet their needs, is an increasingly vital criterion that affects the design and operational approaches of achieving biological nutrient removal (BNR). Original design of the facility should carefully consider and incorporate sustainable approaches to conservation of materials, minimizing site effects caused by construction and controlling initial costs. Operation of BNR facilities should continuously consider ways to improve sustainability by balancing the sometimes-competing goals of optimizing treatment, conserving energy, reducing chemical use, and minimizing sludge production. (Refer to Chapter 11 for a detailed discussion of optimization of BNR systems.)

2.5 Safety

Safety is always of paramount importance in the operation of any water resource recovery facility (WRRF). One of the biggest operational safety concerns for NDN facilities is the use of methanol or other hazardous chemicals as a supplemental carbon source. (See Section 4.7 for additional safety information for methanol use.)

3.0 PROCESS CONFIGURATIONS

3.1 Modified Ludzack–Ettinger Process

3.1.1 Process Description

In this configuration, nitrate in the mixed liquor recycle and RAS flow streams are mixed with influent wastewater and reduced to nitrogen gas in a pre-denitrification reactor upstream of the main aeration basin. An earlier process configuration, the Ludzack–Ettinger process, was initially developed to take advantage of the carbonaceous substrate available in influent wastewater by placing the anoxic zone upstream of the nitrification reactor, as shown in Figure 6.1a (Ludzack and Ettinger, 1962). However, the total nitrogen removal efficiency of the Ludzack–Ettinger process is limited by the quantity of nitrate recycled back to the anoxic zone in RAS flow. In response to that limitation, Barnard developed an improvement to the Ludzack–Ettinger process, identified as the modified Ludzack–Ettinger (MLE) process, which adds the recirculation of mixed liquor recycle (MLR) from the end of the aeration tank to the beginning of the anoxic tank, as shown in Figure 6.1b (Barnard, 1974).

a)

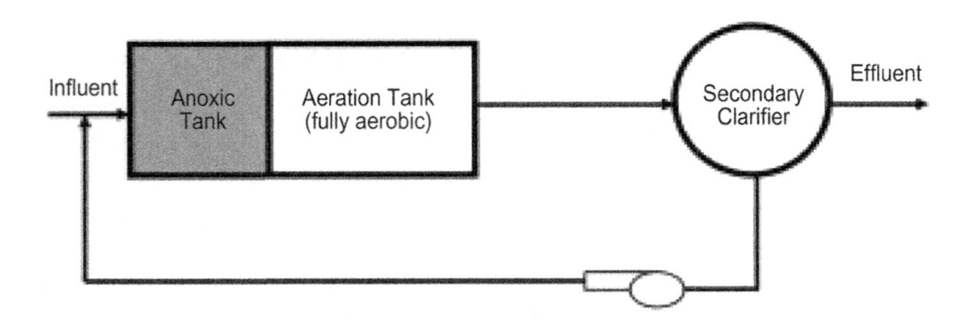

b)

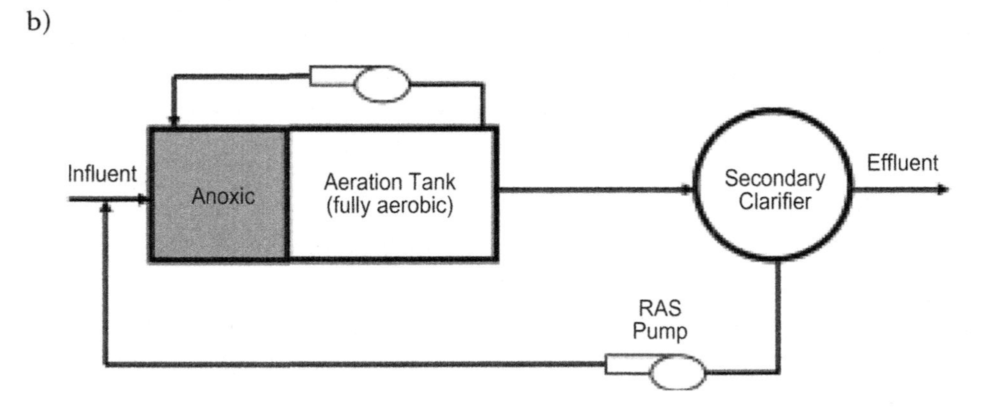

Figure 6.1 (a) Ludzack–Ettinger and (b) MLE process schematics.

Figure 6.2 depicts the amount of denitrification that can be achieved in a pre-denitrification reactor of an MLE process based on MLR flow as a percentage of influent flow. It is important to note that Figure 6.2 indicates denitrification that can be achieved if other factors such as substrate availability or kinetic limitations do not limit performance.

3.1.2 Expected Process Performance

Effluent total nitrogen from an MLE process is typically in the range of 8 to 10 mg/L. This is a function of the relationship presented in Figure 6.2, typical process influent TKN in the 40- to 50-mg/L range, and the common determination that a MLR flow of approximately 4 times influent flow is typically the feasible upper limit.

3.1.3 Operational Features

Table 6.1 lists key operating parameters of the MLE process. The presence of any dissolved oxygen in the anoxic reactor is detrimental to the denitrification rate.

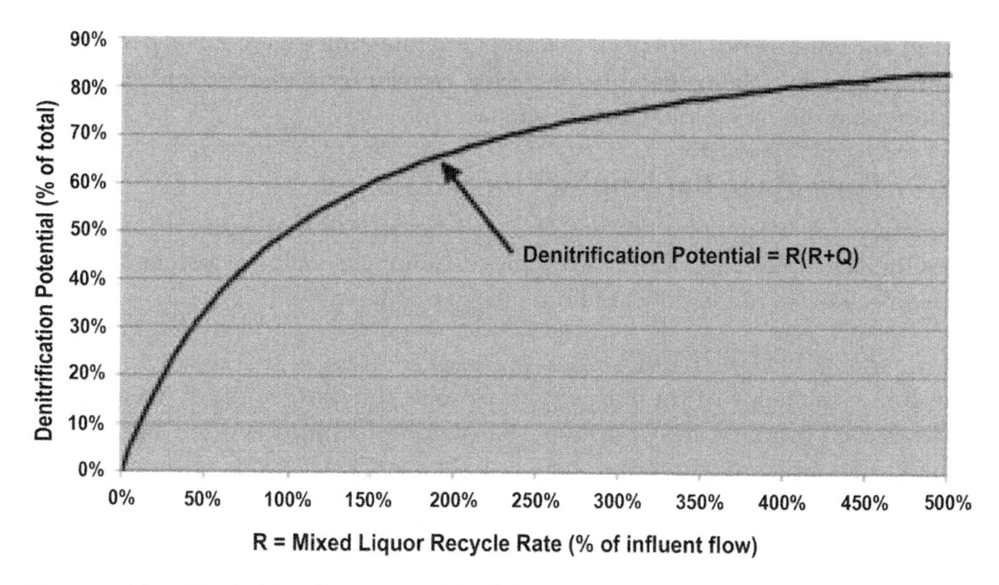

FIGURE 6.2 Denitrification potential of the MLE process.

Therefore, dissolved oxygen inputs from sources such as MLR and back-mixing from the aerobic reactor need to be minimized. The maximum effective MLR will be dependent on the dissolved oxygen included in the MLR flow and the oxygen demand (COD or BOD) of the influent wastewater.

Weaker-strength wastewater and/or higher MLR dissolved oxygen concentrations may limit the maximum beneficial MLR rate to less than 200%; higher-strength wastewater and low MLR dissolved oxygen concentrations may allow beneficial use of MLR rates of 400% or higher. Often, MLR rates greater than 400%

TABLE 6.1 Key operating parameters of the MLE process.

Reactor	Parameter	Rationale
Anoxic	Dissolved oxygen	Will reduce denitrification rate.
		Inadequate load limits denitrification.
	Nitrate	High nitrate recycled to aerobic zone may cause filamentous bulking.
Aerobic	Mixed liquor recycle	Controls nitrate load.
		High dissolved oxygen may inhibit upstream denitrification.
	Dissolved oxygen	Low dissolved oxygen may inhibit nitrification.
	Alkalinity, pH	Nitrification consumes alkalinity.

of influent flow are not beneficial to improving total nitrogen removal because of insufficient substrate availability, dissolved oxygen recirculation, and substrate dilution that reduces denitrification kinetics.

3.2 Four-Stage Bardenpho Process

The four-stage Bardenpho process, shown schematically in Figure 6.3, incorporates the principles used for the MLE process, but also includes a second anoxic zone to achieve a higher level of total nitrogen removal.

3.2.1 Process Description

The first two stages of the Bardenpho process function similarly to an MLE process. The primary anoxic zone in Bardenpho facilities is often sized large enough to consistently accommodate at least a 400% MLR rate without nitrate "bleed-through". This primary anoxic zone removes most nitrate generated in the process (per the relationship shown in Figure 6.2), and the secondary anoxic zone, located outside the MLE "loop", provides denitrification for that portion of flow that is not recycled to the primary anoxic zone. The fourth reactor zone in the Bardenpho process is an aerobic or reaeration reactor and serves to strip any nitrogen gas formed in the second anoxic zone, increase the dissolved oxygen concentration before secondary clarification, improve activated sludge flocculation, minimize phosphorus release in the clarifier, and reduce effluent turbidity.

3.2.2 Expected Process Performance

The Bardenpho process can generally achieve greater than 80% (and often better than 90%) total nitrogen removal (U.S. EPA, 1993; 2010a). With the addition of supplemental carbon in the second anoxic zone, effluent total nitrogen from a Bardenpho process can be low, limited only by refractory dissolved organic nitrogen and the organic nitrogen associated with effluent total suspended solids (TSS). The ability to achieve effluent total nitrogen in this range depends on the

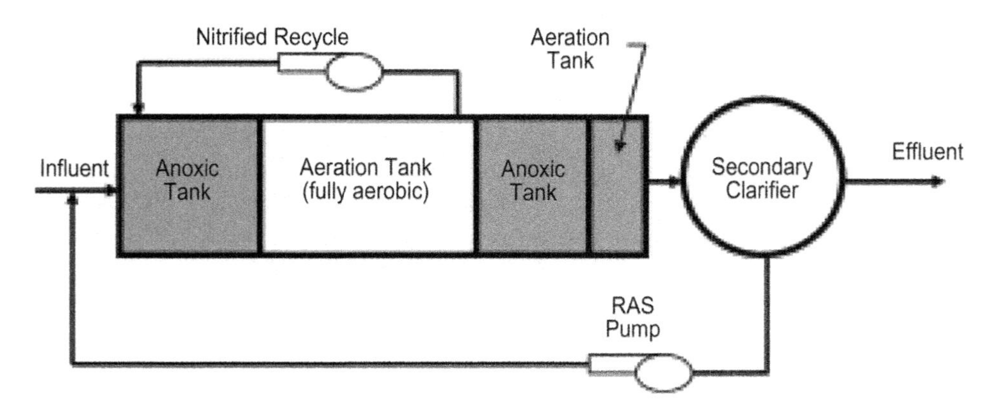

FIGURE 6.3 Four-stage Bardenpho process schematic.

ratio of oxidizable nitrogen to carbon in the influent to the biological process and use of supplemental carbon.

3.2.3 *Operational Features*

The Bardenpho process was designed with many reactor configurations, including plug flow, complete-mix, and oxidation-ditch reactors. In the United States, the systems often use an oxidation ditch as the MLE portion of the system, with separate complete-mix reactors for the secondary anoxic and secondary aerobic (reaeration) zones. Table 6.2 lists the key monitoring requirements at different stages of the four-stage Bardenpho process (U.S. EPA, 1993).

3.3 Sequencing Batch Reactors

Sequencing batch reactors (SBRs) represent a variable-volume, suspended growth treatment technology that uses time sequences to perform various treatment operations that continuous treatment processes conduct in different tanks. The first activated sludge processes were actually SBRs, but the lack of automation forced a shift toward continuous treatment systems that required less operator intervention. Modern automatic monitoring and control systems have resulted in SBRs being a commonly used nitrogen-removal process.

3.3.1 *Process Description*

Sequencing batch reactors proceed through a series of phases for every cycle and will typically complete 4 to 6 cycles per day per SBR tank for domestic

TABLE 6.2 Key operating parameters of the Bardenpho process.

Reactor	Parameter	Rationale
First anoxic	Dissolved oxygen	Will reduce denitrification rate.
		Inadequate load limits denitrification.
	Nitrate	High nitrate recycled to aerobic zone may cause filamentous bulking.
Aerobic	Mixed liquor recycle	Controls nitrate load.
		High dissolved oxygen may inhibit upstream denitrification.
	Dissolved oxygen	Low dissolved oxygen may inhibit nitrification.
Second anoxic	Nitrate	High nitrification in aerobic zone may overwhelm endogenous denitrification capacity, resulting in NO^3 in effluent.
	Dissolved oxygen	High dissolved oxygen will inhibit endogenous denitrification.

wastewater treatment. Typically, 50 to 75% of the liquid volume of the SBR containing settled biomass is retained at the end of every cycle, although a higher percentage may be necessary depending on effluent requirements. The minimum liquid level determines the volume of sludge inventory that can be retained in the SBR reactor and the maximum volume of influent that can be accommodated for any individual cycle. Sequencing batch reactors function in the following four basic phases:

1. Fill—wastewater is added to the SBR, raising the liquid level from the minimum level to a depth that corresponds to the amount of influent received during the fill phase time period. The mixing and aeration treatment steps generally commence during the fill phase;
2. React—biological processes are performed, including various aeration and nonaerated mixing regimes, depending on process goals. In this way, SBRs can be operated to mimic the process conditions of a MLE, four-stage Bardenpho, or other configuration;
3. Settle—aeration and mixing are terminated and biomass is allowed to settle; and
4. Decant—clarified effluent is removed from the basin, aeration and mixing are off, and biomass is wasted as necessary.

For nitrogen removal, the fill and react phases are broken down into static fill, mixed fill, and mixed react. Carbon oxidation and nitrification occur in the aerobic react phase and denitrification will happen in the anoxic fill and react phase. Denitrification results from selecting static fill, mixed fill, and mixed react periods that are long enough to allow use of all dissolved oxygen, thus creating anoxic conditions.

3.3.2 *Expected Process Performance*

With sufficient volume to provide for the necessary reaction time, SBRs can achieve greater than 90% total nitrogen removal. A survey was conducted of 10 SBR systems in the Northeast performing nitrogen removal (Young et al., 2008) that showed that facility effluent total nitrogen levels ranged from 2.5 to 9.5 mg/L. These total nitrogen levels were all achieved without the use of supplemental carbon, although it should be noted that the facilities were operating well below design loads.

3.3.3 *Operational Features*

The greatest challenges related to SBRs are related to the significant hydraulic grade-line drop through the system, the difficulty removing floating material from the SBR tanks, and intermittent, high-rate decant that generally warrants equalization before downstream processes such as filtration and disinfection. Figure 6.4 illustrates the basic phases of an SBR cycle. Sequencing batch reactor

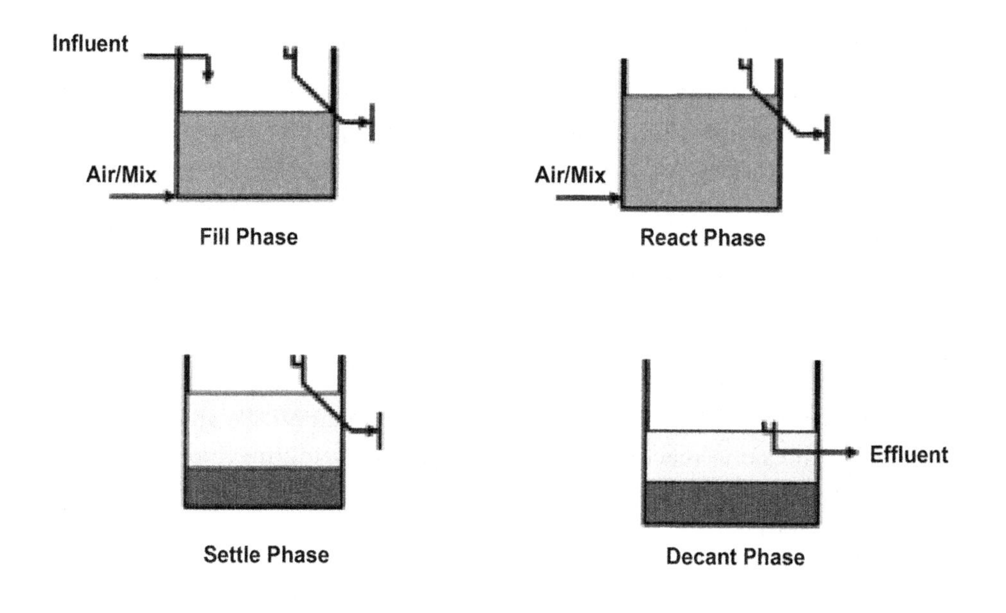

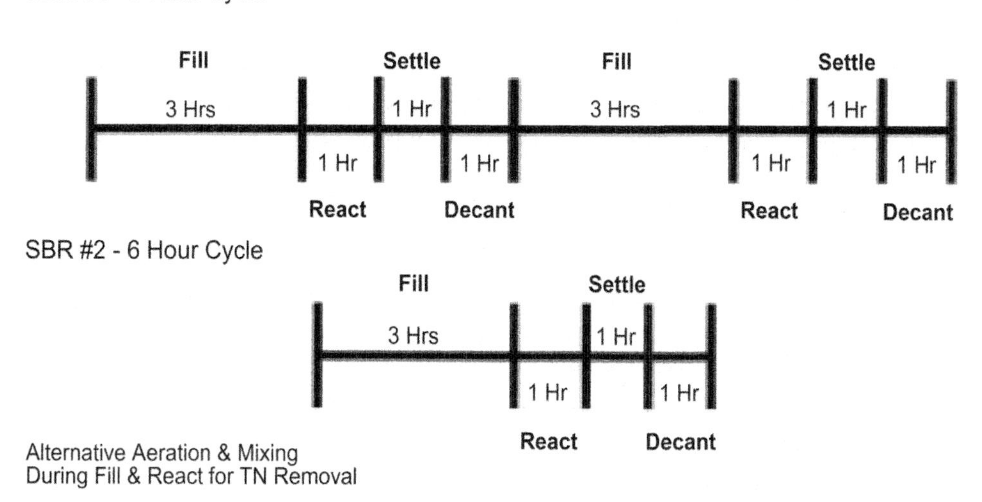

FIGURE 6.4 Sequencing batch reactor process phases and example time sequence.

control systems permit system operation to be configured to mimic almost any other suspended growth reactor configuration. Sequencing batch reactors can be configured and operated to act as a single sludge nitrification system, a multiphased cyclic aeration process, an MLE process, or a Bardenpho system. Figure 6.4 also presents a potential operating strategy for a high level of nitrogen removal.

3.4 Simultaneous Nitrification and Denitrification

Biological reduction of the nitrogen content in wastewater relies primarily on two mechanisms: aerobic nitrification and anoxic denitrification. Generally, the two treatment activities are carried out in physically separated oxic and anoxic treatment zones. However, when the dissolved oxygen concentration is less than 1.0 mg/L, it has been demonstrated that heterotrophs can denitrify and autotrophs can nitrify in the same reactor. This is termed the *simultaneous nitrification and denitrification* (SND) process.

3.4.1 *Process Description*

Studies by Trivedi and Heinen (2000) have revealed that NDN can occur simultaneously in the same reactor, arising from the following mechanisms in suspended growth systems (Barnard et al., 2004; Kaempfer et al., 2000, Satoh et al., 2003; Stensel, 2001):

- Presence of microscopic anoxic and oxic zones within sludge flocs—aeration promotes the dissolution of oxygen into water. The dissolved oxygen subsequently diffuses into the flocs and, along the diffusion path, is consumed by organisms. The diffusion and consumption cause a gradient of dissolved oxygen concentration in the flocs and, at suitably low dissolved oxygen, create flocs that nitrify in the oxic outer layer and denitrify in the anoxic inner core. It is important to note that similar processes have been demonstrated to occur across the dissolved oxygen gradient in denitrifying biofilm processes. Operational approaches to target specific SND performance have not been firmly established. A discussion of SND in biofilms is not within the scope of this manual;
- Presence of macroscopic anoxic and oxic zones within bioreactors—such zones are commonly formed as a result of nonhomogeneous mixing and aeration, particularly in bioreactors with surface aerators in which dissolved oxygen is high near the aerators and low or zero away from the aerators; and
- Presence of novel microorganisms—Robertson et al. (1988) reported that *Thiosphaera pantotropha* could simultaneously nitrify and denitrify under aerobic conditions. In addition, several bacteria were found to perform aerobic denitrification (Davies et al., 1989).

Oxidation ditches, as described in Section 3.6, are often configured to provide SND. Different configurations of oxidation ditches may provide low dissolved oxygen zones or channels as part of the reactor or sections that operate in a swing mode to vary with daily load conditions. Different oxidation-ditch systems use either an aeration disc for oxygen transfer and mixing, turbine aerators, or a combination of vertical-drum mixers and diffused air to maintain low dissolved oxygen around the entire oxidation ditch.

3.4.2 *Expected Process Performance*

Because of the lack of dependence on mixed liquor recycle for achieving denitrification, use of SND can result in higher total nitrogen removal than MLE systems, although accomplishing SND typically requires longer SRT and larger bioreactor volumes than staged processes. Tight dissolved oxygen control must be maintained because SND can be lost if the dissolved oxygen concentration is allowed to climb too high or drop too low.

3.4.3 *Operational Features*

This process offers an advantage by potentially eliminating the MLR stream and reducing energy use because of increased oxygen-transfer efficiency. However, Jenkins et al. (2003) found the process to use influent carbon less efficiently than the MLE or Bardenpho processes and determined that low dissolved oxygen conditions resulted in filamentous bulking. However, Schuyler et al. (2009) found no or few low dissolved oxygen filaments when the dissolved oxygen was consistently maintained below 0.4 mg/L.

3.5 Cyclic Aeration

Development of cyclic aeration technology began in Australia as early as 1965 (Goronszy, 1979). Many activated sludge reactor basins are easily converted from continuous aeration to cyclical aeration by cycling mechanical aerators on and off with timers or by alternating diffused aeration zones by electrically or pneumatically actuated valves.

3.5.1 *Process Description*

The following process features of cyclic aeration systems have been identified:

- Short cycle times (4 hours and less), introduction of biological selector zones using transverse partial baffle walls, fill sequence regulation for filamentous sludge bulking control and cyclic aeration, and a means of regulating the kinetics of oxidation–reduction potential depletion for enhanced biological phosphorus removal (Goronszy, 1979);
- Process control using in-basin respiration rates, which allowed permanent control of the metabolic activity of the biomass and, consequently, changed the principles of process operation from a time-based control to a demand-oriented process control;
- A clear water withdrawal system for high-rate decanting of up to 2.5 m of solids-free effluent within a short time from basins having surface areas of up to 8000 m^2 (i.e., multiple basins and multiple decanters); and
- Establishment of adjustable normal- and high-flow operating protocols.

3.5.2 *Expected Process Performance*

A properly operating cyclic aeration process typically achieves 85 to 90% total nitrogen removal.

3.5.3 *Operational Features*

A key factor that will likely determine if cyclic aeration can function successfully is if the existing system has the ability to maintain a sufficient aerobic SRT to be able to consistently nitrify with intermittent aeration. If the aerobic SRT (total SRT multiplied by the percentage of time aerating) is more than is necessary to maintain nitrification, the cyclical aeration process can reduce effluent total nitrogen.

3.6 Oxidation Ditches

3.6.1 *Process Description*

Oxidation-ditch processes use looped channels that provide a continuous circulation of wastewater and biomass. They are often referred to as "closed-loop" or "racetrack" reactors and are equivalent in flow configuration to a conventional plug flow system with high internal recycle rates (exceeding 250 to 300% of influent flow) without needing a recycle pumping system. The high recycle ratio is implemented by configuring the entire channel flow to recirculate in a closed loop, with the influent and effluent flow representing only a small fraction of the main channel flow. Because of the high recycle rate, oxidation ditches can be efficient NDN systems with properly configured anoxic and aerobic zones. Anoxic and aerobic zones in oxidation ditches are implemented by establishing a dissolved oxygen profile along the flow path rather than with discrete compartments. The resulting intermediate low dissolved oxygen environments create conditions where simultaneous NDN can occur, as described in Section 3.4. Figure 6.5 presents a typical oxidation-ditch configuration with an external secondary clarifier.

A phased isolation ditch system is a continuous-flow activated sludge process in which the main treatment phases of the process are isolated into separate

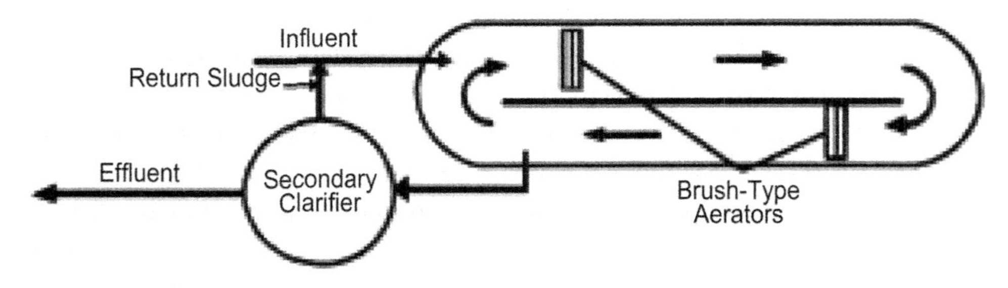

FIGURE 6.5 Oxidation-ditch process schematic.

oxidation ditches. Process conditions within the oxidation ditches will alternate or phase between oxic, anoxic, and/or settling.

3.6.2 *Expected Process Performance*

Oxidation ditches provide up to 90% total nitrogen removal. Most oxidation ditches operate at 6 to 10 mg/L effluent total nitrogen (Sen et al., 1990).

3.6.3 *Operational Features*

Aeration can be provided with horizontal brush aerators, vertical-shaft mechanical aerators that impart a horizontal liquid velocity, or diffused aeration with submersible or vertical drum mixers. Typically, aeration is provided in a few locations within the continuous loop to create variations in dissolved oxygen concentration along the length of the loop, ranging from high dissolved oxygen just downstream of the aerator to low or no dissolved oxygen just upstream of the aerator.

3.7 Step-Feed Configuration

Use of a step-feed activated sludge process configuration with anoxic zones (step-feed BNR) has been a popular method to achieve nitrogen removal without internal recirculation and minimal capital upgrades in existing treatment facilities in the United States. This process entails feeding a portion of the influent to multiple points downstream of the front end of the reactor; each of the feed points has an anoxic zone for denitrification.

3.7.1 *Process Description*

The step-feed configuration with anoxic zones provides conditions for denitrification of nitrate produced in the aerobic zones using the organic carbon in the step-fed wastewater without the need for internal nitrate recirculation and external carbon addition (Larrea et al., 2001). Typically, there are three or four "passes", although some facilities have as many as six passes. Small, mixed, anoxic zones are provided at each feed point, bringing raw wastewater carbon to the nitrate created in the previous pass. This conversion, or retrofit, is well suited for activated sludge facilities consisting of multiple passes in the aeration tanks. In general, by splitting the flow to several influent feed locations and directing RAS to the beginning of the first pass, a higher system SRT is achieved than in a plug flow system with the same basin volume. The increase in SRT can be obtained without increasing the mixed liquor suspended solids (MLSS); as such, solids loading to the clarifiers is not increased.

3.7.2 *Expected Process Performance*

Implementation of a step-feed BNR system for the removal of total nitrogen at several treatment facilities has proven successful in achieving effluent total nitrogen concentrations in the 6 to 8 mg/L range (deBarbadillo et al., 2002; Fillos et al., 1996; Goodwin et al., 2005).

3.7.3 *Operational Features*

Control of influent flow splitting is imperative for a successfully operating step-feed BNR system and can affect step-feed BNR system optimization in several ways, depending on which aspect of the BNR process is limiting. When the BNR system is fully nitrifying, denitrification may be limiting. Total nitrogen removal can be increased by adjusting the flow-splitting regime to minimize the quantity of unused carbon in the anoxic zones of certain passes. The RAS flow also has an effect on the performance of step-feed systems. Typically, RAS is discharged to the beginning of the first pass. If settling characteristics allow, increasing or reducing the RAS flow can affect total nitrogen removal. At lower RAS flows, the solids are more concentrated, which results in higher MLSS concentrations in the early passes and an increased SRT. A postanoxic zone is sometimes included to obtain lower effluent total nitrogen concentrations.

Minimizing carryover of dissolved oxygen from the last oxic zone of one pass to the anoxic zone of the next pass optimizes anoxic zone performance and reduces the readily biodegradable carbon consumed during reduction of residual dissolved oxygen. This must also be considered for plug flow BNR systems; however, it is even more critical for step-feed systems because of the need to minimize dissolved oxygen entering the anoxic zones of each pass.

Some facilities have experienced frothing events while operating in step-feed BNR mode at a high MLSS concentration (Chandran et al., 2003). Foam-control considerations for step-feed systems are similar to those associated with plug flow systems except that foaming can be more problematic with high MLSS concentrations in the early passes. Additional passes and zones in a step-feed system complicate control of foaming because there are more locations for the foam to become trapped in the aeration tanks. The New York City Department of Environmental Protection found surface wasting, surface chlorination, and RAS chlorination effective in controlling froth (Chandran et al., 2003).

3.8 Integrated Fixed-Film Activated Sludge

3.8.1 *Process Description*

As described in Chapter 4, integrated fixed-film activated sludge (IFAS) processes are wastewater treatment systems that incorporate some type of media within a suspended growth activated sludge process to provide a fixed surface area for biofilm growth to supplement the suspended-phase solids inventory. The principal behind the IFAS process is to expand treatment capacity or to upgrade the level of treatment by supplementing the suspended biomass by growing additional biomass (biofilm) on media contained within the aeration tank. The additional biomass allows a higher effective rate of treatment within the existing process tanks, thus making other tank volume available to incorporate anoxic zones for denitrification or anaerobic zones for biological phosphorus removal within the same reactor volume. Advantages of using this type of process to expand or upgrade WRRFs are as follows:

- Additional biomass for treatment without increasing solids loading on final clarifiers;
- Higher-rate treatment, thus allowing greater treatment in a smaller space;
- Simultaneous NDN;
- Similar operation to conventional activated sludge (CAS);
- Improved resistance to inhibitory compounds and washout;
- Increased robustness in colder temperatures because of an increased total SRT; and
- "Self-regulating" biofilm responds to process requirements and sudden load changes.

The IFAS systems also have disadvantages compared to suspended growth systems, including high capital costs for media and other system components, issues with media handling, and—most importantly—the potential vulnerability to media loss because of hydraulic limitations. Leaf et al. (2011) discuss issues with media loss and recommend methods for reducing risks.

Although media types vary, they can be grouped into two categories: moving bed and fixed media. Moving bed media include sponge cuboids and plastic wheels that are designed to remain in suspension and mixed using conventional aeration. Examples of plastic and sponge-wheel-type media are shown in Figure 6.6.

FIGURE 6.6 Plastic wheels (moving bed) (courtesy of I. Kruger Inc.) and sponge cuboids (courtesy of M2T Technologies, Inc.).

Fixed media include rope or looped strand media installed in cages or frames or plastic structured sheet media supported by spaced piers, which are placed within an activated sludge basin. Examples of fixed-type media are shown in Figure 6.7.

All types of IFAS media produce the same result of increasing the capacity of the treatment process within the existing reactor volume, with each type having specific advantages and disadvantages. These are summarized in Table 3.5 of Chapter 3.

3.8.2 *Expected Process Performance*

Integrated fixed-film activated sludge installations effectively increase the treatment capacity of an activated sludge process within the existing tank volume without the need for additional secondary clarifiers. The actual increase in nitrification is directly related to the fill fraction or surface area of IFAS media that are added to a basin and the additional biomass that can be grown on the media. At the Broomfield, Colorado, WRRF, which was the site of the first installation of IFAS in North America, the addition of IFAS media doubled the nitrification capacity of the facility from 15 mL/d (4 mgd) up to 30 mL/d (8 mgd) within the same reactor volume (Rutt et al., 2006). Although optimal fill fractions of floating media are site dependent, they are in the range of 30 to 60% fill by volume. The Broomfield, Colorado, WRRF was originally expanded up to 30 mL/d (8 mgd) using a fill fraction of 48% during phase 1. However, the fill fraction was reduced to 30% for phase 2 (up to 45 mL/d [12 mgd]) based on full-scale operation data that were collected during phase 1 (Phillips et. al., 2008).

Simultaneous nitrification and denitrification has been observed in many IFAS facilities because low dissolved oxygen in the inner layers of the biofilm results in anoxic conditions. The degree of reliable denitrification that occurs in

FIGURE 6.7 Rope-type media (fixed) (courtesy of Entex Technologies, Inc.) and plastic structured sheet with pier support (fixed) (courtesy of Brentwood Industries).

these inner layers is difficult to anticipate. As a result, this SND is often considered a "bonus" of IFAS processes rather than an operational requirement.

3.8.3 Operational Features

3.8.3.1 Reactor Configuration

The IFAS process schematic may be configured similarly to many of the process configurations described earlier in this chapter (e.g., MLE and Bardenpho), with media installed in the aerobic zone(s) of the process. Often, the most efficient configuration is to have more than one aerobic reactor in series, with no media installed in the first aerobic zone. The media provide the most benefit in achieving nitrification, and using the first zone as a carbonaceous BOD removal zone, followed by a media-filled zone for nitrification, increases the nitrification rate of the biofilm.

3.8.3.2 Media Retention, Influent Screens, and Foam Management

Integrated fixed-film activated sludge systems require process modifications that are unique to the media type being considered. Floating media require the installation of wedge-wire media retention screens on top of (Figure 6.8) and through the cell dividing walls (see Figure 4.4, Chapter 4) to prevent floating media from entering the secondary clarifiers and other treatment processes. The cylindrical

FIGURE 6.8 Top-of-wall media retention screens (courtesy of T.Z. Osborne Water Reclamation Facility, Greensboro, North Carolina).

screens installed through the wall are designed to encourage contact with floating plastic media to agitate and clean any accumulated biogrowth or debris off of the screens and floating media. The top of the wall screens retains the media while allowing the passing of foam and floatables on to the secondary clarifiers for removal. Although fixed-type media do not require retention screens, they have media-specific fine influent screen requirements to prevent debris from accumulating on the fixed-media structures within the basins.

Influent screening and foam management are important components to consider in IFAS systems. In most applications of free-floating media, media retention screens are a wedge-wire configuration with a smaller effective passing orifice than influent bar screens that can allow long, fibrous debris to pass through the bars. Floatables and debris that pass through the headworks screens can accumulate on media retention screens and create an operational challenge. Therefore, installing finer headworks screens or a debris removal method within aeration basins should be considered. Similarly, the top of wall screens can impede the passing of foam on to the secondary clarifiers. Foam-causing organisms that are retained can proliferate and cause increased foaming compared to conventional systems. Therefore, foam removal and management needs to be evaluated when considering an IFAS system.

3.8.3.3 Aeration Systems and Dissolved Oxygen Concentrations

The biological treatment feature of an IFAS system is a biofilm grown on fixed media. To achieve equal performance to a suspended growth system and effectively activate most of the biomass within the layers of the biofilm, residual dissolved oxygen concentrations of 3 to 4 mg/L need to be maintained (Stricker et al., 2009). Aeration system requirements vary between IFAS media types. Although less efficient than fine-bubble diffusers, coarse-bubble diffusers may be required if a method for removing the media to perform diffuser maintenance is not available. Coarse-bubble aeration also increases agitation of the media and reduces debris accumulation and excess biogrowth on the exterior of media carriers. Operation costs can, therefore, be much greater than suspended growth systems that operate at lower bulk dissolved oxygen concentrations and with more efficient aeration systems. Operating at higher dissolved oxygen concentrations can also interfere with efficient denitrification if oxygen is recycled into an anoxic zone.

3.8.3.4 Media Management

Maintenance requirements for aeration systems or other equipment can require a basin to be taken out of service, thereby requiring the removal of IFAS media. Fixed-type media are often installed in modules that can be removed with cranes or other lifting equipment. Some floating media such as sponge-type cuboids can be pumped to an adjacent basin or holding tank. Other types of floating media such as plastic carriers are more difficult to remove from a basin. Less-maintenance-intensive aeration systems such as coarse-bubble diffusers should be considered when a media management method is not readily available. Proper media management systems should be evaluated and coordinated with each manufacturer being considered in an IFAS application. A summary of the benefits and drawbacks of each IFAS media type is provided in Table 3.5 of Chapter 3.

3.9 Membrane Bioreactor

3.9.1 Process Description

Membrane bioreactors (MBRs) are a form of activated sludge treatment that use membrane filters instead of clarifiers to separate the mixed liquor in the bioreactor from treated effluent. The most common form of MBRs uses a membrane tank, which has suction-type membranes submerged directly in the mixed liquor in a tank separated from the bioreactor. The treated water or "permeate" is pulled through the outside of the membranes and transported inside the membranes ("outside-in" using low-pressure pumps). The first version of the MBR and several current versions use pressure-type ("inside-out") membranes located outside the bioreactor, with current versions using larger-diameter membranes to avoid plugging issues associated with the early pressure-membrane MBR. Membranes used are either in the microfiltration (0.1- to 0.4-μm openings) or ultrafiltration (0.04- to 0.1-μm openings) categories. There are two basic types of suction membranes that are used: hollow fiber and flat panel.

With final clarifier solids loading not being a concern, MBRs are able to operate at high MLSS concentrations. Typical MBR design is for 8000 mg/L MLSS in the bioreactor and a maximum of 10 000 mg/L in the membrane tank. Because of the emphasis on energy use, there is a trend to design MBRs for lower MLSS concentrations because of the lower oxygen-transfer efficiency at higher MLSS (if space requirements allow for the construction of larger bioreactors).

Coarse-bubble aeration and periodic chemical cleaning with chlorine and acid are used to maintain the design flow (flux) rate through the membranes. The energy required for aeration of the membrane tank is one of the biggest disadvantages of an MBR, resulting in continuing advancements designed to reduce this energy requirement by using more efficient aeration systems.

Current thinking in MBR design is to size the bioreactor for an SRT that it is sufficient to produce a fully nitrified effluent. Because the SRT required for full nitrification varies with wastewater temperature, the design SRT for an MBR will vary similar to any activated sludge process designed for full nitrification.

Membrane integrity is monitored using online turbidity meters. If a membrane becomes ruptured, the effluent turbidity will suddenly increase. Most membrane systems tend to be "self healing" to a degree; however, membrane repair or replacement may be necessary if a permanent increase in turbidity occurs. Membrane life is typically 5 to 10 years or even longer.

Nitrogen removal configurations for MBR systems are generally the same as those for conventional systems, as described previously, although they typically have somewhat different recycle processes and rates. The one significant difference is related to the RAS rate in an MBR. When designed for typical MLSS values of 8000 mg/L in the bioreactor, a RAS rate of 4 times the influent flow is required to keep the solids concentration below the required 10 000 mg/L in the membrane tank. Because this is a typical mixed liquor recirculation rate for a BNR configuration for total nitrogen removal, the RAS stream of the MBR can also provide the necessary mixed liquor recirculation for nitrogen removal in a BNR process.

The RAS flow from the membrane tank can contain high concentrations of dissolved oxygen (typically from 2 to 6 mg/L) because of membrane scouring. This dissolved oxygen will reduce the soluble BOD necessary for the denitrification process. For this reason, one (of many) design and operating approaches involves returning the RAS stream of an MBR to the aeration tank, depleting the dissolved oxygen in that reactor, and then using separate mixed liquor recycle pumps for the return of the lower dissolved oxygen mixed liquor to the anoxic zones, similar to conventional plants. This also lowers aeration requirements in the bioreactor by using the dissolved oxygen in RAS for process oxygen. What is lost with this approach are the more concentrated solids in the membrane tank and RAS, which results in a higher solids concentration in the anoxic zones, thereby increasing nitrogen removal efficiency.

Because of these factors, design and operational troubleshooting for nitrogen-removal MBR facilities should be based on process modeling to properly evaluate all of these factors. Figure 6.9a shows the simplest configuration of a nitrogen-removal MBR that is typically appropriate for meeting limits of 8 mg/L total nitrogen. Figure 6.9b shows the nitrogen-removal MBR with internal recycle that is capable of achieving slightly lower total nitrogen limits, depending on the dissolved oxygen in the membrane tank. If dissolved oxygen in the membrane tank is 6 mg/L or higher, this configuration will provide slightly better total nitrogen removal; however, at 4 mg/L or less of dissolved oxygen in the membrane tank (as is achievable in some of the more modern MBR designs), the single RAS configuration will provide equivalent total nitrogen removal; therefore, the cost of operating a separate mixed liquor recycle pump would not be justified.

3.9.2 *Expected Process Performance*

Because of the high degree of filtration, MBRs can achieve effluent total nitrogen concentrations equivalent to that achievable with denitrification filters without the use of external carbon sources. Second-stage anoxic zones following the aerated portion of the bioreactor are required to achieve these low-effluent total nitrogen concentrations.

a)

Influent Anoxic Aeration Membrane Tank Effluent

WAS

b)

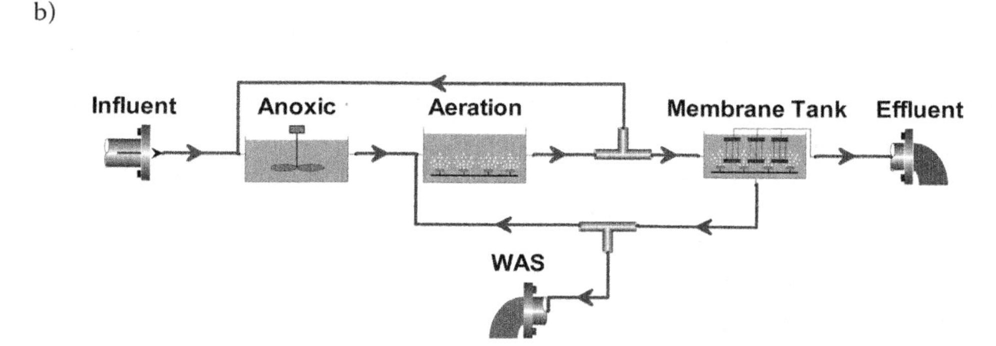

FIGURE 6.9 (a) Basic total nitrogen removal MBR configuration and (b) separate mixed liquor recycle total nitrogen removal MBR configuration (shown in BioWin™ and developed by EnviroSim Associates Ltd.).

3.9.3 *Operational Features*

The most significant operational issue with MBRs is the potential for membrane biofouling. The higher the MLSS concentration and the higher the flux rate, the greater the potential for fouling. Additionally, as the potential for fouling increases, the frequency of cleanings that are required increases. The need to clean a membrane is determined by the pressure differential across the membrane. The initial buildup of solids on the membrane surface improves performance; however, as the solids continue to accumulate and the flowrate is increased to compensate for increased pressure loss across the membrane, the operation can become unstable. Attempting to increase the flowrate further compresses the solids on the surface, thereby increasing the pressure drop again.

There are several operational changes that can be implemented to minimize biofouling. Fouling may be reduced by increasing the turbulence induced by the coarse-bubble aeration system beneath the membranes, which reduces the thickness of the biofilm. There is evidence that the primary cause of biofouling are extracellular polymeric substances (EPS) that form on the surface of the bacteria that comprise the biofilm. Extracellular polymeric substances tend to clog the smaller pores in the cake layer surrounding the membrane. Reducing the amount of EPS reduces filtration resistance. The amount of EPS is influenced by SRT, with the degree of EPS formation increasing at low and high SRTs. Improvements can also be realized by adding a flocculant, which creates larger particles in the cake layer, thereby providing larger pores in the biofilm and reducing filtration resistance.

There are several cleaning cycles required to maintain MBR operation. Typically, there is a frequent but short duration backwash using permeate or filtered wastewater to flush solids from the membrane surface. Periodically, a more extensive backwash is performed with chemical cleaning agents. A more extensive cleaning operation in which the membrane is removed from service and washed in a chemical bath is typically only required once every 6 to 12 months depending on the waste and operating conditions. Most backwash and cleaning operations can be automated.

Because of the trapping nature of the membrane filtration process, foaming is one of the most serious operational issues with MBR facilities. Excessive foam can result in a large amount of old sludge in the system and the release of cellular nitrogen and phosphorus into the effluent. The most reliable method of preventing excessive foaming is incorporating a surface wasting method to the design and operation of the MBR. Skimming pipes or weirs are used in areas where the mixed liquor passes through a narrow channel, such as the channels between the bioreactor and the membrane tank. Either all or a part of the waste activated sludge (WAS) is then taken from the surface of the tank, preventing the buildup of foam. Because of the high solids concentration in an MBR, downstream solids handling processes will be loaded with similar concentrations as that of WAS from a conventional facility.

3.10 Process Comparison Summary

Table 6.3 presents a summary of the process configurations described in this section.

4.0 RELATED SYSTEMS AND EQUIPMENT

4.1 Primary Clarification

Operation of primary clarifiers is important when operating nitrogen removal facilities because the overall capacity of the nitrification process and the efficiency of the denitrification process are directly related to the amount of COD in the influent to the bioreactor. Primary clarification will remove 30 to 50% of raw wastewater COD without the addition of coagulant chemicals and more than 80% COD with the addition of coagulant chemicals. This reduced COD loading

TABLE 6.3 Process comparison summary.

Process	Features, Benefits, and Drawbacks
Modified Ludzack–Ettinger	Two-stage process: anoxic–aerobic; internal mixed liquor recycle recycles nitrate from the aerobic zone to the anoxic zone at a rate of 3 to 5 times the influent flow; achieves moderate total nitrogen removal (effluent total nitrogen: 6 to 8 mg/L); low tank volume requirement compared to other multiple-stage nitrogen removal processes.
Four-stage Bardenpho	Four-stage process: anoxic 1, aerobic 1, anoxic 2, and reaeration; internal mixed liquor recycle recycles nitrate from aerobic 1 to anoxic 1 at a rate of 3 to 5 times the influent flow; achieves excellent total nitrogen removal (effluent total nitrogen: 2 to 4 mg/L); larger footprint required compared to fewer stage flow through nitrogen removal processes.
Sequencing batch reactor	Single reactor tanks that cycle between aerobic and anoxic conditions to achieve nitrogen removal; wastewater influent cycles between reactor tanks to balance influent organic and nitrogen loading; each reactor cycles between 4 to 6 phases depending on the application and nutrient limits required; secondary clarifiers not required; secondary settling occurs within the SBR reactor; relatively large reactor volumes; typical application: WRRFs with a capacity range of 3.8 to 38 ML/d (0.1 to 10 mgd).
Cyclic aeration	Process used in conjunction with an SBR system to achieve nitrification and denitrification within the same reactor volume.
Oxidation ditches	Bioreactors configured in a carousel, where flow continually flows around the ditch with influent and effluent weirs controlling flow based on water-level elevations; continual flow of nitrate from the effluent to the influent end of the ditch eliminates the need for internal recycle that is common with other flow-through nitrogen removal processes; achieves moderate total nitrogen removal (effluent total nitrogen: 3 to 7 mg/L).
Step feed	Includes the addition of a portion of influent flow at multiple points in a flow-through treatment process; addition of carbon from influent wastewater eliminates the need for internal recycle; lower volume requirements compared to CAS.
Integrated fixed-film activated sludge	Supplemental process used to increase the treatment capacity of an existing BNR process; addition of fixed or floating media to the aerobic zones of bioreactors to encourage biofilm growth and increase the solids inventory; most applicable for WRRF locations that are space limited, subject to a wide range of influent loadings, colder climates, or subject to toxic shock because of inhibitory substances in the influent; higher operations costs compared to CAS.
Membrane bioreactor	Solids separation process applicable to any flow-through BNR process; eliminates the need for secondary clarification; typical MLSS: 8000 to 12 000 mg/L; smaller bioreactor volumes because of increased MLSS; higher operations costs compared to CAS; nitrogen removal similar to CAS.

will increase the capacity of the aerobic process (nitrification). However, typically less than 10% of total nitrogen is removed during primary clarification; as such, the COD-to-total-nitrogen ratio is significantly lowered by primary clarification, which reduces the efficiency of the denitrification process. These effects must be considered in the management of primary clarifier operation.

4.2 Secondary Clarifiers

Based on empirical measurements by McCarty (1970), typical effluent TSS from biological treatment units contain approximately 12% nitrogen and 2.2% phosphorus. This translates into 1.2 mg/L of total nitrogen for every 10 mg/L of effluent TSS. With the exception of some of the biological filter processes and the MBR, the biological nitrogen removal processes presented herein depend on the efficient capture of biological solids by secondary clarifiers. It is critical that secondary clarifiers are operated to achieve a consistently high level of treatment to both protect the water environment and avoid the high cost of downstream tertiary treatment. Although many of the basic principles of suspended and fixed growth secondary clarification are the same, the units have some key differences related to upstream biological processes.

4.2.1 *Clarifiers in Suspended Growth Systems*

Suspended growth (activated sludge) reactors and clarifiers must be designed and operated as a coordinated system to provide a consistently high level of performance. It is crucial that the operations staff responsible for process control decisions enhance their understanding of clarifier capacity and operating procedures. Incorporation of nitrogen removal to an activated sludge facility will typically result in improved settling characteristics because of the selection of floc-forming organisms and the suppression of filamentous organisms in the anoxic zones. With the exception of non-nitrifying facilities, BNR facilities will result in lower concentrations of nitrate in the clarifier and decreased problems associated with denitrification in the sludge blanket of the clarifier. Generally, reducing nitrate to less than 10 mg/L will help minimize denitrification issues in the sludge blanket.

The Water Environment Research Foundation Clarifier Research Technical Committee's activated sludge secondary clarifier evaluation protocol defines a step-by-step procedure for evaluating secondary clarifier capacity and troubleshooting poor clarifier performance (WERF, 2001). Poor activated sludge secondary clarifier performance has been associated with one or more of the following conditions (WERF, 2001):

- Sludge blanket denitrification,
- Sludge particle flocculation problems,
- High sludge blankets caused by solids overload, and
- Poor clarifier hydraulics.

With the possible exception of clarifier hydraulics, operators have corrective action measures that can be used to address these conditions. The actual measures undertaken depend on site-specific conditions that exist at the time of the problem. (Refer to Appendix A for a troubleshooting and process optimization guide.)

4.2.2 *Clarifiers in Biofilm Systems*

Secondary clarifiers for biofilm systems have not experienced the same level of interest or research as those for activated sludge systems. Two of the conditions identified as challenges to suspended growth system clarifier performance, flocculation and hydraulic problems, have been identified as the primary cause of problems with biofilm system clarifier performance as well (WEF, 2000). For moving bed biofilm reactor (MBBR) systems, dissolved air flotation clarifiers are commonly used as an alternative to conventional clarifiers because of the light nature of the MBBR floc. (Refer to Chapter 4 for a detailed discussion of this topic.)

4.3 Aeration Systems

In the context of nitrogen removal, the fundamental purpose of aeration systems is to provide the process oxygen necessary for nitrification as the first step of the nitrogen removal process. The sensitivity of nitrifying organisms to environmental conditions makes this the "weak link" in the nitrogen removal process and maintaining sufficient dissolved oxygen in the bioreactor is critical to ensuring a healthy population of nitrifiers. Additionally, the nitrification rate is directly related to the dissolved oxygen concentration, with the nitrification rate dropping quickly as the dissolved oxygen concentration drops below 1 mg/L. With the discovery of nitrifying organisms such as *Nitrospira* that can efficiently nitrify at lower dissolved oxygen concentrations, the traditional belief that dissolved oxygen must be maintained at 2 mg/L or greater is being challenged. This is a critical concern because the implementation of nitrogen removal in WRRFs has caused a fundamental shift in operational control strategies of aeration systems. Whereas historically the emphasis has been on maintaining an excess of dissolved oxygen in the bioreactor, this philosophy can be detrimental to the efficiency of denitrification processes. This is because of the reliance of denitrification processes on internal recycle streams within the bioreactor.

With up to 4 times the influent flow typically being recirculated to anoxic zones, the concentration of dissolved oxygen at the point in the bioreactor where the recirculation stream is being taken must be kept to a minimum to prevent "oxygen poisoning" of the anoxic zone. This can be done by either incorporating a dissolved oxygen control zone (i.e., a zone designed to reduce dissolved oxygen in the recycle stream) or providing a mixer at the end of the bioreactor to allow the operator to not have to depend on the aeration system for mixing at a point where oxygen demand is low.

4.4 Anoxic Zone Mixing

Mixing of anoxic zones and the portions of aerobic zones where aeration is decreased below the minimum value required for mixing is a critical feature of nitrogen removal processes. It is required to keep solids in suspension and/or "degassing" mixed liquor before secondary clarification. The following are general guidelines to consider with anoxic zone mixing:

- Mixers should be configured such that no oxygen is entrained into the mixed liquor,
- Position of mixers should minimize localized backflow into the preceding zones, and
- Baffling and reactor configuration must be considered in the mixing evaluation (WEF et al., 2009).

Mixing can be accomplished by a variety of means and equipment including submersible horizontal propeller-type mixers, pumped-jet mixing, large-bubble compressed air, vertical turbine mixers, and hyperboloid mixers. Anoxic zone mixing requirements vary for each application and are specific to the reactor configuration and process requirements. Mixing solutions should be carefully selected in coordination with a qualified vendor. The features of each mixing type are discussed in the following subsections.

4.4.1 *Submersible Horizontal Propeller*

Submersible horizontal propeller mixers are installed on the sidewalls of an anoxic zone reactor and rotate to produce turbulence and keep the MLSS in suspension (Figure 6.10). Propeller mixing has been successfully used in anoxic zones for many years and is less expensive than other types of commonly used mixing technologies. However, submersible horizontal propellers often have higher maintenance requirements because of submerged motor and debris accumulation on the propellers. Operation costs for propeller mixing are also higher than other mixing technologies because of the higher power requirements, depending on the reactor configuration.

4.4.2 *Jet Mixing*

Jet mixing consists of a mixing pump that can be submerged in the basin or located on top of the basin that discharges to a header containing a series of jet nozzles to produce turbulence and keep solids in suspension. The advantage of jet mixing is that the pumps can be installed outside of the basin, requiring no routine maintenance within the tank. Operators are often already familiar with

FIGURE 6.10 Submersible propeller mixer.

maintenance requirements because the pumps used in jet-mixing systems are common to other processes in a WRRF.

4.4.3 *Large-Bubble Compressed Air*

A relatively new technology that has had success in mixing anoxic zones is large-bubble compressed air mixing. Large-bubble mixing systems consist of a compressor, receiving tank, solenoid valve panel(s), and large-bubble diffuser grid(s). Solenoid valves operate in a defined sequence pattern to release pulses of air into a distribution header with large-bubble diffusers. The diffusers produce large bubbles ranging from marble to softball size that keep the MLSS in suspension (Figure 6.11). Although bubbles are introduced into nonoxygenated zones, anoxic environments are maintained that are fully compatible with BNR requirements. Full-scale oxygen-transfer studies have been conducted between large-bubble systems and conventional submersible propeller-type mixers to confirm that the use of large bubbles does not impede the anoxic reaction because of oxygen entrainment (U.S. EPA, 2010b). Large-bubble mixing systems have lower power consumption and

FIGURE 6.11 Large-bubble mixing (U.S. EPA, 2010a).

maintenance requirements than other forms of mixing. Capital costs of large-bubble mixing systems are often higher than other conventional mixing alternatives, although the more separate zones that require mixing, the more cost effective large-bubble systems become. Actual power requirements and operation costs are site specific and should be carefully evaluated for each application.

4.4.4 *Vertical Turbine*

Vertical turbine mixers are mounted on platforms above the anoxic zone reactor and consist of a drive motor and gearbox, vertical shaft, and impeller (or turbine). Flow is directed toward the floor of the tank, where it is redirected up the sides of the reactor to mix the tank contents. Compared to submersible-style mixers, vertical turbine mixers offer the advantage of having the motor and electrical components located above the tank rather than being submerged.

4.4.5 *Hyperboloid*

Hyperboloid mixers are vertical mixers with a drive motor installed on top of the tank connected to a vertical shaft and hyperboloid-shaped mixer body that is installed relatively close to the tank bottom. Hyperboloid mixers offer the advantages of reduced power consumption compared to other vertical and horizontal propeller- or turbine-type mixers and a reduction in operation and maintenance requirements. The hyperboloid shape and configuration also reduces the potential for debris accumulation.

4.5 Internal Recycle Pumping

Internal recycle pumping is critical to denitrification and BNR systems. It is used to return nitrate to the upstream anoxic zone to use influent carbon in the raw wastewater to fuel denitrification and alkalinity recovery, as discussed in Chapter 5. The denitrification process also produces a net oxygen demand reduction because carbon (COD or BOD) is being degraded under anoxic conditions rather than aerobically with aeration energy. Internal recycle rates are dependent on the process goals and configuration. Two types of pumps are commonly used for internal recycle: axial flow propeller pumps and submersible nonclog pumps.

4.5.1 *Axial Flow Propeller Pumps*

Axial flow propeller-type pumps are known as "high flow–low head" pumps and represent the most common type of internal recycle pump. These pumps typically operate in a head range of 0.9 to 1.8 m (3 to 6 ft). Axial flow pumps are less costly and have lower power consumption and operation costs compared to submersible nonclog pumps because of the low head requirement. However, axial flow pumps have been known to have ragging issues in facilities where only coarse (6-mm [.25-in.]) or greater) influent screens are installed or where ragging is known to be an operational issue.

4.5.2 *Submersible Nonclog Pumps*

Submersible nonclog-type pumps have been used in wastewater applications for many years in influent pumping, RAS, and internal recycle because of their ability to pass solids that are found in raw wastewater and sometimes in mixed liquor. Submersible nonclog pumps typically operate in a head range of 5 to 15 m (15 to 35 ft). Therefore, their power consumption and operating costs are higher than axial flow propeller pumps. Submersible nonclog pumps have the advantage of passing solids and typically have less maintenance requirements than axial flow propeller pumps in an application where ragging issues are known to exist. The pumping properties of submersible nonclog pumps also offer greater flexibility in flow variation when a variable frequency drive (VFD) is added compared to an axial flow propeller pump. With VFD, operators can adjust the flow as needed based on process requirements and environmental conditions.

4.6 Staging

Staging is used in nitrogen removal systems through the implementation of baffles for a variety of purposes, such as the following:

- Separation of aerated and unaerated zones,
- Separation of diffuser zones in a tapered aeration aerobic zone, and
- Nitrified mixed liquor recycle pump baffle to reduce the amount of oxygen recycled to the anoxic zone (WEF et al., 2009).

Baffles can be constructed of a variety of materials including cast-in-place and precast concrete, high-density polyethylene, stainless and painted carbon steel, wood, fiber-reinforced plastic, and hanging curtains. Baffles used to separate reactor and diffuser zones should be placed such that they result in minimal head loss across the baffle. Operation and maintenance of diffusers and other reactor components should be considered when installing baffle walls. Baffle layout should not result in capture and trapping of surface scum. Drain ports are needed in the bottom of the baffle to allow for complete draining of the tank for maintenance purposes so that no water is retained between baffled zones. Openings installed in baffle walls should be sized for operator access between zones without having to exit each baffled zone through the top of the reactor on a ladder or stairwell.

4.7 Supplemental Carbon Systems

The process requirement for the provision of supplemental carbon to improve denitrification is described in detail in Chapter 5. Once it is determined that supplemental carbon is required to meet process goals, the next step is to determine which of many available sources of supplemental carbon are best suited for a particular application (WERF, 2010).

Components of supplemental carbon storage and feed systems are similar to most other chemicals used in wastewater treatment processes: one or more bulk storage tanks, one or more chemical feed pumps and associated piping, and a means of controlling chemical dose. Two of the most commonly used sources of supplemental carbon (methanol and ethanol) are hazardous and highly flammable; therefore, storage and feed systems must be carefully designed, constructed, maintained, and operated to comply with all necessary safety standards and procedures.

Methanol has historically been the most economical alternative carbon source and is typically used to increase denitrification rates in second-stage anoxic zones and denitrification filters. Handling of methanol and designing facilities for feeding methanol must take into account the dangers associated with its use. Methanol represents an explosion hazard and is toxic when ingested. Methanol should be stored in mild or stainless steel tanks; plastic, galvanized, and aluminum tanks are not suitable. Grounding is important to protect against static electricity. Carbide-tipped clamps (to ensure good contact through paint) and dip-tube filling are generally used to guard against ignition from static electricity. Proper venting of tanks is important because of the volatility of methanol, and tanks should be painted white to reduce methanol vapors.

Piping for these chemicals must comply with Section 30 of the National Fire Protection Association Standards and all indoor spaces through which the piping passes must be suitably equipped with fire-detection and prevention systems. Prolonged or repeated breathing of methanol vapors should be avoided at all times. Proper ventilation is required to ensure safe working conditions. The type of ventilation will depend on such factors as dead air spaces, temperature, convection currents, and wind direction and must be considered when determining equipment location, type, and capacity. If mechanical ventilation is used, spark-proof fans should be implemented. Because of the explosive nature of methanol, smoking and welding around methanol facilities must be avoided. Other, less hazardous options include glycerin-based products and acetic acid (refer to Chapter 5).

4.8 Froth Control and Scum Removal

Froth, foam, and scum can be common nuisances at WRRFs. Scum and foam accumulation occurs because of a variety of causes including grease, soaps, scum, detergents, and other surfactants in the influent and trapping of foam-causing bacteria in biological reactors (Metcalf and Eddy, 2003). The following techniques are commonly used for foam- and scum-control measures in the operation of BNR systems:

- Careful placement of over and under baffles, inverted chimneys to direct the flow to the bottom of the following basin, and peak-flow overtopping weirs, as appropriate, in the partitions between the various zones of a bioreactor. There should be no back-mixing from the aerobic to the anoxic zone, and foam should spill freely from the anoxic to aerobic zones. Partitions within any zones should be just below the surface to avoid contact areas for scum to grow;
- Implementation of continuous selective wasting for the preferential removal of foam and scum organisms from the aeration basin as part of the WAS stream. Recycling of this foam to the facility should be minimized because it will reseed the foaming organisms;
- Elimination of dead ends, dead corners, and other quiescent zones in channels or bioreactors where there is a potential for foam and scum to accumulate;
- Installation of an effective scum removal system on primary and secondary clarifiers;
- Avoidance of opportunities for recycling foam and scum organisms to the mainstream treatment train from sidestream solids processing facilities;
- Use of chlorine sprays, as necessary, at localized points of foam and scum collection and/or accumulation to kill these organisms and prevent them from causing problems in either mainstream or sidestream treatment processes. Early microscopic detection is important to minimize the extent of the problem. Alternatively, there are some polymers that could be sprayed on the foam; and

- Effective destruction of foam and scum removed from primary and secondary processes in the solids handling process through digestion, incineration, composting, or other solids handling methods.

4.9 Alkalinity Addition

Alkalinity is a measure of a wastewater's acid-neutralizing capacity. Alkalinity and pH are closely related and are important in BNR facilities. The nitrification process consumes alkalinity, which, in turn, causes the wastewater pH to drop. As the pH drops, the rate of nitrification can decrease and stop at a pH of approximately 6. Alkalinity supplementation may be needed to support nitrification in some WRRFs or to support nitrification at certain times of the year. Alkalinity consumption for both nitrification and phosphorus precipitation must be accounted for to determine if alkalinity supplementation is necessary.

When alkalinity supplementation is required, there are several chemicals that can be used. The preferred chemical is influenced by local conditions, local chemical prices, and operator preferences. Chemicals that can be used as an alkalinity supplement include the following:

- Sodium hydroxide (caustic soda) ($NaOH$);
- Calcium hydroxide (lime) [$Ca(OH)_2$];
- Calcium oxide (quick lime) (CaO);
- Magnesium hydroxide [$Mg(OH)_2$];
- Sodium carbonate (soda ash) (Na_2CO_3); and
- Sodium bicarbonate ($NaHCO_3$).

4.9.1 *Sodium Hydroxide*

Sodium hydroxide (commonly called *caustic* or *caustic soda*) is used for alkalinity supplementation because of its ease of handling. Sodium hydroxide is often not the lowest-cost chemical for alkalinity addition; however, it is easy to use, and annual maintenance costs for the caustic storage and feed system are much lower. Many utility operators believe that the ease of handling of sodium hydroxide far outweighs the higher cost of the chemical. Sodium hydroxide can be purchased at a 50%, by weight, solution strength or, in some locations, it is available at 20 or 25%, by weight, solution strengths. It is classified as a strong base and, if overdosed, can raise the pH much higher than expected. An automatic system to control chemical dose based on pH can be used to prevent overdosing and optimize chemical usage. Once the liquid temperature drops below 12.8 °C, sodium hydroxide will begin to crystallize out of solution. Heat-tracing systems may be used on the storage tank and piping system components to prevent crystallization. It is difficult to redissolve crystallized sodium hydroxide. If sodium hydroxide is delivered to the site at 50% strength and then diluted on-site with facility water or potable water, scaling will occur at the point of mixing. The local pH will rise well above pH 10, and calcium carbonate will form immediately.

4.9.2 *Calcium Hydroxide*

Calcium hydroxide, or hydrated lime, has been slaked by the manufacturer and is sold as a dry material. Hydrated lime must be slurried before use. The slurry tank will be susceptible to scaling if facility water or potable water is used because the hardness in these waters will precipitate. Calcium hydroxide is typically prepared in a 3 to 5%, by weight, slurry. A small fraction of the calcium hydroxide will dissolve and raise the pH to approximately 12, which is high enough to be dangerous to operators. Calcium hydroxide should be added at a point of turbulence to ensure sufficient mixing.

4.9.3 *Quicklime*

Before being used, calcium oxide, or quicklime, must be slaked, a process in which the calcium oxide is mixed with water and allowed to react. Slaker operating temperatures are typically between 120 and 180 °C and the elevated temperatures increase the amount of scaling that can form. Quicklime is typically cheaper than hydrated lime. However, slaking is a labor-intensive process and operation costs must be considered in a life cycle analysis.

4.9.4 *Magnesium Hydroxide*

Magnesium hydroxide is increasingly being used in BNR processes because it will not raise the pH above approximately 10.5, which correlates to its precipitation point range of pH 10.2 to 10.5. However, the effects of magnesium addition on other facility processes such as struvite formation in anaerobic digestion operations must be carefully considered. The costs are regionally driven; therefore, before selecting magnesium hydroxide for long-term use, delivery costs should be investigated. Magnesium hydroxide is generally sold as a slurry, although a dry product can also be purchased. Scaling will begin above pH 8.

4.9.5 *Sodium Carbonate*

Sodium carbonate or soda ash is generally available as a dry product; therefore, it must be dissolved on-site. The maximum pH level of a sodium carbonate solution is approximately 12.

4.9.6 *Sodium Bicarbonate*

Sodium bicarbonate use is not widespread because other chemicals are easier to handle and cost less. Sodium bicarbonate is generally available as a dry product and, therefore, must be dissolved on-site. The maximum pH level of a sodium bicarbonate solution is 8.3.

5.0 REFERENCES

Barnard, J. L. (1974) *Cut N and P without Chemicals. Water Wastes Eng.*, **11,** 41–44.

Barnard, J.; Steichen, M.; deBarbadillo, C. (2004) Interaction Between Aerator Type and Simultaneous Nitrification and Denitrification. *Proceedings of the 77th Annual Water Environment Federation Technical Exhibition and Conference* [CD-ROM]; New Orleans, Louisiana, Oct 2–6; Water Environment Federation: Alexandria, Virginia.

Chandran, K.; Pape, R.; Ezenekwe, I.; Stinson, B.; Anderson, J. (2003) Froth Control and Prevention Strategies for Step-Feed BNR at NYC WPCP. *Proceedings of the 76th Annual Water Environment Federation Technical Exhibition and Conference* [CD-ROM]; Los Angeles, California, Oct 11–15; Water Environment Federation: Alexandria, Virginia.

Davies, K. J. P.; Lloyd, D.; Boddy, L. (1989) The Effect of Oxygen on Denitrification in *Paracoccus denitrificans* and *Pseudomonas aeruginosa. J. Gen. Microbiol.,* **135**, 2445–2451.

deBarbadillo, C.; Carrio, L.; Mahoney, K.; Anderson, J.; Passarelli, N.; Streett, F.; Abraham K. (2002) Practical Considerations for Design of a Step-Feed Biological Nutrient Removal System. *Florida Water Resour. J.*, **54,** (1), 18–35.

Fillos, J.; Diyamandoglu, V.; Carrio, L.; Robinson, L. (1996) Full-Scale Evaluation of Biological Nitrification/Denitrification in a Step-Fed Activated Sludge Process. *Water Environ. Res.*, **68** (2), 131–142.

Goodwin, S.; Johnson, B.; Daigger, G.; Crawford, G. (2005) A Comparison Between the Theory and Reality of Full-Scale Step-Feed Nutrient Removal Systems. *Proceedings of the European Federation of Clean Air and Environmental Protection Associations (Warsaw, Poland) General Assembly Meeting and Conference;* Krakow Poland, May 28–June 1.

Goronszy, M. C. (1979) Intermittent Operation of the Extended Aeration Process for Small Systems. *J.—Water Pollut. Control Fed.*, **51**, 274.

Jenkins, D.; Richard, M. G.; Daigger, G. T. (2003) *Manual on the Causes and Control of Activated Sludge Bulking and Foaming*; Lewis Publishers: University City, Missouri.

Kaempfer, H.; Daigger, G.; Adams, C. (2000) Characterization of the Floc Micro-Environment in Dispersed Growth Systems. *Proceedings of the 73rd Annual Water Environment Federation Technical Exposition and Conference* [CD-ROM]; Anaheim, California, Oct 14–18; Water Environment Federation: Alexandria, Virginia.

Leaf, W. R.; Boltz, J. P.; McQuarrie, J. P.; Menniti, A.; Daigger, G.; Adams, C. (2011) Overcoming Hydraulic Limitations of the Integrated Fixed-Film Activated

Sludge (IFAS) Process. *Proceedings of the 84th Annual Water Environment Federation Technical Exhibition and Conference* [CD-ROM]; Los Angeles, California, Oct 15–19; Water Environment Federation: Alexandria, Virginia.

Ludzack, F. T.; Ettinger, M. B. (1962) Controlling Operation to Minimize Activated Sludge Effluent Nitrogen. *J.—Water Pollut. Control Fed.*, **34,** 9.

McCarty, P. L. (1970) Phosphorus and Nitrogen Removal in Biological Systems. *Proceedings of the Wastewater Reclamation and Reuse Workshop;* Lake Tahoe, California, June 25–27; p 226; University of California: Berkeley, California.

Metcalf and Eddy, Inc. (2003) *Wastewater Engineering, Treatment and Reuse,* 4th ed.; Tchobanoglous, G.; Burton, F. L.; Stensel, H. D., Eds.; McGraw-Hill: New York.

Phillips, H. M.; Maxwell, M.; Johnson, T.; Barnard, J.; Rutt, K.; Seda, J.; Corning, B.; Grebenc, J. M.; Love, N.; Ellis, S. (2008) Optimizing IFAS and MBBR Designs Using Full-Scale Data. *Proceedings of the 81st Annual Water Environment Federation Technical Exhibition and Conference* [CD-ROM]; Chicago, Illinois, Oct 18–22; Water Environment Federation: Alexandria, Virginia.

Robertson, L. A.; van Niel, E. W. J.; Torremans, R. A. M.; Gijs Kuenen, J. (1988) Simultaneous Nitrification and Denitrification in Aerobic Chemostat Cultures of *Thiosphaera pantotropha. Appl. Environ. Microbiol.*, **54,** 11, 2812–2818.

Rutt, K; Seda, J; Johnson, C. H. (2006) Two Year Case Study of Integrated Fixed Film Activated Sludge (IFAS) at Broomfield, CO WWTP. *Proceedings of the 79th Annual Water Environment Federation Technical Exhibition and Conference* [CD-ROM]; Dallas, Texas, Oct 21–25; Water Environment Federation: Alexandria, Virginia.

Satoh, H.; Nakamura, Y.; Ono, H.; Okabe, S. (2003) Effect of Oxygen Concentration on Nitrification and Denitrification in Single Activated Sludge Flocs. *Biotechnol. Bioeng.*, **83** (5), 604–607.

Sen, D.; Randall, C.; Grizzard, T. (1990) *Biological Nitrogen and Phosphorus Removal in Oxidation Ditch and High Nitrate Recycle Systems;* Pub. CBP/TRS 47/90; U.S. Environmental Protection Agency, Chesapeake Bay Program: Annapolis, Maryland.

Schuyler, R. G.; Tamburini, S. H.; Staggs, R. (2009) How Low Is too Low? *Water Environ. Technol.*, **21** (6), 32–39.

Stensel, H. D. (2001) Biological Nutrient Removal: Merging Engineering Innovation and Science. *Proceedings of the 74th Annual Water Environment Federation Technical Exhibition and Conference* [CD-ROM]; Atlanta, Georgia, Oct 13–17; Water Environment Federation: Alexandria, Virginia.

Stricker, A.; Barrie, A.; Maas, C. L. A.; Fernandes, W.; Lishman, L. (2009) Comparison of Performance and Operation of Side-By-Side Integrated Fixed-Film and Conventional Activated Sludge Processes at Demonstration Scale. *Water Environ. Res.,* **81** (3), 219–232.

Trivedi, H.; Heinen, N. (2000) Simultaneous Nitrification/Denitrification by Monitoring NADH Fluorescence in Activated Sludge. *Proceedings of the 73rd Annual Water Environment Federation Technical Exposition and Conference* [CD-ROM]; Anaheim, California, Oct 14–18; Water Environment Federation: Alexandria, Virginia.

U.S. Environmental Protection Agency (1993) *Nitrogen Control Manual;* EPA-625/R-93-010; U.S. Environmental Protection Agency: Washington, D.C.

U.S. Environmental Protection Agency (2010a) *Nutrient Control Design Manual;* EPA-600/R-10-100; U.S. Environmental Protection Agency: Washington, D.C.

United States Environmental Protection Agency (2010b) *Evaluation of Energy Conservation Measures for Wastewater Treatment Facilities;* EPA-832/R-10-005; U.S. Environmental Protection Agency: Washington, D.C.

Water Environment Federation (2009) *An Introduction to Process Modeling for Designers;* WEF Manual of Practice No. 31; Water Environment Federation: Alexandria, Virginia.

Water Environment Federation; American Society of Civil Engineers; Environmental and Water Resources Institute (2009) *Design of Municipal Wastewater Treatment Plants,* 5th ed.; WEF Manual of Practice No. 8; ASCE Manuals and Reports on Engineering Practice No. 76; McGraw-Hill: New York.

Water Environment Research Foundation (2001) *WERF/Clarifier Research Technical Committee (CRTC) Protocols for Evaluating Secondary Clarifier Performance;* Water Environment Research Foundation: Alexandria, Virginia.

Water Environment Research Foundation (2010) *Protocol to Evaluate Alternative External Carbon Sources for Denitrification at Full-Scale Wastewater Treatment Plants;* Water Environment Research Foundation: Alexandria, Virginia.

Young, T.; Crosswell, S.; Wendle, J. (2008) Comparison of Nitrogen Removal Performance in SBR Systems. *Proceedings of the 81st Annual Water Environment Federation Technical Exhibition and Conference* [CD-ROM]; Chicago, Illinois, Oct 18–22; Water Environment Federation: Alexandria, Virginia.

Chapter 7

Enhanced Biological Phosphorus Removal

Somnath Basu, Ph.D., P.E., BCEE;
M. Kim Fries, M.A.Sc., P.Eng.;
Nehreen Majed, Ph.D.; and
Annalisa Onnis-Hayden, Ph.D.

1.0 PROCESS FUNDAMENTALS

Enhanced biological phosphorus removal (EBPR) relies on the selection and proliferation of a microbial population capable of storing orthophosphate in excess of its biological growth requirements. This group of organisms, referred to as *polyphosphate-accumulating organisms* (PAOs), relies on operational conditions that impose a selective advantage for them while putting other groups at a temporary disadvantage with respect to access to food. Once this is achieved by implementation of special design and operational conditions (e.g., alternative anaerobic and aerobic cycles), PAOs gain the selective advantage to grow and function, resulting in excessive accumulation of orthophosphate in activated sludge. With proper biomass wasting, phosphate removal can then be achieved.

Enhanced biological phosphorus removal research has clearly shown that PAOs are a subset of heterotrophs. Several mechanisms have been proposed to explain the enhanced uptake of phosphorus by microorganisms in wastewater. It has been shown that, for biological phosphorus removal to occur in water resource recovery facilities (WRRFs), biomass first needs to pass through an oxygen- and nitrate-free phase (i.e., an anaerobic phase) before entering a phase where an electron acceptor is present (i.e., an anoxic phase where nitrate is present or an aerobic phase where oxygen is present). The oxygen- and nitrate-free phase can be achieved in a separate reactor, the first section of a plug flow reactor, or a part of a sequencing batch reactor cycle (for schemes for EBPR, refer to Chapter 9).

When wastewater enters the anaerobic phase, the PAOs accumulate carbon sources (volatile fatty acids [VFAs]) as an internal polymer called *polyhydroxyalkanoate* (PHA). The main forms of these PHAs are poly-beta-hydroxybutyrate

(PHB) and poly-beta-hydroxyvalerate (PHV). The energy to store this polymer is obtained from the breakdown of glycogen and hydrolysis of polyphosphate bonds. Polyphosphates are formed by a series of high-energy bonds; PAOs can subsequently obtain energy by breaking these bonds. Because polyphosphate is broken down to orthophosphate for energy supply, the phosphate concentration in the anaerobic phase increases. The ability of PAOs to accomplish anaerobic VFA uptake and to store PHA polymers is the main mechanism through which they gain selective advantage in EBPR systems.

The anaerobic phase needs to be followed by an oxygen or nitrate-rich phase, that is, an anoxic phase (anoxic phosphorus removal) or an aerobic phase (aerobic phosphorus removal). During this phase, the stored PHB is consumed, generating energy for growth and uptake of orthophosphate from the liquid phase and generating energy and carbon for replenishment of the glycogen and polyphosphate pools.

Figure 7.1 presents the process schematic and concentration profiles of the mean measurable components for EBPR operated under anaerobic–aerobic conditions, while Figure 7.2 shows a schematic representation of the significant biochemical processes involved.

Under these conditions, the orthophosphate concentration decreases. Most importantly, because the amount of biomass containing large amounts of polyphosphate is increasing under these conditions (PAOs are able to store up to 15% of their dry weight [Bond et al., 1999]), a net phosphorus removal occurs with the wasted sludge.

Initially, little attention was paid to reactions occurring in the anaerobic phase and aerobic uptake was thought to occur because of stress caused to the organisms

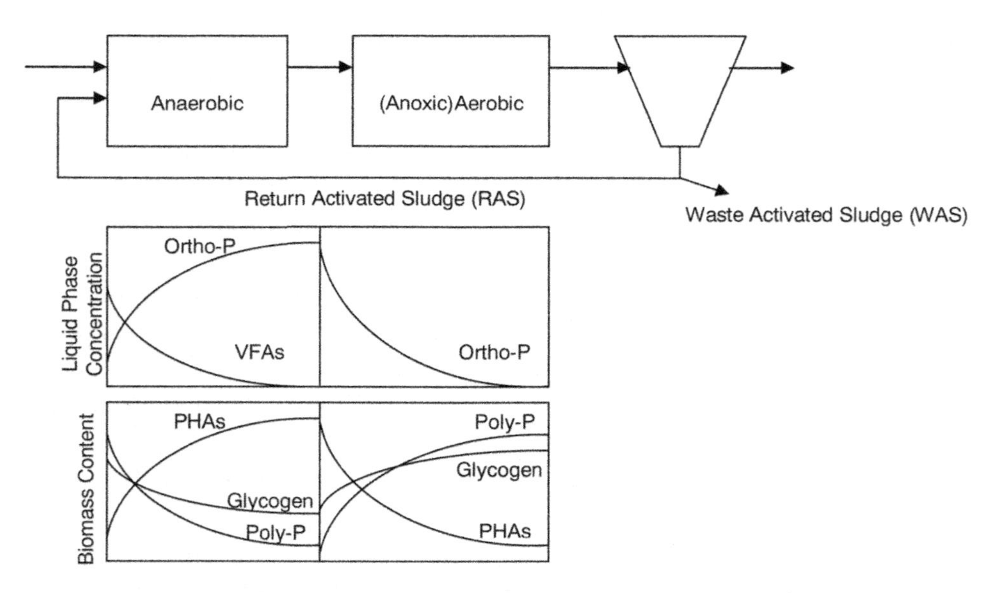

FIGURE 7.1 Typical concentration profiles observed in a generic EBPR system.

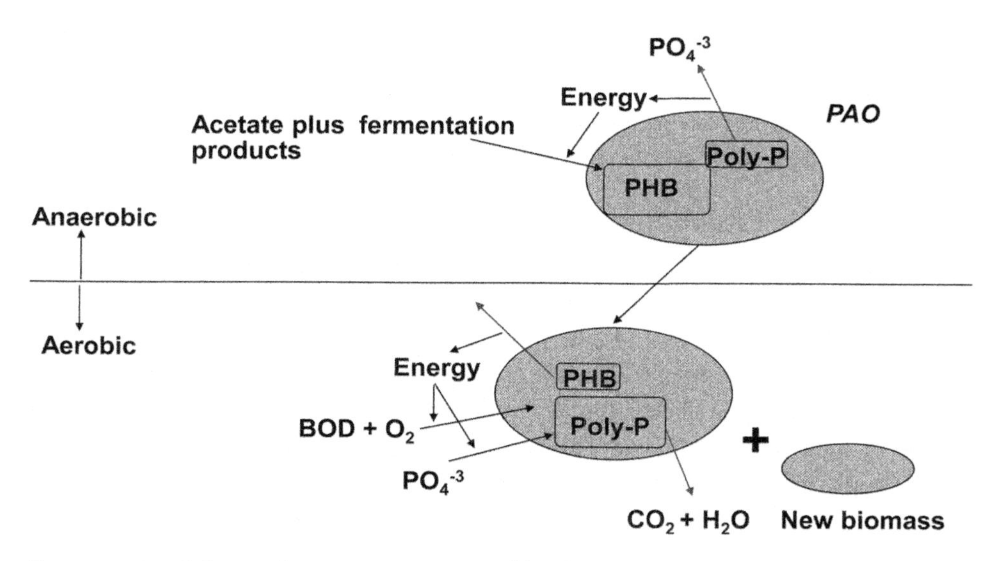

FIGURE 7.2 Schematic representation of biological phosphorus removal in an activated sludge system.

during the different stages of operation. As such, nearly all attention was initially focused on aerobic reactions and optimization of that phase. Considerably more insight has now been gained in the metabolisms underlying the EBPR process, which is presented in Section 3.0 of this chapter.

The anaerobic phase was believed to provide a unique, positive environment for PAOs, enabling them to reserve the necessary amount of carbon to themselves without having to compete with other microorganisms (Matsuo et al., 1992). Later research provided evidence that other organisms, termed *glycogen-accumulating organisms* (GAOs), are also able to store carbon sources (see Section 2.3) and can compete with PAOs in such systems.

It has been assumed that the availability of VFAs, with acetate as the main constituent, is a prerequisite for EBPR. In the absence of these components, fermentation of readily biodegradable carbon sources under anaerobic conditions is necessary. Results of later research also indicated good phosphorus removal with direct use of readily biodegradable material (see Section 4.1).

There is also evidence that EBPR can occur under continuously aerated "aerobic" conditions, where phosphorus and the carbon source (acetate) are temporarily separated and not added simultaneously (Ahn et al., 2007). This occurs with anaerobic–aerobic EBPR systems.

1.1 Aerobic Phosphorus Removal

In the early days of EBPR research, nearly all attention was focused on aerobic processes. A link between anaerobic processes and aerobic processes was not recognized. Aerobic phosphorus, then called "overplus" or "luxury" uptake (Levin and Shapiro, 1965), was supposed to result from stress conditions because of the dynamic feeding of activated sludge plants. It is now recognized that aerobic processes, use of PHA, replenishment of glycogen, and uptake of phosphorus are linked to anaerobic processes.

1.2 Anoxic Phosphorus Removal

Initially, it was believed that PAOs lacked the ability to denitrify and, hence, aerobic conditions were considered essential for the proliferation of these microorganisms for the effectiveness of phosphorus removal. This hypothesis was initially supported by observations that EBPR was negatively affected by nitrate entering the anaerobic phase, that a lower denitrification rate occurred compared to systems with non-PAO heterotroph organisms, and that a net phosphorus release typically occurred rather than an uptake under anoxic conditions (Barker and Dold, 1996). Later interpretations of earlier results and investigations, however, proved that at least a fraction of the PAOs can accumulate phosphate under anoxic conditions (Barker and Dold, 1996; Kuba et al., 1993, 1994). Enhanced biological phosphorus removal can, therefore, take place under anoxic conditions (in the presence of nitrate) as long as no readily biodegradable substrates are present (Oehmen et al., 2007; Oehmen et al., 2010; Seviour et al., 2003).

The use of nitrate rather than oxygen in PAO sludges is advantageous for several reasons. The availability of organic substrates in wastewater can limit both biological phosphorus removal and denitrification. With anoxic phosphorus removal, the same organics can be used for nitrate and phosphorus removal. This double use of carbon source will result in reduced sludge production; the use of nitrate as an electron acceptor for at least a portion of the phosphate uptake will also reduce oxygen demand (Carvalho et al., 2007; Kuba et al., 1996).

Additionally, Kuba et al. (1993) observed a clear difference in growth yield between anaerobic–aerobic sludges and anaerobic–anoxic ones, that is, 0.35 and 0.25 mg suspended solids/mg carbonaceous oxygen demand (COD), respectively. Using a two-sludge system, that is, denitrifying PAOs (DPAOs) and nitrifiers completely separated in two sludges with recirculation of the nitrified supernatant from the nitrifying stage to the anoxic stage, Kuba et al. (1996) observed the required COD to be up to 50% less than that for conventional aerobic phosphorus and nitrogen removal systems. Oxygen requirements and sludge production decreased to about 30 and 50%, respectively. Kuba et al. (1993) also observed a lower sludge volume index value, indicating better sludge settleability compared to anaerobic–aerobic systems.

Experimental evidence indicates that two different populations of PAOs exist in EBPR systems (Kern-Jespersen and Henze, 1993): PAOs that can only use oxygen as a terminal electron acceptor and DPAOs that can use both oxygen and nitrate as terminal electron acceptors. Filipe and Daigger (1999) stated that DPAOs have a disadvantage when competing with PAOs because of a lower thermodynamic efficiency of anoxic growth compared to aerobic growth.

Kuba et al. (1994) and Chuang et al. (1996) provided evidence that DPAOs can use both internally stored PHAs and external substrate for denitrification. When external substrate is used, phosphorus release is observed, whereas phosphorus uptake is expected when internally stored PHAs are degraded. Both mechanisms occur concurrently. According to Chuang et al. (1996), the kinetic competition observed is determined by the polyphosphate content of the microorganisms. With increased polyphosphate content, an increased specific phosphorus release rate is observed. More details on the microorganisms present in anoxic phosphorus removal and the metabolic pathways and mechanisms involved in anoxic phosphorus removal are presented in the following sections.

2.0 MICROBIOLOGY OF ENHANCED BIOLOGICAL PHOSPHORUS REMOVAL

There has been interest in the microbiology of the process and, as a result, several reviews on the microbiology of EBPR systems have been published (e.g., Blackall et al. [2002]; Forbes et al. [2009]; Oehmen et al. [2007]; Seviour et al. [2003]; and Seviour and McIlroy [2008]). Bacteria that are involved directly or indirectly in the EBPR process are fermentative (acid-producing) bacteria, PAOs, and GAOs.

2.1 Fermentative (Acid-Producing) Bacteria

Short-chain VFAs are absorbed by PAOs and stored as PHBs within the cell of these microorganisms. Volatile fatty acids are generated from complex particulate and soluble organic molecules during the fermentation process, which consists of the following three steps: hydrolysis, acidogenesis, and acetogenesis. These three steps are the first of four involved in the anaerobic digestion of complex organics.

Bacteria involved in hydrolysis and fermentation are not well investigated in full-scale EBPR plants, but recent studies show that hydrolysis seems to be carried out by relatively few and specialized species (Nielsen et al., 2012). Several filamentous bacteria are involved such as *Microthrix*, which produce lipase and consume long-chain fatty acids (Nielsen et al., 2009), *Chloroflexi*, and the epiphytic bacteria, *Candidatus epiflobacter*, which produce proteases and consume amino acids (Xia et al., 2007). Nielsen et al. (2012) found that fermenting bacteria constitute a large fraction (20%) of the microbial community in EBPR and are dominated by *Firmicutes* and *Actinobacteria*.

2.2 Polyphosphate-Accumulating Organisms

Based on cultivation experiments, *Gammaproteobacteria*, of the genus, *Acinetobacter*, were traditionally believed to be the only PAOs. However, it has become clear today that *Acinetobacter* can accumulate polyphosphate but do not possess the aforementioned PAO metabolism. Furthermore, cultivation-independent methods and quantitative fluorescent in situ hybridization (FISH) have demonstrated that the relative abundance of *Acinetobacter* in EBPR systems was dramatically overestimated because of cultivation biases, further confirming that *Acinetobacter* is not important to EBPR.

Clone library analyses followed by FISH instead indicated that *Rhodocyclus*-related *Betaproteobacteria,* given the name *Candidatus Accumulibacter phosphatis,* were more likely candidates (Hesselmann et al., 1999). However, not all clone libraries from EBPR processes have detected these populations, which is likely attributable to difficulties in extracting DNA from such heavily capsulated clustered cells. *Accumulibacter* have frequently showed the previously defined PAO phenotype in both laboratory- and full-scale EBPR systems (Crocetti et al., 2000; Kong et al., 2004).

Combining FISH with 4′,6-diamidino-2-phenylindole (DAPI) staining for detection of polyphosphates has revealed that *Accumulibacter* are not the only PAO populations in EBPR systems (He et al., 2008; Onnis-Hayden et al., 2011; Wong et al., 2005). In fact, evidence shows that *Accumulibacter and Beta-proteobacteria* often comprise only a relatively small proportion of cells accumulating polyphosphate, especially in full-scale facilities.

Many full-scale EBPR facility communities contain high numbers of polyphosphate staining and as-yet uncultured *Tetrasphaera*-related *Actinobacteria* (Kong et al., 2007); however, this class of bacteria did not assimilate acetate or synthesize PHAs anaerobically (Kong et al., 2005). According to Seviour and McIlroy (2008), any population that accumulates more phosphorus than it requires for growth and that stains positively for polyphosphate should be considered as a putative PAO, regardless of whether it synthesizes PHAs anaerobically or not.

Several studies have indicated that *Accumulibacter* are the significant PAO populations in both denitrifying and conventional EBPR processes (Carvalho et al., 2007; Kong et al., 2004; Zeng, Saunders, Yuan, Blackall, and Keller, 2003), emphasizing the metabolic versatility of these bacteria. However, differences in physiologies among *Accumulibacter* strains may exist. Carvalho et al. (2007), for example, suggested that rod-shaped *Accumulibacter* used nitrate, nitrite, and oxygen as electron acceptors for EBPR, whereas the more common *Accumulibacter* morphotype of large coccal cells used nitrite and oxygen. Further work determined that *Accumulibacter* populations established in a laboratory-scale bioreactor under nitrite-reducing conditions could not reduce nitrate (Guisasola et al., 2009). The differences in nitrate reduction activities suggested finer phenotypic and ecological differences among members of the *Accumulibacter* lineage (Flowers et al., 2009).

2.3 Glycogen-Accumulating Organisms

Enhanced biological phosphorus removal facilities often experience upsets and chemical phosphorus removal is used as a backup to polish their effluents (Seviour et al., 2003). In some instances, external disturbances such as high rainfall, excessive nitrate loading to the anaerobic reactor, or nutrient limitation may explain these process upsets. In other instances, different studies have linked such upsets to the appearance of a bacterial population known generically as *GAOs* (Oehmen et al., 2007; Seviour et al., 2003).

Under anaerobic conditions, GAOs have the ability to assimilate substrates like acetate and, under aerobic conditions, to use these to synthesize intracellular PHAs. Like PAOs, GAOs are thought to metabolize this stored PHA; however, like PAOs, they synthesize intracellular glycogen instead of polyphosphate. Glycogen-accumulating organisms are viewed as potential competitors of PAOs for anaerobic substrate uptake and, thus, are a likely cause of EBPR failure (Cech and Hartman, 1993; Liu, Mino, Nakamura, and Matsuo, 1996; Oehmen et al., 2007; Seviour et al., 2003). Therefore, it is recognized that maintaining conditions favoring the proliferation of PAOs over GAOs is critical to the stability of the EBPR process (Gu et al., 2008).

Because GAOs always seem to be present but are suppressed as a minority in well-functioning EBPR processes (Cech and Hartman, 1993; Matsuo, 1994), GAOs may possibly function as a scavenger for soluble COD and may become the dominating population in deteriorated EBPR processes (Liu, Mino, Nakamura, and Matsuo, 1996). Indeed, as GAOs are able to use different mechanisms for anaerobic uptake of substrates, they will be in a favorable position whenever the internal polyphosphate reserve of PAOs is depleted because of external unfavorable conditions (e.g., nitrate inhibition, unfavorable growth conditions, or unknown factors).

Several factors have been identified that influence competition between PAOs and GAOs. These include influent COD-to-bioavailable-phosphorus ratio (as milligrams per liter of influent COD per milligrams per liter of influent phosphorus), solids retention time (SRT), substrate type, hydraulic retention time (HRT), temperature, pH, dissolved oxygen, and feeding strategy (Filipe et al., 2001a, 2001b; Oehmen, Vives, Lu, Yuan, and Keller, 2005; Oehmen, Yuan, Blackall, and Keller, 2005; Rodrigo et al., 1996; Whang and Park, 2006).

Finally, the precise identity of these GAOs is still largely unknown. This is a significant reason for the difficulty in fully comprehending the role of different GAOs in EBPR system failure.

3.0 METABOLISM, METABOLIC PATHWAYS, AND METABOLIC REACTIONS

3.1 Metabolic Pathways for Polyphosphate-Accumulating Organisms in the Anaerobic Phase

Most researchers have focused on anaerobic metabolism because it serves as a selector for the enrichment of PAOs, anaerobic carbon uptake, and storage as intracellular compounds involve complicated and unresolved biochemical dynamics.

Fuhs and Chen (1975) were the first to postulate that PHB may function as a carbon storage reservoir and suggested that the anaerobic phase is necessary to encourage fermentation of complex organic compounds to produce low-molecular-weight VFAs. In the early years of EBPR research, PHB was already recognized as a storage polymer in the anaerobic phase of the EBPR process through staining techniques followed by microscopic observation (Buchan, 1983). Later, Comeau et al. (1986) analytically verified that the PHB-like polymer contains 3-hydroxybutyrate (3HB) and 3-hydroxyvalerate as monomeric building units. Satoh et al. (1992) further revealed that the storage polymer consists of four monomeric units and is now generally referred to as *PHA*.

At earlier stages of investigation into EBPR, polyphosphate was considered the sole energy source for PAOs; however, glycogen was also later found to provide energy during anaerobic substrate uptake and storage as PHAs. According Mino et al. (1998), the energy-requiring processes under anaerobic conditions are as follows: transport of external substrates into the cell, conversion of substrates to PHA and related metabolism, and maintenance or endogenous respiration. According to Smolders et al. (1994a), transport of acetate into the cell is thermodynamically influenced by pH. Based on a number of experiments within the pH range of 5.5 to 8.5, Smolders et al. (1994a) showed that the phosphorus-release-to-acetate-uptake ratio varies between 0.25 and 0.75 P-mol/C-mol, with lower ratios resulting at lower pH.

3.1.1 Source of Reducing Equivalent

An uncertainty that exists in explaining EBPR metabolism is the source of reducing power generation that is used for anaerobic PHA formation. Glycolysis and tricarboxylic acid (TCA) cycle are the significant catabolic pathways that are considered as the means to provide reducing power for PAOs.

3.1.1.1 Glycolysis

Glycolysis is a common biochemical pathway that is functional in many aerobic and anaerobic microorganisms and can be divided into two stages (Madigan et al., 2003). Stage 1 of glycolysis is a series of preparatory rearrangements or reactions that do not involve oxidation–reduction (redox) and do not release energy, but lead to production of two molecules of glyceraldehyde 3-phosphate from glucose. In stage 2, oxidation–reduction occurs, where nicotinamide adenine dinucleotide is converted to hydroxylamine reductase ($NADH_2$), energy is produced in the form of adenosine triphosphate, and pyruvate is produced. In anaerobic organisms, pyruvate can be converted into various products via fermentation, while, in aerobic respiration, it is metabolized via the TCA cycle after decarboxylation to acetyl-CoA. A schematic representation of a glycolytic pathway for an energy-transfer process is presented in Figure 7.3.

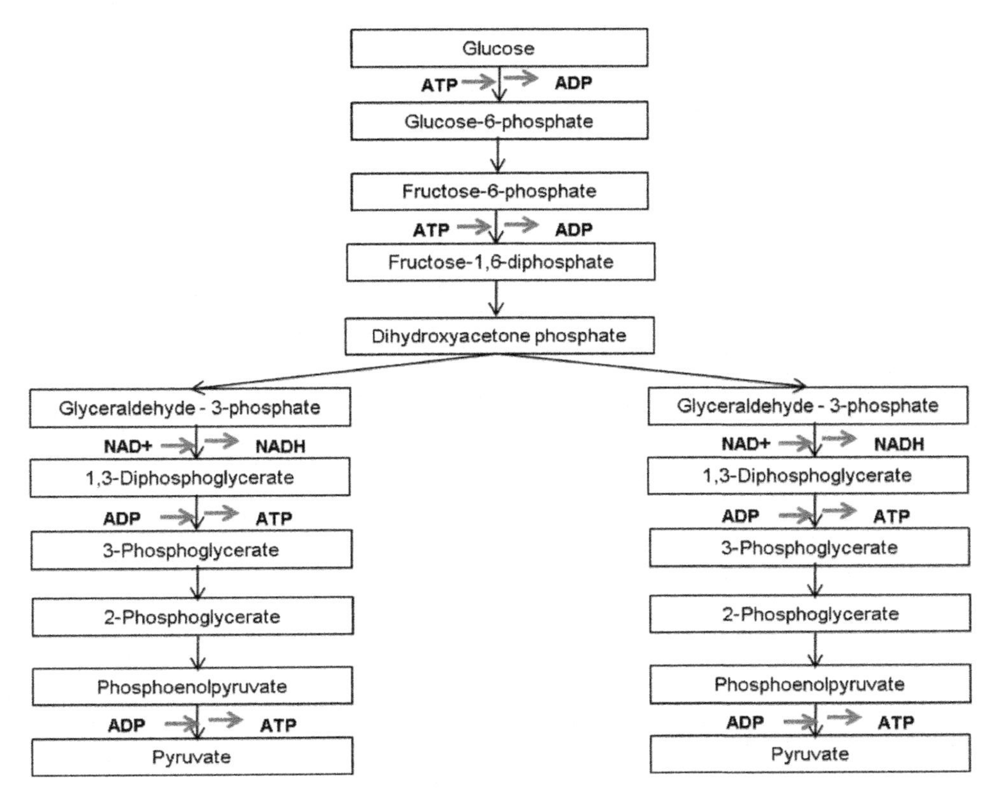

FIGURE 7.3 Glycolytic pathway and glycolysis.

3.1.1.2 Tricarboxylic Acid Cycle

The TCA cycle, also known as the *citric acid cycle* or *Krebs cycle*, is a common pathway for oxidizing all metabolic fuels (Figure 7.4). It consists of a series of enzyme-catalyzed chemical reactions that are typically used by aerobic organisms for cellular respiration (Madigan et al., 2003). The cycle begins with acetyl-CoA transferring its two-carbon acetyl group to the four-carbon acceptor compound, oxaloacetate, to generate a six-carbon tricarboxylic acid, citrate. Citrate enters a series of reactions where two carbons are released as carbon dioxide (CO_2) and the remaining four carbons are regenerated as oxaloacetate, which can then be used in the cycle again. The energy from these reactions is transferred to other metabolic processes as electrons by $NADH_2$. The $NADH_2$ generated in the TCA cycle may later donate its electrons in oxidative phosphorylation to drive ATP synthesis.

3.1.1.3 Glyoxylate Pathway and Partial Tricarboxylic Acid Cycle

A modification of the TCA cycle is the glyoxylate pathway (Figure 7.5). The glyoxylate pathway is a cyclic pathway that converts two acetyl-CoA into one molecule

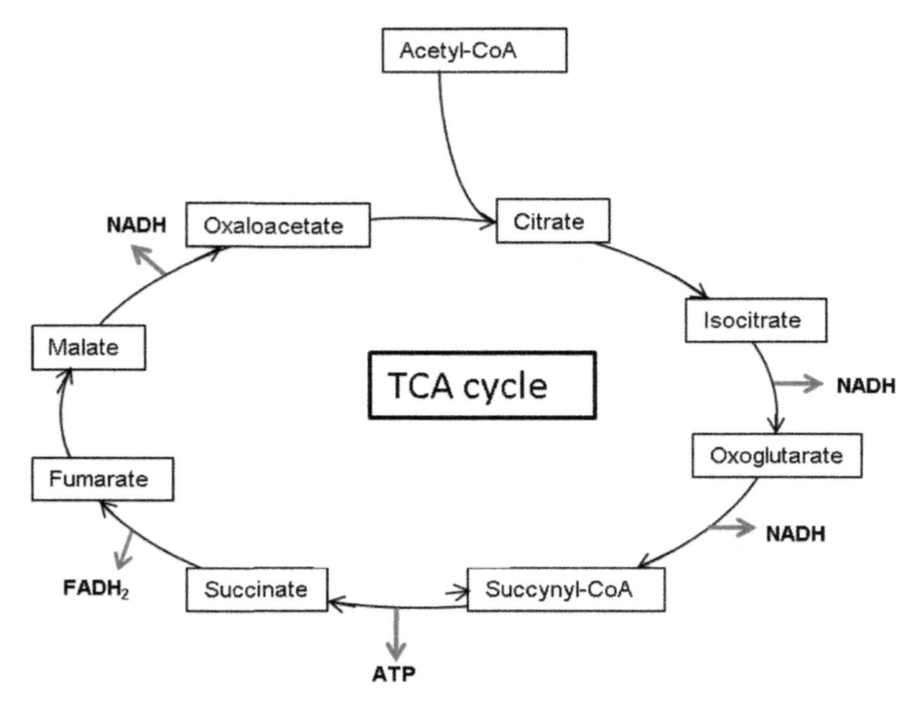

FIGURE 7.4 Full TCA cycle (Zhou et al., 2010).

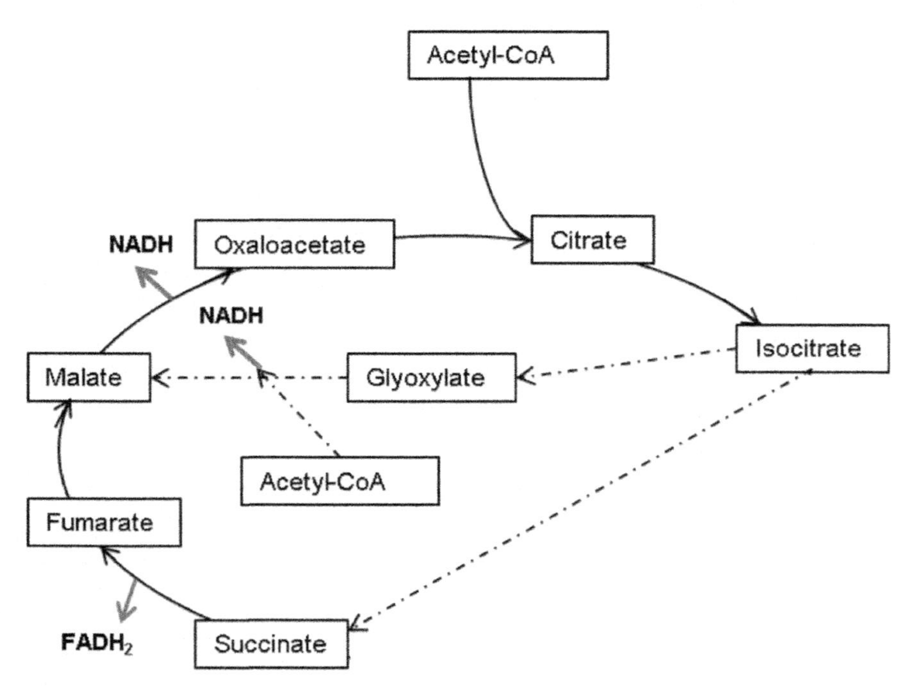

FIGURE 7.5 Partial TCA plus glyoxylate cycle (Oehmen et al., 2007; Zhou et al., 2010).

of succinate. The pathway uses some of the same enzymes as the TCA cycle, but it bypasses the reactions in which carbon dioxide is produced during the TCA cycle. The process occurs in the glyoxysome, which is a specialized organelle that carries out both the oxidation of fatty acids to acetyl-CoA and the use of acetyl-CoA in the glyoxylate pathway (Zhou et al., 2010).

3.1.2 *Source of Reducing Equivalent—Comeau–Wentzel Model Versus Mino Model and Experimental Evidence*

Initially, two biochemical models were proposed to explain the origin of the reducing equivalents produced anaerobically by PAOs. The Comeau–Wentzel model (Comeau et al., 1986; Wentzel et al., 1986) assumed that the reducing equivalents needed to reduce acetyl-CoA to PHB were produced by the TCA cycle; it was also suggested that this pathway would provide sufficient NADH (formed from a metabolic reaction of NAD to compensate for the amount used by PHB synthesis). Later, Mino et al. (1987) proposed that intracellular carbohydrate and glycogen consumption happens through glycolysis, which was reflected by the decrease in total carbohydrate concentration under anaerobic conditions. Thus, intracellular carbohydrate was proposed to be the supplier of reducing power and the provider of additional acetyl-CoA for PHB synthesis. Then, Smolders et al. (1994a) developed the anaerobic metabolic model based on the Mino model and demonstrated that the anaerobic stoichiometry could be successfully explained by the metabolism based on intracellular glycogen. Since then, glycolysis is generally accepted to be the main source of reducing power for anaerobic conversion of VFAs to PHAs by PAOs.

Involvement of glycogen in EBPR was later confirmed by Pereira et al. (1996) and Maurer et al. (1997) via nuclear magnetic resonance (NMR) techniques. Perreira et al. (1996) also suggested that the TCA cycle was functioning anaerobically since being labeled as carbon dioxide, derived directly from labeled acetate, and detected during the anaerobic phase and that the reducing power supplied by glycolysis was estimated to be inadequate to explain the total amount required for PHA production. Recently, Zhou et al. (2009) used an EBPR reactor highly enriched with *Acccumulibacter* PAOs and demonstrated with 13C-labeled acetate that *Accumulibacter* can use both glycogen and TCA cycle for anaerobic-reducing power generation. Pijuan et al. (2008) compared the anaerobic stoichiometry from full-scale EBPR WRRFs with the predictive models from other researchers and concluded that EBPR metabolic models should incorporate both glycolysis and TCA cycle for the production of reducing power in PAOs when acetate is the substrate, whereas the involvement of TCA cycle was suggested to be low or negligible when propionate is used as the carbon source.

Louie et al. (2000) used metabolic inhibitors in combination with NMR technique to obtain detailed and specific involvement of the TCA cycle reactions.

They found out that succinate oxidation to fumarate by succinate dehydrogenase was a significant step in the PHA synthesis pathway, which requires a terminal electron acceptor with a greater redox potential than the fumarate-succinate couple. This was considered to be the reason that the complete TCA cycle could not be functional anaerobically (Mino et al., 1998). This problem was later addressed by Hesselmann et al. (2000), who proposed a modified succinate-propionate pathway combined with an incomplete TCA cycle in which some acetyl-CoA proceeds through the oxidative pathway of the TCA cycle (forward) and later gets converted to propionyl-CoA via the methylmalonyl-CoA pathway, while a portion of pyruvate also proceeds through the reductive pathway of the TCA cycle via oxaloacetate. The involvement of partial TCA cycle was also supported by Brdjanovic, Slamet, van Loosdrecht, Hooijmans, Alaerts, and Heijnen (1998) and Yagci et al. (2003), although they suggested the activity of the glyoxylate pathway.

A metagenomic study by Martin et al. (2006) with enriched *Accumulibacter* PAOs revealed the presence of functional enzymes of both full TCA cycle and split TCA cycle in addition to those of glycolysis. A proteomics study by Wilmes et al. (2008) suggested that the activity of split TCA cycle or glycoxylate pathway in PAOs could be enabled anaerobically.

3.1.3 *Glycolytic Pathway Followed by Polyphosphate-Accumulating Organisms*

Exactly which pathway is used to convert glycogen to pyruvate has been a matter of question. Different research groups have tried to identify whether PAOs use the Embden–Meyerhof–Parnas (EMP) pathway or the Entner–Doudoroff pathway for glycolysis. The former pathway produces 3 mol of ATP/mol of glycosyl unit (monomeric unit of glycogen) degraded, whereas the latter pathway produces 2 mol of ATP (Mino et al., 1998). Previous results based on NMR techniques have suggested the Entner–Doudoroff (Hesselman et al., 2000; Maurer et al., 1997) and EMP (Erdal, 2002) pathways.

The polyphosphate kinase gene (ppk1) has also been used as a genetic marker to study the diversity of *Accumulibacter*; based on this gene, two main types (Type I and Type II) can be identified. Each type is comprised of several distinct clades (He et al., 2007). Flowers et al. (2009) suggested that Clade IIA can use nitrite generated by Clade IA as a denitrification intermediate for phosphorus uptake because genes enabling nitrite reduction to N_2 were found in the genome of Clade IIA (Martin et al., 2006). This metagenomics study demonstrated that *Accumulibacter* Clade IIA contained genes of the EMP, but the key genes of the Entner–Doudoroff pathway were not present. Through a study of metaproteomics, Wexler et al. (2009) also showed that EMP enzymes were present in Type I and Type II *Accumulibacter* sludge. These contrasting findings suggest that different strains of *Accumulibacter* could possess and thus manifest different metabolic behaviors given that *Accumulibacter* is the dominant type of PAO found in both full-scale and laboratory-scale EBPR systems.

3.2 Metabolic Pathways for Polyphosphate-Accumulating Organisms in the Aerobic Phase

Polyphosphate-accumulating organisms grow aerobically on the anaerobically stored PHAs. Then, stored PHAs are used as the carbon and energy source to replenish the polyphosphate and glycogen pools. Consequently, intracellular polyphosphate and glycogen increases, stored PHAs decrease, and soluble orthophosphate is taken up from the solution, leading to removal of phosphorus from the solution. These aerobic biotransformations in PAOs are similar in all of the proposed models, in which catabolism proceeds through the TCA cycle. The degradation of PHB and PHV leads to acetyl-CoA, the former entering the TCA at the oxaloacetate level and the latter entering at the succinyl-CoA level. Both are used as the carbon and energy source for biomass growth, while a portion of ATP is used for external phosphate uptake and regeneration of polyphosphate. The original model proposed by Comeau et al. (1986) was modified according to the Mino model to include glycogen formation and, hence, some carbon and energy were assigned for the replenishment of glycogen. Only one structured metabolic model has been established on aerobic metabolism of PAOs by Smolders et al. (1994b), who linked aerobic carbon and phosphorus transformations to the oxygen consumption rate with a system of equations. Growth of PAOs under aerobic conditions is a rather slow process, the maximum growth rate being about $0.04\ h^{-1}$ (Smolders et al., 1995). The fact that significant parts of intracellular PHAs are spent to accumulate polyphosphate and glycogen aerobically and not merely for cell growth in PAO metabolism is a unique characteristic of these organisms (Kortstee et al., 2000).

3.3 Metabolic Pathways for Denitrifying Polyphosphate-Accumulating Organisms in the Anoxic Phase

Under anoxic conditions, where nitrate is available as an electron acceptor, similar aerobic carbon metabolism has been observed (Kuba et al., 1997) because a part of PAOs has been found to use nitrate as an electron acceptor (Kern-Jespersen and Henze, 1993; Kuba et al., 1993). Under anoxic conditions, PAOs also recover intracellular glycogen and replenish polyphosphate by consuming stored PHAs and taking up external orthophosphate (Kuba et al., 1993). However, the energy production efficiency with nitrate expressed in terms of mole of ATP per mole of NADH is estimated to be 40% lower than that with oxygen (Kuba et al., 1994). Therefore, the rate of phosphorus uptake by PAOs under anoxic conditions is generally lower than under aerobic conditions. Consequently, a 20% lower cell yield value was reported for an anaerobic–anoxic EBPR process than for an anaerobic–aerobic EBPR process (Murnleitner et al., 1997).

3.4 Metabolic Pathways for Glycogen-Accumulating Organisms in the Anaerobic Phase

Another population group that has been found to be relevant and to have an effect on EBPR process performance are GAOs (Liu, Mino, Nakamura, and Matsuo, 1996). Similar to PAOs, GAOs can uptake VFAs in anaerobic condition for PHA formation while using glycogen as the primary source of energy rather than polyphosphate cleavage like PAOs. Glycogen-accumulating organisms can also oxidize PHAs aerobically, leading to biomass growth and glycogen replenishment as PAOs. Because GAOs compete with PAOs for VFAs without contributing to phosphorus removal, their presence may affect EBPR systems (Satoh et al., 1994). Different studies have highlighted several factors such as pH (Filipe et al., 2001b; Oehmen, Vives, Lu, Yuan, and Keller, 2005), temperature (Lopez-Vazquez et al., 2008; Whang and Park, 2006), substrate type (Oehmen et al., 2006; Oehmen, Yuan, Blackall, and Keller, 2005; Pijuan et al., 2004), and influent carbon-to-phosphorus ratio (C/P) (Liu et al., 1997) that could potentially favor PAOs over GAOs if they are optimally incorporated. However, the combined effects of specified factors on the competition between PAOs and GAOs are still not fully known.

Because glycogen is believed to be the main source of energy for GAOs, it is used in these organisms to a much larger extent than PAOs. Additionally, because of this higher glycogen demand to produce ATP, reducing equivalents are produced in excess of those required for reduction of acetyl-CoA to PHB. Thus, NADH must also be consumed to maintain the redox balance in this process, which was proposed via the reduction of pyruvate to propionyl-CoA (Liu et al., 1994; Satoh et al., 1994). Pyruvate, which is an intermediate of glycolysis, would enter the left branch of the TCA at the oxaloacetate level (Figure 7.4), be converted to succinyl-CoA, and then further converted to propionyl-CoA through the methylmalonyl-CoA pathway. Thus, when acetate is consumed, GAOs produce both acetyl-CoA and propionyl-CoA anaerobically, which leads GAOs to store approximately 75% PHB and 25% PHV (C-mol basis) (Filipe et al., 2001a; Zeng, van Loosdrecht, Yuan, and Keller, 2003) with only a small fraction of poly-3-hydroxy-2-methylvalerate produced (Filipe et al., 2001a; Satoh et al., 1994). It is still unclear if the full TCA cycle is active in GAOs for anaerobic metabolism because of the limited number of research that has been carried out on the metabolism of GAOs.

Satoh et al. (1994) hypothesized that GAOs follow the EMP pathway for glycolysis because of the absence of the Entner–Doudoroff specific glucose-6-phosphate-dehydrogenase enzyme in GAO-enriched sludge. However, NMR techniques by Lemos et al. (2007) also implicated the Entner–Doudoroff pathway through which partial carbon cycling occurs for gluconeogenesis. Saunders et al. (2007) studied the anaerobic VFA uptake by GAOs and observed that GAOs generate proton motive force by a combination of the efflux of protons through ATPase, at the expense of ATP, and the reductive TCA cycle enzyme, fumarate-reductase. The aerobic metabolism of GAOs is similar to that of PAOs, except that there is no polyphosphate accumulation like PAOs. The degradation of both PHB and PHV produces the energy required for biomass growth, glycogen replenishment, and cell maintenance (Filipe et al., 2001a; Zeng, van Loosdrecht, Yuan, and Keller, 2003).

Although it is generally accepted that PAOs and GAOs are different organisms (Mino et al., 1998), it has not been definitively proven that the theoretical metabolic pathways of PAOs and GAOs are components of a metabolism in one group of organisms or unique metabolisms in separate groups, especially because the majority of the pathways of carbon transformations are similar and involve intracellular storage polymers. Table 7.1 summarizes the reactions and processes

TABLE 7.1 Reactions or processes involved in anaerobic, anoxic, and aerobic phases of EBPR and the respective populations involved.

	Anaerobic	Anoxic	Aerobic
Reactions	e^- donor—acetate	e^- donor—acetate/PHB	e^- donor—acetate/PHB
Populations	e^- acceptor—none	e^- acceptor—nitrate/ nitrite	e^- acceptor—NO_3/O_2
1	Acetate uptake *PAO, GAO*	Acetate uptake *PAO, GAO, DPAO, denitrifiers, heterotrophs*	Acetate uptake *PAO, GAO, DPAO, heterotrophs*
2	PHB formation *PAO, GAO*	PHB use/formation *PAO, GAO, DPAO*	PHB use *PAO, GAO, DPAO*
3	Polyphosphate degradation, orthophosphate release *PAO*	Polyphosphate degradation, orthophosphate release *PAO*	
4	Glycogen use *PAO, GAO*	Glycogen use/formation *PAO, DPAO, GAO*	Glycogen formation *PAO, DPAO, GAO*
5		Polyphosphate formation from PHB, orthophosphate uptake *DPAO*	Polyphosphate formation from PHB, orthophosphate uptake *PAO, DPAO*
6	Fermentation *Heterotrophs/non-PAOs*	Denitrification *DPAO, denitrifiers*	Nitrification *Nitrifiers*
7	Secondary phosphorus release, decay *PAO, GAO*	Secondary phosphorus release, decay *PAO, GAO, DPAO, denitrifiers, heterotrophs*	Secondary phosphorus release, decay *PAO, GAO, DPAO, nitrifiers, denitrifiers, heterotrophs*

involved in anaerobic, anoxic, and aerobic phases in EBPR and the populations that are involved in the respective processes.

3.5 Stoichiometry of the Process

Because isolation of PAOs or GAOs has not been possible so far, metabolic models have been proposed by defining the reaction stoichiometry using assumed biochemical pathways (as described previously). Oehmen et al. (2007) summarized that models establish a set of reactions to describe the metabolism, each with known stoichiometry but unknown reaction rates. The consumption and/or production rate of each compound involved in the reactions either as an initial substrate (e.g., acetate), a final product (e.g., PHAs), or an intermediate (e.g., ATP and NADH) is then expressed as a function of the rates of the reactions involving this compound, with the use of the reaction stoichiometry established. By assuming that the reaction intermediates, including ATP and NADH, do not accumulate in bacterial cells, the substrate consumption and product formation rates are expressed with a minimum number of independent reaction rates. The theoretical stoichiometry is based on the yield coefficient for these independent reaction rates, which are verified by measuring the consumption rates of a set of substrates and the production rates of a set of products. When the number of measured rates is larger than the number of independent reaction rates, equations involving only known stoichiometric coefficients and measured variables are obtained, forming independent checks of theoretical stoichiometry (Oehmen et al., 2007).

Smolders et al. (1994a, 1994b; 1995) described the anaerobic and aerobic stoichiometries of the reactions in PAOs, which are summarized in Tables 7.2 and 7.3.

TABLE 7.2 Anaerobic reactions of PAOs (Smolders et al., 1994a).

No.	Process	Reaction
1	Acetate uptake and storage as PHB	$CH_2O + (0.5+\alpha_1^a)\ ATP + 0.25\ NADH \Rightarrow CH_{1.5}O_{0.5} + H_2O$ Acetate PHB
2	Polyphosphate degradation for ATP production	$HPO_3 + H_2O \Rightarrow \alpha_2^b ATP + H_3PO_4$ Polyphosphate
3a	Reducing power/NADH generation in TCA cycle	$CH_2O + (0.5 + \alpha_1)\ ATP + H_2O \Rightarrow 2\ NADH + CO_2$
3b	Reducing power/NADH generation from degradation of glycogen	$CH_{10/6}O_{5/6} + 0.17\ H_2O \Rightarrow 0.67\ CH_{1.5}O_{0.5} + 0.33\ CO_2 +$ Glycogen $0.5\ NADH + 0.5\ ATP$

[a] α_1 = the amount of ATP required for active transport of acetate across cell membrane; varies between 0 and 0.5 mol ATP/C-mol acetate depending on pH.

[b] α_2 = the amount of ATP required for polyphosphate release, which is typically found as 1 mol ATP/P-mol released.

TABLE 7.3 Aerobic reactions of PAOs (Smolders et al., 1994b).

No.	Process	Reaction
1	Poly-beta-hydroxybutyrate catabolism	$CH_{1.5}O_{0.5} + 1.5\ H_2O \Rightarrow 2.25\ NADH + 0.5\ ATP + CO_2$ PHB
2	Oxidative phosphorylation	$NADH_2 + 0.5\ O_2 \Rightarrow H_2O + \delta^a ATP$
3	Biomass synthesis from PHB	$1.27\ CH_{1.5}O_{0.5} + 0.2\ NH_3 + 0.015\ H_3PO_4 + (K^b + m_{ATP}{}^c/\mu)\ ATP + 0.385\ ATP \Rightarrow CH_{2.09}O_{0.54}N_{0.20}P_{0.15} + 0.615\ NADH_2 + 0.27\ CO_2$
4a	Phosphate transport	$H_3PO_4{}^{out} + 1/\varepsilon^d\ NADH_2 + 1/2\varepsilon\ O_2 \Rightarrow H_3PO_4{}^{in} + 1/\varepsilon\ H_2O$
4b	Polyphosphate synthesis	$H_3PO_4{}^{in} + \alpha_3{}^e\ ATP \Rightarrow HPO_3 + H_2O$ Polyphosphate
5	Glycogen production	$4/3\ CH_{1.5}O_{0.5} + 5/6\ ATP + 5/6\ H_2O \Rightarrow CH_{10/6}O_{5/6} + 1/3\ CO_2 + NADH_2$

[a] δ = the amount of ATP produced per electron pair.
[b] K = polymerization coefficient.
[c] m_{ATP} = maintenance coefficient.
[d] ε = transport coefficient.
[e] α_3 = the amount of ATP required for the synthesis of polyphosphate.

There are experimentally determined, model-predicted, and validated values for all the biochemical coefficients that are included in the tables.

3.6 Enhanced Biological Phosphorus Removal Kinetics

As summarized in Oehmen et al. (2007), reaction kinetics in EBPR are typically modeled using Monod-type kinetics. The measured variables for verifying metabolic models of PAOs and GAOs have typically been the rates of VFA consumption, glycogen consumption and regeneration, PHA production and consumption, phosphate release and uptake, oxygen consumption, and ammonium uptake (Filipe et al., 2001a, 2001b; Smolders et al., 1995; Yagci et al., 2003). In some instances, rates of carbon dioxide and/or proton production have also been used (Oehmen, Zeng, Yuan, and Keller, 2005; Smolders et al., 1994a; Zeng, van Loosdrecht, Yuan, and Keller, 2003). Table 7.4 presents an extract from Oehmen et al. (2010) of the metabolic model predictions for kinetic rates using EMP and Entner–Doudoroff pathways for PAOs. Table 7.5 shows some ranges of the EBPR kinetic rates that previous studies have obtained.

4.0 FACTORS INFLUENCING ENHANCED BIOLOGICAL PHOSPHORUS REMOVAL

In this section, external factors influencing the EBPR process will be discussed. Although initial research mainly focused on aerobic processes, it has gradually

TABLE 7.4 Comparison of anaerobic metabolic model predictions using the EMP and Entner–Doudoroff pathways for PAOs (Oehmen et al., 2010).

Carbon source	Pathway	P/VFA[a] P-mol/C-mol	Glycogen/VFA[b] (C-mol/C-mol)	PHA/VFA[c] (C-mol/C-mol)	CO_2/VFA (C-mol/C-mol)
Acetate	EMP	$0.25 + \alpha_1$	0.5	1.33	0.17
Acetate	Entner–Doudoroff	$0.33 + \alpha_1$	0.5	1.33	0.17
Propionate	EMP	$0.17 + \alpha_1$	0.33	1.22	0.11
Propionate	Entner–Doudoroff	$0.22 + \alpha_1$	0.33	1.22	0.11

[a]Anaerobic phosphorus released with respect to acetate uptake.
[b]Anaerobic glycogen degradation with respect to acetate uptake.
[c]Anaerobic PHA production with respect to acetate uptake.

become clear that good phosphorus removal activity can only be obtained when anaerobic microorganisms are subjected to conditions favoring storage of sufficient carbon sources to be used under aerobic conditions with simultaneous uptake of orthophosphate. First, attention will be focused on carbon sources used by responsible microorganisms and how these carbon sources influence phosphorus release, formation of PHAs, and the influence on overall EBPR performance. Other external factors that can influence EBPR performance are C/P, secondary release, dissolved oxygen, pH, HRT, SRT, presence of nitrate and nitrite, and temperature.

4.1 Short-Chain Volatile Fatty Acids and Non-Short-Chain Volatile Fatty Acids as Carbon Sources

4.1.1 *Effect of Carbon Source on Phosphorus Release*

Availability of carbon is a primary factor for controlling the EBPR process because the most favorable substrate would cause the optimum release of phosphorus in the anaerobic phase. The EBPR process tends to use substantially more influent COD than the conventional activated sludge process, which is likely because EBPR organisms accumulate storage products that require carbon alone for synthesis (Harper and Jenkins, 2003). The form of COD, or the form of carbon that is available to organisms, is another crucial factor determining the performance of the EBPR process. Carbonaceous oxygen demand must have a sufficient portion of short-chain VFAs, or a readily biodegradable fraction that ferments into VFAs in sewer lines or in fermentation units, to achieve EBPR (Randall et al., 1992). Many laboratory-scale EBPR reactors have been successfully operated with acetate as the significant carbon source. If the retention time of wastewater in collection systems is long enough for fermentation, a significant portion of organic matter in the wastewater will be fermented to soluble VFAs (mainly acetic acid) before being transported to the facility (Mino et al., 1998). Thus, acetate is an

Table 7.5 Enhanced biological phosphorus removal kinetic rates as reported by researchers.

Reference	EBPR system	Phosphorus-release rate mg-P/gVSS·h	Phosphorus-uptake rate mg-P/gVSS·h	P/VFA (P-mol/C-mol)	Glycogen/VFA (C-mol/C-mol)	PHA/VFA (C-mol/C-mol)
Smolders et al. (1994a)	Laboratory-scale SBR			0.25–0.75		
Chuang et al. (1996)	Pilot scale	4–18.6 (TSS)				
Liu et al. (1997)	Laboratory-scale SBR	3–155				
Saunders et al. (2003)	6 Australian facilities			0.29–0.51		
Schuler and Jenkins (2003)	Laboratory-scale SBR			0.015–0.93	0.08–1.19	0.79–1.33
Gu et al. (2008)	6 U.S. facilities	5.6–31.9	2.4–9.7	0.27–0.75		
He et al. (2008)	6 U.S. facilities	11.1–49.5	1.9–50.8	0.36–0.76		
Lopez-Vazquez et al. (2008)	7 Dutch facilities	9.6–20.9	6.2–19.2	0.15–0.20		

important substrate for EBPR. Propionate has also been gaining significant interest in recent studies as a complementary substrate in EBPR; certain long-chain VFAs have also been used. Table 7.6 shows the types of VFAs typically observed in fermented wastewater.

4.1.2 *Effect of Carbon Source on Storage Products*

When acetate is the only carbon source available, 3HB represents the significant component in the PHA that is formed (Satoh, 1992; Smolders et al., 1994a); this is then called *PHB*. Experimental studies with PAO- and GAO-enriched sludge have revealed that PAOs produce mainly PHB when fed with acetate, with little PHV production (generally less than 10% [C-mol basis]) (Mino et al., 1998; Satoh, 1992; Smolders et al., 1994a). As shown in Figure 7.6 (Satoh et al., 1994), the composition of PHAs for both acetate and propionate uptake and the fractions include hydroxy butyrate, hydroxy methyl butyrate, hydroxyvalerate, and hydroxy methyl valerate.

4.1.3 *Effect of Carbon Source on Enhanced Biological Phosphorus Removal Performance*

While carbon availability in the form of readily biodegradable COD or VFAs is critical to successful EBPR, carbon type, particularly relative to process stability, is also important (Gebremariam et al., 2011). Choice of substrate or carbon source is critical to EBPR process economics as influenced by the cost of the carbon source and selective use of the carbon source of PAOs against GAOs (Puig et al., 2008). Metabolism of GAOs was first reported by Cech and Hartman (1990), and the main experimental variable in their study, leading to the discovery of GAOs, was feed composition. Glycogen-accumulating organisms were found to be developed in a reactor fed with acetate and glucose, whereas no GAOs were found to be present in the reactor fed with acetate only. This implied that glucose is a preferred substrate for GAOs. Acetate-fed EBPR systems have demonstrated good EBPR performance with PAO enrichment; however, proliferation of GAOs

TABLE 7.6 Volatile fatty acids typically found in fermented wastewater.

Volatile fatty acid	Chemical formula	Phosphorus uptake/ VFA COD consumed
Acetic acid	CH_3-COOH	0.37
Propionic acid	CH_3-CH_3-COOH	0.10
Butyric acid	CH_3-CH_3-CH_3-COOH	0.12
Isobutyric acid	CH_3-CH_3-COOH-CH_3	0.14
Valeric acid	CH_3-CH_3-CH_3-CH_3-COOH	0.15
Isovaleric acid	CH_3-CH_3-COOH-CH_3-CH_3	0.24

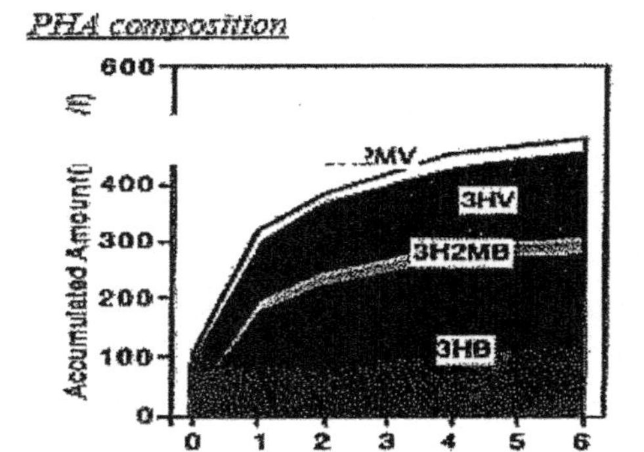

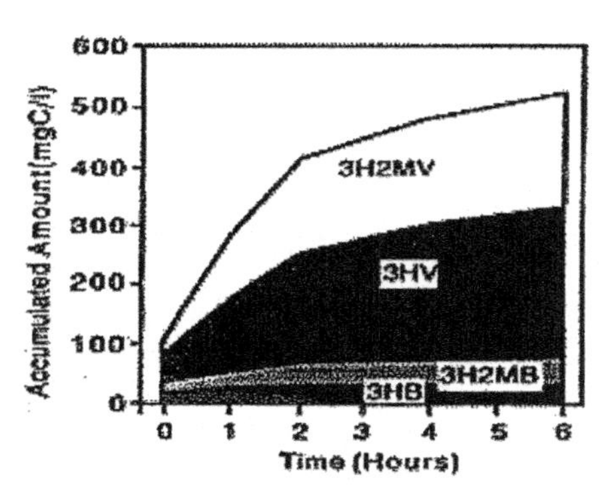

FIGURE 7.6 Polyhydroxyalkanoate composition for acetate update (left) and propionate uptake (right).

is possible with acetate as the substrate if other operational conditions (e.g., higher COD-to-phosphorus ratio [COD/P] and lower pH) become favorable for GAOs.

Although acetate is the most prevalent VFA that is used as substrate in EBPR facilities, propionate can also be present in substantial quantities (up to 45% of the total) (Burow et al., 2007). Most EBPR studies use acetate as the substrate; however, there has been an increased interest in the effect of propionate on EBPR performance. A number of laboratory studies have used a strategy of alternative feed between acetate and propionate to obtain highly enriched PAO cultures as those evidences suggest that *Competibacter* GAOs are less competitive with PAOs for propionate uptake compared to acetate uptake alone (Saunders et al., 2007; Zhang et al., 2007; Zhou et al., 2008). Thus, propionate is commonly accepted as an efficient substrate to select against GAOs, especially *Competibacter*. Thomas (2008) reported increased stability of a full-scale facility in Australia when propionate was supplied in an adequate amount relative to acetate. However, this strategy was contradicted by later studies in which some *Alphaproteobacteria* GAOs were observed to be more efficient in propionate uptake compared to acetate uptake (Kong et al., 2006; Oehmen, Yuan, Blackall, and Keller, 2005).

4.2 Effect of Influent Carbon-to-Phosphorus Ratio

4.2.1 *Effect of Influent Carbon-to-Phosphorus Ratio on Enhanced Biological Phosphorus Removal Populations*

For the removal of each phosphorus unit, there is a stoichiometric requirement of COD. Above this required carbon loading, at a higher than stoichiometrically required COD-to-phosphorus ratio, an advantage is provided to the non-phosphorus-removing population (GAOs). Previous studies on sequencing batch reactors (SBRs) showed that higher ratios of influent COD to phosphorus create a phosphorus-limiting condition, thus favoring GAO proliferation. Oehmen et al. (2007) stated that PAOs tend to dominate at a COD/P of 10 to 20 mg COD/mg P, whereas GAOs tend to dominate at a COD/P greater than 50 mg COD/mg P. In the intermediate levels of COD/P, both populations can coexist. Gu et al. (2008) postulated that, to a certain extent, this coexistence of GAOs with PAOs might be beneficial to the EBPR process in that GAOs could potentially act as buffer against higher carbon loading and prevent the system from deteriorating. However, the presence of GAOs also presents the system with such vulnerability that, because of some operational fluctuations, if the overall conditions become favorable to GAOs, the system might collapse. Thus, maintenance of optimum COD/P and good control over operational conditions are required to use the competition for substrate between PAOs and GAOs to provide a positive outcome. Any operational condition that results in a reduction of COD before the biological treatment system must be closely monitored to allow an appropriate COD/P in the EBPR system's feed. Seasonal variations in biochemical oxygen demand (BOD)-to-COD ratios and wastewater VFA content must be closely investigated as part of the wastewater characterization preceding EBPR design and as part of the process control strategy of an operating facility.

4.2.2 *Effect of Influent Carbon-to-Phosphorus Ratio on Enhanced Biological Phosphorus Removal Performance*

Municipal wastewater fermented in the collection system is generally a good source of VFAs for EBPR operation. Advanced primary treatment practiced at some treatment facilities must be looked at carefully because, at facilities where EBPR is going to be implemented, adequate quantities of organic material must be supplied to support PAO functions. In some instances where sufficient carbon substrate is not available, carbon augmentation of biological reactor influent is practiced. Facility recycles also have significant bearing in EBPR systems operation because they can also contain some VFAs that can be sufficient to ameliorate the phosphorus functions. The influent COD/P or BOD/P (influent COD/P or influent C/P) is crucial for proper design and operation of phosphorus removal systems. Whether a system is limited by COD (or BOD) or phosphorus determines the extent to which PAOs can function, and the amount of excess phosphorus can be taken up from the solution. Previous research has shown that the influent COD/P correlated well with the EBPR biomass total phosphorus content and phosphorus removal functions (Kisoglu et al., 2000; Liu et al., 1997; Schuler and Jenkins, 2003). Because EBPR systems rely on enrichment of PAOs to be able to select the phosphorus accumulators, the system must be fed with at least a minimum amount of phosphorus that can be taken up in correlation with the amount of substrate COD available in the feed. Figures 7.7 and 7.8 show the effect of the influent COD/P on mixed liquor PAO enrichment and sludge phosphorus content.

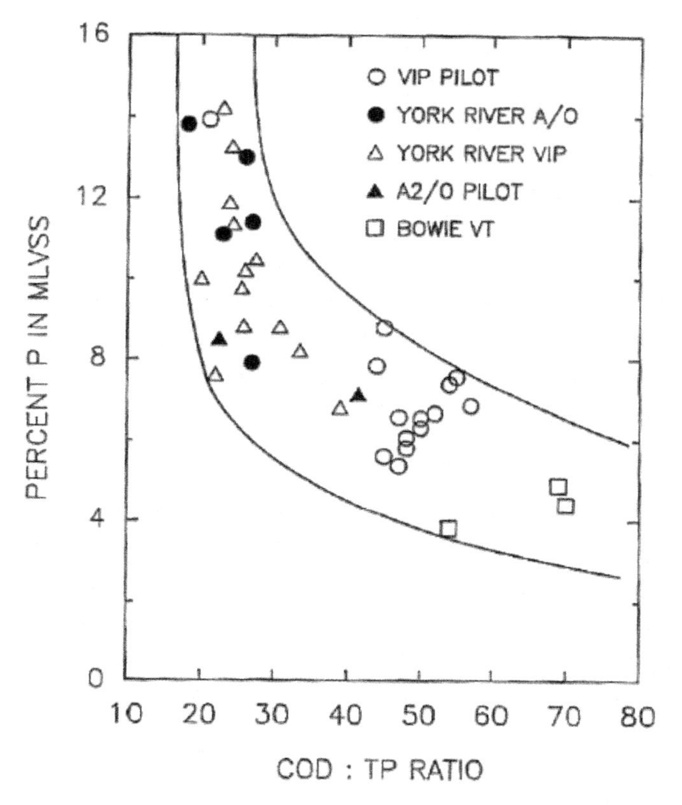

FIGURE 7.7 Effect of COD-to-total-phosphorus ratio on mixed liquor PAO enrichment and phosphorus storage at different EBPR facilities.

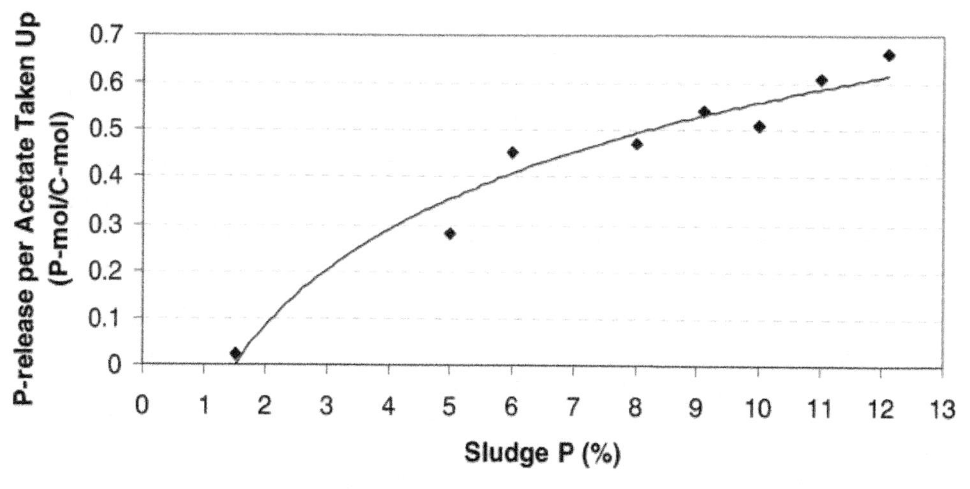

FIGURE 7.8 Data collected by Liu et al. (1997) using pilot-scale EBPR facilities fed with VFAs at various feed COD-to-phosphorus ratios.

4.3 Secondary Release of Phosphorus

As described in previous sections, release of phosphorus under anaerobic conditions with simultaneous uptake of acetate, and its storage as PHB, is called *primary release* (Barnard, 1984). However, phosphorus is also released from stored polyphosphate under anaerobic conditions even in the absence, or inadequate presence, of VFAs for the microorganisms to derive sufficient energy for maintaining their metabolic activities. This is called *secondary release* of phosphorus (Barnard and Scruggs, 2003). Secondary release takes place at a much slower rate than primary release. Phosphorus resulting from the secondary release mechanism is not removed under a subsequent aerobic or anoxic environment because of the unavailability of stored PHB. Therefore, secondary release should be avoided as much as possible because it adversely affects the performance of an EBPR process.

A common location for secondary release of phosphorus in an EBPR facility is in the secondary clarifier(s). If the sludge blanket in a secondary clarifier is deep, or the sludge is not removed from the clarifier bottom on a regular basis, the resulting anaerobic environment leads to secondary release.

Another possible location is in the upfront anaerobic compartment of an EBPR activated sludge facility. If the anaerobic compartment volume is too large and HRT is much longer than the time needed for fermentation of the incoming COD, then secondary release can take place in the anaerobic compartment. On the other hand, if the compartment volume is too small, then the incoming COD may not have adequate HRT for fermentation to acetic acid, which is a precursor to the formation of PHB as the storage product. In this instance, the stored PHB may not be adequate to supply the energy necessary for subsequent aerobic uptake of orthophosphate over and above that released by the primary mechanism. An inadequate amount of PHB limits the capacity of the PAO cells to uptake all the released orthophosphate in the subsequent aerobic or anoxic compartment.

Organic matter contained in wastewater can get ample opportunity for fermentation when it is conveyed to the facility through a large collection system that provides a long retention time. However, the effectiveness of fermentation depends on a number of factors, including temperature, wastewater characteristics, and characteristics of the sewer system (refer to Section 5.2 for a detailed discussion).

Conversely, fresh wastewater may not be completely fermented if the combined retention time in the collection system and the anaerobic compartment is inadequate for complete fermentation of incoming COD. This was experienced in several facilities in South Africa during the early stages of development of the EBPR process (Uys, 1984). Such situations may be resolved by providing a prefermenter to process the settled primary solids and adding the resulting VFAs to the anaerobic compartment. Some facilities with inadequate influent VFAs also resort to direct addition of acetic acid in the anaerobic compartment.

4.4 Influence of External pH

The pH affects EBPR activity. At a pH lower than 5.4, EBPR activity is completely inhibited, while EBPR mechanisms can operate in the pH range of 8.5 to 9, according to Reddy (1998). Recent research by Filipe et al. (2001a, 2001b) documented an optimum pH range of 7.0 to 7.5 for PAOs, with decreasing activity at lower pH values. In the following subsections, the influence of pH on anaerobic and aerobic reactions and on the overall EBPR process is presented.

4.4.1 *Effect of pH on Anaerobic Reactions*

Smolders et al. (1995) observed a direct linear relationship between pH and phosphorus release to the acetate-uptake ratio in EBPR systems (Figure 7.9). In their study, the phosphorus release acetate-uptake ratio was found to increase from 0.25 to 0.75 mol-P/mol-C when pH increased from 5.5 to 8.2. However, while an increase in phosphorus release and the phosphorus release rate with increasing pH was observed, the acetate-uptake rate showed no pH dependency (Figure 7.10).

In reality, a pH of less than 6 should never be allowed to occur in a biological nutrient removal (BNR) facility. Liu, Mino, Matsuo, and Nakamura (1996) observed a more complex pH dependency of the acetate-uptake rate. Between pH 5 and 6.5, the acetate-uptake rate increased linearly from 0 to about 50 mg C/g volatile suspended solids (VSS)/h. Between pH 6.5 and 8, no pH dependency was observed, which was in accordance with Smolders et al. (1994a). Above pH 8, the acetate-uptake rate started to decrease. Release rates significantly lower than those

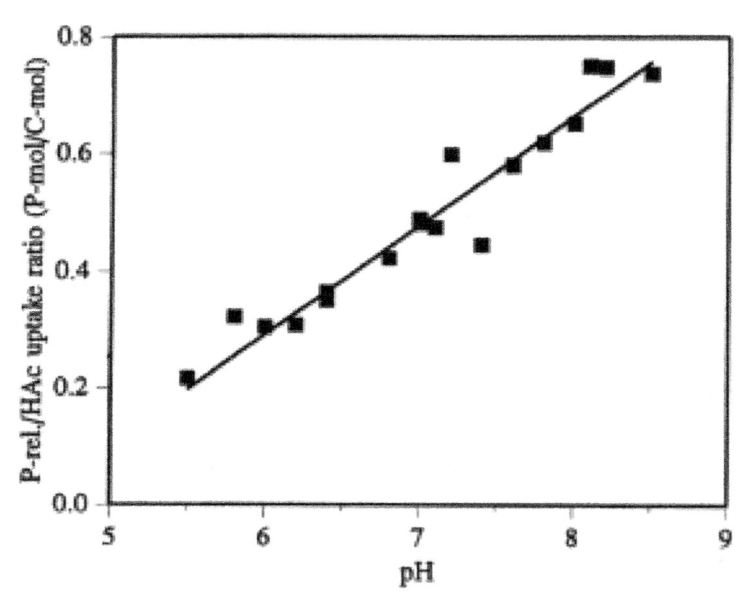

FIGURE 7.9 Phosphate release to acetate-uptake ratio as a function of pH (Smolders, 1995).

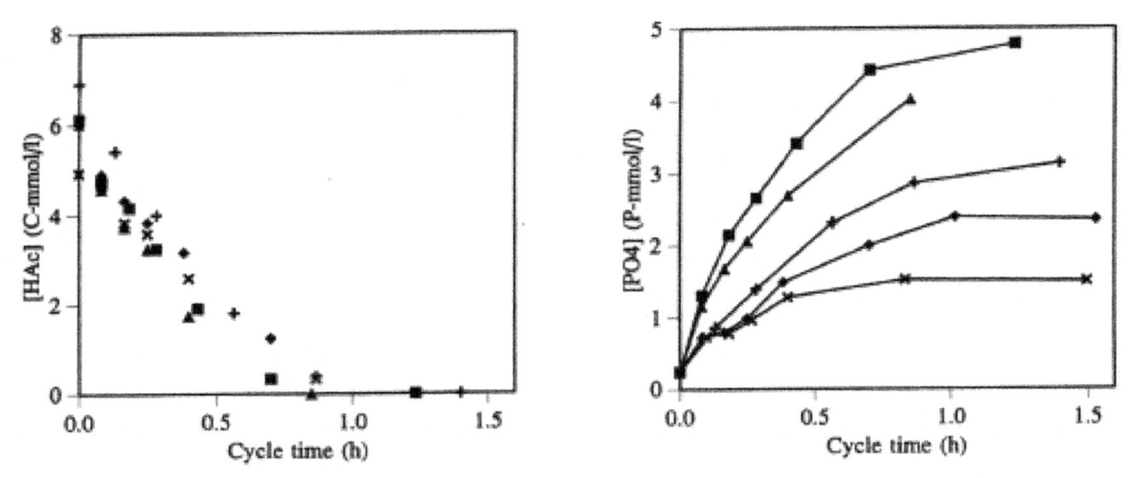

FIGURE 7.10 Acetate uptake and phosphorus release at different pH values. Average initial acetate concentration 6 C-mmol/L, MLSS 3.2 g/L, VSS 2.2 g/L, pH 5.8 (×), pH 6.4 (◆), pH 7 (+), pH 8.2 (▲), and pH 8.6 (■) (Smolders, 1995).

reported by Smolders et al. (1994a) were observed for orthophosphate. Between pH 5.0 and 6.5, the orthophosphate release rate ranged from 20 to about 50 mg P/g VSS/h; higher than pH 6.5, this value continuously increased. From pH 8.5 and higher, a decrease in orthophosphate release rate was observed (Figure 7.11).

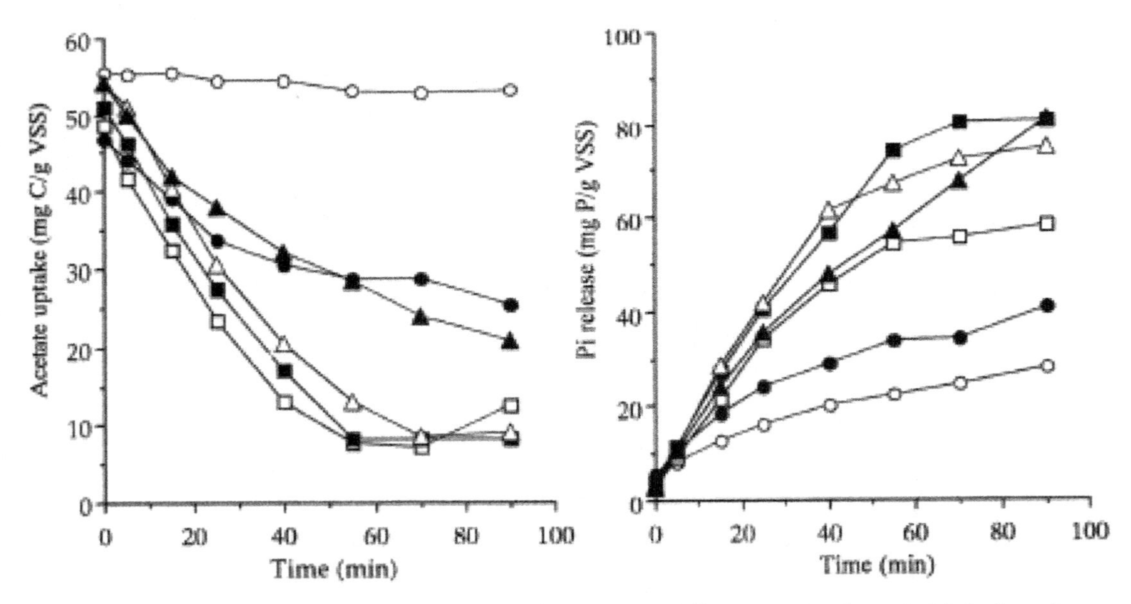

FIGURE 7.11 Acetate uptake and phosphorus release at different pH values. Initial phosphorus content of sludge 12 w%, pH 5 (○), pH 5.7 (●), pH 6.5 (□), pH 7.1 (■), pH 7.8 (△), and pH 8.6 (▲) (Liu, Mino, Matsuo, and Nakamura, 1996).

More recent studies have shown that the rate of acetate uptake is independent of pH between pH 6.5 and 8, which is the case with PAOs (in accordance with both Smolders et al. [1994a] and Liu, Mino, Matsuo, and Nakamura [1996]), but strongly dependent on pH for GAOs. The rate of acetate uptake by GAOs is significantly decreased when the pH of the medium is increased (Filipe et al., 2001a, 2001b).

4.4.2 *Effect of pH on Aerobic Reactions*

Aerobically, a series of batch tests has shown that phosphorus uptake, PHA use, and biomass growth were all inhibited by a low pH (6.5). The batch tests suggested that a higher aerobic pH (7 to 7.5) would be more beneficial for PAOs (Filipe et al., 2001a).

4.4.3 *Effect of pH on Enhanced Biological Phosphorus Removal Performance*

Lower than pH 6.9, the EBPR process has been shown to decline in efficiency (WEF et al., 2005). This may be attributable to competition with GAOs. Filipe et al. (2001b) found that GAOs grow faster than PAOs at a pH of less than 7.25. This is attributable to the rate at which PAOs and GAOs can take up VFAs under anaerobic conditions; that is, GAOs take up acetate faster at a pH lower than 7.25, whereas PAOs are faster at a pH higher than this value (Filipe et al., 2001b). At higher pH, the energy needed for acetate uptake is higher; therefore, PAOs, having an extra energy source in polyphosphorus compared to GAOs, can use it to meet this higher energy demand and successfully outcompete GAOs for VFAs (Filipe et al., 2001b).

Several studies have, indeed, shown higher phosphorus removal when the anaerobic and/or aerobic pH level was increased (from pH 7 to 7.5 to 8.5) (Schuler and Jenkins, 2002; Serafim et al., 2002). The reason for the improved performance was hypothesized to be from a shift in microbial competition from GAOs to PAOs. This hypothesis has been supported by assessing population changes in the microbial community. In a study by Zhang et al. (2005), a complete loss of phosphate-removing organisms was observed when the pH was decreased from 7.0 to 6.5; at the same time, a shift in the microbial population was also observed. With the decrease in pH, the phosphorus removal efficiency went from 99.9% of phosphorus to 17% of phosphorus in just 14 days after the pH was lowered to pH 6.5. Clearly, the results show that pH strongly influences PAO–GAO competition, and suggest pH as a performance parameter that can be manipulated to favor PAOs to optimize phosphorus removal.

In some studies, a decrease in the VFA uptake, phosphorus release, and phosphorus-uptake rates has been observed at pH values above 8.0 (Liu, Mino, Matsuo, and Nakamura, 1996; Oehmen, Vives, Lu, Yuan, and Keller, 2005; Schuler and Jenkins, 2002). Therefore, an upper limit for pH should also be considered.

It should also be noted that the fraction of phosphorus removed via biologically induced chemical precipitation increases with increasing pH (Maurer et al., 1999).

4.5 Influence of Oxidation–Reduction Potential

Use of oxidation–reduction potential (ORP) to determine the aerobic or anaerobic state of wastewater has found widespread application. Measurement of ORP in a wastewater environment is not precise because of the limitation of the electrolytic cells to respond completely to most of the species undergoing biochemical oxidation and reduction reactions. Because of this uncertainty in obtaining precise values, ranges of ORP values have been established that are widely used in wastewater treatment practice as indicators of many important microbiological processes, including EBPR. Typically, ORP values of −50 to −200 mV are accepted as the regime for anaerobic polyphosphate breakdown, whereas +50 to +150 mV is considered as the range for aerobic phosphorus uptake.

4.6 Dissolved Oxygen

Few studies have tried to quantify the effect of dissolved oxygen on EBPR processes. Brdjanovic, van Loosdrecht, Hooijmans, Mino, Alaerts, and Heijnen (1998) demonstrated that when dissolved oxygen is too high, phosphorus uptake stops because of a gradual depletion of PHB. If organic substrate is introduced to the system, phosphorus release takes place; however, the released phosphorus cannot be taken up fully again because the PHB content limits the uptake rate. The authors suggest that this phenomenon might explain the observed deterioration of EBPR activity after heavy rainfall or on weekends. Griffiths et al. (2002) also hypothesized that the dissolved oxygen concentration affects competition between PAOs and GAOs, therefore affecting EBPR performance. In this study, the dissolved oxygen concentration was adjusted in numerous full-scale WRRFs and changes in process performance were identified. The abundance of PAOs and tetrad-forming organisms (TFOs) in the sludge were assessed using staining techniques. Poor phosphorus removal performance and a high number of TFOs were observed more frequently at high dissolved oxygen concentrations of 4.5 to 5.0 mg/L, while dissolved oxygen concentrations of approximately 2.5 to 3.0 mg/L seemed to correlate with a greater abundance of PAOs. When a stream containing high dissolved oxygen (i.e., 5 mg/L) is returned from the aerobic zone to the anaerobic zone, it may adversely affect EBPR performance regardless of PAO and GAO competition.

Further study is necessary to clarify the effect of dissolved oxygen concentration on PAO–GAO competition. This is increasingly relevant because there is a general trend today to operate full-scale BNR facilities at lower dissolved oxygen concentrations (0.5 to 1.5 mg/L) to both reduce aeration costs and improve nitrogen removal through simultaneous nitrification and denitrification.

4.7 Temperature

Effects of temperature on the efficiency and kinetics of EBPR systems have been investigated for the past few decades, with studies yielding contradictory results. Early researchers such as Barnard et al. (1985) and Ekama et al. (1984) reported that EBPR efficiency was unchanged over the range of 5 to 24 °C. Other studies by Beatons et al. (1999) and Choi et al. (1998) showed that cold temperatures adversely affect EBPR performance.

Contradictory to previous findings, Helmer and Kunst (1997) and Erdal (2002) reported that, despite slowing reaction rates, EBPR performance can be significantly greater at 5 °C compared to 20 °C. This shows that better system performance can be achieved as a result of reduced competition for substrate in the anaerobic zones and increased population of PAOs. The phosphorus content of EBPR biomass achieved by Erdal (2002) was up to 50% VSS in the end of the aerobic zone. Figure 7.12 illustrates the importance of acclimation and the resulting improved cold temperature operation of an EBPR system.

Good EBPR performance can be achieved as long as total SRT values of 16 and 12 days are provided for 5 and 10 °C, respectively. System performance was not affected for SRTs between 16 and 24 days and 12 and 17 days for 5 and 10°C, respectively. Low temperatures can also lower phosphorus uptake, although this is not an issue in well-operated and properly acclimatized facilities (WEF et al., 2005).

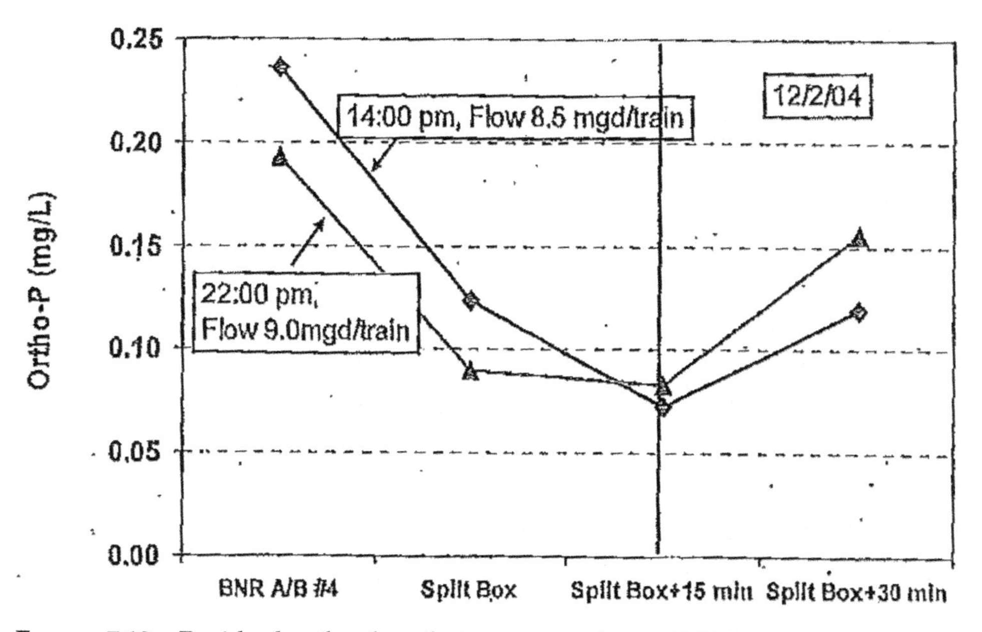

FIGURE 7.12 Residual orthophosphate measured at a BNR basin, downstream split box, and after it was further aerated for 30 minutes via laboratory aeration testing (Gu et al., 2006).

High SRT operations increase endogenous glycogen use, thereby consuming the available reducing power used for PHA formation in anaerobic stages. Glycogen metabolism was found to be the most rate-limiting step in EBPR biochemistry at temperatures below 15 °C. The pilot EBPR systems removed phosphorus until the complete shutdown of glycogen use and replenishment was observed. Despite the presence of available energy sources (polyphosphate and PHA), shutdown of the glycogen metabolism was the significant reason for washout to occur. The shutdown of glycogen use through the anaerobic stage in washout SRTs prevented acetate use and PHA formation. While PAOs wash out of the system, ordinary heterotrophs can continue to grow, using the acetate passing through the anaerobic stages unconsumed and into aerobic stages.

High temperatures adversely affect phosphorus removal because of the increasing effect of GAOs. Modeling studies have shown that GAOs can predominate at higher temperatures because of their increased ability to uptake acetate at those temperatures compared to PAOs (Whang et al., 2002). At temperatures greater than 28 °C, phosphorus removal will generally be impaired, apparently by the predominance of GAOs.

Li et al. (2010) investigated temperature effects on intracellular absorption and extracellular phosphorus removal by extracellular polymeric substances (EPS) in EBPR processes with the help of a laboratory-scale SBR. Studies conducted at temperatures of 5, 15, and 25 °C revealed that lower temperatures were favorable to removal by EPS, primarily because of precipitation as magnesium phosphate, the maximum of which occurred at 5 °C. Maximum intracellular absorption was observed at 15 °C.

Based on the aforementioned studies, a general conclusion can be made that higher temperatures are detrimental to EBPR performance, primarily because of a shift in population from PAOs to GAOs. The EBPR process is favored by, or is indifferent to, low temperatures down to 5 °C.

4.8 Solids Retention Time

Solids retention time is one parameter that can influence EBPR performance and the competition between PAOs and GAOs; however, only a few studies have explored the relationship between SRT and EBPR performance.

It has been demonstrated that an increase in SRT can lead to the decrease of biomass yield and excess sludge discharged, which reduces the phosphorus removed by discharging excess sludge (U.S. EPA, 1987). However, Randall et al. (1992) found that phosphorus content in biomass increases, but that phosphorus removal efficiency does not change as SRT increases. Tremblay et al. (1999) also addressed the fact that PAOs take predominant roles in EBPR systems at a long SRT. Additionally, Barnard (1983) reported that SRT plays a smaller role in phosphorus removal in practice than is expected in EBPR. Mamais and Jenkins (1992) showed that there is a washout SRT for all temperatures higher than 10 to 30 °C.

Erdal et al. (2003, 2006) investigated mechanisms leading to washout or cessation of EBPR activity before other heterotrophic functions halt. After examining underlying biochemical methods, they showed that the main effect of system SRT in EBPR systems is on PHA and glycogen polymerization reactions.

A more recent study by Li et al. (2008) reported performance degradation and worse sludge settleability when the SRT was increased from 8 to 16 days. In summary, there are still a lot of contradictions about the effect of SRT on EBPR performance.

There are also few studies that have tried to determine the effect of SRT on the competition between PAOs and GAOs. Seviour et al. (2003) reported that GAOs could successfully compete with PAOs at a long SRT, which resulted in the decrease of phosphorus removal in the EBPR system. Rodrigo et al. (1999) concluded that shorter SRTs are beneficial for PAOs after observing that the EBPR biomass activity decreased as the SRT was extended, suggesting that GAOs may tend to dominate at longer SRTs.

In an acetate-fed laboratory-scale reactor operated at 30 °C and pH 7.5, Whang and Park (2006) observed a switch in the dominant microbial population from an enriched-GAO to an enriched-PAO culture when lowering the applied SRT from 10 to 3 days. Through a model-based analysis, Whang et al. (2007) inferred that, under the operating conditions applied by Whang and Park (2006), GAOs had a lower net biomass growth rate than PAOs and, therefore, were outcompeted after the SRT was shortened. However, those studies do not provide further details about the effect of SRT on the population dynamics and microbial and biochemical mechanisms involved. Moreover, considering the temperature of 30 °C applied by Whang and Park (2006), their observations may not represent the scenario at ambient temperatures similar to those at full-scale facilities. Although studies regarding the minimum anaerobic and aerobic SRTs of PAOs are available in literature (Brdjanovic, Van Loosdrecht, Hooijmans, Alaerts, and Heijnen, 1998; Matsuo, 1994), no data concerning the effect of SRT on GAO cultures have been reported. However, Onnis-Hayden et al. (2011) recently reported the low abundance and absence of the most commonly found GAOs (*Competibacter*-type) in a full-scale system operating at an SRT of less than 4 days.

4.9 Hydraulic Retention Time

Both anaerobic and aerobic retention times affect phosphorus uptake and storage by PAOs. Sufficient time should be allowed for formation of the VFA and its storage as PHA, with simultaneous release of orthophosphate. If the anaerobic retention time is too short, the phosphorus uptake in the aerobic zone will be lower than achievable because the stored PHA in the anaerobic zone is insufficient. The effects of HRT in both anaerobic and aerobic zones are discussed in the following subsections.

4.9.1 *Anaerobic Hydraulic Retention Time*

Excessive retention in an anaerobic environment gives rise to secondary release of phosphorus. Optimum anaerobic HRT is typically between 0.25 and 1 hour (Barnard and Fothergill, 1998). The hypothesis of secondary release and the effect of anaerobic HRT on this phenomenon were supported by the findings of Stephens and Stensel (1998) based on results of bench-scale SBR tests. However, results obtained from a recent study by Coats et al. (2011) contradicted the earlier findings. The authors conducted tests on bench-scale SBRs consisting of feed, anaerobic, aerobic, settle, and decant steps in each operating cycle. The wastewater was obtained from the influent of a Moscow, Idaho, WRRF. Settled primary sludge was fermented externally to feed the SBR reactors as the source of VFA. The feed consisted of 90% wastewater and 10% fermented liquor. Three reactors were operated with an overall HRT of 12 hours, with an anaerobic HRT of 1, 2, and 3 hours. Coats et al. (2011) observed that the rate of phosphorus release decreased with increasing anaerobic HRT, with the reactor having 3-hour anaerobic HRT exhibiting about half of the rate of the reactor with 1-hour HRT. During the subsequent aerobic step, however, complete removal of phosphorus was achieved by each of the three SBRs. Interestingly, the phosphorus uptake by the reactors with 2- and 3-hour anaerobic HRT were faster (less than 30 minutes) compared to the one with 1-hour anaerobic HRT (more than 30 minutes). Each reactor completely consumed the VFAs and demonstrated no secondary release of phosphorus.

Based on this study, the authors concluded that long anaerobic HRT does not necessarily adversely affect EBPR performance. Additionally, longer anaerobic HRT enriches the concentration of PAOs in the microbial population. These results contradict observations made by earlier researchers and, therefore, warrant further study.

4.9.2 *Aerobic Hydraulic Retention Time*

Enhanced biological phosphorus removal performance is a function of aerobic HRT. Although long aeration time facilitates uptake of orthophosphate by PAOs, excessive aeration time can be detrimental because it induces secondary release (Stephens and Stensel, 1998). In a full-scale study conducted at the Las Vegas Water Pollution Control Facility (Las Vegas, Nevada), Gu et al. (2006) investigated the effect of HRT on soluble phosphorus concentration in the treated effluent from the BNR trains. The authors observed significantly lower (about half, or less) phosphorus concentration at the secondary clarifier effluent compared to the aeration basin effluent. By extending the aerobic HRT by 15 minutes downstream of the secondary clarifier splitter box, an even lower orthophosphate concentration (less than 0.10 mg/L) could be achieved in the final effluent. However, further extension of the aerobic HRT by 30 minutes resulted in a rise of orthophosphate concentration because of the onset of secondary release (Figure 7.12). The extended aeration condition for the secondary clarifier was simulated by aerating the clarifier splitter box samples by 15 and 30 minutes, respectively.

In a bench-scale experimental SBR study, Stephens and Stensel (1998) reported that, after extending the aeration step from 3 hours to 7 hours in a given cycle, phosphorus release and phosphorus uptake were not adversely affected during the same cycle. However, the rate of phosphorus release in the anaerobic step and the rate of uptake in the aerobic step in the subsequent cycle were much slower.

Extending the aerobic HRT beyond a critical point induces secondary release as a result of endogenous decay and cell lysis. Both anaerobic and aerobic HRT affect EBPR performance. Neethling et al. (2005) observed that a ratio of 3 to 4 for aerobic to anaerobic HRT led to optimal EBPR performance.

4.10 Nitrate

4.10.1 Decrease of Enhanced Biological Phosphorus Removal Performance

Poor phosphorus removal under nitrate-rich conditions in the anaerobic zone has been attributed to the disruption of anaerobic conditions by nitrate (Barnard, 1976), consumption of fatty acids by denitrifying non-polyphosphate heterotrophs (Barker and Dold, 1996), and the inhibition of PAOs by nitrite as a result of incomplete denitrification (Jiang et al., 2006; Zhou et al., 2007). Previous studies indicated that there are two types of PAOs based on their ability to use nitrate and/or nitrite as electron acceptors; these are referred to as *DPAOs* as opposed to aerobic PAOs (non-DPAOs) (Kern-Jespersen and Henze, 1993; Kuba et al., 1993). It was suggested that incorporation of an anoxic zone in EBPR can take advantage of DPAOs that can use the intracellular PHA's storage for simultaneous removal of nitrogen and phosphorus, thereby leading to reduced aeration requirements and sludge production of the system (Kuba et al., 1993. However, little is currently known about the factors that govern their abundance and functions.

4.10.2 Proliferation of Denitrifying Polyphosphate-Accumulating Organisms

There is controversy regarding whether DPAOs are different from conventional aerobic PAOs or if DPAOs and PAOs are the same organisms that have the capability to induce different enzymes under different electron acceptor conditions (Gebremariam et al., 2011). A number of researchers have hypothesized the former based on observations of lower phosphorus uptake under anoxic vs aerobic conditions (Freitas et al., 2005; Kerrn-Jespersen and Henze, 1993; Kuba et al., 1993; Shoji et al., 2003; Smolders et al., 1995). A number of earlier studies, through a comparison of EBPR activities (e.g., phosphate-uptake rates) and the abundance of *Accumulibacter*-like PAOs (quantified via FISH methods), indicated that only a fraction of *Accumulibacter* PAOs are able to use nitrate (DPAO phenotype) and that there are likely two different subgroups of PAOs possessing different denitrification capabilities (Kong et al., 2004; Zeng, Saunders, Yuan, Blackall, and Keller, 2003). He et al. (2007) investigated the phylogeny of *Accumulibacter* PAOs with finer-resolution FISH probes and identified different subclusters within the *Accumulibacter* group (differentiated by 16s rRNA nucleotides). A recent study from the same group (Flowers et al., 2009) demonstrated that different subgroups within the *Accumulibacter* PAOs (Clade IA and IIA) have different nitrate reduction capabilities, which was later confirmed by Oehmen et al. (2010). Additional evidence supporting the hypothesis that there are two separate subgroups of *Accumulibacter* that possess different abilities to use nitrate and nitrite as the electron acceptor is the metagenomic study by Martin et al. (2006), which revealed that the particular *Accumulibacter*-like PAOs (*Accumulibacter* Clade IIA) enriched in their laboratory-scale acetate-fed EBPR reactor lacked the gene that encodes for the nitrate reductase enzyme.

Factors that affect the enrichment of DPAOs vs non-DPAOs are not fully understood. Carvalho et al. (2007) and Oehmen et al. (2010) showed that carbon source type may affect the enrichment of DPAOs vs non-DPAOs because they observed different denitrifying performance and identified different *Accumulibacter* clades within two SBR–EBPR systems fed with acetate and propionate as carbon sources, respectively. Another factor that was suggested to possibly affect DPAO activities is the level of carbon source in the anoxic condition, which affects the kinetic competition between heterotrophic denitrifiers and denitrifying PAOs under anoxic conditions (Ahn et al., 2002; Chuang et al., 1996).

4.11 Septage and Sidestreams

Sidestreams consist of return streams from thickeners, sludge tanks (digested and undigested), and dewatering devices. Among these, streams returning after digestion of sludge are the most concentrated in nutrients. Approximately 60% of phosphorus contained in the biomass is released as orthophosphate upon anaerobic digestion (Phillips et al., 2006). In EBPR facilities, where the waste activated sludge (WAS) contains high concentrations of stored phosphorus, these streams can have high phosphorus concentrations. The characteristics of these streams are highly variable (CCNY, 1994; U.S. EPA, 1994). When recycled and blended with influent for treatment as a combined stream, the resulting loads can upset EBPR performance because of altered carbon-to-phosphorus ratios in the influent to the bioreactor system, as described in Section 4.2.

Because of this high variability and the strength of the streams, it is important to control the introduction of these streams in the facility. This can be addressed through an equalization tank, which helps dampen the large fluctuations in concentrations and minimizes any adverse effect on the treatment process, or by treating the combined recycle stream separately to remove phosphorus by a separate sidestream treatment before returning it to the facility. In many instances, sidestreams from several facilities in a region are collected and transported to a larger facility capable of sidestream treatment.

The reduction of phosphorus in a separate sidestream facility in not generally recommended for centrate streams from anaerobically digested processes (Phillips et al., 2006). This is primarily attributable to low concentrations of readily biodegradable chemical oxygen demand (rbCOD) in anaerobically digested sludge because most VFAs are converted to methane in the digester. The VFA concentration of digested sludge centrate may be inadequate to support phosphorus removal by a separate sidestream EBPR process. Chemical precipitation with the help of a ferric chloride ($FeCl_3$) or alum [$Al_2(SO_4)_3$, xH_2O] solution is the preferred option if a separate sidestream treatment is planned for return streams containing large fractions of centrate. Another option is removal of phosphorus as struvite ($Mg \cdot NH_4 \cdot PO_4, 6H_2O$) using processes such as Ostara. Chapter 8 provides a detailed discussion of chemical precipitation.

4.12 Recalcitrant Phosphorus

Dissolved acid hydrolyzable fractions and organic phosphorus fractions of influent phosphorus can adversely affect the effluent soluble phosphorus concentration. Pilot studies at the Westerly Wastewater Treatment Facility in Marlborough, Massachusetts (Lancaster and Madden, 2008), determined that only 35 to 39% of these phosphorus species are removed by EBPR. The remaining part is biologically recalcitrant, which also did not respond to any commercially available physicochemical tertiary treatment process. Based on these findings, the authors recommended that the presence of these species of phosphorus in the influent should be considered while setting practical limits of phosphorus in the treated effluent.

5.0 FERMENTATION

Fermentation of complex organics to short-chain VFA end products provides a key input to EPBR. Effective EPBR requires VFA concentrations as acetic acid that are 5 to 7 times the concentration of the influent phosphorus (as phosphorus). The following subsections summarize critical considerations that must be taken into account to ensure that these fermentation products are available to the EPBR process.

5.1 Fermentation Process Fundamentals

As stated previously, short-chain VFAs are absorbed by PAOs and stored as PHBs within the cell of these microorganisms. Volatile fatty acids are generated from complex particulate and soluble organic molecules during the fermentation process, which consists of the following three steps: hydrolysis, acidogenesis, and acetogenesis. A schematic of the various processes involved in anaerobic metabolism is shown in Figure 7.13.

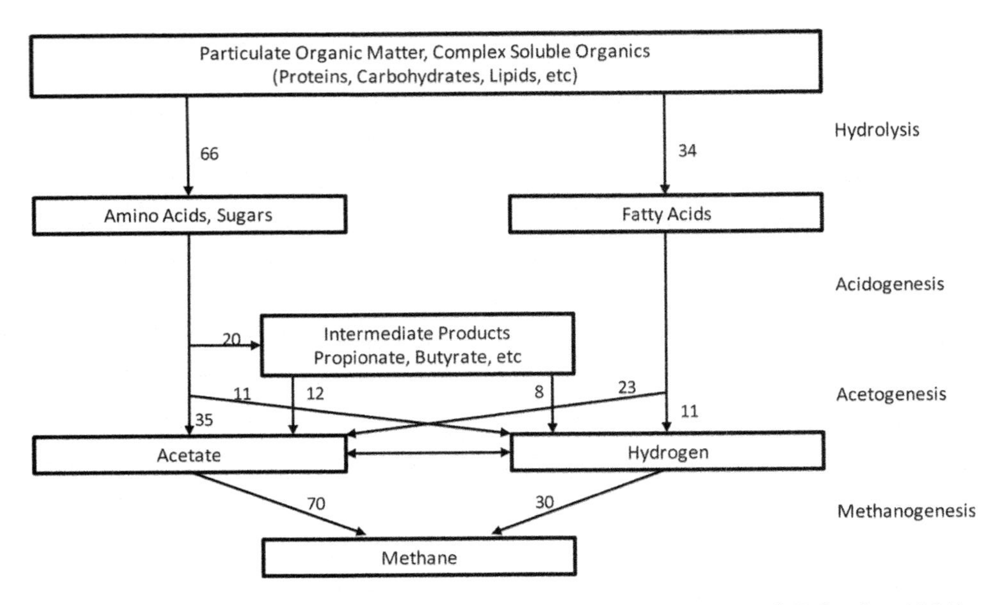

FIGURE 7.13 Four phases of anaerobic fermentation (Gujer and Zehnder, 1983).

5.1.1 *Hydrolysis*

Hydrolysis is an enzyme-mediated reaction in which a chemical bond of the complex organic molecule is cleaved and ionized water (H^+ and OH^-) is inserted to the two ends of the severed bond. Through successive hydrolysis reactions, carbohydrates, lipids, and other carbon-based molecules are ultimately converted to glucose or other similar, shorter-chain organic molecules. Proteins and other nitrogen-based complex organic molecules are hydrolyzed to amino acids.

Hydrolysis is mediated by extracellular enzymes and is the first step of heterotrophic bacteria in preparation for substrate transfer across the cell membrane to support chemoheterotrophic metabolism. These hydrolytic enzymes, or hydrolase, are exerted by microorganisms found external to the surface of heterotrophic bacteria and are critical to cell metabolism. Different hydrolase are specific to the substrate bonds upon which they are effective. Burgess and Pletschke (2008) published a review of the importance of hydrolase in wastewater sludge treatment.

Hydrolysis occurs in aerobic, anoxic, and anaerobic environments. The rate appears to be dependent on the oxidizing environment; however, there is some evidence to suggest that the presence of electron acceptors is of minimal importance in complex mixtures such as activated sludge (Goel et al., 1998).

5.1.2 *Acidogenesis*

The second stage of short-chain VFA production is acidogenesis, in which complex soluble organics are converted to VFAs, hydrogen, alcohols, ketones, and carbon dioxide. Again, this process is catalyzed by extracellular enzymes that facilitate the biochemical reactions necessary for the conversion. Various degradation pathways are potentially available, ultimately leading to the predominant VFA produced being acetic acid. Not only do these reactions make the carbon substrate more amenable to cross-membrane transport, energy is produced through the cleavage of bonds that activates the enzymes.

5.1.3 *Acetogenesis*

A companion process to acidogenesis is acetogenesis, wherein longer-chain VFAs and alcohols are converted to acetic acid and hydrogen by acetogenic bacteria. Generally, this conversion is achieved by obligate anaerobic bacteria that generate elemental hydrogen and acetic acid from the degradation of the more complex molecules. Because these bacteria are inhibited by high concentrations of elemental hydrogen, they only exist in a synergistic relationship with bacteria that consume hydrogen ions, such as non-VFA consuming, sulfur-reducing bacteria (*Desulfovibrio* sp.) or methanogens that convert hydrogen (H_2) and carbon dioxide to methane. Although a few species of acetogenic bacteria have been isolated (Lorowitz et al., 1989; McInerney et al., 1981), there are difficulties in growing these organisms in pure culture because of inhibition caused by hydrogen accumulations.

The key to effective fermentation is truncating the anaerobic conversion process so that methanogenesis, which is the conversion of acetic acid and hydrogen to methane, does not occur. Generally, SRT is controlled in acid-fermentation processes so that methanogenic bacteria are washed out of the system. However, it is also key that hydrogen is removed so that it does not inhibit acetogenic bacteria. Additionally, it is critical that other strains of bacteria that do not consume acetic acid but do scavenge the hydrogen for energy, such as *Desulfovibrio* sp. noted previously, coexist in a fermentation reactor.

Sulfur-reducing bacteria (SRB) preferentially use sulfate as their electron acceptor, converting it to sulfide. Not all SRB use hydrogen as the electron donor in their metabolism. A number of these strains (*Desulfobacter, Desulfobulbus, Desulfococcus,* etc.) use VFAs as the electron donor, with the end products being sulfide and carbon dioxide (Bruser et al., 2000). These bacteria will consume short-chain VFAs generated by acidogenesis and acetogenesis. Middleton and Lawrence (1977) studied the kinetics of acetic-acid-consuming SRB and determined that the minimum SRT for these bacteria would be temperature dependent, varying from 2.9 days at 20 °C to 4.7 days at 10 °C. Given that these minimum SRTs are similar to those used in fermentation, consumption of VFAs by these bacteria is an issue that needs further study.

Sulfur-reducing bacteria outcompete methanogenic bacteria for acetate in most instances. They also outcompete other syntropic bacteria for propionate (Chen et al., 2008). However, SRB only convert propionate to acetate; hence, SRB are important microorganisms that participate in acetogenesis reactions.

5.2 Naturally Occurring Volatile Fatty Acid Production in Wastewater Collection Systems

Fermentation occurs in collection systems and primary treatment before an EPBR secondary treatment system. Furthermore, there are some industrial discharges that will inherently include high naturally occurring VFA concentrations that will persist in the wastewater until entry into the WRRF.

Fermentation of wastewater organics occurs in a wastewater collection system (Vollertsen et al., 2011). The degree of complex COD removal depends on the characteristics of the wastewater, the characteristics of the collection system that affect wastewater oxygen concentration (extent, slope, etc.), the presence of force mains, wastewater temperature, and the presence of other inhibiting substances. Collection system fermentation is dynamic because these characteristics fluctuate because of external influences. For instance, conversion of complex organics to short-chain VFAs will be substantially reduced during wet weather events because of the reduced retention time in the system, higher rates of reaeration caused by increased turbulence, and lower wastewater temperatures that often accompany wet weather flows.

In communities where the weather is relatively warm, fermentation in the upstream wastewater collection system often generates sufficient VFAs to sustain EPBR, especially where the retention time in the collection system is relatively long. For communities in more temperate climates, VFAs may have to be generated within the treatment process to augment those arriving from the collection system.

There are few effective operating parameters that can be manipulated in the collection system to enhance VFA production. Some facilities have used upstream trunk sewers as equalization basins to reduce flow variations delivered to the facility. Retention times in these ad hoc basins are often sufficient to cause fermentation, especially as solids deposited along the bottom during relatively quiescent times have SRTs greater than liquid retention times. However, VFA production from this type of system is inconsistent because, during high-flow periods, solids are flushed from the system and HRTs are short.

One interaction of primary importance to VFA production in collection systems is odor control. Chemicals added to inhibit sulfide production or to combine with sulfides so they do not evolve into a gaseous state seriously affect fermentation in sewers. Oxidizing chemicals such as chlorine, oxygen, potassium permanganate, nitrates, or hydrogen peroxide are often added to wastewater flows to suppress sulfate reduction. The oxidizing conditions created by these chemical additions inhibit fermentation, especially those reactions that are obligate anaerobic.

The other method used to control odor emissions is the addition of compounds that will precipitate sulfides, that is, the addition of ferrous, or ferric, salts, predominantly. However, the addition of ferrous salts to wastewater also results in some precipitation of VFAs in ferro-organic compounds, which can inhibit acetogenesis (Ahmed et al., 2001) and, hence, the production of acetic acid.

In many WRRFs, EBPR processes suffer because of the inadequacy of VFAs in the influent. To maintain EBPR without interruption, those facilities resort to the addition of acetic acid from external sources or self-generation of VFAs by fermenting parts of primary sludge in dedicated tanks that are used as fermenters. Primary sludge is a ready source of carbon substrate that is not accompanied by substantial amounts of nitrogen and phosphorus. Short-chain VFA production by fermenting this material has been used by a large number of facilities to facilitate EBPR. While primary sludge is commonly used as the source of VFAs, secondary solids also have been used by a few facilities. Typically, facilities that are not equipped with primary clarifiers resort to the fermentation of return activated sludge (RAS) or WAS as an alternative source. A detailed discussion of various types of fermenters and their operating parameters is presented in Chapter 9.

6.0 TROUBLESHOOTING ENHANCED BIOLOGICAL PHOSPHORUS REMOVAL

Enhanced biological phosphorus removal is one of the most widely studied biological wastewater treatment processes because of its role in controlling nutrients discharged to the aquatic environment in a manner that is cost effective and free from chemical usage. Although many full-scale facilities consistently demonstrate excellent phosphorus removal performance under stringent discharge limits, they are not free from occasional process upsets, potentially resulting in permit violations. From this standpoint, it is essential to analyze the causes of such upsets and initiate urgent troubleshooting measures to minimize the risk of such permit violations and to restore the facility to the normal running condition. In this section, three case studies are presented as examples of problem analyses and solutions at selected full-scale treatment facilities. Chapter 17 is devoted to case studies only and presents the details of eight operating facilities. The purpose of the case studies presented in this chapter is to demonstrate how the underlying principles of EBPR, as described in the preceding sections, have been used by facilities to achieve their respective nutrient control goals.

6.1 Kelowna Wastewater Treatment Plant, British Columbia, Canada

The bioreactor system at the Kelowna Wastewater Pollution Control Centre (42 mL/d) in British Columbia, Canada, was designed and operated as a five-stage Bardenpho process to achieve high levels of nitrogen and phosphorus removals required to comply with the permit limits of 6 mg/L total nitrogen and 0.25 mg/L total phosphorus. Oldham and Stevens (1984) conducted a study to investigate the reasons for failure of the facility to achieve the desired total phosphorus concentration in the final effluent. At the time of this study, total flow was handled by two parallel, multicell, plug flow reactor trains with an HRT of 24 hours each. The authors observed that the total phosphorus concentration reached about 0.15 mg/L at about 80% of the total length of the aerobic stage, after which it climbed above 1 mg/L through the rest of the treatment train and in the final effluent.

After changing the mode of operation to only one train in line, with an HRT in the bioreactor system of 11 hours, Oldham and Stevens (1984) achieved effluent total phosphorus of ~0.1 mg/L. It was revealed that, in the two-train operation COD oxidation, nitrification and EBPR were complete in the front 80% and denitrification was occurring in the tail end of the aerobic stage, leaving no nitrate for denitrification in the anoxic stages. Secondary release of phosphorus was taking place in the absence of any dissolved oxygen or nitrate in the second anoxic zone, resulting in a higher than expected total phosphorus concentration in the effluent. After reducing the HRT in the one-train operating mode, the denitrification reaction was moved to the second anoxic zone. This, combined with reaeration in the final stage, helped to stop the secondary release, which demonstrates that excessive operating HRT may be detrimental to effective EBPR performance.

A detailed description of this facility, including its design details and operating experience, is presented in Chapter 17. A discussion of the steps taken by the facility based on a multiyear process monitoring and control program is also included in Chapter 17.

6.2 Black River Falls Wastewater Treatment Plant, Wisconsin

This case study was conducted in a small (~0.76 mL/d) community WRRF in Black River Falls, located in the east-central part of Wisconsin (Basu et al., 2008). The biological treatment process includes both nitrogen and phosphorus removal in an oxidation ditch, with brush-type aerators. The RAS returns to the outer ditch and mixes with the influent. The facility has occasionally experienced EBPR process failure despite maintaining dissolved oxygen concentrations of 3 to 4 mg/L in the aerobic zone and the presence of adequate VFAs in the influent. Microbiological analysis of the mixed liquor indicated healthy populations of PAOs and absence of GAOs. However, the biomass indicated proliferation of filamentous microorganisms, primarily Type 0092. The root cause of this was traced back to the operation under excessively long SRT (~30 days), which caused high effluent phosphorus because of secondary release and poor settleability, resulting in carryover of excessive solids with the effluent. Gradual reduction of SRT to the level still adequate for nitrification enabled elimination of the bulking problem and restoration of a normal EBPR process.

6.3 Eagles Point Wastewater Treatment Plant, Cottage Grove, Minnesota

The 30-mL/d Eagles Point Wastewater Treatment Plant in Cottage Grove, Minnesota, was started up in September 2002. The treatment train includes screening and grit removal, primary settling, secondary treatment, and secondary clarification.

The biological treatment process has been designed for both phosphorus and nitrogen removal, with the flexibility of operating in plug flow or step-feed modes. Primary solids are fermented in a gravity thickener, as needed, to supplement for VFAs to support EBPR. Lindeke and Barnard (2005) reported on the experience gained from operating the facility for the first 2.5 years since its startup.

During the period of this study, the average influent BOD, total suspended solids (TSS), ammonia-nitrogen, and total phosphorus were 248, 220, 27.2, and 6.8 mg/L, respectively. Corresponding effluent characteristics were 4, 4, 2.4, and 0.68, respectively. Operating mixed liquor suspended solids and SRT of the facility were maintained at 2000 to 2500 mg/L and 6.8 days (for two aeration tanks together), respectively.

The operational experience revealed three potential challenges to EBPR performance of this facility. These were seasonal unavailability of sufficient VFAs, carryover of excessive solids with the clarifier effluent, and operation of the biological reactor, as discussed in the following paragraphs.

Volatile fatty acid concentrations (expressed in terms of COD) across the primary sedimentation tank at the facility increased by about 50% annually because of fermentation in the sludge blanket. The VFA species were primarily acetic acid, with little propionic acid and no heavier acids. However, the fermenter overflow demonstrated 5 times as much VFA concentration in terms of COD as that in the primary effluent, with propionic acid about half that of acetic acid and a minor presence of isomers of butyric and valeric acids. During this study, facility staff monitored influent VFAs generated in the collection system as a function of wastewater temperature. The data presented in Figure 7.14 demonstrate the strong influence of temperature on influent VFAs, which can also potentially affect the

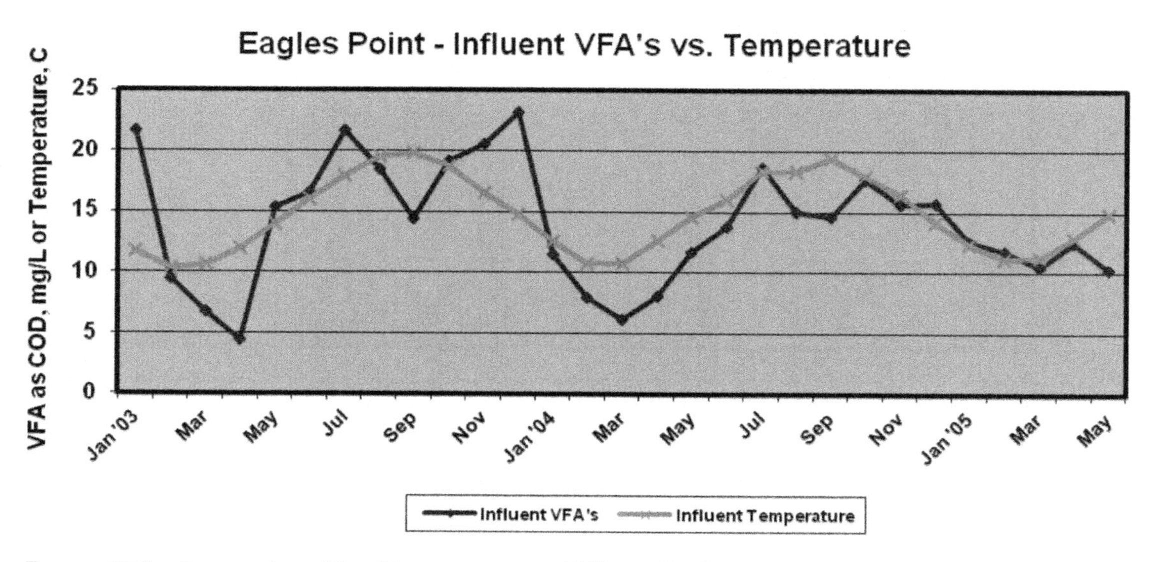

FIGURE 7.14 Seasonal profile of temperature vs VFAs at Eagles Point Wastewater Treatment Plant.

EBPER performance of the facility because of its dependence on the presence of adequate amounts of VFAs. Accordingly, the operating staff supplemented any deficiency by adding fermenter overflow to the influent during winter months.

The facility effluent TSS spiked several times during the study period because of process upsets, as shown in Figure 7.15. Twice, it was caused by the proliferation of microorganisms that caused poor settling in secondary clarifiers as a result of difficulty of formation of biological flocculation. Two other times, the primary reason was operational changes. Effect of effluent TSS on the effluent total phosphorus concentration is clearly demonstrated in this figure.

The facility demonstrated superior EBPR performance during the plug flow mode of operation of the bioreactor compared to the step-feed mode. This is primarily attributable to the fact that during the step-feed mode, primary effluent containing rbCOD is introduced to the reactor through multiple inlet points along its length. This results in a significant part of VFAs being directed to different points along the biological reactor, with the front end, which needs VFAs most for EBPR, receiving only a small fraction. This creates much more of a likelihood of VFA deficit for EBPR in the step-feed mode than in the plug flow mode of operation. On the other hand, during plug flow operation all VFAs enter the front end of the reactor with the primary effluent. Figure 7.16 demonstrates poor

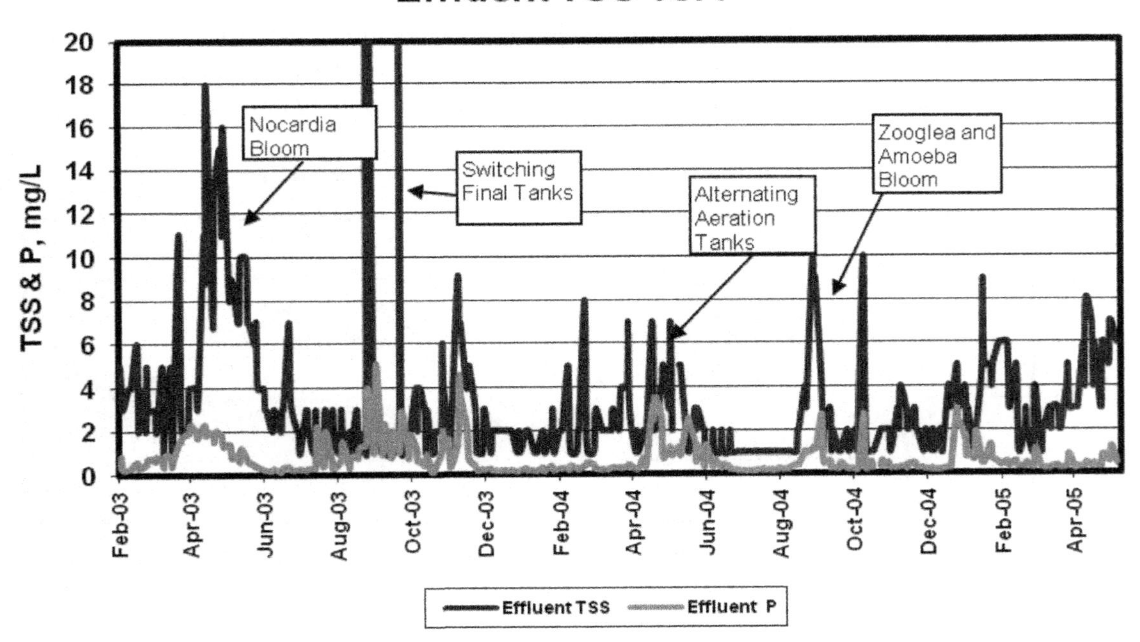

FIGURE 7.15 Effect of effluent TSS on effluent total phosphorus at Eagles Point Wastewater Treatment Plant.

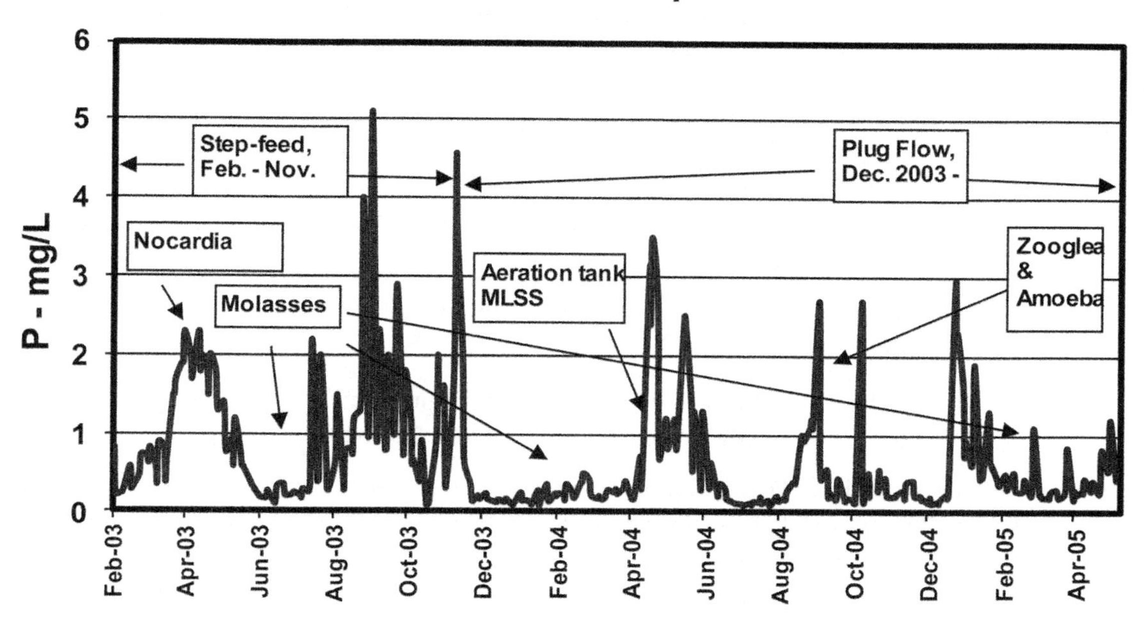

FIGURE 7.16 Effects of step-feed vs plug flow operation of the bioreactor on effluent total phosphorus at Eagles Point Wastewater Treatment Plant.

performance of EBPR processes between August and December 2003 operating in step-feed mode. After this period, the reactor was changed to a plug flow mode of operation.

Under the circumstances of EBPR failure because of VFA deficiency, facility staff dosed rbCOD in the form of molasses, which supplemented VFAs by fermentation and restored EBPR. The periods of molasses dosing and the effect on effluent total phosphorus are clearly shown in Figure 7.16.

The case studies presented herein demonstrate that there is no single solution to resolve all the problems leading to EBPR failure. Indeed, each situation requires careful analysis on a case-by-case basis for effective troubleshooting.

7.0 REFERENCES

Ahmed, Z.; Ivanov, V.; Hyun, S.-H.; Cho, K. M.; Kim, I. S. (2001) Effect of Divalent Iron on Methanogenic Fermentation of Fat-Containing Wastewater. *Environ. Eng. Resour.*, **6** (3), 139–146.

Ahn, J.; Schroeder, S.; Beer, M.; McIlroy, S.; Bayly, R. C.; May, J. W.; Vasiliadis, G.; Seviour, R. J. (2007) Ecology of the Microbial Community Removing Phosphate from Wastewater under Continuously Aerobic Conditions in a Sequencing Batch Reactor. *Appl. Environ. Microbiol.*, **73**, 2257–2270.

Ahn, J.; Daidou, T.; Tsuneda, S.; Hirata, A. (2002) Transformation of Phosphorus and Relevant Intracellular Compounds by a Phosphorus-Accumulating Enrichment Culture in the Presence of Both the Electron Acceptor and Electron Donor. *Biotechnol. Bioeng.*, **79**, 83–93.

Barker, P. S.; Dold, P. C. (1996) Denitrification Behavior in Biological Excess Phosphorus Removal Activated Sludge Systems. *Water Res.*, **30** (4), 769–780.

Barnard, J. L. (1984) Activated Primary Tanks for Phosphate Removal. *Water SA*, **10** (3), 121–126.

Barnard, J. L. (1976) A Review of Biological Phosphorus Removal in the Activated Sludge Process. *Water SA*, **2** (3), 136–144.

Barnard, J. L. (1983) Design Consideration Regarding Phosphate Removal in Activated Sludge Plants. *Water Sci. Technol.*, **15**, 319–328.

Barnard, J. L.; Fothergill, S. (1998) Secondary Phosphorus Release in Biological Phosphorus Removal Systems. *Proceedings of the 71st Annual Water Environment Technical Exposition and Conference*; Orlando, Florida, Oct 3–7; Water Environment Federation: Alexandria, Virginia.

Barnard, J. L.; Scruggs, C. L. (2003) Biological Phosphorus Removal. *Water Environ. Technol.*, **15** (2), 27–33.

Barnard, J. L.; Stevens, G. M.; Leslie, P. J. (1985) Design Strategies for Nutrient Removal Plant. *Water Sci. Technol.*, **17** (11/12), 233–242.

Basu, S.; Wiesner, T.; Page, J. (2008) Failure and Subsequent Restoration of EBPR Capability of a Small Community Wastewater Treatment Plant. *Proceedings of the 81st Annual Water Environment Federation Technical Exhibition and Conference* [CD-ROM]; Chicago, Illinois, Oct 18–22; Water Environment Federation: Alexandria, Virginia.

Beatons, D.; Vanrolleghem, P. A.; van Loosdtrecht, M. C. M.; Hosten, L. H. (1999) Temperature Effects in Bio-P Removal. *Water Sci. Technol.*, **39** (1), 215–225.

Blackall, L. L.; Crocetti, G.; Saunders, A. M.; Bond, P. L. (2002) A Review and Update of the Microbiology of Enhanced Biological Phosphorus Removal in Wastewater Treatment Plants. *Antonie Van Leeuwenhoek Int. J. Gen. Molecular Microbiol.*, **81** (1/4), 681–691.

Bond, P. L.; Keller, J.; Blackall, L. L. (1999) Anaerobic Phosphate Release from Activated Sludge with Enhanced Biological Phosphorus Removal. A Possible Mechanism of Intracellular pH control. *Biotechnol. Bioeng.*, **63** (5), 507–515.

Brdjanovic, D.; Slamet, A.; van Loosdrecht, M. C. M.; Hooijmans, C. M.; Alaerts, G. J.; Heijnen, J. J. (1998) Impact of Excessive Aeration on Biological Phosphorus Removal from Wastewater. *Water Res.*, **32**, 200–208.

Brdjanovic, D.; van Loosdrecht, M. C. M.; Hooijmans, C. M.; Alaerts, G. J.; Heijnen, J. J. (1998) Minimal Aerobic Sludge Retention Time in Biological Phosphorus Removal Systems. *Biotechnol. Bioeng.*, **60**, 326–332.

Brdjanovic, D.; van Loosdrecht, M. C. M.; Hooijmans, C. M.; Mino, T.; Alaerts, G. J.; Heijnen, J. J. (1998) Effect of Polyphosphate Limitation on the Anaerobic Metabolism of Phosphorus-Accumulating Microorganisms. *Appl. Microbiol. Biotechnol.*, **50**, 273–276.

Bruser, T.; Lens, P. N. L.; Truper, H. G. (2000) Chapter 3—The Biological Sulphur Cycle. In *Environmental Technologies to Treat Sulfur Pollution*; IWA Publishing: London.

Buchan, L. (1983) Possible Biological Mechanism of Phosphorus Removal. *Water Sci. Technol.*, **15**, 87–103.

Burgess, J. E.; Pletschke, B. I. (2008) Hydrolytic Enzymes in Sewage Sludge Treatment: A Mini-Review. *Water SA*, **34** (3), 343–350.

Burow, L. C.; Kong, Y.; Nielsen, J. L.; Blackall, L. L.; Nielsen, P. H. (2007) Abundance and Ecophysiology of *Defluviicoccus* spp., Glycogen-Accumulating Organisms in Full-Scale Wastewater Treatment Processes. *Microbiology*, **153**, 178–185.

Carvalho, G.; Lemos, P. C.; Oehmen, A.; Reis, M. A. M. (2007) Denitrifying Phosphorus Removal: Linking the Process Performance with the Microbial Community Structure. *Water Res.*, **41**, 4383–4396.

Cech, J. S.; Hartman, P. (1993) Competition between Polyphosphate and Polysaccharide Accumulating Bacteria in Enhanced Biological Phosphate Removal Systems. *Water Res.*, **27** (7), 1219–1225.

Cech, J. S.; Hartman, P. (1990) Glucose Induced Break Down of Enhanced Biological Phosphate Removal. *Environ. Technol.*, **11** (7), 651–656.

Chen, Y.; Cheng, J. J.; Creamer, K. S. (2008) Inhibition of Anaerobic Digestion Process: A Review. *Bioresour. Technol.*, **99** (10), 4044–4064.

Choi, E.; Rhu, D.; Yun, Z.; Lee, E. (1998) Temperature Effects on Biological Nutrient Removal System with Municipal Wastewater. *Water Sci. Technol.*, **37** (9), 219–226.

Chuang, S. H.; Ouyang, C. F.; Wang, Y. B. (1996) Kinetic Competition between Phosphorus Release and Denitrification on Sludge under Anoxic Conditions. *Water Res.*, **30** (12), 2961–2968.

City College of New York (1994) Chapter 2—Characterization and Treatment of New York City Sludge Dewatering Centrate. In *Recycles from Centrate in New York City Wastewater Treatment Plants*; City College of New York: New York.

Coats, E. R.; Watkins, D. L.; Brinkman, C. K.; Loge, F. R. (2011) Effect of Anaerobic HRT on Biological Phosphorus Removal and the Enrichment of Phosphorus Accumulating Organisms. *Water Environ. Res.*, **83** (5), 461–469.

Comeau, Y.; Hall, K. J.; Hancock, R. E. W.; Oldham, W. K. (1986) Biochemical Model for Enhanced Biological Phosphorus Removal. *Water Res.*, **20** (12), 1511–1521.

Crocetti, G. R.; Hugenholtz, P.; Bond, P. L.; Schuler, A.; Keller, J.; Jenkins, D.; Blackall, L. L. (2000) Identification of Polyphosphate-Accumulating Organisms and Design of 16S rRNA-Directed Probes for their Detection and Quantitation. *Appl. Environ. Microbiol.*, **66**, 1175–1182.

Ekama, G.; Marais, G.; Siebritz, I. (1984) Biological Excess Phosphorus Removal. In *Theory, Design, and Operation of Nutrient Removal Activated Sludge Processes;* Water Research Commission: Pretoria, South Africa.

Erdal, U. G.; Erdal, Z. K.; Randall, C. W. (2003) The Competition between PAOs (Phosphorus Accumulating Organisms) and GAOs (Glycogen Accumulating Organisms) in EBPR (Enhanced Biological Phosphorus Removal) Systems at Different Temperatures and the Effects on System Performance. *Water Sci. Technol.*, **47**, 1–8.

Erdal, U. G.; Erdal, Z. K.; Randall, C. W. (2006) The Mechanism of Enhanced Biological Phosphorus Removal Washout and Temperature Relationships. *Water Environ. Res.*, **78** (7), 710–750.

Erdal, Z. K. (2002) *The Biochemistry of EBPR: Role of Glycogen in Biological Phosphorus Removal and the Impact of the Operating Conditions on the Involvement of Glycogen;* Department of Civil and Environmental Engineering, Virginia Polytechnic Institute and State University: Blacksburg, Virginia.

Erdal, U. G. (2002) The Effects of Temperature on System Performance and Bacterial Community Structure in a Biological Phosphorus Removal System. Ph.D. Thesis, Virginia Polytechnic Institute and State University: Blacksburg, Virginia.

Filipe, C. D. M.; Daigger, G. T.; Grady, C. P. L. (2001a) A Metabolic Model for Acetate Uptake under Anaerobic Conditions by Glycogen Accumulating Organisms: Stoichiometry, Kinetics, and the Effect of pH. *Biotechnol. Bioeng.*, **76** (1), 17–31.

Filipe, C. D. M.; Daigger, G. T.; Grady, C. P. L. (2001b) pH as a Key Factor in the Competition Between Glycogen-Accumulating Organisms and Phosphorus-Accumulating Organisms. *Water Environ. Res.*, **73** (2), 223–232.

Flowers, J. J.; He, S.; Yllmaz, S.; Noguera, D. R.; McMahon, K. D. (2009) Denitrification Capabilities of Two Biological Phosphorus Removal Sludges Dominated by Different *"Candidatus Accumulibacter"* Clades. *Environ. Microbiol. Reports*, **1** (6), 583–588.

Forbes, C. M.; O'Leary, N. D.; Dobson, A. D.; Marchesi, J. R. (2009) The Contribution of "Omic"-Based Approaches to the Study of Enhanced Biological Phosphorus Removal Microbiology. *FEMS Microbiol. Ecol.*, **69** (1), 1–15.

Freitas, F.; Temudo, M.; Reis, M. A. M. (2005) Microbial Population Response to Changes of the Operating Conditions in a Dynamic Nutrient-Removal Sequencing Batch Reactor. *Bioprocess Biosyst. Eng.*, **28** (3), 199–209.

Fuhs, G. W.; Chen, M. (1975) Microbiological Basis of Phosphate Removal in the Activated Sludge Process for the Treatment of Wastewater. *Microbiol. Ecol.*, **2** (2), 119–138.

Gebremariam, S. Y.; Beutel, M. W.; Christian, D.; Hess, T. F. (2011) Research Advances and Challenges in the Microbiology of Enhanced Biological Phosphorus Removal—A Critical Review. *Water Environ. Res.*, **83** (3), 195–219.

Goel, R.; Mino, T.; Satoh, H.; Matsuo, T. (1998) Enzyme Activities under Aerobic and Anaerobic Conditions in Activated Sludge Sequencing Batch Reactor. *Water Res.*, **32**, 7.

Griffiths, P. C.; Stratton, H. M.; Seviour, R. J. (2002) Environmental Factors Contributing to the "G Bacteria" Population in Full-Scale EBPR Plants. *Water Sci. Technol.*, **46** (4-5), 185–192.

Gu, A. Z.; Hughes, T.; Fisher, D.; Swartzlander, D.; Dacko, B.; Ellis, W. G.; He, S.; McMahon, K. D.; Neethling, J. B.; Wei, H. P.; Chapman, M. (2006) The Devil Is in the Details: Full Scale Optimization of the EBPR Process at the City of Las Vegas WPCF. *Proceedings of the 79th Annual Water Environment Federation Technical Exhibition and Conference* [CD-ROM]; Dallas, Texas, Oct 22–25; pp 5110–5130.

Gu, A. Z.; Saunders, A.; Neethling, J. B.; Stensel, H. D.; Blackall, L. L. (2008) Functionally Relevant Microorganisms to Enhanced Biological Phosphorus Removal Performance at Full-Scale Wastewater Treatment Plants in the United States. *Water Environ. Res.*, **80**, 688–698.

Guisasola, A.; Qurie, M.; Vargas, M. D.; Casas, C.; Baeza, J. A. (2009) Failure of an Enriched Nitrite-DPAO Population to Use Nitrate as an Electron Acceptor. *Process Biochem.*, **44** (7), 689–695.

Gujer, W.; Zehnder, A. J. B. (1983) Conversion Process in Anaerobic Digestion. *Water Sci. Technol.*, **15** (8/9), 127–167.

Harper, W. F.; Jenkins, D. (2003) The Effect of an Initial Anaerobic Zone on the Nutrient Requirements of Activated Sludge. *Water Environ. Res.*, **75** (3), 216–224.

He, S.; Gall, D. L.; McMahon, K. D. (2007) "Candidatus Accumulibacter" Population Structure in Enhanced Biological Phosphorus Removal Sludges as Revealed by Polyphosphate Kinase Genes. *Appl. Environ. Microbiol.*, **73**, 5865–5874.

He, S.; Gu, A. Z.; McMahon, K. D. (2008) Progress Toward Understanding the Distribution of Accumulibacter Among Full-Scale Enhanced Biological Phosphorus Removal Systems. *Microbial Ecol.*, **55**, 229–236.

Helmer, C.; Kunst, S. (1997) Low Temperature Effects on Phosphorus Release and Uptake by Microorganisms in EBPR Plants. *Water Sci. Technol.*, **37** (4-5), 531–539.

Hesselmann, R.; Werlen, C.; Hahn, D.; van Der Meer, J.; Zehnder, A. (1999) Enrichment, Phylogenetic Analysis and Detection of a Bacterium that Performs Enhanced Biological Phosphate Removal in Activated Sludge. *Syst. Appl. Microbiol.*, **22**, 454–465.

Hesselmann, R. P. X.; von Rummell, R.; Resnick, S. M.; Hany, R.; Zehnder, A. J. B. (2000) Anaerobic Metabolism of Bacteria Performing Enhanced Biological Phosphate Removal. *Water Res.*, **34**, 3487–3494.

Jiang, Y.; Wang, B.; Wang, L.; Chen, J.; He, S. (2006) Dynamic Response of Denitrifying Poly-P Accumulating Organisms Batch Culture to Increased Nitrite Concentration as Electron Acceptor. *J. Environ. Sci. Health,* **41** (11), 2557–2570.

Kern-Jespersen, J. P.; Henze, M. (1993) Biological Phosphorus Uptake under Anoxic and Aerobic Conditions. *Water Res.*, **27**, 617–624.

Kisoglu, Z.; Erdal, U. G.; Randall, C. W. (2000) The Effect of COD/TP Ratio on Intracellular Storage Materials, System Performance and Kinetic Parameters in a BNR System. *Proceedings of the 73rd Annual Water Environment Federation Technical Exposition and Conference* [CD-ROM]; Anaheim, California, Oct 14–18; Water Environment Federation: Alexandria, Virginia.

Kong, Y. H.; Nielsen, J.; Nielsen, P. (2004) Microautoradiographic Study of Rhodocyclus-Related Polyphosphate-Accumulating Bacteria in Full-Scale Enhanced Biological Phosphorus Removal Plants. *Appl. Environ. Microbiol.*, **70** (9), 5383–5390.

Kong, Y. H.; Ong, S. L.; Ng, W. J.; Liu, W.-T. (2002) Diversity and Distribution of a Deeply Branched Novel Proteobacterial Group Found in Anaerobic-Aerobic Activated Sludge Processes. *Environ. Microbiol.*, **4** (11), 753–757.

Kong, Y. H.; Xia, Y.; Nielsen, J. L.; Nielsen, P. H. (2006) Ecophysiology of a Group of Uncultured Gammaproteobacterial Glycogen-Accumulating Organisms in Full-Scale Enhanced Biological Phosphorus Removal Wastewater Treatment Plants. *Environ. Microbiol.*, **8** (3), 479–489.

Kong, Y. H.; Nielsen, J. L.; Nielsen, P. H. (2005) Identity and Ecophysiology of Uncultured Actinobacterial Polyphosphate-Accumulating Organisms in Full-Scale Enhanced Biological Phosphorus Removal Plants. *Appl. Environ. Microbiol.*, **71** (7), 4076–4085.

Kong, Y., Ong, S. L.; Ng, W. J.; Liu. W.-T. (2002) Diversity and Distribution of a Deeply Branched Novel Proteobacterial Group Found in Anaerobic-Aerobic Activated Sludge Processes. *Environ. Microbiol.*, **4**, 753–757.

Kortstee, G. J. J.; Appeldoorn, K. J.; Bonting, C. F. C.; van Niel, E. W. J.; van Veen, H. W. (2000) Recent Developments in the Biochemistry and Ecology of Enhanced Biological Phosphorus Removal. *Biochem.* (Moscow), **65** (3), 332–340.

Kuba, T.; Smolders, G.; van Loosdrecht, M. C. M.; Heijnen, J. J. (1993) Biological Phosphorus Removal from Waste-Water by Anaerobic–Anoxic Sequencing Batch Reactor. *Water Sci. Technol.*, **27** (5-6), 241–252.

Kuba, T.; van Loosdrecht, M. C. M.; Heijnen, J. J. (1996) Phosphorus and Nitrogen Removal with Minimum COD Requirement by Integration of Denitrification in a Two-Sludge System. *Water Res.*, **30** (7), 1702–1710.

Kuba, T.; Wachtmeister, A.; van Loosdrecht, M. C. M.; Heijnen, J. J. (1994) Effect of Nitrate on Phosphorus Release in Biological Phosphorus Removal Systems. *Water Sci. Technol.*, **30** (6), 263–269.

Lancaster, C. B.; Madden, J. E. (2008) Not So Fast: The Impact of Recalcitrant Phosphorus on the Ability to Meet Low Phosphorus Limits. *Proceedings of the 81st Annual Water Environment Federation Technical Exhibition and Conference* [CD-ROM]; Chicago, Illinois, Oct 18–22; Water Environment Federation: Alexandria, Virginia; pp 3531–3543.

Lemos, P. C.; Dai, Y.; Yuan, Z.; Keller, J.; Santos, H.; Reis, M. A. M. (2007) Elucidation of Metabolic Pathways in Glycogen-Accumulating Organisms with in Vivo C-13 Nuclear Magnetic Resonance. *Environ. Microbiol.*, **9** (11), 2694–2706.

Levin, G. V.; Shapiro, J. (1965) Metabolic Uptake of Phosphorus by Waste Water Organisms. *J.—Water Pollut. Control Fed.*, **37** (6), 800–821.

Li, N.; Ren, N.-Q.; Wang, X.-H.; Kang, H. (2010) Effect of Temperature on Intracellular Phosphorus Absorption and Extra-Cellular Phosphorus Removal in EBPR Process. *Bioresour. Technol.*, **101** (15), 6265–6268.

Li, N.; Wang, X.; Ren, N.; Zhang, K.; Kang, H.; Youa, S. (2008) Effects of Solid Retention Time (SRT) on Sludge Characteristics in Enhanced Biological Phosphorus Removal (EBPR) Reactor. *Chem. Biochem. Eng. Q.*, **22** (4), 453–458.

Lindeke, D.; Barnard, J. (2005) The Role and Production of VFA's in a Highly Flexible BNR Plant. *Proceedings of the 78th Annual Water Environment Federation Technical Exhibition and Conference* [CD-ROM]; Washington, D.C., Oct 29–Nov 2; Water Environment Federation: Alexandria, Virginia.

Liu, W.-T., Mino, T.; Nakamura, K. ; Matsuo, T. (1994) Role of Glycogen in Acetate Uptake and Polyhydroyalkanoate Synthesis in Anaerobic-Aerobic Activated Sludge with a Minimized Polyphosphate Content. *J. Fermentation Bioeng.*, **77** (5), 535–540.

Liu, W.-T.; Mino, T.; Matsuo, T.; Nakamura, K. (1996) Biological Phosphorus Removal Processes—Effect of pH on Anaerobic Substrate Metabolism. *Water Sci. Technol.*, **34** (1-2), 25–32.

Liu, W.-T.; Mino, T.; Nakamura, K.; Matsuo, T. (1996) Glycogen Accumulating Population and its Anaerobic Substrate Uptake in Anaerobic-Aerobic Activated Sludge without Biological Phosphorus Removal. *Water Res.*, **30** (1), 75–82.

Liu, W.-T.; Nakamura, K.; Matsuo, T.; Mino, T. (1997) Internal Energy-Based Competition Between Polyphosphate- and Glycogen-Accumulating Bacteria in Biological Phosphorus Removal Reactors—Effect of P/C Feeding Ratio. *Water Res.*, **31** (6), 1430–1438.

Lopez-Vazquez, C. M.; Song, Y. I.; Hooijmans, C. M.; Brdjanovic, D.; Moussa, M. S.; Gijzen, H. J.; van Loosdrecht, M. C. M. (2008) Temperature Effects on the Aerobic Metabolism of Glycogen Accumulating Organisms. *Biotechnol. Bioeng.*, **101** (2), 295–306.

Lorowitz, W. H.; Zhao, H.; Bryant, M. P. (1989) *Syntrophomonas wolfei* Subspecies *saponavida* Subsp. *nov.*, a Long Chain Fatty-Acid-Degrading Subspecies *Wolfei* Subsp *nov.*; and Amended Descriptions of the Genus and Species. *Int. J. System. Microbiol.*, **39**, 122–126.

Louie, T. M.; Mah, T. J.; Oldham, W.; Ramey, W. D. (2000) Use of Metabolic Inhibitors and Gas Chromatography/Mass Spectrometry to Study Poly-beta-hydroxyalkanoates Metabolism Involving Cryptic Nutrients in Enhanced Biological Phosphorus Removal Systems. *Water Res.*, **34** (5), 1507–1514.

Madigan, M. T.; Martinko, J. M.; Parker, J. (2003) Brock Biology of Microorganisms, 10th ed.; Pearson Education: Upper Saddle River, New Jersey.

Mamais, D.; Jenkins, D. (1992) The Effects of MCRT and Temperature on Enhanced Biological Phosphorus Removal. *Water Sci. Technol.*, **26** (5/6), 955–965.

Martin, H. G.; Ivanova, N.; Kunin, V.; Warnecke, F.; Barry, K. W.; McHardy, A. C.; Yeates, C.; He, S. M.; Salamov, A. A.; Szeto, E.; Dalin, E.; Putnam, N. H.; Shapiro, H. J.; Pangilinan, J. L.; Rigoutsos, I.; Kyrpides, N. C.; Blackall, L. L.; McMahon, K. D.; Hugenholtz, P. (2006) Metagenomic Analysis of Two Enhanced Biological Phosphorus Removal (EBPR) Sludge Communities. *Nature Biotechnol.*, **24** (10), 1263–1269.

Matsuo, Y. (1994) Effect of the Anaerobic Solids Retention Time on Enhanced Biological Phosphorus Removal. *Water Sci. Technol.*, **30** (6), 193–202.

Matsuo, T.; Mino, T.; Sato, H. (1992) Metabolism of Organic Substances in Anaerobic Phase of Biological Phosphate Uptake Process. *Water Sci. Technol.*, **25** (6), 83–92.

Maurer, M.; Abramovich, D.; Siegrist, H.; Gujer, W. (1999) Kinetics of Biologically Induced Phosphorus Precipitation in Waste-Water Treatment. *Water Res.*, **33** (2), 484–493

Maurer, M.; Gujer, W.; Hany, R.; Bachmann, S. (1997) Intracellular Carbon Flow in Phosphorus Accumulating Organisms from Activated Sludge Systems. *Water Res.,* **31** (4), 907–917.

McInerney, M. J.; Bryant, M. P.; Hespell, R. B.; Costerton, J. W. (1981) *Syntrophomonas wolfei gen. nov.* sp. *nov.,* An Anaerobic Syntrophic, Fatty Acid-Oxidizing Bacterium. *Appl. Environ. Microbiol.,* **41** (4), 1029–1039.

Middleton, A. C.; Lawrence, A. W. (1977) Kinetics of Microbial Sulfate Reduction. *J.—Water Pollut. Control Fed.,* **49** (7), 1659–1670.

Mino, T.; Arun, V.; Tsuzuki, Y.; Matsuo, T. (1987) Effect of Phosphorus Accumulation on Acetate Metabolism in the Biological Phosphorus Removal. In *Biological Phosphate Removal from Wastewaters;* Ramadori, R., Ed.; Pergamon Press: Elmsford, New York, pp 27–38.

Mino, T.; van Loosdrecht, M. C. M.; Heijnen, J. J. (1998) Microbiology and Biochemistry of the Enhanced Biological Phosphate Removal Process. *Water Res.,* **32** (11), 3193–3207.

Murnleitner, E.; Kuba, T.; van Loosdrecht, M. C. M.; Heijnen, J. J. (1997) An Integrated Metabolic Model for the Aerobic and Denitrifying Biological Phosphorus Removal. *Biotechnol. Bioeng.,* **54** (5), 434–450.

Neethling, J. B.; Blake, B.; Benisch, M.; Gu, A. Z.; Stephens, H.; Stensel, D.; Moore, R. (2005) Factors Influencing the Reliability of Enhanced Biological Phosphorus Removal; Report 01-CTS-3; Water Environment Research Foundation: Alexandria, Virginia.

Nielsen, P. H.; Kragelund, C.; Seviour, R. J.; Nielsen, J. L. (2009) Identity and Ecophysiology of Filamentous Bacteria in Activated Sludge. *FEMS Microbiol. Rev.,* **33** (6), 969–998.

Nielsen, J. L.; Nguyen, H.; Meyer, R. L.; Nielsen, P. H. (2012) Identification of Glucose-Fermenting Bacteria in a Full-Scale Enhanced Biological Phosphorus Removal Plant by Stable Isotope Probing. *Microbiology,* **158** (7), 1818–182.

Oehmen, A.; Carvalho, G.; Freitas, F.; Reis, M. A. M. (2010) Assessing the Abundance and Activity of Denitrifying Polyphosphate Accumulating Organisms through Molecular and Chemical Techniques. *Water Sci. Technol.,* **61** (8), 2061–2068.

Oehmen, A.; Lemos, P. C.; Carvalho, G.; Yuan, Z. G.; Keller, J.; Blackall, L. L.; Reis, M. A. M. (2007) Advances in Enhanced Biological Phosphorus Removal: From Micro to Macro Scale. *Water Res.,* **41** (11), 2271–2300.

Oehmen, A.; Saunders, A. M.; Vives, M. T.; Yuan, Z. G.; Keller, H. (2006) Competition Between Polyphosphate and Glycogen Accumulating Organisms in Enhanced Biological Phosphorus Removal Systems with Acetate and Propionate as Carbon Sources. *J. Biotechnol.,* **123** (1), 22–32.

Oehmen, A.; Vives, M. T.; Lu, H. B.; Yuan, Z. G.; Keller, J. (2005) The Effect of pH on the Competition Between Polyphosphate-Accumulating Organisms and Glycogen-Accumulating Organisms. *Water Res.,* **39** (15), 3727–3737.

Oehmen, A.; Yuan, Z. G.; Blackall, L. L.; Keller, J. (2005) Comparison of Acetate and Propionate Uptake by Polyphosphate Accumulating Organisms and Glycogen Accumulating Organisms. *Biotechnol. Bioeng.,* **91** (2), 162–168.

Oehmen, A.; Zeng, R. J.; Yuan, Z. G.; Keller, J. (2005) Anaerobic Metabolism of Propionate by Polyphosphate-Accumulating Organisms in Enhanced Biological Phosphorus Removal Systems. *Biotechnol. Bioeng.,* **91** (1), 43–53.

Oldham, W. K.; Stevens, G. M. (1984) Initial Operating Experience of a Nutrient Removal Process (Modified Bardenpho) at Kelowna, British Columbia. *Can. J. Civil Eng.,* **11** (3), 474–479.

Onnis-Hayden, A.; Majed, N.; Schramm, A.; Gu, A. Z. (2011) Process Optimization by Decoupled Control of Key Microbial Populations: Distribution of Activity and Abundance of Polyphosphate-Accumulating Organisms and Nitrifying Populations in A Full-Scale IFAS-EBPR Plant. *Water Res.,* **45** (13), 3845–3854.

Pereira, H.; Lemos, P. C.; Reis, M. A. M.; Crespo, J. P. S. G.; Carrondo, M. J. T.; Santos, H. (1996) Model for Carbon Metabolism in Biological Phosphorus Removal Processes Based on in Vivo C-13-NMR Labelling Experiments. *Water Res.,* **30** (9), 2128–2138.

Phillips, H. M.; Kobylinski, E.; Barnard, J.; Wallis-Lage, C. (2006) Nitrogen and Phosphorus Rich Streams: Managing the Nutrient Merry-Go-Round. *Proceedings of the 79th Annual Water Environment Federation Technical Exhibition and Conference* [CD-ROM]; Dallas, Texas, Oct 21–25; Water Environment Federation: Alexandria, Virginia; pp 5282–5304.

Pijuan, M.; Oehmen, A.; Baeza, J. A.; Casas, C.; Yuan, Z. (2008) Characterizing the Biochemical Activity of Full-Scale Enhanced Biological Phosphorus Removal Systems: A Comparison with Metabolic Models. *Biotechnol. Bioeng.,* **99** (1), 170–179.

Pijuan, M.; Saunders, A. M.; Guisasola, A.; Baeza, J. A.; Casas, C.; Blackall, L. L. (2004) Enhanced Biological Phosphorus Removal in a Sequencing Batch Reactor Using Propionate as the Sole Carbon Source. *Biotechnol. Bioeng.,* **85** (1), 56–67.

Puig, S.; Coma, M.; Monclús, H.; van Loosdrecht, M. C. M.; Colprim, J.; Balaguer, M. D. (2008) Selection Between Alcohols and Volatile Fatty Acids as External Carbon Sources for EBPR. *Water Res.,* **42** (3), 557–566.

Randall, C. W.; Barnard, J. L.; Stensel, H. D., Eds. (1992) *Design and Retrofit of Wastewater Treatment Plants for Biological Nutrient Removal*; Vol. 5, Water Quality Management Library, Technomic Publishing: Lancaster, Pennsylvania.

Reddy, M. (1998) *Biological and Chemical Systems for Nutrient Removal;* WEF Special Publication; Water Environment Federation: Alexandria, Virginia.

Rodrigo, M. A.; Seco, A.; Ferrer, J.; Penya-Roja, J. M. (1999) The Effect of the Sludge Age on the Deterioration of the Enhanced Biological Phosphorus Removal Process. *Environ. Technol.,* **20** (10), 1055–1063.

Rodrigo, M. A.; Seco, A.; Penya-Roja, J. M.; Ferrer, J. (1996) Influence of Sludge Age on Enhanced Phosphorus Removal in Biological Systems. *Water Sci. Technol.,* **34** (1-2), 41–48.

Satoh, H.; Mino, T.; Matsuo, T. (1992) Uptake of Organic Substrates and Accumulation of Polyhydroxyalkanoates Linked with Glycolysis of Intracellular Carbohydrates under Anaerobic Conditions in the Biological Excess Phosphate Removal Processes. *Water Sci. Technol.,* **26** (5-6), 933–942.

Satoh, H.; Mino, T.; Matsuo, T. (1994) Deterioration of Enhanced Biological Phosphorus Removal by the Domination of Microorganisms without Polyphosphate Accumulation. *Water Sci. Technol.,* **30** (6), 203–211.

Saunders, A. M.; Mabbett, A. N.; McEwan, A. G.; Blackall, L. L. (2007) Proton Motive Force Generation from Stored Polymers for the Uptake of Acetate Under Anaerobic Conditions. *FEMS Microbiol. Lett.,* **274** (2), 245–251.

Saunders, A. M.; Oehmen, A.; Blackall, L. L.; Yuan, Z.; Keller, J. (2003) The Effect of GAOs (Glycogen Accumulating Organisms) on Anaerobic Carbon Requirements in Full-Scale Australian EBPR (Enhanced Biological Phosphorus Removal Process) Plants. *Water Sci. Technol.,* **47** (11), 37–43.

Schuler, A. J.; Jenkins, D. (2002) Effects of pH on Enhanced Biological Phosphorus Removal Metabolisms. *Water Sci. Technol.,* **46** (4–5), 171–178.

Schuler, A. J.; Jenkins, D. (2003) Enhanced Biological Phosphorus Removal from Wastewater by Biomass with Different Phosphorus Contents, Part I: Experimental Results and Comparison with Metabolic Models. *Water Environ. Res.,* **75** (6), 485–498.

Serafim, L. S.; Lemos, P. C.; Levantesi, C.; Tandoi, V.; Santos, H.; Reis, M. A. M. (2002) Methods for Detection and Visualization of Intracellular Polymers Stored by Polyphosphate-Accumulating Microorganisms. *J. Microbiol. Methods,* **51** (1), 1–18.

Seviour, R. J.; McIlroy, S. (2008) The Microbiology of Phosphorus Removal in Activated Sludge Processes—The Current State of Play. *J. Microbiol.,* **46** (2), 115–124.

Seviour, R. J.; Mino, T.; Onuki, M. (2003) The Microbiology of Biological Phosphorus Removal in Activated Sludge Systems. *FEMS Microbiol. Rev.,* **27** (1), 99–127.

Shoji, T.; Satoh, H.; Mino, T. (2003) Quantitative Estimation of the Role of Denitrifying Phosphate Accumulating Organisms in Nutrient Removal. *Water Sci. Technol.*, **47** (11), 23–29.

Smolders, G. J. F.; van der Meij, J.; van Loosdrecht, M. C. M.; Heijnen, J. J. (1995) A Structured Metabolic Model for the Anaerobic and Aerobic Stoichiometry and Kinetics of the Biological Phosphorus Removal Process. *Biotechnol. Bioeng.*, **47** (3), 277–287.

Smolders, G. J. F. (1995) A Metabolic Model of the Biological Phosphorus Removal—Stoichiometry, Kinetics and Dynamic Behaviour. Ph.D. Thesis, Delft University of Technology, Delft, Netherlands.

Smolders, G. J. F.; van der Meij, J.; van Loosdrecht, M. C. M.; Heijnen, J. J. (1994a) Model of the Anaerobic Metabolism of the Biological Phosphorus Removal Process—Stoichiometry and pH Influence. *Biotechnol. Bioeng.*, **43** (6), 461–470.

Smolders, G. J. F.; van der Meij, J.; van Loosdrecht, M. C. M.; Heijnen, J. J. (1994b) Stoichiometric Model of the Aerobic Metabolism of the Biological Phosphorus Removal Process. *Biotechnol. Bioeng.*, **44** (7), 837–848.

Stephens, H. L.; Stensel, D. (1998) Effect of Operating Conditions on Biological Phosphorus Removal. *Water Environ. Res.*, **70** (3), 362–369.

Tremblay, A.; Tyagi, R. D.; Urampalli, S. R. Y. (1999) Practice Periodical of Hazardous, Toxic, and Radioactive Waste. *Management*, **10**, 183.

U.S. Environmental Protection Agency (1987) *Design Manual: Phosphorus Removal;* EPA-625/1-87-001; U.S. Environmental Protection Agency: Washington, D.C.

U.S. Environmental Protection Agency (1994) Decentralized Systems Technology Fact Sheet Septage Treatment Disposal; Municipal Technology Branch, U.S. Environmental Protection Agency: Washington, D.C.

Uys, R. (1984) Municipal Chemist, Witbank RSA; Personal Communication with J. L. Barnard.

Vollertsen, R. E.; Hvilved-Jacobsen, J.; Nielsen, T. (2011) Anaerobic Transformations of Organic Matter in Collection Systems. *Water Environ. Resour.*, **83** (6), 532–540.

Water Environment Federation; American Society of Civil Engineers; Environmental and Water Resources Institute (2005) *Biological Nutrient Removal (BNR) Operation in Wastewater Treatment Plants;* WEF Manual of Practice No. 29; ASCE/EWRI Manuals and Reports on Engineering Practice No. 109; McGraw-Hill: New York.

Wentzel, M. C.; Lotter, L. H.; Loewenthal, R. E.; Marais, G. v. R. (1986) Metabolic Behaviour of *Acinetobacter spp.* in Enhanced Biological Phosphorus Removal—A Biochemical Model. *Water SA*, **12** (4), 209–224.

Wexler, M.; Richardson, D. J.; Bond, P. L. (2009) Radiolabelled Proteomics to Determine Differential Functioning of *Accumulibacter* during the Anaerobic and Aerobic Phases of a Bioreactor Operating for Enhanced Biological Phosphorus Removal. *Environ. Microbiol.*, **11** (12), 3029–3044.

Whang, L. M.; Filipe, C. D. M.; Park, J. K. (2007) Model-Based Evaluation of Competition between Polyphosphate- and Glycogen-Accumulating Organisms. *Water Res.*, **41** (6), 1312–1324.

Whang, L. M.; Park, J. K. (2002) Competition between Polyphosphate- and Glycogen-Accumulating Organisms in Biological Phosphorus Removal Systems—Effect of Temperature. *Water Sci. Technol.*, **46** (1-2), 191–194.

Whang, L. M.; Park, J. K. (2006) Competition between Polyphosphate- and Glycogen-Accumulating Organisms in Enhanced-Biological-Phosphorus-Removal Systems: Effect of Temperature and Sludge Age. *Water Environ. Res.*, **78** (1), 4–11.

Wilmes, P.; Andersson, A. F.; Lefsrud, M. G.; Wexler, M.; Shah, M.; Zhang, B.; Hettich, R. L.; Bond, P. L.; VerBerkmoes, N. C.; Banfield, J. F. (2008) Community Proteogenomics Highlights Microbial Strain-Variant Protein Expression within Activated Sludge Performing Enhanced Biological Phosphorus Removal. *ISME J.*, **2** (8), 853–864.

Wong, M.; Mino, T.; Seviour, R.; Onuki, M.; Liu, W. (2005) In situ Identification and Characterization of the Microbial Community Structure of Full-Scale Enhanced Biological Phosphorous Removal Plants in Japan. *Water Res.*, **39** (13), 2901–2914.

Xia, Y.; Kong, Y. H.; Nielsen, P. H. (2007) In situ Detection of Protein-Hydrolysing Microorganisms in Activated Sludge. *FEMS Microbiol. Ecol.*, **60** (1), 156–165.

Yagci, N.; Artan, N.; Cokgor, E. U.; Randall, C. W.; Orhon, D. (2003) Metabolic Model for Acetate Uptake by a Mixed Culture of Phosphate- and Glycogen-Accumulating Organisms under Anaerobic Conditions. *Biotechnol. Bioeng.*, **84** (3), 359–373.

Zeng, R. J.; Saunders, A. M.; Yuan, Z. G.; Blackall, L. L.; Keller, J. (2003) Identification and Comparison of Aerobic and Denitrifying Polyphosphate-Accumulating Organisms. *Biotechnol. Bioeng.*, **83** (2), 140–148.

Zeng, R. J.; van Loosdrecht, M. C. M.; Yuan, Z. G.; Keller, J. (2003) Metabolic Model for Glycogen-Accumulating Organisms in Anaerobic/Aerobic Activated Sludge Systems. *Biotechnol. Bioeng.*, **81** (1), 92–105.

Zhang, T.; Liu, Y.; Fang, H. H. P. (2005) Effect of pH Change on the Performance and Microbial Community of Enhanced Biological Phosphate Removal Process. *Biotechnol. Bioeng.*, **92** (2), 173–182.

Zhang, C.; Chen, Y.; Liu, Y. (2007) The Long-Term Effect of Initial pH control on the Enrichment Culture of Phosphorus- and Glycogen-Accumulating Organisms with a Mixture of Propionic and Acetic Acids as Carbon Sources. *Chemosphere,* **69** (11), 1713–1721.

Zhou, Y.; Pijuan, M.; Oehmen, A.; Yuan, Z. G. (2010) The Source of Reducing Power in the Anaerobic Metabolism of Polyphosphate Accumulating Organisms (PAOs)—A Mini-Review. *Water Sci. Technol.,* **61** (7), 1653–1662.

Zhou, Y.; Pijuan, M.; Yuan, Z. (2007) Free Nitrous Acid Inhibition on Anoxic Phosphorus Uptake and Denitrification by Poly-Phosphate Accumulating Organisms. *Biotechnol. Bioeng.,* **98** (4), 903–912.

Zhou, Y.; Pijuan, M.; Zeng, R. J.; Lu, H.; Yuan, Z. (2008) Could Polyphosphate-Accumulating Organisms (PAOs) be Glycogen Accumulating Organisms (GAOs)? *Water Res.,* **42** (10-11), 2361–2368.

Zhou, Y.; Pijuan, M.; Zeng, R. J.; Yuan, Z. G. (2009) Involvement of the TCA Cycle in the Anaerobic Metabolism of Polyphosphate Accumulating Organisms (PAOs). *Water Res.,* **43** (5), 1330–1340.

8.0 SUGGESTED READINGS

Oehmen, A.; Yuan, Z.; Blackall, L. L.; Keller, J. (2004) Short-Term Effects of Carbon Source on the Competition of Polyphosphate Accumulating Organisms and Glycogen Accumulating Organisms. *Water Sci. Technol.,* **50** (10), 139–144.

Rassie, W. H.; Pretorius, W. A. (2001) A Review of Characterisation Requirements for In-Line Prefermenters Paper 1: Wastewater Characterisation. *Water SA,* **27** (3), 405–412.

Snoeyink, V. L.; Jenkins, D. (1980) *Water Chemistry;* Wiley & Sons: New York.

Vollertsen, J.; Petersen, G.; Borregaard, V. R. (2005) Hydrolysis and Fermentation of Activated Sludge to Enhance Biological Phosphorus Removal. *Proceedings of the IWA Specialized Conference on Nutrient Management in Wastewater Treatment Processes and Recycle Streams;* Krakow, Poland, Sept 19–21; International Water Association: London.

Chapter 8

Chemical Precipitation of Phosphorus

Jurek Patoczka, Ph.D., P.E.; Maria Inneo; and Susan Hansler, M.A.Sc., P.Eng.

1.0 INTRODUCTION

Chemical precipitation of phosphorus is a method frequently used as a primary phosphorus removal process or in conjunction with biological removal processes as a supplemental, polishing, or backup process. Chemical addition is often the phosphorus removal method of choice for small- and medium-size facilities, where implementation of a dedicated, enhanced biological phosphorus removal (EBPR) process is not practical or economical. This is particularly the case when an existing treatment facility has to meet new phosphorus limits and the configuration of the existing biological treatment process does not lend itself easily to an EBPR conversion. Because the EBPR process can be subject to periods of inferior performance or upsets, it is almost always supported by standby or polishing chemical addition facilities. In practice, whenever an effluent phosphorus limit has to be consistently met, chemical addition is almost always practiced, at least in a standby mode.

An advantage of chemical precipitation is that the phosphorus removed in this fashion remains fixed in the waste sludge as inorganic precipitate and is not readily released during sludge storage, thickening, or anaerobic digestion. At facilities with EBPR, phosphorus release from the sludge is common, particularly in oxygen-deficient conditions. Consequently, recycle streams from sludge processing operations can return a significant phosphorus load to the head of the facility, particularly when anaerobic digestion is practiced. At such facilities, the sludge processing train and return streams must be carefully designed and operated because uncontrolled releases of phosphorus can offset the purpose of the

EBPR process. When chemical phosphorus precipitation is practiced, chemically bound phosphorus in sludge is not subject to such release.

Use of chemicals for phosphorus removal has a number of disadvantages, which are discussed in more detail in the subsequent sections. Some of these are increased sludge generation (Section 6.1), alkalinity depletion (Section 6.3), an increase in total dissolved solids (TDS) (Section 6.4), increased inert concentration in mixed liquor with related potential effects on ultimate biosolids disposal (Section 3.2), and potential effects on other processes such as UV disinfection (Section 6.7).

2.0 PHOSPHORUS FORMS IN RAW WASTEWATER AND IN THE EFFLUENT

The principal form of phosphorus removed by chemical addition is soluble orthophosphate ions. These ions represent the significant form of phosphorus in raw municipal wastewater. Biological treatment processes will convert most of the remaining phosphorus forms (polyphosphates, or condensed phosphates, and organic phosphorus) into orthophosphates while assimilating and incorporating some of the available phosphorus to the biomass (approximately 1 mg/100 mg of 5-day biochemical oxygen demand [BOD_5] removed in the biological process). Effluent from the activated sludge, or from any other biological process, will contain residual, soluble phosphorus primarily in the orthophosphate form, with a relatively minor contribution from refractory or recalcitrant dissolved organic phosphates (rDOP). For practical purposes, the contributions from rDOP are only of concern where stringent limits (at the level of 0.1 mg/L total phosphorus or less) are required (Section 5.1). Consequently, soluble phosphorus in the effluent from biological processes, including in the mixed liquor as it exits the aeration basin, is amenable to almost complete removal by chemical precipitation. In addition to soluble phosphorus, effluent from any treatment process will contain phosphorus in particulate form, either as inorganic precipitate or as part of biological cells. In a typical situation in which chemical is added to the activated sludge system, the effluent suspended solids will contain phosphorus in both forms enmeshed together into the floc (see Section 3.3 for a more detailed discussion of this topic).

Facility recycle streams from solids processing operations, particularly when anaerobic digestion is used and particularly for facilities practicing EBPR, could be rich in phosphorus released from sludge, as discussed in more detail in Section 4.0 of Chapter 2. The effect of the phosphorus load from recycle streams should be considered when designing and operating chemical phosphorus removal processes.

Chemical addition to the primary clarifier, which is known as *chemically enhanced primary treatment* (CEPT), or *pre-precipitation*, is frequently used for enhanced particulate and colloidal BOD_5 and total suspended solids (TSS) removal in the primary clarifiers to lower organic and solids loading on the downstream biological process. Chemicals commonly used in CEPT treatment (aluminum and iron salts) are the same ones used for orthophosphate precipitation. Consequently, most of the phosphorus present in raw wastewater in soluble, particulate, or colloidal form can be removed by chemical precipitation in the primary clarifier, potentially leading to phosphorus deficiency in downstream biological treatment processes. (For a detailed discussion of phosphorus sources and forms of phosphorus present in the influent and effluent, refer to Chapter 2 of this manual or Section 2.0 of Chapter 7 of Water Environment Federation's [WEF's] *Nutrient Removal* [WEF, 2010]).

3.0 PRINCIPLES OF CHEMICAL PHOSPHORUS REMOVAL

3.1 Mechanism of Phosphorus Removal

The basic mechanism of chemical phosphorus removal is precipitation followed by a solids separation step such as sedimentation and/or filtration. Phosphorus precipitation is the transformation of the soluble phosphorus present in wastewater as orthophosphate anion (PO_4^{3-}) to an insoluble chemical compound (salt) and the removal of these insoluble precipitates by sedimentation or filtration. Such a transformation occurs when a chemical agent bearing a proper cation, such as aluminum (Al^{3+}) or ferric (Fe^{3+}), is added to the wastewater and reacts with the soluble phosphorus to form an insoluble orthophosphate salt. In practice, more complex reactions are taking place, which involve co-precipitation and surface adsorption. These mechanisms are discussed in more detail in *Nutrient Removal* (WEF, 2010).

Precipitate formed in this way, together with other forms of particulate phosphorus present in the wastewater, can be separated from the wastewater by clarification or filtration. The phosphorus precipitated using chemicals is removed from the wastewater with sludge (i.e., either combined with waste primary or biological sludge or as a separate chemical sludge).

3.2 Fate of Chemicals Added During Treatment

The application dosages of chemicals needed to remove phosphorus to low levels are typically in excess of a stoichiometric requirement, as discussed in more detail in Section 4.0. When chemical is added to the mixed liquor at the effluent from the aeration basin (or influent to the final clarifier), which is the most common application point, any excess chemical that does not bind with orthophosphates will precipitate nevertheless. This precipitate will be in the form of aluminum hydroxide (for aluminum-based chemicals) and ferric hydroxide (for ferric-based salts), although, in reality, more complex hydrated-oxides mixtures will also be formed. These inorganic precipitates, only partially saturated with orthophosphates, will be enmeshed with biological floc and settle in the final clarifier. Thus, the excess, unused chemicals will be incorporated to the mixed liquor and returned to the head of the facility with the return activated sludge (RAS). Here, these chemicals will be available to react with phosphorus entering the aeration basin with raw wastewater, reducing overall chemical use.

The precipitate accumulating in aeration basins will result in an increase in the basin inert solids concentration. The effect of an increased inert content of mixed liquor should be taken into account when designing and operating the activated sludge system. In particular, when operating the system at a given sludge age (such as that required for nitrification), inert solids from chemical addition will increase the equilibrium mixed liquor suspended solids (MLSS) concentration necessary to maintain the target sludge age. Conversely, if the same MLSS concentration is being maintained, the resulting sludge age will be lower after initiation of chemical addition. Refer to Section 6.1 for calculations of different chemicals' contribution to the inert content of MLSS.

3.3 Role of Solids Separation

Because effluent permit limits for phosphorus are commonly expressed as total phosphorus, precipitation of soluble phosphorus (orthophosphates) into a particulate form is only part of the job of phosphorus removal. Indeed, the solids separation step (clarification and/or filtration) must be capable of removing effluent TSS with associated particulate phosphorus forms to levels consistent with the effluent phosphorus limit.

The contribution of particulate phosphorus is particularly important for facilities with low effluent limits because even seemingly low effluent TSS can carry a significant phosphorus contribution (Figure 8.1). An additional consideration is that, in systems with a dedicated phosphorus removal process either by chemical addition to the activated sludge or by EBPR, the phosphorus content in MLSS and effluent TSS are increased. While the phosphorus content of MLSS is approximately 2% in a municipal conventional activated sludge facility, the content increases to 4 to 5% for a facility with a stringent phosphorus limit, regardless of whether the phosphorus is incorporated to the sludge by precipitation or by EBPR. For facilities operating at a low sludge age and with an influent phosphorus concentration higher than that for typical municipal wastewater, this could even be much higher.

Figure 8.1 shows that, for total phosphorus limits of 0.1 mg/L, effluent TSS should be no more than 1 mg/L. Presently, this can only be consistently achieved by membrane filtration (as in a membrane bioreactor [MBR]), multistage filtration, or by well-performing conventional filtration supported by polymer use. For limits of 0.5 to 1 mg/L total phosphorus, conventional filtration or even well-performing clarification (possibly supported by polymer addition) will be adequate, assuming that soluble orthophosphates are kept to a minimum by

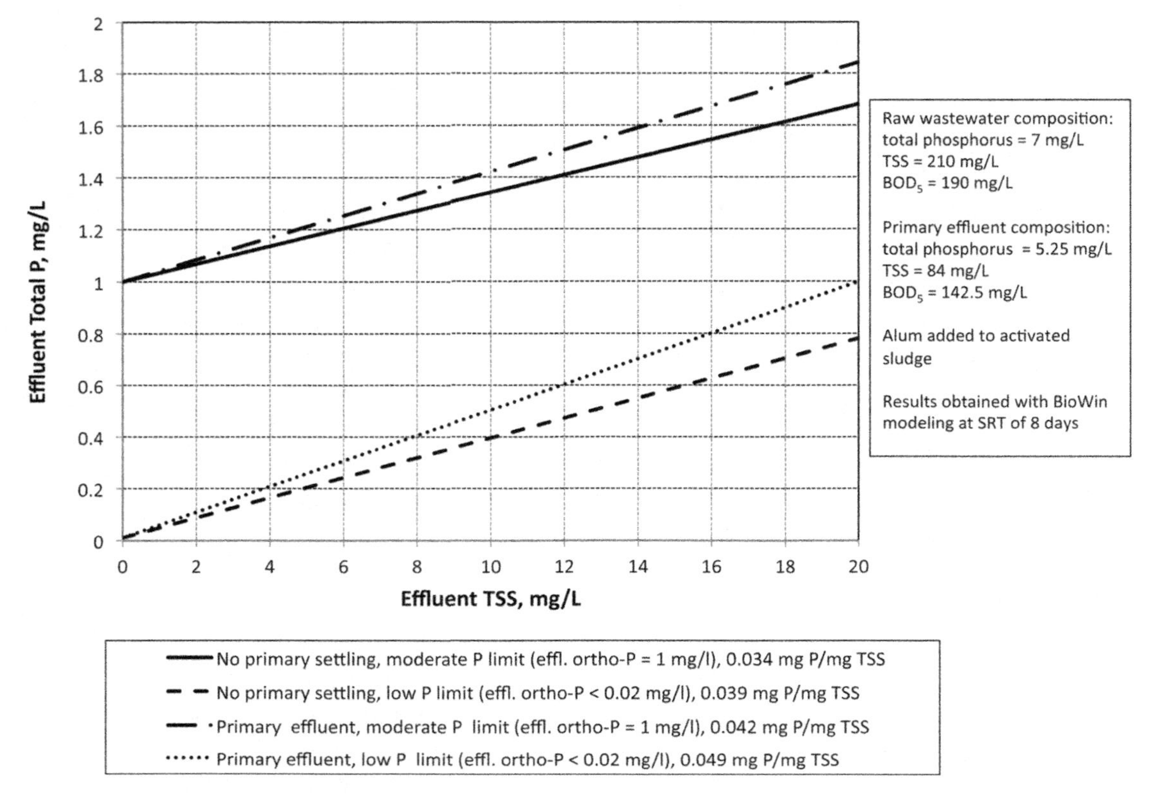

FIGURE 8.1 Contribution of effluent TSS to total phosphorus in the effluent for different treatment scenarios.

adequate chemical dose (or a well-functioning EBPR). As Figure 8.1 illustrates, for systems with primary clarification, which increases the mixed liquor volatile suspended solids (VSS) fraction of the MLSS, the phosphorus content in the MLSS and effluent TSS is even higher.

4.0 CHEMICALS USED FOR PHOSPHORUS PRECIPITATION

4.1 Introduction

In general, cations of the following metals can be used for the precipitation of phosphorus (orthophosphates) from wastewater:

- Aluminum,
- Iron,
- Calcium, and
- Magnesium.

Most commonly utilized are various chemicals containing aluminum (mainly alum) and iron (mainly ferric chloride), with calcium and magnesium being much less common. The following sections will discuss the different chemicals available, the precipitation process, and sample calculations.

4.2 Aluminum Salts

4.2.1 *Overview of Aluminum-Based Chemicals*

There are several aluminum compounds that are used in the wastewater industry for phosphorus removal, including aluminum sulfate, polyaluminum chloride (PACl), and sodium aluminate. The most widely used chemical is aluminum sulfate, commonly known as *alum*, although PACl is gaining popularity because of its ease of handling and lowering effect on pH. Polyaluminum chloride was originally used in water treatment when enhanced solids removal was a treatment objective. Because the strength and composition of sodium aluminate and PACl products vary considerably, efficacy of a specific product considered for an application should be established and compared to alternative products in jar tests, as discussed in Section 4.6. Sodium aluminate in a dry form or in solutions of various strengths was also originally developed for water treatment applications and is typically used when additional alkalinity is required.

Solid alum could be supplied as a dry, powdery chemical ("filter alum"), with an average molecular composition of $A1_2(SO_4)_3 \cdot 14H_2O$. This is a hydrated aluminum sulfate salt that indicates the presence of a defined number of water molecules in the structure of the dry salt crystal under normal conditions. However, alum for wastewater treatment is commonly delivered as a 49% solution (indicating that 49 parts of dry alum are combined with 51 parts of water, by weight). By convention, the alum dose applied in the treatment process is expressed in equivalent units of weight of dry alum (including water of crystallization) per volume of water (wastewater). Properties of alum and other aluminum-based chemicals are summarized in Table 8.1. The effect of aluminum-based chemicals on sludge generation, alkalinity consumption, and the increase in TDS are discussed in Sections 6.1, 6.3, and 6.4, respectively.

Regardless of the chemical, the active precipitating agent is aluminum ion, which combines with orthophosphate ions to form aluminum phosphate, as follows:

$$Al^{3+} + (PO_4)^{3-} \rightarrow AlPO_4 \qquad (8.1)$$

In reality, precipitation of phosphorus with aluminum and iron salts is a more complex process involving co-precipitation with aluminum hydroxide, hydrated aluminum oxides, and surface adsorption. This makes stoichiometric (based on chemistry of the reaction) calculations approximate; however, eq 8.1 will be used as the basis of dose calculations for illustrative purposes.

Table 8.1 Properties of aluminum-based chemicals used for phosphorus precipitation.

Name	Chemical formula	Molecular weight	Aluminum metal contents, % by weight	Specific density		Weight ratio of dry chemical ($Al_2(SO_4)_3 \cdot 14H_2O$) for stoichiometric phosphorus precipitation (g of chemical/g of P)
				kg/L	lb/gal	
Alum, dry	$A1_2(SO_4)_3 \cdot 14H_2O$	594	9.1	0.6 to 1.1	5 to 9.5	9.6
Alum, 49% solution*	$A1_2(SO_4)_3 \cdot 14H_2O$	594	4.4	1.33	11.1	9.6
Sodium aluminate, anhydrous (powder)	$NaAlO_2$	82	33	0.72	6.0	2.64
Sodium aluminate, trihydrate (granular)	$Na_2O \cdot Al_2O_3 \cdot 3H_2O$	218	25	1.02	8.51	3.52
Sodium aluminate, 20 to 45% solution (could vary significantly)	$NaAlO_2$	82	Varies	Varies	Varies	
Polyaluminium chloride, solutions of various strengths	$Al_nCl_{(3n-m)}(OH)_m$ example: $Al_{12}Cl_{12}(OH)_{24}$	Varies	Varies	Varies	Varies	

*49% alum solution has a dry alum ($Al_2(SO_4)_3 \times 14H_2O$) content of 0.647 kg/L (5.4 lb/gal) and an aluminum metal content of 0.059 kg/L (0.492 lb/gal).

Equation 8.1 indicates that it will take 1 mol of aluminum ion to react with 1 mol of phosphate and, thus, a 1-to-1-mol ratio of aluminum to phosphorus is required for this reaction. Because the molecular weight of aluminum is 27 and the molecular weight of phosphorus is 31, the weight ratio of aluminum to phosphorus in eq 8.1 is 0.87 to 1, as shown in the following calculation:

$$(27 \text{ g Al}/1 \text{ mol A1})/(31 \text{ g P}/1 \text{ mol P}) = 0.87/1 \text{ or } 0.87:1$$

With aluminum sulfate, there are two molecules of aluminum per molecule of alum; therefore, the stoichiometric weight ratio is

$$594 \text{ g } Al_2\ (SO_4)_3 \cdot 14\ H_2O/2/31 \text{ g P} = 9.6:1.$$

A larger amount of aluminum salt is required for actual operation than the chemistry of the reaction (stoichiometry) predicts. The amount of excess chemical required increases as the target residual phosphorus concentration decreases. The excess aluminum salt will precipitate as aluminum hydroxide, with a simplified reaction as follows:

$$A1^{3+} + 3OH^- \rightarrow A1(OH)_3 \tag{8.2}$$

Therefore, sludge that is generated at the point of chemical application will include a mixture of aluminum hydroxide and aluminum phosphate generated according to eqs 8.1 and 8.2. Equation 8.2 also illustrates the consumption of alkalinity (pH reduction) encountered when applying alum. These effects are discussed in more detail in Section 6.3.

An alternative aluminum-based chemical available for phosphorus removal is PACl, with a general composition of $Al_nCl_{(3n-m)}(OH)_m$. Polyaluminum chloride products also include an admixture of polyaluminum chlorohydrate and vary in the degree of acid neutralization (amount of caustic added, or basicity) and polymerization and aluminum content. In the water treatment industry, PACl is used as a more effective coagulant for TSS removal than alum and is also being used for phosphorus removal from wastewater. Polyaluminum chloride use generally does not depress wastewater pH, although its effects on pH depend on the exact formulation. Polyaluminum chloride use for phosphorus removal is gaining popularity, particularly at smaller facilities, because it could eliminate the need for use of caustic to increase pH when straight alum is used.

Sodium aluminate can also serve as a source of aluminum for the precipitation of phosphorus. It is generated by reacting aluminum hydroxide with caustic soda. The chemical formula for sodium aluminate is $Na_2Al_2O_4$ or $NaAlO_2$. One commercial form of sodium aluminate is granular trihydrate, which may be written as $Na_2O \cdot Al_2O_3 \cdot 3H_2O$ and contains approximately 46% Al_2O_3 or 25% Al. It is also being sold as a solution in various strengths. In contrast to alum, which reduces wastewater pH, a rise in pH may be expected upon addition of sodium aluminate to wastewater. Consequently, in situations where caustic addition would be required to counter pH depression caused by alum addition, application of sodium aluminate may eliminate the need for handling and dosing two different chemicals.

4.2.2 *Determination of Dose for Aluminum-Based Chemicals*

An approximate dose of alum (or other aluminum-based chemicals) required for a particular application can be calculated based on calculations presented in this section (including eqs 8.4 and 8.5 in Section 4.2.3). Refinement of the chemical application rate could then be performed in full scale based on the results obtained because the system response is fairly quick (Section 5.6).

Jar tests may also be performed to determine the chemical dosage for phosphorus removal in atypical applications (e.g., unusual wastewater composition) and to assess alkalinity destruction and pH depression from dosing of aluminum-based chemicals (and iron-based chemicals). Jar testing will also aid in determining requirements for supplemental alkalinity addition that may be necessary to maintain the required pH for nitrification (see also Section 6.3 for a discussion of the effects of pH and alkalinity). Jar test results may only be an approximation of the dose requirements and the chemical's effect on pH because many aspects of full-scale, continuous chemical addition may not be well duplicated in jar tests. However, jar tests are well suited for comparative testing of efficiency and costs of alternative chemicals (see also Section 4.6). Because of the inherent variability of wastewater, jar tests should be repeated on several different samples.

A design dose curve for chemical phosphorus removal using aluminum ions was developed using literature and pilot-facility data (Figure 8.2). Data in the literature include laboratory-scale test data and data from sites that are operating with chemical phosphorus removal (Gates et al., 1990). The full data set includes results for a range of pH values (mostly 6.5 to 7.5), temperatures, and wastewater characteristics. The effluent soluble phosphorus concentration is labeled "residual soluble P" on the logarithmic x-axis. The molar ratio for metal ion dose to soluble phosphorus removed is labeled "M_{dose}/soluble $P_{removed}$ (mol/mol)" on the y-axis. It is important to note that the curves apply to the soluble portion of the phosphorus only. The curve used to fit the data is based on the following equation:

$$y = a/(1 + be^{-cXe}) \tag{8.3}$$

Where

y = moles aluminum required per mole soluble phosphate removed,
a = 0.8,
b = −0.95,
c = 1.9, and
Xe = target effluent soluble phosphorus concentration (mg/L).

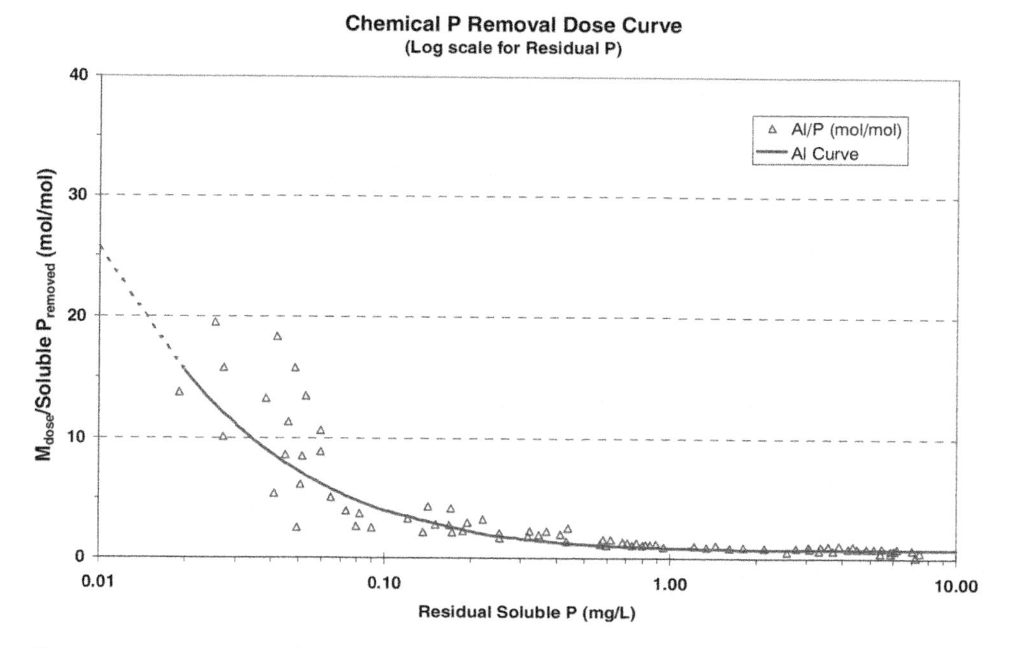

FIGURE 8.2 Ratio of aluminum (Al^{3+}) dose to phosphorus removed as a function of residual orthophosphate concentration (Gates et al., 1990).

Because eq 8.3 was derived as a best-fit formula for empirical data, it is valid only for a limited range of residual soluble phosphorus concentrations (approximately 0.1 to 0.8 mg/L). For residual phosphorus concentrations above 0.8 mg/L, the value of "y" should be assumed to be 1.0 (i.e., no excess over the stoichiometric amount indicated).

In general, the dose response found in the literature is quite variable and the data scatter increases at residual soluble phosphorus concentrations below 0.1 mg/L. Because of this variability, pilot-scale testing is recommended to determine the actual chemical dose required to reach a low-targeted effluent soluble phosphorus concentration. While jar tests will provide the approximate dose required and are effective for comparison of efficacy of alternative chemicals, the actual dose required in the full-scale application, particularly when applied to an activated sludge process, could be different. This is attributable to the full-scale effects and recirculation of the excess chemicals, as discussed in Section 3.2, which are not duplicated in jar tests. The solids separation step, which plays a critical role in achieving a low effluent phosphorus concentration (Section 3.3), cannot be well simulated in jar tests, emphasizing the need for pilot (or full-scale) testing.

Table 8.1 includes the molecular weight for the different chemicals mentioned in this section. A sample calculation is provided in this section to determine the dose required to precipitate soluble phosphate. Following the detailed, step-by-step derivation, simplified formulas for both International System of Units and U.S. customary units are provided. The following conditions are assumed:

- Influent facility flowrate is 37 850 m^3/d (10 mgd),
- A residual (influent) soluble phosphorus concentration at the point of chemical application is 3.0 mg/L, and
- A residual (effluent) soluble phosphorus concentration of 0.4 mg/L is required.

The amount of soluble phosphorus to be removed in kilograms (pounds) per day is as follows:

$$\text{Phosphorus} = (3 \text{ to } 0.4 \text{ mg/L}) \times 37\,850 \text{ m}^3\text{/d} \times 1\,000 \text{ L/m}^3 \times 1 \text{ kg}/1\,000\,000 \text{ mg} = 98.4 \text{ kg/d}$$

or

$$(3 \text{ to } 0.4) \text{ mg/L} \times 10 \text{ mgd} \times 8.34 \text{ (lb/mg} \times \text{L/mg)} = 217 \text{ 1b/d of soluble phosphorus to be removed.}$$

This is equivalent to (98.4 kg/d)/(31 kg/kg-mol P) = 3.17 kg-mol or (217 lb/d)/(31 lb/lb-mol P) = 7.0 lb-mol phosphorus to be removed.

The amount of aluminum ions required is calculated from eq 8.3 as follows:

$$Y = 0.8/(1 - 0.95e^{(-1.9)(0.4)}) = 1.44 \text{ kg-mol Al/kg-mol soluble phosphorus removed (or lb-mol Al/lb-mol P).}$$

Consequently, the amount of aluminum ions required per day is

$$3.17 \text{ kg-mol P} \times 1.44 \text{ kg-mol Al/kg-mol P} = 4.57 \text{ kg-mol Al}$$
$$\text{or } 7.0 \text{ lb-mol P} \times 1.44 \text{ lb-mol Al/lb-mol P} = 10.1 \text{ lb-mol Al.}$$

As the molecular weight of the Al ion is 27, this is converted to mass of aluminum ions per day, as follows:

$$4.57 \text{ kg-mol Al} \times 27 \text{ kg Al/kg-mol Al} = 123 \text{ kg of aluminum ions per day}$$
$$\text{or } 10.1 \text{ lb-mol Al} \times 27 \text{ lb Al/lb-mol Al} = 272 \text{ lb of aluminum ions per day.}$$

As indicated in Table 8.1, dry alum contains 9.1% aluminum. Consequently, the amount of dry alum required for this application will be

$$123 \text{ kg Al}/0.091\text{kg Al/kg alum} = 1\,356 \text{ kg alum}$$
$$\text{or } 272 \text{ lb Al}/0.091 \text{ lb Al/lb alum} = 2\,988 \text{ lb of dry alum per day.}$$

If 49% alum solution is used, it contains 0.059 kg/L (or 0.492 lb/gal) of aluminum ions (Table 8.1). Thus, the amount of liquid alum required for this application will be

$$123 \text{ kg Al}/0.059 \text{ kg Al/L} = 2\,092 \text{ L}$$
$$\text{or } 273 \text{ lb Al}/0.492 \text{ lb Al/gal} = 553 \text{ gal of solution per day.}$$

The liquid alum required per day to reduce soluble phosphorus from 3.0 to 0.4 mg/L in the aforementioned example is 2092 L/d (553 gpd). The alum dose equivalent to this alum use rate may be calculated from a factor of 0.647 kg/L (5.4 lb/gal) of alum in the 49% solution, as listed in Table 8.1, as follows:

$$2092 \text{ L/d} \times 0.647 \text{ kg/L}/37\,850 \text{ m}^3\text{/d} = 0.0358 \text{ kg/m}^3$$
$$0.0358 \text{ kg/m}^3 \times 1\,000\,000 \text{ mg/kg}/1000 \text{ L/m}^3 = 35.8 \text{ mg/L alum dose.}$$

In U.S. customary units, calculation for this 10-mgd facility will be as follows:

$$553 \text{ gpd} \times 5.4 \text{ lb/gal}/8.34 \text{ (lb/mg} \times \text{L/mg)}/10 \text{ mgd} = 35.8 \text{ mg/L alum dose.}$$

4.2.3 *Summary Dose Formulas for Alum*

The aforementioned calculations can be abbreviated to the following overall formula for calculation of the application rate of 49% alum solution in International System of Units:

$$A = (0.0118)(Xi - Xe)(Q)/[1 - 0.95 \times \exp(-1.9 \times Xe)] \quad (8.4)$$

Where

A = 49% alum solution application rate (L/d),
Xi = soluble phosphorus concentration at the application point (mg/L),
Xe = target effluent soluble phosphorus concentration (mg/L), and
Q = facility flow (m^3/d).

In U.S. customary units, the formula is as follows:

$$A = (11.8)(Xi - Xe)(Q)/[1 - 0.95 \times \exp(-1.9 \times Xe)] \quad (8.5)$$

Where

A = 49% alum solution application rate (gpd),
Xi = soluble phosphorus concentration at the application point (mg/L),
Xe = target effluent soluble phosphorus concentration (mg/L), and
Q = facility flow (mgd).

The aforementioned dose calculations are approximate. Many site-specific factors such as mixing conditions (Section 5.3), application point (Section 5.4), and wastewater chemistry and temperature will affect the actual dose of chemical required to accomplish treatment objectives. This is particularly true at residual soluble phosphorus concentrations below 0.1 mg/L.

The aforementioned calculations were based on a single application point; however, the optimum operating mode may involve dual application points (i.e., chemical addition to the primary clarifiers and to the secondary treatment process), which could result in overall savings in chemical consumption and cost (Section 5.4).

4.2.4 *Other Considerations for Aluminum-Based Chemicals*

A significant amount of sludge is produced when aluminum salts are added to the process to remove phosphorus. The quantity of sludge and handling considerations are discussed in Sections 6.1 and 6.2. Alkalinity and pH effects of aluminum sulfate addition are discussed in Section 6.3. Aluminum sulfate and, to a lesser extent, sodium aluminate and PACl, will increase TDS in the system. These effects are discussed in more detail in Section 6.4.

Aluminum compounds (and ferric compounds, discussed later) are mildly acidic and, therefore, storage and handling issues are of concern. Fiber-glass-reinforced plastic or polyethylene tanks can be used to store any of the aluminum compounds. Recommended metering pumps include solenoid, peristaltic, and diaphragm types. Carrier water should be avoided, if possible, because it will result in a more neutral pH; in addition, aluminum hydroxide may precipitate, causing plating in the chemical feed lines. If it is necessary to add carrier water for mixing or dilution, then it should be added as close to the injection point as possible to minimize plating effects. The point of addition of chemical should be in an accessible location above the water level so the delivery rate can be verified with a "bucket and stopwatch" method. For mixing recommendations, refer to Section 5.3. Pump heads should be constructed of polyvinyl chloride (PVC). Piping, valves, and fittings should be PVC or chlorinated PVC. Where feasible, chemical storage and handling equipment should be designed to be compatible with alternative chemicals to provide flexibility when availability and prices change. Storage and delivery piping should be heat-traced, where necessary, to prevent crystallization (49% alum will start to crystallize at −1 °C [30 °F]).

Facility personnel should wear personal protection equipment (PPE) when handling chemicals. The PPE should include, but not be limited to, gloves, respirators, goggles, aprons, and face shields, and should be worn when working or handling any aluminum salt solutions.

4.3 Iron Salts

4.3.1 *Overview of Iron-Based Chemicals*

The most common iron compounds used for phosphorus precipitation from wastewater are trivalent ferric (Fe^{3+}) salts, chiefly ferric chloride ($FeCl_3$), and, sometimes, ferric sulfate [$Fe_2 (SO4)_3$]. Because ferric salts addition to preliminary treatment facilities or to primary clarifiers is sometimes used for odor control, ferric may be a chemical of choice for phosphorus removal as it could serve dual purposes at facilities with a need to control odor in the liquid train. Table 8.2 includes general information for the different chemicals mentioned in this section. Chemistry of phosphorus precipitation with ferric salts is similar to that of aluminum compounds. Bivalent ferrous (Fe^{2+}) salts, primarily in the form of spent pickle liquor from metal surface cleaning with acids, are sometimes used as an inexpensive alternative, when locally available. Because pickle liquor can contain metal contaminants, it is important to ensure that these contaminants do not have an adverse effect on the sludge generated or facility effluent quality. It is not clear if the use of spent pickle liquor will continue to be a common practice in wastewater treatment because alternative methods of metal surface cleaning have recently been developed.

Ferrous salts (such as those found in pickle liquor) react with phosphates and precipitate rather poorly (WEF, 2010); therefore, they should only be added to the aeration basin, where they will first be oxidized to ferric salts. It should be noted that some additional oxygen demand will be created in the aeration basin to oxidize ferrous ions to ferric ions. Ferrous ions should never be added to the final clarifier for phosphorus removal because excess or unreacted ferrous ions will carry over into the disinfection system to consume chlorine and form a precipitate, contributing to effluent TSS. Furthermore, if a UV disinfection system is used, iron added to final clarifier may interfere with UV absorbance and foul the lamp sleeves, increasing the frequency of lamp cleaning (Section 6.7).

Effects of iron-based chemicals on sludge generation, alkalinity consumption, and TDS increase is discussed in Sections 6.1, 6.3, and 6.4, respectively (see also Table 8.3). Ferric salts can be added at various treatment facility locations, similar

TABLE 8.2 Properties of iron-based chemicals used for phosphorus precipitation.

Name	Chemical formula	Molecular weight	Iron metal contents, % by weight	Weight ratio of dry chemical for stoichiometric phosphorus precipitation (g of chemical/g of P)
Ferric chloride, dry	$FeCl_3$	162.5	34.5	5.24
Ferric chloride, 37% solution*	$FeCl_3$	162.5	12.8	5.24
Ferric sulfate, dry	$Fe_2(SO_4)_3$	400	28	6.45
Ferrous chloride, dry	$FeCl_2$	127	79	6.14
Ferrous sulfate, dry	$Fe(SO_4)_2$	152	37	7.36

*37% ferric chloride solution has a specific density of 1.36 kg/L (11.4 lb/gal), a dry $FeCl_3$ content of 0.504 kg/L (4.2 lb/gal), and an iron metal content of 0.173 kg/L (1.44 lb/gal).

TABLE 8.3 Sludge generation and TDS increase factors from the use of selected chemicals (Patoczka, 2006).

Chemical*/process	TSS increase factor (*F*), kg per kg (or mg/L per mg/L) of chemical added	TDS increase factor, kg per kg (or mg/L per mg/L) of chemical added
Alum for stoichiometric phosphorus precipitation (as alum orthophosphate), without neutralization	0.411	0.165
Excess alum (precipitating as aluminum hydroxide), without neutralization	0.263	0.485
Typical alum application for chemical phosphorus removal (at 3:1 alum to phosphorus stoichiometric rate), without neutralization	0.312	0.378
Typical alum application for chemical phosphorus removal (at 3:1 alum to phosphorus stoichiometric rate), with full neutralization with caustic	0.312	0.533
Ferric chloride precipitating as ferric orthophosphate	0.929	0.071
Excess ferric (precipitating as ferric hydroxide)	0.658	0.655
Typical ferric application for chemical phosphorus removal (at 3:1 ferric to phosphorus stoichiometric rate), without neutralization	0.748	0.460
Typical ferric application for chemical phosphorus removal (at 3:1 ferric to phosphorus stoichiometric rate), with full neutralization with caustic	0.748	0.745
pH adjustment with caustic	0	0.575
pH adjustment with sulfuric acid	0	0.980

*Alum dose expressed as $Al_2(SO_4)_3 \times 14H_2O$; ferric as $FeCl_3$.

to aluminum-based chemicals. Ferric ion combines with orthophosphate ions to form ferric phosphate, as follows:

$$Fe^{3+} + PO_4^{3-} \rightarrow FePO_4 \quad (8.6)$$

The reaction between ferrous ions and phosphate ions can be written as follows:

$$3\ Fe^{2+} + 2\ PO_4^{3-} \rightarrow Fe_3(PO_4)_2 \quad (8.7)$$

In reality, precipitation of phosphorus with ferric salts is a more complex process (as is the case with aluminum compounds) involving co-precipitation with ferric hydroxide and hydrated ferric oxides and surface adsorption. This makes stoichiometric calculations approximate; however, eq 8.6 will be used as the basis of dose calculations for illustrative purposes.

Equation 8.6 indicates that it will take 1 mol of ferric ion to react with 1 mol of phosphate and, thus, a 1-to-1-mol ratio of iron to phosphorus is required for this reaction. Because the molecular weight of iron is 56 and the molecular weight of phosphorus is 31, the weight ratio of iron to phosphorus in eq 8.6 is 1.8 to 1, as shown in the following calculation:

$$(56 \text{ g Fe}/1 \text{ mol Fe})/(31 \text{ g P}/1 \text{ mol P}) = 1.81/1 \text{ or } 1.8:1$$

With ferric chloride, there is one molecule of ferric per molecule of ferric chloride; therefore, the stoichiometric weight ratio of ferric chloride to phosphorus is

$$(162.5 \text{ g } FeCl_3)/31 \text{ g P} = 5.24:1.$$

Similarly calculated chemical stoichiometric weight ratios for other ferric chemicals are provided in Table 8.2.

Bench-, pilot-, and full-scale studies have shown that considerably higher stoichiometric quantities of chemicals are typically necessary to meet phosphorus-removal objectives as a result of competing hydroxide and sulfide reactions. If iron salts are added to the primary clarifier, any sulfides presents will first react with the ferric or ferrous ion. Consequently, if a facility influent has significant levels of sulfide, then the iron dosing must be higher for the same amount of phosphorus removed in the primary clarifier.

The excess chemical required increases as the target residual phosphorus concentration decreases. The excess ferric salt will precipitate as ferric hydroxide, with a simplified reaction as follows:

$$Fe^{3+} + 3\,OH^- \rightarrow Fe(OH)_3 \tag{8.8}$$

Therefore, sludge that is generated at the point of ferric application will include a mixture of ferric hydroxide and ferric phosphate generated according to eqs 8.6 and 8.8. Equation 8.8 also illustrates the consumption of alkalinity (pH reduction) encountered when applying ferric. These effects are discussed in more detail in Section 6.3.

4.3.2 *Determination of Dose for Iron-Based Chemicals*

An approximate dose of ferric salt required for the particular application can be determined based on calculations presented in this section, including eqs 8.10 and 8.11 presented in Section 4.3.3). Refinement of the chemical application rate can then be performed in full scale based on the results obtained because the system response is fairly quick (Section 5.6).

For comments on performing jar tests to determine chemical dose, refer to the beginning of Section 4.2.2 and to Section 4.6. A design dose curve for chemical phosphorus removal using ferric ions was developed using literature and pilot-facility data (Figure 8.3). Literature data include laboratory-scale test data and

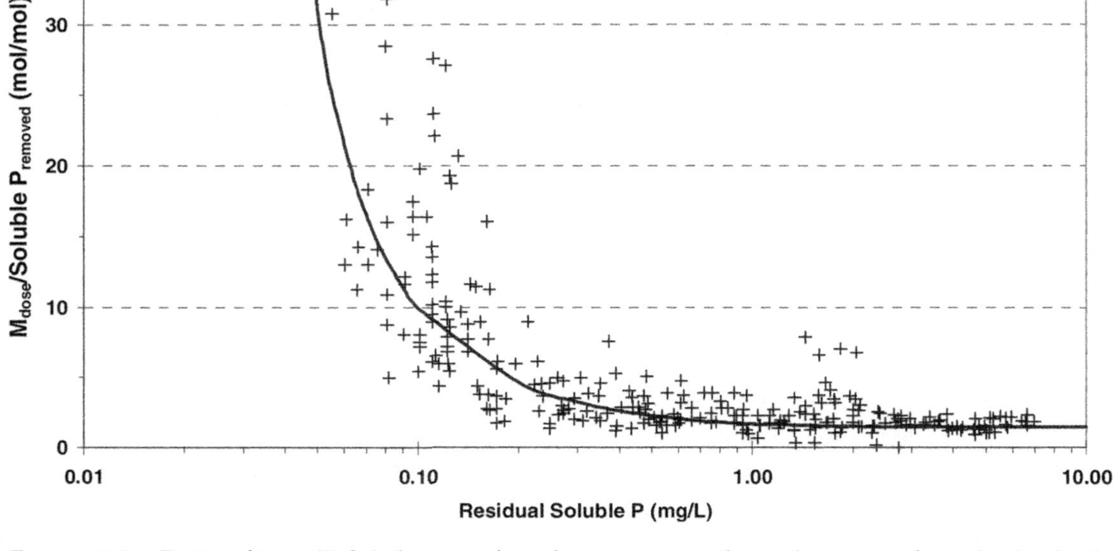

FIGURE 8.3 Ratio of iron (Fe^{3+}) dose to phosphorus removed as a function of residual soluble orthophosphate concentration (Luedecke et al. [1988], and data from the Blue Plains Wastewater Treatment Plant, Washington, D.C.).

data from sites that are operating with chemical phosphorus removal (Luedecke et al., 1988). The data set includes results for a range of pH values (mostly 6.5 to 7.5), temperatures, and wastewater characteristics. The effluent soluble phosphorus concentration is labeled "residual soluble P" on the logarithmic x-axis. The molar ratio for metal ion dose to soluble phosphorus removed is labeled "M_{dose}/soluble $P_{removed}$ (mol/mol)" on the y-axis. It is important to note that the curves apply to the soluble portion of the phosphorus only. The curve used to fit the data is based on the following equation:

$$y = a/(1 + b \times e^{-cXe}) \tag{8.9}$$

Where

y = mole iron required per mole soluble phosphate removed,
a = 1.48,
b = −1.07,
c = 2.25, and
Xe = target effluent soluble phosphorus concentration at the chemical application point (mg/L).

In general, the data at lower residual soluble phosphorus concentrations are more scattered. Because of this variability, bench- or pilot-scale testing is recommended at each facility to determine the actual molar dose required to reach the targeted effluent soluble phosphorus concentration.

The molar dose for phosphorus precipitation is based on the desired final effluent soluble phosphorus concentration rather than the starting phosphorus concentration. For example, to meet a 0.5-mg/L soluble phosphorus concentration, a 2.7-mol ratio of ferric ion to phosphorus or a weight ratio of 4.1 g Fe^{3+}/g P is required. To remove 2.5 mg/L P (from 3 to 0.5 mg/L), an iron dose of 10.25 mg/L Fe^{3+} is required. It is important to determine if a value greater than the influent soluble phosphorus concentration should be used because of the potential for solubilization of the particulate phosphorus, which would increase the soluble phosphorus concentration above measured influent concentrations.

As noted previously, using dual application points may yield the optimum operating point with respect to chemical dose and sludge production. Using the same concentrations as those previously cited, if the phosphorus concentration were to be reduced to 1 mg/L in the primary clarifier, with additional iron added to the aeration basin to achieve a final effluent of 0.5 mg/L soluble phosphorus, less iron will be used in total. To achieve 1 mg/L of soluble phosphorus out of the primary clarifier requires a molar ratio of 1.67 to 1 or a weight ratio of 3 g Fe^{3+}/g P. Therefore, to remove 2 mg/L of phosphorus across the primary clarifier, an iron dose of 6 mg/L is required. To remove the remaining 0.5 mg/L of soluble phosphorus across the secondary treatment system requires a molar ratio of 2.27 to 1 or a weight ratio of 4.1 g Fe^{3+}/g P, which equates to an iron dose of 2.05 mg/L Fe^{3+}. The total iron dose to meet a residential soluble phosphorus concentration of 0.5 mg/L is, therefore, $6 + 2.05 = 8.05$ mg/L Fe^{3+} as opposed to 10.25 mg/L Fe^{3+} if all of the phosphorus is removed at one time. There is a 20% savings in chemical use and a reduction in the overall chemical sludge production. It should be noted that this example has neither taken any credit for phosphorus that would be removed biologically across the secondary treatment system nor has it accounted for potential solubilization of particulate phosphorus. Actual dosages must be fine tuned in the field to account for these issues. A similar relationship exists when aluminum is used for phosphorus precipitation.

A sample calculation is provided in this section to determine the ferric chloride dose required to precipitate soluble phosphorus. Following the detailed, step-by-step derivation, simplified formulas for both International System of Units and U.S. customary units are provided. The following conditions are assumed:

- Influent facility flowrate is 37 850 m^3/d (10 mgd),
- Soluble phosphorus influent concentration to the facility is 3.0 mg/L, and
- A residual primary effluent phosphorus concentration of 1 mg/L is required.

The amount of soluble phosphorus to be removed in kilograms per day (pounds per day) is as follows:

$$\text{Phosphorus} = (3 - 1\ \text{mg/L}) \times 37\,850\ \text{m}^3\text{/d} \times 1000\ \text{L/m}^3 \times 1\ \text{kg}/1\,000\,000\ \text{mg} = 75.7\ \text{kg/d}$$

or

$$(3 - 1)\ \text{mg/L} \times 10\ \text{mgd} \times 8.34\ (\text{lb/mg} \times \text{L/mg}) = 167\ \text{lb/d of soluble phosphorus to be removed.}$$

This is equivalent to (75.7 kg/d)/(31 kg/kg-mol P) = 2.44 kg-mol P or (167 lb/d)/(31 lb/lb-mol P) = 5.38 lb-mol of P to be removed.

The amount of ferric ions required can be calculated from eq 8.9 as follows:

$$y = 1.48/(1 - 1.07e^{(-2.25)(1)}) = 1.67\ \text{kg-mol Fe/kg-mol}$$

(or 1.67 lb-mol Fe/lb mol) of soluble phosphorus removed.

Consequently, the amount of ferric ions required per day is

2.44 kg-mol P × 1.67 kg-mol Fe/kg-mol P = 4.07 kg-mol Fe
or 5.38 lb-mol P × 1.67 lb-mol Fe/lb-mol P = 8.99 lb-mol Fe.

The molecular weight of the Fe ion is 56, which can be converted to mass of ferric ions per day, as follows:

4.07 kg-mol Fe × 56 kg Fe/kg-mol Fe = 228 kg
or 8.99 lb-mol Fe × 56 lb Fe/lb-mol Fe = 503 lb/d of ferric ions.

As indicated in Table 8.2, dry ferric chloride contains 34.5 % ferric. Consequently, the amount of dry ferric required for this application will be

228 kg Fe/0.345 = 661 kg, or 503 lb Fe/0.345 =
1459 lb/d of dry ferric chloride.

If 37% ferric chloride solution is used, it contains 0.173 kg/L of ferric ions (Table 8.2). Thus, the amount of ferric solution required for this application will be

228 kg Fe/0.173 kg/L = 1318 L
or 503 lb Fe/1.44 = 349 gpd of solution.

The 37% ferric chloride solution required per day to reduce the soluble phosphorus from 3.0 to 1.0 mg/L in the aforementioned example is 1318 L/d

(349 gpd). The ferric dose equivalent to this ferric chloride use rate can be calculated from the factor listed in Table 8.2 of 0.504 kg/L (4.2 lb/gal) of ferric chloride in 37% solution, as follows:

$$1318 \text{ L/d} \times 0.504 \text{ kg/L}/37\,850 \text{ m}^3\text{/d} = 0.0176 \text{ kg/m}^3$$
$$0.0175 \text{ kg/m}^3 \times 1\,000\,000 \text{ mg/kg}/1000 \text{ L/m}^3 = 17.6 \text{ mg/L}$$

In U.S. customary units, calculation for this 10-mgd facility will be as follows:

$$349 \text{ gpd} \times 4.2 \text{ lb/gal}/8.34 \text{ (lb/mg} \times \text{L/mgd)}/10 \text{ mgd} = 17.6 \text{ mg/L}$$

4.3.3 *Summary Dose Formulas for Ferric Chloride*

The aforementioned calculations can be abbreviated to the following overall formula for calculation of the application rate of 37% ferric chloride solution in International System of Units:

$$A = (0.0155)(Xi - Xe)(Q)/[1 - 1.07 \times \exp(-2.25 \times Xe)] \tag{8.10}$$

Where

A = 37% ferric chloride solution application rate (L/d),
Xi = soluble phosphorus concentration at the application point (mg/L),
Xe = target effluent soluble phosphorus concentration (mg/L), and
Q = facility flow (m^3/d).

In U.S. customary units, the formula is as follows:

$$A = (15.5)(Xi - Xe)(Q)/[1 - 1.07 \times \exp(-2.25 \times Xe)] \tag{8.11}$$

Where

A = 37% ferric solution application rate (gpd),
Xi = soluble phosphorus concentration at the application point (mg/L),
Xe = target effluent soluble phosphorus concentration (mg/L), and
Q = facility flow (mgd).

It should be recognized that the aforementioned dose calculations are approximate. Many site-specific factors such as mixing conditions (Section 5.3), application point (Section 5.4), wastewater chemistry, and temperature, discussed in more detail later in this chapter, will affect the actual dose of chemical required to accomplish treatment objectives. This is particularly true at low residual soluble phosphorus concentrations (below 0.1 mg/L).

The aforementioned calculations were based on a single application point; however, the optimum operating mode may involve dual application points (i.e., chemical addition to the primary clarifiers and to the secondary treatment process), which could result in savings in chemical consumption and cost (Section 5.4).

4.3.4 *Other Considerations for Iron-Based Chemicals*

As with aluminum-based compounds, there will be a significant amount of sludge produced when iron salts are added to the process to remove phosphorus. The quantity of sludge and handling considerations are discussed in Sections 6.1 and 6.2, respectively. Alkalinity and pH effects of ferric salts are discussed in Section 6.3. Metal salts will also increase TDS in the system. These effects are discussed in more detail in Section 6.4. A potentially significant drawback of using ferric is its potential effect on UV disinfection, as discussed in Section 6.7. Tank and pipe material, pump type, and safety considerations are all similar to the aluminum-based options discussed in Section 4.2.4.

4.4 Lime

Lime has historically been used to increase alkalinity, remove phosphorus, and improve removal efficiencies across primary clarifiers. More recently, use of lime at water resource recovery facilities for phosphorus removal has become less common, presumably because of lime-handling difficulties and the availability of more effective chemicals.

Because high pH ($>$10) is needed for phosphorus precipitation with lime, lime addition for the purpose of phosphorus removal is not compatible with biological treatment as the required pH is too high for microorganisms. However, lime-treated primary effluent may potentially be accepted by the subsequent biological treatment step without neutralization (or with minimal neutralization) because carbon dioxide generated during aerobic treatment will lower the pH. Nitrification, if practiced, could consume alkalinity (i.e., lower the pH) as well.

In addition to the aforementioned use in primary clarifiers, lime addition for phosphorus removal may potentially be used in sidestream treatment (such as in the Phostrip process) or as a post-treatment step with a subsequent pH adjustment. However, use of lime for phosphorus removal is rare. The residual phosphorus concentration typically achieved with lime treatment is approximately 1 mg/L. The dose of lime required for phosphorus removal is typically governed by the alkalinity of the wastewater because lime will first react with bicarbonate before precipitating as hydroxyapatite. That dose is approximately 1.5 times that of total alkalinity in milligrams per liter as $CaCO_3$ (Sedlak, 1991). The hydroxyapatite has a variable composition; however, an approximate equation for its formation can be written as follows, assuming, in this instance, that the phosphate present is the hydrogen phosphate ion (HPO_4^{2-}):

$$3\ HPO_4^{2-} + 5\ Ca^{2+} + 4\ OH^- \rightarrow Ca_5\ (OH)(PO_4)_3 + 3\ H_2O \qquad (8.12)$$

Use of lime for phosphorus removal results in generation of inert solids, the quantity of which is governed mainly by wastewater alkalinity. The following two issues should also be considered when lime is added to the treatment process: (1) relatively high calcium concentrations in the process water can inhibit VSS destruction in digesters and (2) high calcium content in the final sludge may not be advantageous to certain soils if the sludge is ultimately used for soil amendments.

Lime is either gravity fed or pumped to the point of application. Materials of construction for lime systems are carbon steel or PVC. Personnel should wear PPE when handling chemicals; the PPE should include, but not be limited to, gloves, respirators, goggles, aprons, and face shields and should be worn when working or handling chemical solids or slurries.

4.5 Magnesium Hydroxide

Another alternative chemical available to precipitate phosphorus is magnesium hydroxide, although its use in the main liquid train is presently limited because of its high cost. Magnesium hydroxide raises the pH to precipitate phosphorus and, therefore, yields similar results as lime addition, although chemical handling issues are not as significant as those for lime.

In anaerobic digesters, magnesium in the presence of ammonia and orthophosphate promotes formation of magnesium ammonium phosphate (struvite), which is known to cause severe clogging of digester piping. Magnesium addition is used in dedicated treatment processes recently developed for removal and recovery of phosphorus and ammonia from sludge recycle streams (e.g., the Pearl® process by Ostara). This is discussed in more detail in Chapter 12.

Magnesium hydroxide is received in liquid form and can be used similar to ferric or aluminum addition. Tank materials are typically constructed of fiberglass reinforced plastic (FRP) or polyethylene. Piping, valves, and fittings are typically constructed of PVC or other compatible plastic material.

4.6 Proprietary Formulations

A number of proprietary formulations for phosphorus removal have been marketed in recent years. While the exact composition of these products is typically not known, they are likely based on a combination of some of the chemicals previously discussed in this chapter. The products variously claim to be easier to handle, generate less sludge, be less sensitive to lower wastewater temperatures, and/or have a lesser effect on facility pH. While claims of lower sludge generation should be treated with skepticism, the other claims may be valid. This is attributable to the fact that these formulations typically contain some form of neutralizing alkali agent, which increases pH and alkalinity and makes the chemicals less corrosive, as in the case of sodium aluminate and PACl discussed earlier in this chapter.

However, this convenience typically comes at a higher chemical cost. The recommended method of evaluating the expected benefits of using alternative chemicals is to determine the dose of each chemical needed to achieve the required phosphorus removal in side-by-side jar tests. Such jar tests should include evaluation of the required dose of a neutralizing agent. The dose requirements established in the comparative jar tests can then be combined with unit costs of alternative chemicals to determine overall application costs. Only then can the relative costs of using different chemicals be compared to other factors, such as ease of handling and need for any additional neutralizing chemicals, and an informed decision be made. For small facilities, use of specialized or proprietary formulations may sometimes be justified; however, at larger facilities, chemical costs will likely be an overriding consideration.

4.7 Water Treatment Sludge

Water treatment residuals typically contain a large fraction of aluminum or ferric salts, which may have residual capacity to bound and adsorb phosphorus. Such sludges can be used as cost-effective materials to reduce soluble phosphorus in wastewater. Although fresh alum was found to be more efficient at phosphorus removal than spent alum sludge (Georgantas et al., 2006), alum sludge may be a reasonable substitute for alum because of its low cost and high availability. It should also be noted that the efficiency of alum sludge to remove phosphorus decreases through aging. More than 90% phosphorus removal from wastewater has been observed at sufficiently high spent-water-treatment sludge application doses (Asada et al., 2010; Ippolito et al., 2011; Mortula and Gagnon, 2007). Naturally, use of spent water treatment facility sludge for wastewater treatment transfers the burden of water treatment residual handling to the wastewater facility and could result in a large increase of waste sludge generated at the facility.

4.8 Role of Polymers

Organic polymers do not remove or precipitate soluble phosphorus on their own to any significant extent. However, their use can significantly improve the solids separation process by coagulating colloids and increasing the size and compactness of the chemical and biological solids, thus increasing their settling velocity.

Use of metal salts for phosphorus removal can create, at least initially, fine precipitate (floc) or even colloids. Flocculation and settling of such particulates can be greatly improved by the application of a polymer. The use of polymer can be particularly effective when low-effluent phosphorus levels are required. In some high-rate applications used for phosphorus removal, such as several patented sand-ballasted flocculation and sedimentation processes, the use of polymer at a proper dose is critical to process performance.

When used, the polymer addition point should be as far downstream from the point of addition of the metal salts as practical and should be located in a place where adequate mixing is available or can be created. In many typical configurations of the activated sludge process, metal salts are added at the outlet of the aeration basin, with polymer addition at the final clarifier splitter box.

Adequate mixing, which is critical to polymer effectiveness, can be provided by simple air agitation.

When added to mixed liquor, the polymer dose required is typically in the range of 0.5 to 1 mg/L for dry polymers, which typically have approximately 90% active content. For emulsion polymers, with only 30 to 40% active content, the required dose will be about 2.5 to 3 times higher (as expressed in the weight of the emulsion, as supplied).

It should be noted that, at some activated sludge facilities, polymer addition is continually added to final clarifiers to improve their performance. Other facilities use polymer addition to final clarifiers in a standby mode during storm flows.

When metal salts are added to the primary clarifier (CEPT), polymer addition can significantly increase phosphorus, TSS, and particulate biochemical oxygen demand (BOD) removal compared to that obtained by coagulant alone. The optimal polymer dose in CEPT treatment, or when applied in a tertiary phosphorus removal process, will typically be lower for a mixed liquor application than mentioned previously.

Typical polymer systems require stainless steel or FRP storage or aging tanks. Polyvinyl chloride piping, valves, and fittings are typically used for polymer service. Because polymers are sensitive to shear and have higher viscosity than most metal salts, progressive cavity pumps are often recommended for polymer service.

Polymer does not have the chemical handling issues associated with most other chemicals. Generally, slips and falls are the most common hazard when handling a polymer solution. Care should be taken when changing the type of the polymer from anionic to cationic based (or vice versa) because a thorough cleaning of polymer storing and handling equipment with solvents may be needed to prevent formation of scale-type deposits.

5.0 PROCESS CONSIDERATIONS

5.1 Limits of Technology

The minimum achievable effluent phosphorus concentration depends on a number of factors, including efficiency of the solids separation process, as discussed in Section 3.3. In general, soluble orthophosphate concentrations in wastewater can be reduced to below 0.01 mg/L with a sufficiently high chemical dose (Smith et al., 2008; U.S. EPA, 2007). Dual-step filtration processes are particularly suited to achieving low concentrations because they provide extended contact time for precipitation, coagulation, and separation of fine particulates initially generated after chemical addition. Microfiltration or ultrafiltration (e.g., in an MBR) while achieving a reliable separation of suspended matter may not be as effective in lowering soluble orthophosphate concentration to the limits of solubility because contact time in the filtration step is limited (WEF, 2010). In some effluents, a measurable concentration (as much as 0.05 mg/L) of non-reactive rDOP may be present (refer to Chapter 11).

In practice, taking into account process and wastewater chemistry variability, the potential presence of measurable rDOP, the effect of any residual particulate phosphorus, and any imperfections in the solids separation step, a monthly average effluent total phosphorus concentration below 0.05 mg/L is generally achievable at a well-designed and operated full-scale facility. In a survey of advanced treatment facilities (U.S. EPA, 2007), the range of monthly average phosphorus concentrations for facilities having a phosphorus limit of 0.1 mg/L or less typically was approximately 0.03 to 0.09 mg/L.

It should be noted that analyzing residual phosphorus at low levels by the standard colorimetrical method is uncertain. Concentration readings from a spectrophotometer calibration curve should be verified manually (as opposed to relying on a regression equation), with calibration data developed at the low end of the concentration range.

5.2 Effect of pH

The minimum solubility of ferric and aluminum orthophosphates in wastewater is in the acidic pH range of 3 to 5 (WEF, 2010), with the exact optimum pH likely being site specific depending on a number of factors, including wastewater chemistry. At pH greater than 7, the residual dissolved orthophosphate concentration starts to increase for the same chemical dose applied. Considering that most facilities must maintain their effluent pH above 6, the optimal pH for minimizing residual soluble phosphorus concentration will be approximately 6.5 to maintain a safety margin. However, maintaining a particular target pH may not be practical or necessary because low effluent phosphorus concentrations were achieved at facilities operating at a pH as high as 7.5 (Takacs et al., 2006).

The target pH of the precipitation reaction, as discussed in this section, should be measured at the point of the precipitate formation, that is, downstream of the chemical application point or at the solids separation point (clarifier effluent). The final effluent pH or aeration basin pH could be somewhat different.

An important consideration is that some commonly used chemicals, including alum and all iron salts, are acidic and will lower the wastewater pH to a degree depending on the alkalinity of the wastewater and the dose of chemical. For larger chemical doses and/or for wastewater with low alkalinity, use of neutralizing agents such as caustic may be necessary. In some instances, use of ferric or alum may actually lower the pH of the wastewater to around 6.5, at which point use of chemicals for phosphorus removal is likely to be optimal.

5.3 Effect of Mixing and Contact Time

Orthophosphate removal with metal salts significantly improves if vigorous mixing at the point of chemical addition is provided. The benefits of mixing seem to be greater at a higher pH and are more pronounced for alum than for ferric (Sagberg et al., 2006; Szabo et al., 2008). Extending the contact time between the precipitated matter and wastewater allows for additional adsorption of soluble orthophosphates and increases the removal efficiency. Consequently, the recommended chemical addition location should be at a turbulent point as far upstream from the solids separation point (i.e., clarifier or filter) as practical. A convenient location for coagulant addition to an activated sludge train is to a turbulent area at the tail end of the aeration basin (e.g., effluent weir).

When chemical addition to the primary clarifier is contemplated, a suitable addition point may be a turbulent location at a preliminary treatment facility, such as at the screens, or to grit tanks (if available), with conveyance and flow distribution structures upstream of the primary clarifiers providing a good flocculating environment. Vigorous mixing desired at the point of chemical addition could also be enhanced by an additional mixing device, such as an air sparger.

5.4 Points of Addition

Metal salts addition points should be upstream of solids separation steps such as clarification of filtration. Figure 8.4 depicts typical available options. The use of chemicals is most efficient at a tertiary application point, where most primary or biological solids or other constituents that can bind with metal salts are already removed. However, most existing facilities are not equipped with a suitable tertiary treatment process.

Chemicals can be added in front of conventional sand filters; however, this decreases the filter run time (the higher the dose, the shorter the filter run) and the acceptable chemical dose in such an application can be too small for adequate phosphorus removal. In such instances, split-point addition may be most

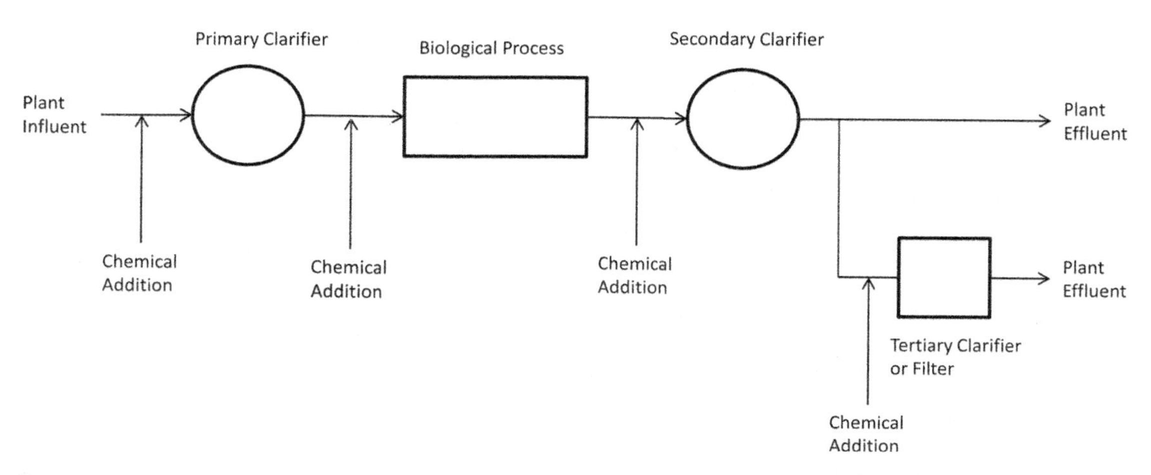

FIGURE 8.4 Chemical phosphorus removal dosing locations.

effective, with part of the chemical going to secondary clarifier influent (aeration basin effluent) and part going to secondary clarifier effluent (filter influent).

At conventional activated sludge facilities, and particularly at smaller, extended aeration facilities, common practice is to add chemicals to a turbulent location in front of final clarifiers such as the effluent weir of the aeration basin or to a clarifier's distribution box.

Application of chemicals for phosphorus removal to the primary clarifier (CEPT) is effective, particularly when biological processes could benefit from lower loading of organics, nitrogen, and TSS. Such practice also minimizes the inert fraction of the mixed liquor and, in case of ferric addition, minimizes the potential effects of chemicals on UV disinfection. Selection of the chemical application point at existing facilities may be limited by the need for good mixing previously discussed, which may not be easy to achieve at all potential locations.

A potentially significant reduction in chemical use could be accomplished by multipoint (or split-point) chemical addition. A commensurable reduction in waste sludge generation and the need for neutralizing chemicals and other process effects can also be realized. Multipoint addition is typically achieved by splitting chemical addition between primary clarifier and aeration basin effluent or between aeration basin effluent and final clarifier effluent (i.e., filter influent).

When phosphorus is removed by chemical addition to tertiary facilities (tertiary clarifiers and/or polishing filters), recycling the resulting chemical sludge to the primary treatment process should be considered, if feasible. Such practice will facilitate use of the residual phosphate bonding capacity of tertiary sludge and could substantially reduce the overall use of chemicals (Takacs et al., 2006).

5.5 Chemical Feed Control

Many smaller and medium-size facilities adding chemicals to the aeration basin operate at a constant chemical flowrate that is manually adjusted periodically, as needed. Diurnal variations in influent phosphorus loading are partially absorbed by large quantities of partially active chemical precipitate present in the aeration basin. Such facilities add chemicals in excess of what is required to meet the permit limit.

Because of the increasing availability and reliability of online probes measuring residual orthophosphate, real-time dose control is becoming feasible and cost effective, even at smaller facilities. A less efficient but simpler option would be to flow-pace the chemical addition with periodic adjustment of the feed ratio based on observed results. Implementation of phosphorus removal process monitoring and dose control is recommended because it lowers chemical costs and mitigates other negative aspects of chemical addition. Chapter 14 provides more details on online analyzers and dose control.

5.6 Response Time and Startup of Chemical Addition

Phosphorus removal by chemical addition can be initiated on demand without the prolonged acclimation period required for some biological processes, such as EBPR. This is typically the case for chemical processes. Some dedicated tertiary processes with a minimal hydraulic retention time (HRT) can lower the effluent phosphorus concentration within minutes of startup. For more conventional applications, the response is more measured and predictable because it depends on HRT in downstream treatment facilities.

Figure 8.5 illustrates results from a test performed at an oxidation-ditch facility without primary clarifiers or tertiary filters (Patoczka, 2008). The alum dose required for achieving the desired effluent total phosphorus level (below 0.5 mg/L) was determined before initiating the test; however, no chemical was added for more than 3 weeks before initiation of the test. As Figure 8.5 shows, the clarifier effluent total phosphorus, originally at 3.7 mg/L, started to decrease almost immediately upon activation of alum addition and dropped below 1 mg/L within 24 hours. In 3 days, the concentration stabilized at 0.35 mg/L.

The delay in system response to initiation of chemical addition to activated sludge is likely attributable to a combination of the followings factors:

- Travel time between the chemical application point and the effluent sampling location. If a final clarifier is only present, this delay will be marginal.

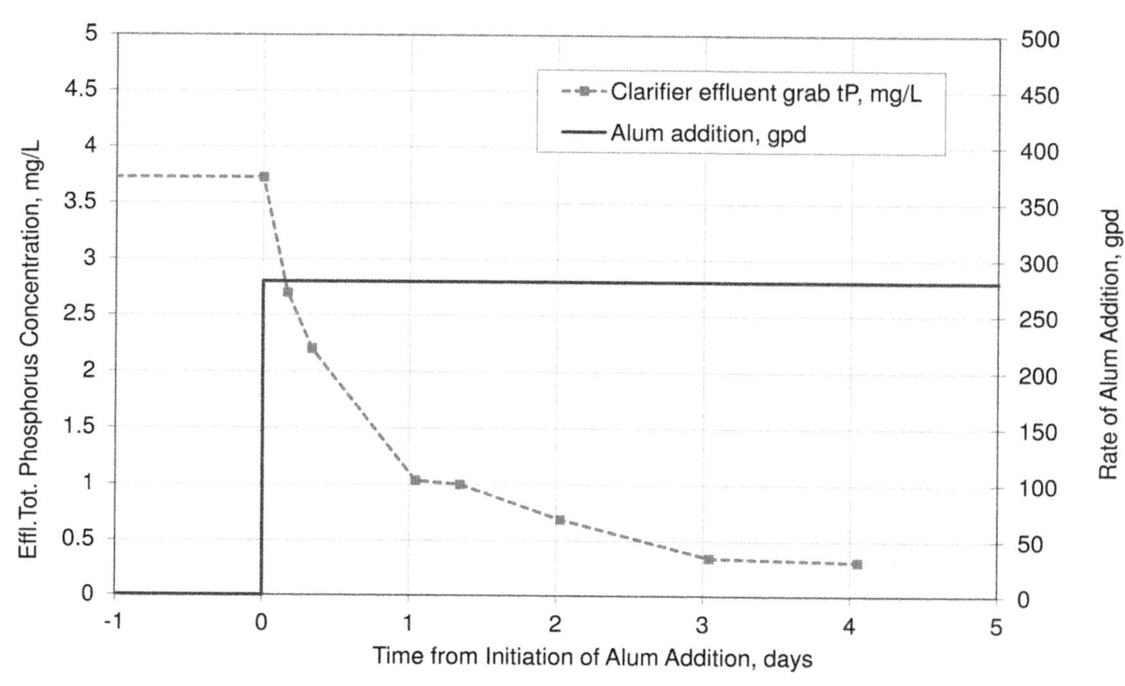

FIGURE 8.5 Response time to initiation of chemical addition.

Any tertiary facilities, chlorine contact tanks, and, particularly, a stabilization lagoon, will add delay commeasurable with HRT of these facilities;

- As discussed in Section 3.2, the excess, unused chemicals applied at the secondary clarifier influent will be returned with RAS to the aeration basin and will bind with phosphorus in the raw wastewater. The time required to reach an accumulation of these chemicals close to a saturation point (steady state) corresponds to several multiples (3 is a good number) of HRT. For an aeration basin with a nominal HRT of 1 day, the time to approach equilibrium conditions will thus be approximately 3 days;
- Metal salts are coagulating agents and will be partially used by biomass and colloids in the aeration basin for coagulation of mixed liquor flocs until a degree of saturation of the biomass inventory is achieved and the mixed liquor at the point of the chemical application is presaturated with chemicals and pre-coagulated; and
- Precipitates resulting from metal salt addition continue to adsorb orthophosphate for several hours or even days, delaying the time to achieve equilibrium.

6.0 EFFECT OF CHEMICAL ADDITION ON FACILITY OPERATIONS

6.1 Sludge Generation

The addition of chemicals for precipitation of phosphorus will generate additional inert chemical sludge at the point of addition, be it in a primary clarifier, activated sludge system, or tertiary application. The amount of additional sludge (on a dry-weight basis) generated by the addition of selected chemicals can be calculated based on the conversion factors provided in Table 8.3. For example, for a 3785-mL/d (1-mgd) facility adding 10 mg/L of alum, the expected additional sludge generation will be 4.11 mg/L, or 15.6 kg/d (33.9 lb/d), based on a 0.411-conversion factor. In many applications, particularly when levels of phosphorus below 1 mg/L are targeted with chemical addition, an excess of chemical over that required by the stoichiometric equation is needed. Because the excess chemical will precipitate in a different form, a different conversion factor will result, as indicated in Table 8.3. As discussed in Section 4.0, the stoichiometric amount of chemical for the removal of 1 mg/L of phosphorus is 9.6 mg/L for alum and 5.2 mg/L for ferric chloride (the chemical dose convention is noted in the footnote in Table 8.3).

Conversion factors listed in Table 8.3 are approximate because simplifying assumptions of precipitate composition were used in their derivation. The addition of coagulating chemicals will increase the capture of colloidal and particulate matter, which may further increase additional sludge generation.

This is particularly applicable when chemicals are added to primary clarifiers (CEPT), where enhanced removal of TSS and BOD will generate even more primary sludge. However, decreased TSS and organic loading to the secondary system will result in a lower waste biological sludge generation. Typical primary clarifier removal rates of 50% for TSS and 30% for BOD_5 may increase to 70 and 50%, respectively, or even more, particularly when polymer addition is also provided.

As discussed in Section 3.2, chemical precipitates accumulating in the aeration basins will result in an increase in the basin inert solids concentration. This effect should be considered when designing a biological treatment system. The equilibrium concentration of inert chemical residue resulting from chemical addition to the activated sludge system can be calculated as follows:

$$\text{MLSSci} = D \times F \times \text{SRT}/\text{HRT} \quad (8.13)$$

Where

MLSSci = concentration of chemical inerts in the mixed liquor arising from chemical addition, mg/L;

D = chemical dose applied to the activated sludge process, mg/L (based on the nominal, forward flow of wastewater);

F = TSS conversion factor for the chemical applied, as listed in Table 8.3;

SRT = solids retention time, days; and

HRT = hydraulic retention time, days.

The following conditions are assumed for the sample calculations described herein:

- Flowrate, 3785 m^3/d (1 mgd);
- Aeration basin volume, 1892 m^3 (500 000 gal);
- Sludge age of the process, 12 days; and
- Ferric (37% solution) application rate, 300 L/d (79.3 gpd).

The HRT of this activated sludge system is equal to the aeration basin volume divided by flow, that is, 1892 m^3/3785 m^3/d = 0.5 days (or 12 hours). The ferric solution application rate is equal to the ferric chloride dose of 40 mg/L (refer to Section 4.3.2 for a calculation example). With 3-mg/L phosphorus concentration in the influent, this represents a stoichiometric ratio of Fe to P of approximately 3 to 1 (again, refer to Sections 4.2.2 and 4.3.2 for exemplary calculations). From Table 8.3, the TSS increase conversion factor (F) for $FeCl_3$ applied at a 3-to-1 ratio is 0.748. Consequently, the inert MLSS concentration is

$$40 \text{ mg/L} \times 0.748 \times 12 \text{ d}/0.5 \text{ d} = 718 \text{ mg/L}$$

The MLSSci calculated from eq 8.13 only accounts for the component of the inert fraction of the mixed liquor resulting from chemical dosing in addition, and above, the typically expected inert MLSS fraction resulting from inert TSS present in the influent or generated during the treatment.

Equation 8.13 can also be used to calculate the amount of additional waste sludge generated by chemical addition. If the facility operated without ferric addition at 3000 mg/L MLSS, the equilibrium MLSS concentration with ferric addition will be 3718 mg/L for the conditions in the aforementioned example. This corresponds to a 24% (3718 mg/L ÷ 3000 mg/L) increase in the dry mass of waste activated sludge generation.

An even more straightforward way of calculating the additional mass of dry-waste inert chemical sludge that is generated is to multiply the chemical use rate expressed in kilograms per day (pounds per day) (see Tables 8.1 and 8.2 for chemical content of typical technical solutions) by the applicable TSS conversion factor from Table 8.3. For example, 49% alum solution has 0.647 kg/L (5.4 lb/gal) of alum (Table 8.1). If a facility is using 200 L/d (52.8 gpd) of the 49% solution, the mass of alum used is 200 L/d × 0.647 kg/L = 129 kg/d (285 lb/d). Assuming a typical application with a 3-to-1 stoichiometric rate, the appropriate TSS conversion factor from Table 8.3 is 0.312. Consequently, this alum addition will result in generation of 129 kg/d × 0.312 = 40.3 kg/d (88.7 lb/d) of inert alum sludge.

When using iron-based coagulants, the presence of iron in the waste sludge disposed of off-site may be beneficial for using the sludge as a soil amendment. Aluminum has no advantage as a soil amendment; therefore, if the final sludge is blended with soil, there may be additional concerns with aluminum content in the sludge.

6.2 Sludge Settling, Thickening, and Dewatering

In addition to increasing the waste solids generation rate on a dry-mass basis, the addition of chemicals such as aluminum and ferric salts will affect sludge settling and dewatering properties. When added to activated sludge, these flocculating chemicals will generally improve solids separation in the secondary clarifier because of the increased capture of fine floc and colloids. Alum and, in particular, iron-based chemicals will increase activated sludge specific density, thereby improving its settling properties (lowering the sludge volume index). This could significantly improve performance of secondary clarifiers, allowing operation at a lower sludge blanket and/or higher MLSS concentration. Ferrous, ferric, and aluminum-based chemicals could also help control activated sludge foaming and possibly inhibit filamentous bulking.

The effect of chemicals is more difficult to predict in sludge thickening and dewatering operations. On several occasions, the Woodland Plant in Morris Township, New Jersey, has initiated and stopped adding alum to remove total phosphorus to approximately 0.7 mg/L. Each time, the facility experienced a decrease in dry solids concentration during gravity belt thickening from approximately 5.5 to 4.5% solids when chemical was being added. Another New Jersey facility, the East Windsor Municipal Utilities Authority, observed a decrease from 3.2 to 2.8% solids concentration from gravity thickeners when alum was used for phosphorus removal.

6.3 Alkalinity and pH

Addition of alum and, particularly, iron compounds will lower the pH of wastewater because these agents are acidic and alkalinity is consumed during precipitation reactions (Section 4.0). As discussed in Section 4.2, some alternative aluminum-based products such as sodium aluminate and PACl will not depress pH or consume alkalinity.

The extent of pH reduction mainly depends on the alkalinity of the wastewater; the higher the alkalinity, the lower the reduction in pH for a given chemical dosage. In some instances of low wastewater alkalinity, addition of an alkaline substance such as sodium hydroxide, soda ash, or lime may be required to maintain acceptable pH. If the facility also nitrifies, alkalinity consumption by phosphorus precipitation must be added to the nitrification alkalinity demand to evaluate the overall effect on the system.

Alum addition consumes alkalinity at a rate of 0.505 mg (as $CaCO_3$) per milligram of alum added, which corresponds to a need for 0.404 mg of caustic (NaOH). Thus, a 3.785-mL/d (1-mgd) facility that applies a 60-mg/L alum dose (227 kg/d or 500 lb/d alum application rate, corresponding to 351 L/d or 92.7 gpd of 49% alum solution, respectively) would require 91.7 kg/d (202 lb/d) of dry caustic to completely recover alkalinity loss induced by alum. This amount of dry caustic corresponds to 120 L/d (31.8 gpd) of 50% caustic solution.

Similarly, ferric chloride addition consumes alkalinity at a rate of 0.923 mg (as $CaCO_3$) per milligram of ferric chloride added, which corresponds to a need for 0.738 mg of caustic (NaOH). Thus, a 3.785-mL/d (1-mgd) facility that applies 30 mg/L of ferric chloride dose (114 kg/d or 250 lb/d dry ferric chloride application rate, corresponding to 226 L/d or 59.5 gpd of 27% ferric chloride solution, respectively) would require 84.1 kg/d (184 lb/d) of dry caustic to completely recover alkalinity loss induced by ferric chloride. This amount of dry caustic corresponds to 111 L/d (29 gpd) of 50% caustic solution.

It is important to note that complete neutralization may or may not be necessary or desired, depending on wastewater alkalinity, effluent pH requirements, and other process considerations such as maintenance of optimum pH for phosphorus precipitation (typically below pH 7) or optimum pH for nitrification (above 7). All technical iron solutions and, in particular, pickle liquor, contain substantial amounts of free sulfuric acid or hydrochloric acid, which will additionally consume alkalinity and suppress the pH of the wastewater, depending on the acid content. As the effectiveness of chemical phosphorus precipitation increases with the lowering of pH (in the typical range of treatment facility operations), some depression of the pH by chemicals may be desirable, providing that the nitrification process is not affected and the effluent remains within the acceptable range of pH.

6.4 Total Dissolved Solids Increase

Addition of chemicals will, in most instances, result in an increase in TDS concentration. Total dissolved solids represents a parameter of growing regulatory concern in many regions. In many areas, regulators are enforcing water quality standards for TDS, which, for fresh waters, typically are 500 mg/L (e.g., New Jersey Water Quality Standards). Discharging high TDS can also be of concern in instances where reclaimed water is used for irrigation purposes or as a cooling water makeup.

However, in most typical applications, an increase in TDS because of chemical addition will be relatively modest. Table 8.3 provides TDS increase factors for selected chemicals. For example, the addition of a relatively high (100-mg/L) dose of alum, which could be sufficient in producing 0.1 mg/L of effluent total phosphorus concentration according to eq 8.2, will increase TDS by 53.3 mg/L based on a factor of 0.533 from Table 8.3 (assuming a 3-to-1 stoichiometric rate). In most situations where chemical addition is lower, TDS effects from using chemicals for phosphorus removal will be relatively small. Table 8.3 also provides TDS increase factors for some other chemicals and processes commonly used in wastewater treatment.

6.5 Biological Phosphorus Removal

While the EBPR process can sometimes be effective in producing low-effluent phosphorus, consistently meeting low-effluent limits could be impaired by periods of substandard performance or upsets. For these reasons, chemical addition is commonly practiced at EBPR facilities as a tertiary polishing step, a backup process, or, more commonly, in a form of simultaneous biological and chemical phosphorus removal by coagulant addition to the activated sludge train (Gebremariam et al., 2011; Neethling et al., 2005). Although metal salts addition to an EBPR process has been shown to improve the overall phosphorus removal efficiency, there are concerns that continuous dosing of coagulants to an activated sludge facility may lower efficiency of the phosphorus release and uptake cycle by competition or inhibition. As chemicals added to activated sludge remain in the system for a period commensurable with sludge age, the effect of a chemical added at a high dose during a period of inferior EBPR performance could linger for a considerable time, delaying recovery of the EBPR system.

Consequently, metals addition to the EBPR process should be practiced carefully, at the lowest necessary dose, with the point of addition being at the effluent from the aeration basin. Where possible, chemical addition to the EBPR activated sludge train should be avoided and, if necessary, chemicals should be applied to the primary clarifier and/or to the tertiary process.

6.6 Anaerobic Digestion

Because of the low solubility of ferrous sulfide (FeS), ferric salt addition directly to anaerobic digesters has been used to control H_2S in the digester gas and for odor control. Lee et al. (2009), Novak and Park (2010), and Yuan and Bandosz (2007) suggested that the addition of iron salts for phosphorus removal in the liquid train will have a similar effect on digester gas quality. Aluminum-based salts are not expected to have this effect.

When metal salts are used in the liquid train for phosphorus removal or ferric is added directly to the digester for odor control, struvite (magnesium ammonium phosphate) formation in the anaerobic digesters is typically prevented. This is because most of the orthophosphate in the waste sludge is bound in a stable precipitate, making it unavailable for struvite formation. There are some reports (e.g., Chen et al. [2008]; Dentel and Gosset [1982], and Monteith and Atkinson [2001]) that at certain aluminum concentrations anaerobic digestion could become inhibited. However, this does not appear to be a problem at metal addition rates typically used for phosphorus control because no instances of full-scale anaerobic digestion inhibition were found at a number of facilities using alum for phosphorus removal.

In summary, when anaerobic digestion is part of the treatment process, the effect of chemicals added for phosphorus removal should be carefully considered. Ferric salts may be preferred over aluminum-based products in such applications, mainly because of their side benefit in odor control. However, under some conditions, ferric addition at a high dose could cause formation of iron scales in digester piping. Separation of biological and chemical sludges should be considered, if practical, to avoid increased inert load to the digester. Additionally, when phosphorus is bound with aluminum or ferric in the waste (and digested) sludge, it becomes largely unavailable as fertilizer, potentially diminishing the value of biosolids derived from sludge. It is important to note that magnesium ammonium phosphate (struvite) is more soluble than aluminum or ferric sulfates. Consequently, this compound recovered from sidestreams by processes such as Pearl (see Chapter 12) may be more suitable as a long-release fertilizer.

6.7 Ultraviolet Disinfection

Effectiveness of UV disinfection is dependent on the delivered UV dose, contact time, liquid film thickness, wastewater absorbance, wastewater turbidity, wastewater chemistry (including hardness, alkalinity, pH, and oxidation–reduction potential), system configuration, and temperature. Fouling of UV lamp sleeves decreases the effective UV dose reaching the wastewater and is the main cause of decreased UV-system disinfection efficiency.

Lu et al. (2012) observed that iron salts have a greater negative effect on UV transmittance in wastewater than other coagulants. The presence of iron in UV-system influent has been documented to be a significant factor in the overall decrease of the disinfection efficacy through the following mechanisms:

- Dissolved iron molecules can absorb UV radiation in critical wavelengths, preventing UV light from reaching target organisms. This may include cationic free iron or iron complexes adsorbed into the residual suspended solids and bacteria flocs (Kozak et al., 2011);
- Iron precipitates can add to the resulting suspended solids of a treated effluent, thereby causing enhanced shielding effects, blocking the transmittance of UV light. Ultraviolet radiation is believed to promote precipitation processes of residual dissolved iron as $Fe(OH)_3$ and other compounds (Kozak et al., 2011; Nessim and Gehr, 2006); and
- Residual iron, particularly at concentrations in excess of 0.5 mg/L, has been found to be the main constituent associated with fouling of UV lamps, although the presence of calcium, magnesium, phosphorus, and organic matter is also important for scale formation. Formation and deposition of precipitates on the lamp's quartz sleeves is promoted by the lamp's high temperature. The precipitate, in addition to ferric salts, could include $CaCO_3$, $CaSO_4$, $MgCO_3$, $MgSO_4$, $Al(OH)_3$, and $Al_2(SO_4)_3$ (Black and Veatch, 2010; Kozak et al., 2011; Nessim and Gehr, 2006; Peng et al., 2005; Sehnaoui and Gehr, 2001; Sheriff and Gehr, 2001).

Operation of the UV system could also be affected by fouling of UV light sensors, causing an unnecessary increase in power to compensate for the errant light-intensity measurements.

With increasing iron concentrations above 0.5 mg/L, particularly in the presence of other scale-forming constituents, scaling may be quite rapid, resulting in decreased effectiveness of UV disinfection in a matter of days or even hours, depending on the concentration of iron and other constituents (Sheriff and Gehr, 2001). It is important to note that the presence of residual ferric at such elevated concentrations is not typical (and not necessary) and the ferric application rate should be reduced. Chemical and mechanical cleaning of the sleeves could cause permanent fouling because of the scratches that trap foulants (Peng et al., 2005). On the other hand, addition of coagulants and/or polymer to the upstream treatment processes could significantly reduce the concentration of dispersed solids and colloidal matter in the UV influent, thus reducing turbidity and improving disinfection performance.

6.8 Potential Phosphorus Deficiency

It is well established that activated sludge bacteria need both nitrogen and phosphorus (sometimes called *macronutrients*) in addition to other microconstituents.

These are typically available in adequate concentrations in municipal wastewater. An aggressive phosphorus removal approach using chemical addition ahead of the biological process, like CEPT, could result in inadequate phosphorus supply for the bacteria.

As discussed in Section 3.3, a conventional activated sludge biomass has a phosphorus content of about 2% (and much more if EBPR is practiced). As a first approximation, a general rule can be used that for each 100 mg/L of BOD_5 consumed by the biological process, 1 mg/L of phosphorus will be needed for biological growth (in addition to 5 mg/L of nitrogen). Consequently, to avoid phosphorus deficiency in the downstream biological treatment process, effluent from a primary clarifier with 50 mg/L of BOD_5 should have residual phosphorus of no less than 0.5 mg/L in soluble orthophosphate form because other forms of phosphorus may not be bioavailable.

Another treatment process that could experience phosphorus deficiency is a separate-stage tertiary denitrification process at facilities required to also meet a tight effluent phosphorus limit. Some facilities, particularly in the Chesapeake Bay watershed, are required to simultaneously meet total nitrogen limits of 3 mg/L and total phosphorus limits as low as 0.18 mg/L. To ensure that denitrifying bacteria have an adequate supply of phosphorus, deBarbadillo et al. (2006) recommended maintaining an orthophosphorus-P (i.e., expressed as phosphorus to NO_x-N (i.e., nitrates and nitrites expressed as nitrogen) ratio of 0.02 or more.

6.9 Safety and Operational Considerations

All chemicals, whether they are a gas, solid, or liquid, require a feeding system to accurately and repeatedly control the amount applied. Effective use of chemicals depends on accurate dosages and proper mixing. The effectiveness of certain chemicals is more sensitive to dosage rates and mixing than that of others. Design of a chemical feed system must consider the physical and chemical characteristics of each chemical used for feeding; minimum and maximum ambient or room temperatures; minimum, average, and maximum wastewater flows; minimum average and maximum anticipated dosages required; and the reliability of feeding devices.

Operators and maintenance personnel should be aware of the hazards and characteristics of the chemicals that are used at a facility. Material safety data sheets and technical specifications provided by suppliers are a good source of this information. Additional resources for design and operation of chemical feed systems are cited in the Suggested Readings section of this Chapter.

7.0 REFERENCES

Asada, L. N.; Sundefeld, G. C., Jr.; Alvarez, C. R.; Filho, S. S.; Piveli, R. P. (2010) Water Treatment Plant Sludge Discharge to Wastewater Treatment Works:

Effects on the Operation of Upflow Anaerobic sludge Blanket Reactor and Activated Sludge Systems. *Water Environ. Res.*, **82** (5), 392–400.

deBarbadillo, C.; Rectanus, R.; Canham, R.; Schauer, P. (2006) Tertiary Denitrification and Very Low Phosphorus Limits: A Practical Look at Phosphorus Limitations on Denitrification Filters. *Proceedings of the 79th Annual Water Environment Federation Technical Exhibition and Conference* [CD-ROM], Dallas, Texas, Oct 21–25; Water Environment Federation: Alexandria, Virginia.

Black and Veatch Corporation (2010) *White's Handbook of Chlorination and Alternative Disinfectants,* 5th ed.; Wiley & Sons: Hoboken, New Jersey.

Chen, Y.; Cheng, J. J.; Creamer, K. S. (2008) Inhibition of Anaerobic Digestion Process: A Review. *Bioresour. Technol.*, **99** (10), 4044–4064.

Dentel, S. K.; Gosset, J. M. (1982) Effect of Chemical Coagulation on Anaerobic Digestibility of Organic Materials. *Water Res.*, **16** (5), 707–718.

Gates, D. D.; Luedecke, C.; Hermanowicz, S. W.; Jenkins, D. (1990) Mechanisms of Chemical Phosphorus Removal in Activated Sludge with Al(III) and Fe(III). *Proceedings of the 1990 Specialty Conference on Environmental Engineering*; American Society of Civil Engineers; Reston, Virginia; p 322.

Gebremariam, S. Y.; Beutel, M. W.; Christian, D.; Hess, T. F. (2011) Research Advances and Challenges in the Microbiology of Enhanced Biological Phosphorus Removal—A Critical Review. *Water Environ. Res.*, **83** (3), 195–219.

Georgantas, D. A.; Matsis, V. M.; Grigoropoulou, H. P. (2006) Soluble Phosphorus Removal Through Adsorption on Spent Alum Sludge. *Environ. Technol.*, **27** (10), 1081–1088.

Ippolito, J. A.; Barbarick, K. A.; Elliot, H. A. (2011) Drinking Water Treatment Residuals: A Review of Recent Uses. *J. Environ. Qual.*, **40** (1), 1–12.

Kozak, J. A.; Lordi, D. T.; Abedin, Z.; O'Connor, C.; Granato, T.; Kollias, L. (2011) The Effect of Ferric Chloride Addition for Phosphorus Removal on Ultraviolet Radiation Disinfection of Wastewater. *Environ. Practice*, **12** (4), 275–284.

Lee, Y.; Baker, S.; Wang L.; Gibson, D.; Amos, J.; Zaleski, A. (2009) Operational Impact of Phosphorus Removal Using Ferric Chloride on Anaerobic Digesters and Dewatering. *Proceedings of the Water Environment Federation Nutrient Removal Specialty Conference*; Washington, D.C., June 28–July 1; Water Environment Federation: Alexandria, Virginia; pp 7411–7418.

Luedecke, C.; Hermanowicz, S. W.; Jenkins, D. (1988) Precipitation of Ferric Phosphate in Activated Sludge: A Chemical Model and its Verification. *Water Sci. Technol.*, **21**, 325–338.

Lu, G.; Li, C.; Zheng, Y.; Deng, A. (2012) Effect of Different Coagulants on The Ultraviolet Light Intensity Attenuation. *Desalination Water Treat.*, **37** (1–3), 302–307.

Monteith, H.; Atkinson, D. (2001) By The Numbers: Process Data Review Reveals the Causes of Digester Inhibition. *Proceedings of the 74th Annual Water Environment Federation Technical Exhibition and Conference* [CD-ROM]; Atlanta, Georgia, Oct 13–17; Water Environment Federation: Alexandria, Virginia; pp 389–400.

Mortula, M. M.; Gagnon, G. A. (2007) Phosphorus Treatment of Secondary Municipal Effluent Using Oven-Dried Alum Residual. *J. Environ. Sci. Health.*, **42**, 1685–1691.

Neethling, J. B.; Bakke, B.; Benish, M.; Gu, A.; Stephens, H.; Stensel, H. D.; Moore, R. (2005) *Factors Influencing the Reliability of Enhanced Biological Phosphorus Removal*; Report No. 01-CTS-3; Water Environment Research Foundation: Alexandria, Virginia.

Nessim, Y.; Gehr, R. (2006) Fouling Mechanisms in a Laboratory-Scale UV Disinfection System. *Water Environ. Res.*, **78** (12), 2311–2323.

Patoczka, J. (2006) TDS and Sludge Generation Impacts from Use of Chemicals in Wastewater Treatment. *Proceedings of the 79th Annual Water Environment Federation Technical Exhibition and Conference* [CD-ROM], Dallas, Texas, Oct 21–25; Water Environment Federation: Alexandria, Virginia.

Patoczka, J. (2009) Unpublished Report Prepared for Bernards Township Sewerage Authority, New Jersey, by Hatch Mott MacDonald, Millburn, New Jersey.

Peng, J.; Qiu, Y.; Gehr, R. (2005) Characterization of Permanent Fouling on the Surfaces of UV Lamps Used for Wastewater Disinfection. *Water Environ. Res.*, **77** (4), 309–322.

Sedlak, R. (1991) *Phosphorus and Nitrogen Removal from Municipal Wastewater;* Lewis Publishers: New York.

Sehnaoui, K.; Gehr, R. (2001) Fouling of UV Lamp Sleeves: Exploring Inconsistencies in the Role of Iron. *Proceedings of the First International Congress on UV Technologies;* Washington, D.C., June 14–16; International UV Association: Ayr, Ontario, Canada.

Sheriff, M.; Gehr, R. (2001) Laboratory Investigation of Inorganic Fouling of Low Pressure UV Disinfection Lamps. *Water Quality Res.*, **36** (1), 71–92.

Smith, S.; Takacs, I.; Murthy.; Daigger, G. T.; Szabo, A. (2008) Phosphate Complexation Model and Its Implications for Chemical Phosphorus Removal. *Water Environ. Res.*, **80** (5), 428–438.

Takacs, I.; Murthy, S.; Smith, S; McGrath, M. (2006) Chemical Phosphorus Removal to Extremely Low Levels: Experience at Two Plants in the Washington, D.C. Area. *Water Sci. Technol.*, **53,** 21.

U.S. Environmental Protection Agency (2007) *Advanced Wastewater Treatment to Achieve Low Concentration of Phosphorus*; EPA 910-R/07-002; U.S. Environmental Protection Agency: Washington, D.C.

Water Environment Federation (2010) *Nutrient Removal*; Manual of Practice No. 34; McGraw-Hill: New York.

Yuan, W.; Bandosz, T. J. (2007) Removal of Hydrogen Sulfide from Biogas on Sludge-Derived Adsorbents. *Fuel*, **86** (17–18), 2736–2746.

8.0 SUGGESTED READINGS

Cabirol, N.; Barraga'n, E. J.; Dura'n, A.; Nayola, A. (2003) Effect of Aluminum and Sulphate on Anaerobic Digestion of Sludge from Wastewater Enhanced Primary Treatment. *Water Sci. Technol.*, **48** (6), 235–240.

Gehr, R.; Wright, H. (1998) UV Disinfection of Wastewater Coagulated with Ferric Chloride: Recalcitrance and Fouling Problems. *Water Sci. Technol.*, **38** (3), 15–23.

Stumm, W.; Morgan, J. J. (1970) *Aquatic Chemistry*; Wiley & Sons: New York.

Water Environment Federation; American Society of Civil Engineers; Environmental and Water Resources Institute (2009) *Design of Municipal Wastewater Treatment Plants*, 5th ed.; Manual of Practice No. 8; ASCE Manuals and Reports on Engineering Practice No. 76; McGraw-Hill: New York.

Water Environment Federation (2007) *Operation of Municipal Wastewater Treatment Plants*, 6th ed.; Manual of Practice No. 11; McGraw-Hill: New York.

Chapter 9

Enhanced Biological Phosphorus Removal Systems

Samuel S. Jeyanayagam, Ph.D., P.E., BCEE;
Ting Lu, Ph.D.; and Steven Reusser, P.E.

1.0 INTRODUCTION

This chapter will build on the fundamentals of enhanced biological phosphorus removal (EBPR) reviewed in Chapter 7 and will primarily cover the components of a functional EBPR system. Many of the topics covered in this chapter are also discussed elsewhere in the manual; therefore, the relevant chapters are cross referenced.

2.0 SYSTEM CONFIGURATION

Several process configurations are available to achieve EBPR with or without concurrent nitrogen removal. These can be categorized as plug flow or completely mixed systems.

2.1 Plug Flow Systems

In theory, plug flow is characterized by fluid particles passing through the reactor with little or no longitudinal mixing and exiting the reactor with their relative positions remaining the same as when they entered. Consequently, all particles have the same retention time. Because of hydraulic inefficiencies, true plug flow is difficult to achieve in full-scale applications. Tanks are designed with high length-to-width ratios or compartmentalized to approximate plug flow conditions. Commonly used process schemes are summarized in Table 9.1 and are briefly described in the following subsections (WEF, 2011).

2.1.1 *Anaerobic and Oxic*

The anaerobic and oxic (A/O) configuration represents the basic EBPR scheme and incorporates the two environmental conditions needed for phosphorus release and uptake reactions. It was originally developed as the Phoredox system in South Africa in 1974 (Barnard, 1974). In North America, it is now known as the A/O process. The anaerobic zone receives secondary influent and return activated sludge (RAS). This provides polyphosphate-accumulating organisms (PAOs) in RAS the first opportunity to use the influent carbon source. This critical step selects PAOs and provides them with a competitive advantage to perform EBPR in the aerobic zone. Chapter 7 contains a detailed description of the EBPR. The main indicator of the presence of a well-established PAO population is the typical release and uptake pattern of the EBPR process described in Chapter 7.

For the A/O process to be effective, it is important to protect the integrity of the anaerobic zone by eliminating dissolved oxygen in the influent and nitrate in RAS. Secondary influent typically has low or no dissolved oxygen. However, upstream unit operations such as preaeration, use of screw pumps, open discharge from centrifugal pumps, and free-fall over weirs can incorporate dissolved oxygen and should be avoided. In nitrifying systems, nitrate in RAS will turn part of the anaerobic zone into an anoxic zone for denitrification in the presence of influent carbon. If this was not considered during sizing of the anaerobic zone, EBPR performance could be affected because of reduced anaerobic volume.

2.1.2 *Anaerobic–Anoxic–Oxic Variations*

Achieving EBPR requires cycling the microbial consortium (sludge) between anaerobic and aerobic conditions for the appropriate length of time. An anoxic zone is added to accomplish simultaneous nitrogen and phosphorus removal. Several process configurations are available and are outlined in Chapter 10.

2.1.3 *PhoStrip*

PhoStrip is a sidestream biological–chemical phosphorus removal process in which a phosphate stripper tank is used. A portion of the phosphate-rich return sludge is diverted to a stripper designed to operate as an anaerobic sludge thickener where phosphate release takes place (Figure 9.1). Once the supernatant is removed, the sludge is returned to the RAS line and back into the aeration basins. The phosphate-rich supernatant is treated with lime in a separate reactor.

TABLE 9.1 Key features of available EBPR alternatives (WEF, 2010).

Process	Zones	Internal recycles	Key features
Anaerobic–oxic or phoredox	Anaerobic–oxic	None	Total phosphorus removal, mainstream anaerobic and oxic zones
PhoStrip	Oxic–anaerobic	None	Total phosphorus removal, mainstream oxic and sidestream anaerobic zones, biological and chemical removal
Anaerobic–anoxic–oxic (A^2/O)	Anaerobic–anoxic–oxic	• Nitrate (oxic to anoxic)	Total phosphorus and total nitrogen removal
University of Cape Town	Anaerobic–anoxic–oxic	• Nitrate (oxic to anoxic) • Anaerobic (anoxic to anaerobic)	Total phosphorus and total nitrogen removal, RAS to anoxic, anaerobic zone protected, larger anaerobic because of lower mixed liquor volatile suspended solids
Modified University of Cape Town	Anaerobic–anoxic1–anoxic2–oxic	• Nitrate (oxic to anoxic1) • Anaerobic (anoxic2 to anaerobic)	Total phosphorus and total nitrogen removal, RAS to anoxic1 greater protection of anaerobic zone, larger anaerobic because of lower MLSS
Virginia Initiative Process	Anaerobic–anoxic1–anoxic2–oxic	• Nitrate (oxic to anoxic1) • Anaerobic (anoxic2 to anaerobic)	Similar to modified University of Cape Town except nitrate recycle mixed with RAS and anaerobic and anoxic zones are staged
Five-stage Bardenpho	Anaerobic–anoxic1–oxic–anoxic2–reaeration	• Nitrate (oxic to anoxic1)	Total phosphorus and total nitrogen removal, second anoxic (anoxic2)
Oxidation ditch	Anaerobic–oxic	None	Total phosphorus and total nitrogen removal, anaerobic zone precedes oxidation ditch, simultaneous nitrification and denitrification
Johannesburg	Preanoxic–anaerobic–oxic	• Nitrate (oxic to anoxic)	Total phosphorus and total nitrogen removal, preanoxic zone for RAS denitrification and protect anerobic zone
Westbank	Preanoxic–anaerobic–anoxic–oxic	• Nitrate (oxic to anoxic)	Total phosphorus and total nitrogen removal, preanoxic zone for RAS denitrification and protect anaerobic zone
Schreiber	Anerobic–oxic (temporal zone distribution)	None	Total phosphorus removal, air is cycled to achieve anerobic–oxic conditions, incidental total nitrogen removal
Sequencing batch reactor	Anerobic–oxic (temporal zone distribution)	None	Total phosphorus removal, air is cycled to achieve anaerobic–oxic conditions, incidental total nitrogen, no clarifiers
Phased isolation ditch (BioDenipho)	Anerobic–oxic (temporal zone distribution)	None	Total phosphorus removal, air is cycled to achieve anerobic–oxic conditions, incidental total nitrogen removal

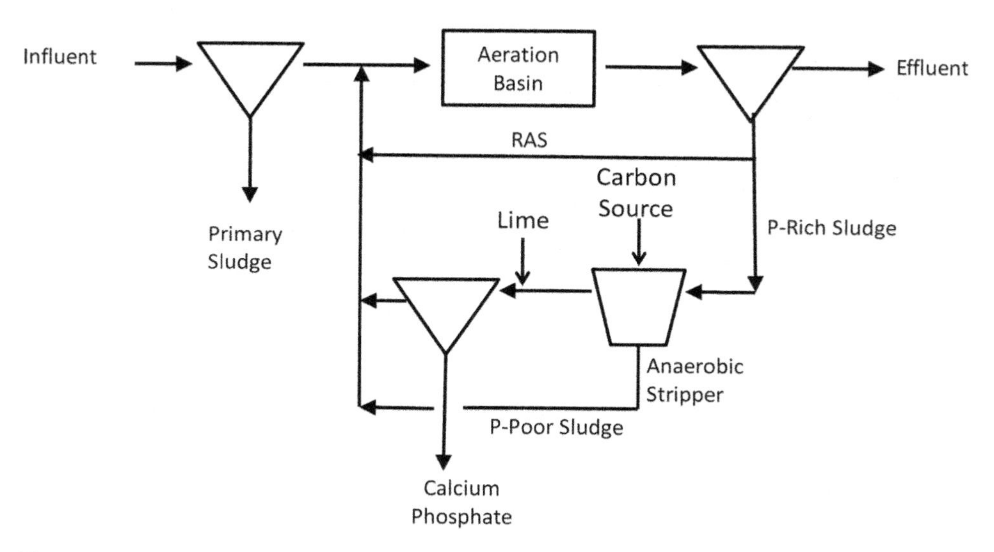

FIGURE 9.1 PhoStrip process.

Phosphorus is then chemically precipitated out of the solution as calcium phosphate in a settler, and the liquid is returned to the head of the system. Unless phosphate recovery is desired, this configuration does not have significant benefits over the biological-only treatment configurations.

2.2 Complete-Mix Systems

In a complete-mix system, as the fluid particle enters the reactor, it is immediately dispersed and uniformly distributed throughout the reactor. Complete-mix conditions are typically established in a round or square reactor. The following subsections present examples of some complete-mix systems.

2.2.1 *Oxidation-Ditch Configurations*

Traditional oxidation ditches are constructed in a racetrack setup in which the secondary influent and mixed liquor are circulated around a center barrier. Vendor-specific variations of the traditional oxidation-ditch concept have been developed. As reported by Daigger and Littleton (2000), full-scale operating experience has shown that simultaneous nitrogen and phosphorus removal can be accomplished in these systems via two mechanisms: (1) biological reactor (bioreactor) mixing patterns that allow the anoxic and anaerobic zones necessary for EBPR to develop, referred to as the *bioreactor macroenvironment*, and (2) development of anoxic and anaerobic zones within the floc, referred to as the *floc microenvironment*. Tight dissolved oxygen control, good mixing, and adequate solids inventory are important to achieving EBPR in oxidation ditches. In some configurations, an anaerobic zone is added upstream of the oxidation ditch to select PAOs.

2.2.2 *Sequencing Batch Reactor*

In a sequencing batch reactor (SBR), the anaerobic and aerobic environments required for EBPR are created in the same tank but not at the same time, which is the case in a plug flow reactor. An anaerobic reaction period can be developed during the fill and initial react periods by removing sufficient nitrate following the aeration period. It is also possible to minimize nitrate concentration by incorporating several alternating aerobic and anoxic periods during the react period.

3.0 FACTORS AFFECTING ENHANCED BIOLOGICAL PHOSPHORUS REMOVAL PERFORMANCE

Enhanced biological phosphorus removal is a complex process encompassing different environmental conditions and complimenting and competing biochemical reactions. This section discusses the many factors that could singly or in combination affect EBPR efficiency.

3.1 Influent Characteristics

The key factor that determines the amount of phosphorus stored in activated sludge is the amount of readily biodegradable organic matter in the anaerobic zone. There must be a large excess beyond that needed to deplete electron acceptors (dissolved oxygen and nitrogen oxide) in the anaerobic zone because the bacteria will preferentially metabolize the organic matter and reduce the amount of stored polyhydroxyalkanoates (PHAs), thereby affecting PAO selection and EBPR performance.

The type of organic compound in the anaerobic zone is also important; it must be soluble and readily biodegradable. Because biodegradability is a function of the length of the carbon chain, short-chain volatile fatty acids (SCVFAs) are ideal sources. In particular, acetic (two carbons) through Valeric acid (five carbons) are considered to be the prototype of readily available organic compounds for EBPR. Although all of these volatile fatty acids (VFAs) are usable for EBPR, their relative effectiveness varies. Of all of the VFAs commonly generated during municipal wastewater fermentation, acetic acid is the most efficient based on mass of phosphorus uptake per carbonaceous oxygen demand (COD) consumed. It is important to note that EBPR cannot be accomplished with one-carbon compounds such as formic acid because their polymerization is thermodynamically unfavorable for the bacteria.

The minimum ratios typically used to indicate the EBPR capability of a system are summarized in Table 9.2 (WEF, 2011). These ratios refer to bioreactor influent and should account for recycle loads and removals in primary clarifiers. Typically, recycle flows represent additional phosphorus load, while primary clarifiers remove 5-day carbonaceous biochemical oxygen demand ($CBOD_5$.) As

TABLE 9.2 Minimum substrate-to-phosphorus requirements for EBPR (WEF, 2010).

Substrate measure	Substrate-to-phosphorus ratio	Remarks
Five-day carbonaceous BOD	25:1	Provides a rough and initial estimate. Based on typically available facility data.
Five-day soluble BOD	15:1	Better indicator than $cBOD_5$.
Carbonaceous oxygen demand	45:1	More accurate than CBOD. Not measured by all facilities.
Volatile fatty acids	7:1 to 10:1	More accurate than COD. Involves specialized laboratory analysis.
Readily biodegradable carbonaceous oxygen demand	15:1	Most accurate. Measures VFA formation potential. Accounts for VFA formation in the anaerobic zone. Specialized laboratory analysis.

a result, primary effluent (bioreactor influent) will contain a lower substrate-to-phosphorus ratio relative to raw influent. The $CBOD_5$-to-total-phosphorus ratio can be used as a first approximation of the adequacy of carbon substrate for EBPR. It is now thought that readily biodegradable COD (rbCOD) is the most accurate measure of substrate availability because it represents influent VFAs and organic compounds that could potentially be fermented to VFAs in the anaerobic zone of the bioreactor. The rbCOD is an estimate of truly soluble COD.

The composition of influent phosphorus affects EBPR efficiency. In particular, corrosion inhibitors used by water treatment facilities can be a significant source of polyphosphate, which is not readily reactive and is difficult to remove. For example, when the Xenia, Ohio, water treatment facility switched its corrosion control chemical from a 25 polyphosphate and 75 orthophosphate formulation to one containing 75 polyphosphates and 25 orthophosphates, the city's two wastewater facilities discharged elevated levels of total phosphorus (Jeyanayagam, 2007). This was corrected by switching the corrosion inhibitor back to the original formulation. It should be noted that the effluent orthophosphate values were consistently low, indicating that the higher effluent total phosphorus was mainly because of increased influent polyphosphate.

Municipal wastewater fermented in the collection system is typically a good source of VFAs for EBPR operation. Seasonal variation of influent VFAs can be tracked by observing the $CBOD_5$-to-COD ratio over a 12-month period. Sewer fermentation is inhibited by dissolved oxygen entrainment, wet weather high flows, the type of odor control strategy used, and cold temperatures, including those from the effect of snowmelt. Although adequate VFAs may be available in the facility influent to meet summer phosphorus limits, sewer system fermentation may not be a reliable and consistent source of carbon substrate throughout the year. In such instances, the anaerobic zone may need to be sized to support additional fermentation. Facility recycles, such as supernatant from primary sludge gravity thickening, also may contain sufficient VFAs for EBPR. Other sources of carbon augmentation include on-site VFA generation via sludge prefermentation, industrial wastewater containing VFAs, or acetic and propionic acids. Often, it may be simpler and more cost effective to provide a standby chemical feed system.

3.2 Integrity of the Anaerobic Zone

The anaerobic zone of an EBPR bioreactor is expected to perform two functions. The primary function is PAO selection, which is a relatively rapid reaction if adequate rapidly biodegradable substrate is available. In some instances, the anaerobic zone is also required to perform a secondary function, that is, VFA generation through fermentation. This is a slower reaction requiring a longer anaerobic hydraulic retention time (HRT).

An anaerobic condition is one in which free (dissolved) and combined oxygen are not available for bacterial metabolism. The oxidation–reduction potential (ORP) can also be used to confirm the presence of anaerobic conditions. Typical ORP values in anaerobic reactors are in the range of –300 mV or less. Field testing, however, is required to establish site-specific ORP for anaerobic conditions. Ideally, anaerobic conditions must be maintained throughout the anaerobic volume for reliable EBPR. Sources of dissolved oxygen include preaeration, influent screw pumps, free-fall over weirs, excessive turbulence, aggressive anaerobic zone mixing, backflow from the aerobic zone, and internal mixed liquor recycle. Typical sources of nitrate include RAS and internal mixed liquor recycle (Jeyanayagam, 2007).

Introduction of nitrate or dissolved oxygen to the anaerobic zone causes a reduction in the actual anaerobic volume. Consequently, the effective anaerobic solids retention time (SRT) and HRT will be reduced. This will decrease the anaerobic contact time between PAOs and the substrate (VFAs), which could potentially compromise phosphorus removal (Jeyanayagam, 2007). In addition, the presence of nitrate and dissolved oxygen will provide competing organisms with access to the substrate. For example, 1.0 mg of nitrate-nitrogen will take up readily biodegradable organics needed for the removal of 0.7 mg of phosphorus by supporting denitrification. Likewise, the presence of 1.0 mg of dissolved oxygen will use up the substrate needed for the removal of 0.3 mg of phosphorus by facilitating normal heterotrophic activity (biochemical oxygen demand oxidation).

3.3 Solids Capture

Effluent total phosphorous consists of two components: soluble phosphorus and particulate phosphorus. A well-designed and operated EBPR system with adequate VFAs can reduce the effluent soluble phosphorus to approximately 0.1 mg/L. Particulate phosphorus, which represents solids-associated phosphorus, depends on effluent total solids and its phosphorus content. For example, if the effluent total suspended solids (TSS) is 10 mg/L (75% volatile suspended solids [VSS]) and the phosphorus content of the mixed liquor is 0.06 mg/mg VSS (6%), then the effluent particulate phosphorus concentration would be 0.45 mg/L. Hence, controlling effluent solids is important to achieve low effluent total phosphorus.

3.4 Recycle Loads

Return streams from sludge operations (e.g., dewatering) can significantly affect EBPR performance. The quantity and quality of these streams vary based on the technology used in solids processing operations. For example, anaerobic digestion is likely to release more phosphorus than aerobic digestion. Sludge dewatering using belt filter presses typically generates 2 times more recycle flow (filtrate) compared to centrifuge dewatering because of the amount of wash water used in the dewatering operation. The total recycle flow can amount to 1 to 3% of the facility influent flow and can reduce process HRT. The recycle nitrogen and phosphorus mass loads can represent 15 to 40% of influent total nitrogen and 5 to 40% of influent total phosphorus loads, respectively. The effect of recycle streams is discussed further in Section 7.0.

3.5 Mixing

The goal of mixing in an anaerobic zone is to keep the mixed liquor suspended solids (MLSS) suspended while minimizing surface turbulence that could transfer oxygen from the atmosphere. Therefore, the minimum power input necessary to keep the solids suspended should be used, and the stirring mechanism should be designed to avoid vortexing. Alternatives include submersible, vertical, and pulsed air mixers. Power input should be sufficient to maintain a velocity of 0.3 m/s (1.0 ft/sec) throughout the zone. Because horizontal velocities can be difficult to quantify in a reactor with a single-point mixing device, a more practical criterion is complete turnover of the cell contents every 20 minutes based on the primary pumping rate of the mixer.

3.6 Baffles

Baffles play an important role in enhancing the EBPR process. The following types of baffles are used in EBPR bioreactors: interzone and intrazone.

3.6.1 *Interzone Baffles*

In EBPR bioreactor configurations, interzone baffles are provided for separating the anaerobic and anoxic or oxic zones to protect the integrity of the anaerobic zone. Interzone baffle design should consider the following:

- Because of density differences, the water surface in the aerobic zone will be considerably higher than the water surface in the nonaerated (anaerobic) zone. This could initiate a backflow of a high dissolved oxygen stream into the anaerobic zone, which will reduce the available substrate for EBPR and could potentially stimulate filamentous growths and cause *nocardiaform* foaming. In addition, foam will be trapped in the zone upstream of the aerated zone because of the increase in water surface;

- Head loss, when available, is the most effective way to segregate zones by providing a drop in water surface across the top baffle. Baffle openings should be sized to ensure a forward flow of 0.15 m/s (0.5 ft/sec) at minimum flow. Fully submerged baffles with the top approximately 25 mm (1 in.) below the water surface would allow the free flow of surface scum and foam without an opportunity to accumulate. Alternatively, when baffles extend well above the water surface, slots should be cut at the top for the scum and foam to pass;
- Bottom baffle openings must be provided to facilitate tank draining. The opening should be small enough to avoid significant forward flow;
- To facilitate tank cleaning, removable walk-through baffle segments should be considered to allow operators to move from zone to zone without having to climb back out; and
- Provisions should be made for confined space activities by including a lifeline and retrieval device.

3.6.2 *Intrazone Baffles*

Intrazone baffles typically are provided within a zone to enhance reaction kinetics within a zone. In an EBPR system, intrazone baffles are provided to improve plug flow conditions. As shown in Figure 9.2, an anaerobic stage divided into discrete stages would result in more rapid substrate uptake kinetics (because of a higher initial food-to-microorganism ratio [F/M]) than a completely mixed zone (Stensel,

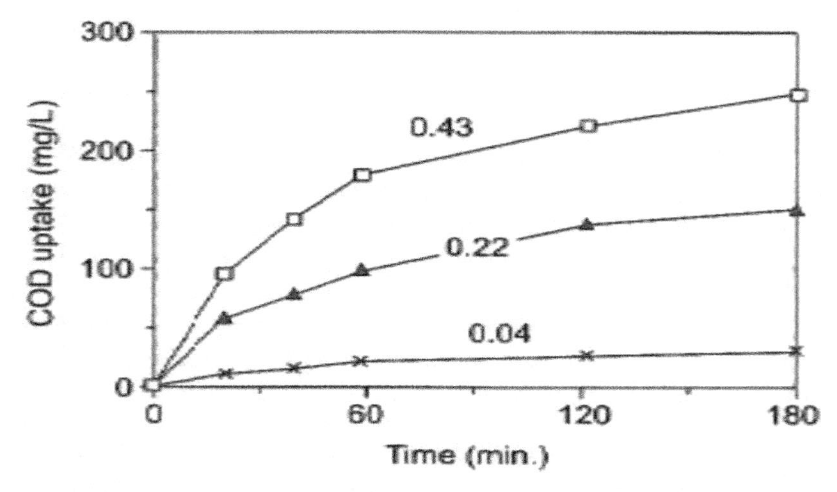

FIGURE 9.2 Effect of initial F/M on anaerobic substrate uptake kinetics (WEF, 2010).

1991). Consequently, the total anaerobic volume can be decreased by staging. Similar benefits have been reported by staging the aerobic zone (Jeyanayagam 2007; Narayanan et al., 2006).

3.7 Supplemental Chemical Addition

Enhanced biological phosphorus removal entails the complex interaction of design and operational factors that, singly or in combination, can affect process efficiency. The flexibility to add metal salts when necessary will ensure process reliability and regulatory compliance. However, indiscriminate use of metal salts could potentially interfere with the recovery of the EBPR process and could facilitate the eventual conversion from backup to exclusive chemical addition. This is because the soluble phosphorus that is chemically precipitated is unavailable for uptake and storage (as energy-rich volutin granules) by PAOs in the anoxic and aerobic zone. Consequently, PAOs will have less internal energy available to them in the anaerobic zone for the uptake of SCVFAs and will lose their competitive advantage over other heterotrophs. This is the reason overdosing with chemical precipitants results in a washout of the EBPR function.

The recommended strategy for chemical addition is to dose only as much metal salt as is needed to precipitate the phosphorus that the PAOs cannot take up. This could be accomplished by measuring the soluble phosphorus (mostly orthophosphate) toward the end of the bioreactor and adding metal salt in direct proportion to it (e.g., a 2-to-1 molar ratio, to be confirmed by jar testing as outlined in Chapter 8). The orthophosphate at the end of the bioreactor represents phosphorus that the organisms are unable to take up at that point in time. The chemical could be added to a well-mixed section of the mixed liquor channel, just upstream of the secondary clarifiers. However, it should be noted that the unreacted metal salt that is returned with RAS may continue to remove phosphorus after chemical addition is terminated. For this reason, a preferred first option might be to chemically treat the sidestream to remove recycled phosphorus before adding supplemental chemical to the EBPR process.

4.0 WASTING STRATEGIES

The parameter of fundamental importance to the design and control of the activated sludge system is SRT. Chapter 7 contains a discussion of SRT requirements for the EBPR process. Operationally, the design SRT governs the mass of sludge wasted daily from the system. In nutrient removal systems, SRT, rather than the F/M or organic loading, should be used for process control. By definition, total system SRT in days is

$$\text{Total SRT} = \frac{\text{Mass of sludge in the bioreactor}}{\text{Mass of sludge wasted per day}} \tag{9.1}$$

In EBPR systems, anaerobic and aerobic SRTs can be calculated by replacing the numerator in the aforementioned equation with the mass of solids in the anaerobic and aerobic zones, respectively. It should be noted that the SRT calculation presented in eq 9.1 considers the solids inventory in the bioreactor only and could underestimate the actual SRT in facilities where significant solids inventory is held outside of the bioreactor in final clarifiers and excessively large mixed liquor and RAS channels. The mass of sludge wasted per day is equal to the daily sludge production. *Design of Municipal Wastewater Treatment Plants* (WEF et al., 2009) contains a discussion of the effect of SRT and wastewater characteristics on daily sludge production.

Available sludge wasting strategies, shown in Figure 9.3, include wasting from the final clarifier underflow (conventional) and hydraulic wasting directly from the bioreactor. The recommended strategy is to waste continuously, when possible.

4.1 Final Clarifier Underflow Wasting (Conventional)

The most common method of sludge wasting is from the return sludge line, sometimes called *conventional wasting*. For this wasting strategy, using the aforementioned mathematical expression, SRT may be computed as follows:

$$\mathrm{SRT} = \frac{V \times X}{(Qw \times Xr + Qe \times Xe)} \tag{9.2}$$

Where

V = bioreactor volume, m^3;

X = MLSS concentration, mg/L;

Qw = waste sludge flowrate from the return sludge line, m^3/d;

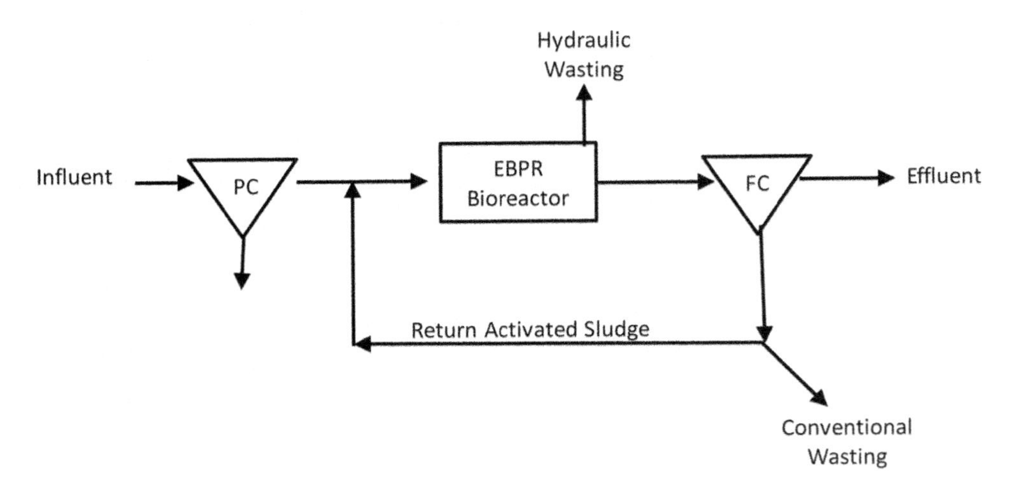

FIGURE 9.3 Alternate wasting strategies.

Xr = RAS solids concentration, mg/L;
Qe = effluent flowrate from secondary clarifier, m^3/d; and
Xe = effluent TSS concentration, mg/L.

If effluent solids are neglected, the wasting rate approximates to

$$Qw \approx \frac{V \times X}{SRT \times Xr} \tag{9.3}$$

4.2 Hydraulic Wasting

The hydraulic control of sludge wasting was first proposed and implemented by Garrett in 1958. In essence, this approach entails wasting MLSS (X) directly from the bioreactor. If effluent solids are ignored, the SRT is calculated as follows:

$$SRT \approx \frac{VX}{QwX} = \frac{V}{Qw} \tag{9.4}$$

Additionally, the waste flowrate from the bioreactor (Qw) is independent of MLSS and RAS characteristics, as follows:

$$Qw \approx \frac{V}{SRT} \tag{9.5}$$

Hydraulic wasting offers the following advantages for the EBPR process:

- Ease of determining the amount of daily wasting by simply dividing the volume of the bioreactor by the desired SRT, as shown in the aforementioned equation;
- Because the solids recycle rate does not enter the calculation of SRT, the RAS flowrate can be chosen to give proper operation of the final clarifier, thereby ensuring that almost all biomass is returned to the bioreactor (Grady et al., 2011);
- Wasting aerobic sludge has a lower tendency to release phosphorus during subsequent handling than underflow from a secondary clarifier. In addition, the wasted MLSS stream can be thickened continuously in a dissolved air floatation unit, which is the preferred method of thickening phosphorus-rich EBPR solids. Use of other thickening methods often requires sludge storage to allow intermittent operation. This practice could potentially lead to secondary phosphorus release during storage; and
- Because of lower solids concentration than clarifiers under flow, continuous wasting is viable and more aligned with continuous biological synthesis and growth in the bioreactor.

The main drawback of hydraulic wasting is the need to size subsequent solids handling equipment for the higher hydraulic load.

Hydraulic wasting also can be used for wasting clarifier underflow if the RAS flow and influent flow are metered. The Seven Springs facility in Madison, Wisconsin, has successfully used hydraulic wasting of underflow with biological phosphorus removal for approximately 15 years. A mass balance on the aeration/clarifier system using influent flow (Qi), RAS flow (Qr), MLSS concentration (X), and RAS concentration (Xr) yields

$$(Qi + Qr)X = QrXr \tag{9.6}$$

or

$$Xr = (Qi + Qr)X/Qi \tag{9.7}$$

Substituting for Xr in eq 9.3, the expression for wasting (Qw) becomes

$$Qw = VQi/[SRT(Qi + Qr)] \tag{9.8}$$

With known influent and recycle flowrates, the waste sludge rate can be set hydraulically for wasting underflow. If the return rate is constant with a varying influent flow, the wasting rate must be varied to maintain a constant solids loading on a downstream thickener. If the return rate is varied proportionally to influent flow, the waste sludge flowrate can be constant throughout the day. This method will work successfully if metering is accurate. And, as long as thickening is not attempted in the final clarifier, secondary release of phosphorus in the clarifier is not likely to be a problem.

5.0 APPLICATION OF MASS-BALANCE PRINCIPLES IN ENHANCED BIOLOGICAL PHOSPHORUS REMOVAL OPERATIONS

Mass balances provide a convenient way to evaluate the performance of a unit operation. Using mass-balance concepts, it is also possible to perform a reality check of facility data. Because of the dynamic nature of facility operating conditions, average data should initially be used for mass-balance calculations. This section presents how mass balances can be used to make operating decisions.

The best way to monitor the oxygen and nitrate returned to the head of EBPR systems is the "mass-balance method". The mass of phosphorus, oxygen, and nitrate can be calculated using the flow and the respective concentration as shown in eq 9.9. Once the mass going in and coming out are calculated for each reactor, the change in the reactor can be calculated from the difference between the two, as follows:

$$\begin{aligned}\text{Mass difference} &= \text{Mass in} - \text{Mass out}\\ &= [C_{in} \times Q_{in}] - [C_{out} \times Q_{out}]\end{aligned} \tag{9.9}$$

Where

M = mass of the specific constituent (lb/d or kg/d),
C = concentration of the specific constituent (mg/L), and
Q = flowrate (mgd or ML/d).

Figure 9.4 illustrates the results of mass-balance calculations performed for a University of Cape Town-type EBPR system that consists of two anaerobic (An), two anoxic (Ax), and three aerobic (Ae) reactors. The figure also illustrates the effect of the mixed liquor recycle from the end of the aeration zone (Ae3) to the first anoxic zone (Ax1). The recycle contains nitrate and oxygen, which contributed to the electron acceptor budget in the anoxic zone.

Once the recycle was taken out of service, anoxic phosphate uptake decreased significantly, leading to an increase of aerobic uptake. By using mass-balance calculations, one can avoid confusion that can be caused by using concentration values as a monitoring tool. Because the recycle flows also have a dilution effect on facility influent at different points of the treatment system, mass loading must be used instead of concentrations.

Nitrate in RAS can be reduced by providing a wide spot in the RAS line to allow the nitrate to dissipate. A small amount of primary effluent may be added to enhance the denitrification rate. Another strategy is to operate at the lowest possible RAS rate. With RAS rates of 50% or less, little nitrate remains when RAS arrives at the anaerobic zone. While it is possible to reduce the RAS nitrate load by promoting denitrification in the final clarifiers, this is not a recommended practice,

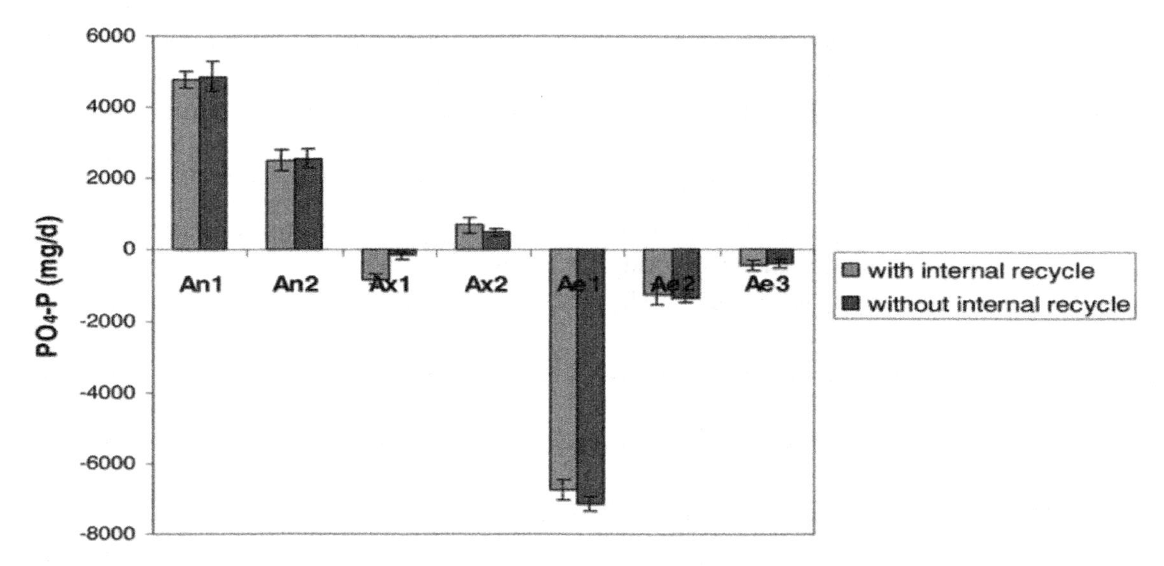

FIGURE 9.4 Phosphorus mass balance performed for phosphorus in a pilot-scale, Virginia Initiative Process-type EBPR system (Erdal et al., 2002). Positive values indicate phosphorus release, whereas negative values indicate phosphorus uptake (WEF et al., 2005).

particularly if the clarifier effluent nitrate-nitrogen is more than approximately 5 mg/L. At these higher nitrate levels, denitrification in the clarifier can release enough nitrogen gas to affect sludge settleability and the ability of the facility to meet TSS permit limits.

6.0 CARBON SOURCES

The availability of adequate carbon is crucial to reliable EBPR. Broadly categorized, carbon sources for EBPR include *internal sources*, which refer to the generation of short-chain fatty acids through sludge fermentation of the influent rapidly biodegradable organic matter, and *external sources*, which refer to the addition of purchased chemicals.

6.1 Internal Carbon Source

Acid fermentation is the process by which complex particulate and soluble substrates present in the sludge and wastewater are aerobically broken down to form SCVFAs. Because they are soluble and readily biodegradable, SCVFAs are a highly effective carbon source for EBPR and denitrification. The fundamental mechanism of fermentation is presented in Chapter 7. In this section, the various primary sludge fermenter configurations are outlined. It should be noted that VFA production also occurs naturally in wastewater collection systems under the right conditions. This topic is also discussed in Chapter 7.

Sludge fermenters are configured to accomplish three basic functions: acid fermentation, solids separation, and elutriation. The available flow sheets are shown in Figure 9.5 and briefly described in the following subsections. Table 9.3 summarizes the differentiating features of the various fermenter configurations.

6.1.1 *Activated Primary*

The primary clarifier is operated as a fermenter by continuously recycling the primary sludge directly to the inlet or through an elutriation tank. Fermentation occurs in the accumulated sludge blanket. Generally, activated primary tanks require purposely designed primary clarifiers that allow sufficient sludge blanket depth to be maintained. The SCVFAs generated in the sludge mass are released into the bulk liquid by elutriation with influent wastewater and conveyed to the EBPR process together with the primary effluent. The sludge inventory and blanket height are controlled by wasting a fraction of the primary sludge to the solids handling system. Barnard (1984) first proposed this fermentation concept. A modified operating strategy, called *alternating activated primaries*, involves two primary clarifiers. Primary sludge is retained and fermented in one and then pumped to another for SCVFA elutriation. Raw wastewater is settled in both tanks at the same time. This design was used at the Baviaanspoort Wastewater

(a) Activated primary

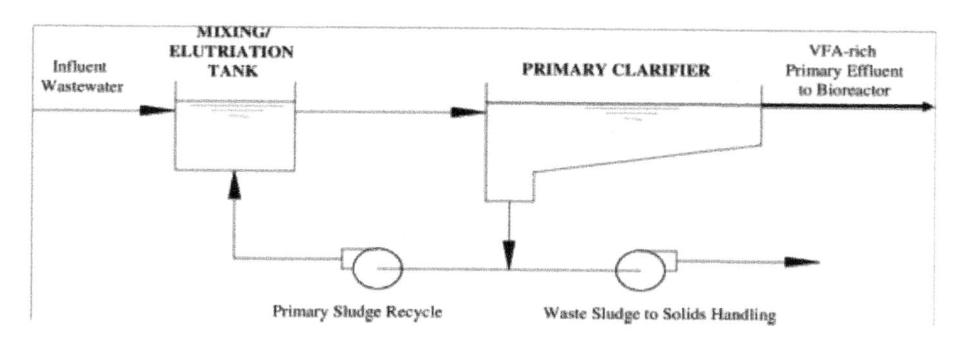

(b) Complete-mix fermenter

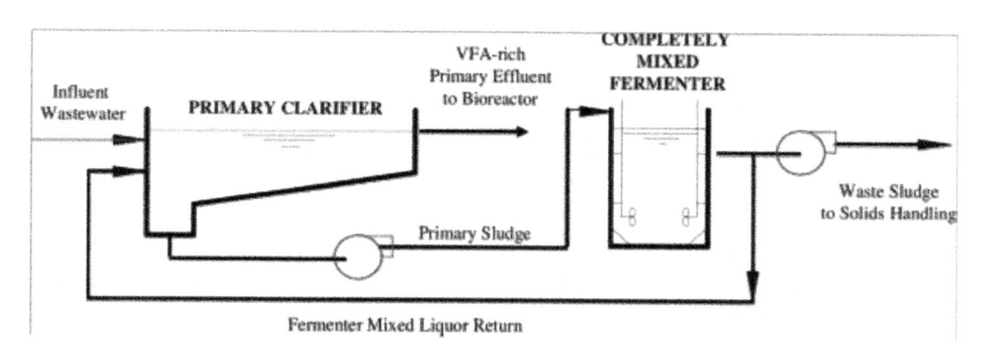

(c) Single-stage static fermenter and thickener

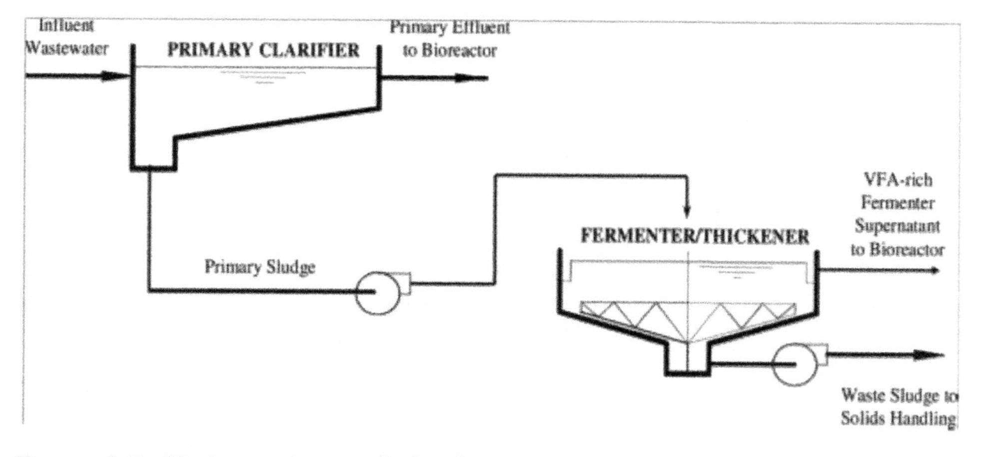

FIGURE 9.5 Various primary sludge fermentation configurations.

(d) Two-stage fixed fermenter and thickener

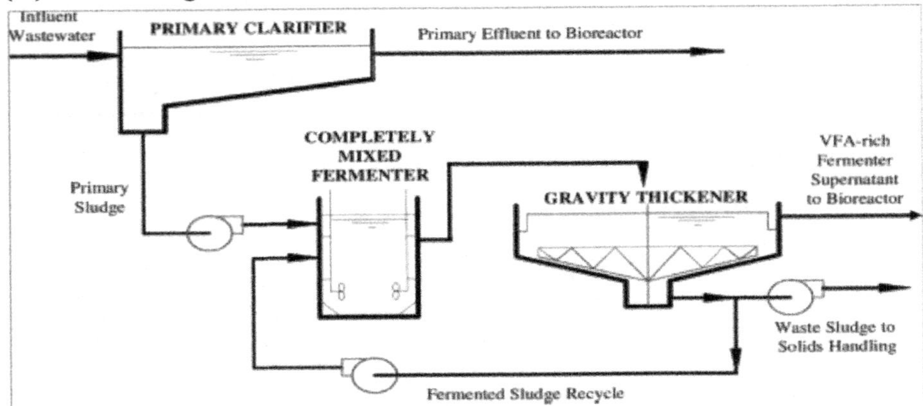

(e) United fermentation thickening fermenter

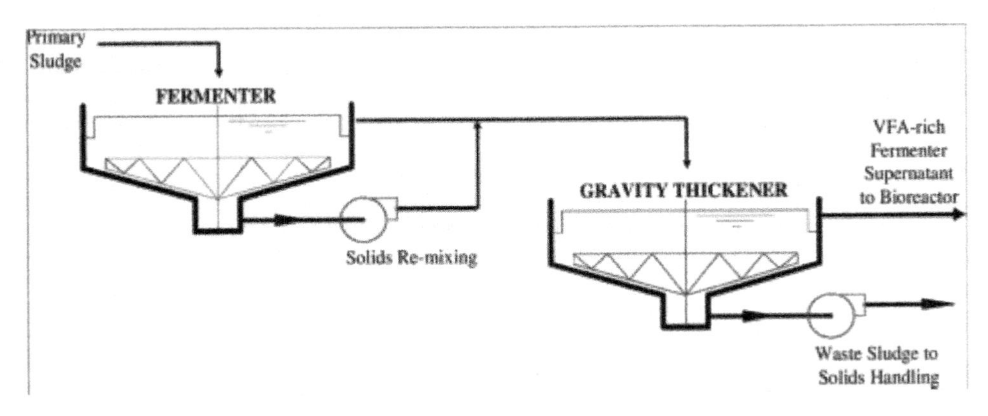

FIGURE 9.5 *(Continued).*

Treatment Works in South Africa and the operation allowed for the process to be reversed (Randall et al., 1992).

6.1.2 *Complete-Mix Fermenter*

Complete-mix fermenters enable the fermentation process to be removed in large part from the primary treatment process, although the solids loadings to primary clarifiers are significantly reduced. Complete mixing of the fermenter also provides better contact between the fermenting solids and the bulk liquid. Primary sludge is continuously pumped to a mechanically mixed tank where acid fermentation occurs. The complete-mix fermenter tank overflow is returned to the primary clarifier inlet. The SCVFAs generated in the fermenter are elutriated though mixing with influent wastewater and conveyed to the EBPR process

TABLE 9.3 Differentiating features of the various fermenter configurations.

Configuration	Fermentation	Solids separation	VFA stream
Active primary	Primary clarifier	Primary clarifier	Primary effluent
Complete mix	Separate completely mixed tank	Primary clarifier	Primary effluent
Single-stage static fermenter/thickener	Gravity thickener	Gravity thickener	Gravity thickener supernatant
Two-stage complete mix fermenter/thickener	Separate completely mixed tank	Gravity thickener	Gravity thickener supernatant
Unified fermentation and thickening	First gravity thickner	Second gravity thickener	Gravity thickener supernatant

together with the primary effluent. The fermenter SRT is controlled by wasting a fraction of the fermented sludge to the solids handling system. The fermenter HRT is controlled either by adjusting the complete-mix tank volume or by adjusting the primary sludge pumping rate. The SCVFAs elutriated in the primary clarifier are conveyed to the EBPR process together with the primary effluent. This fermenter configuration was first proposed by Rabinowitz et al. (1987).

6.1.3 *Single-Stage Static Fermenter and Thickener*

Static fermenters simplify the fermentation process, removing it completely to a separate unit process. In addition, supernatant from this process can be directed to the EPBR bioreactor independent of the primary effluent so that it may be applied where it is most appropriate in the process. This fermenter is a lowly loaded gravity thickener with an increased side-water depth to allow a fermenting sludge inventory to accumulate in the bottom of the thickener. Primary sludge is continuously pumped into a center well and allowed to settle and thicken in the fermenter and thickener. The SCVFA-rich supernatant is conveyed directly to the EBPR bioreactor, where it can be optimally used for EPBR and/or denitrification. A source of nitrate-rich elutriation water is typically added to the incoming primary sludge or directly to the thickener to minimize sulfide and methane formation in the sludge blanket and to help convey SCVFAs to the EBPR bioreactor. The fermenter SRT and sludge inventory are controlled by wasting a portion of the sludge inventory, as thickened sludge, to the solids handling system to maintain a target sludge blanket height. This fermentation concept was first applied at the Kelowna's water resource recovery facility (WRRF) in British Columbia in the early 1980s (Oldham and Stevens, 1984).

6.1.4 *Two-Stage Mixed Fermenter and Thickener*

This configuration combines complete-mix and static fermenters into a single unit process. Primary sludge is pumped to a mechanically mixed tank, where acid fermentation occurs. The complete-mix tank overflows to a gravity thickener for liquid–solids separation. The SCVFA-rich supernatant is directed to the EPBR bioreactor independent of the primary effluent so that it may be applied where it is most appropriate in the process. Furthermore, the complete-mix basin allows intimate contact between the fermenting solids and the bulk liquid, optimizing the transfer of SCVFAs to the supernatant. Primary sludge is continuously pumped from the mainstream primary clarifier to a mechanically mixed tank where acid fermentation occurs. The complete-mix tank overflows to the gravity thickener for liquid–solids separation. Thickened sludge is continuously recycled from the bottom of the thickener to the complete-mix tank inlet. The VFA-rich thickener supernatant is conveyed directly to the EBPR bioreactor, where it can be directed to an anaerobic or anoxic zone to optimally be used for EPBR and/or denitrification. A source of elutriation water is typically added to the gravity thickener inlet or into the thickener blanket to minimize sulfide and methane formation in the sludge blanket and to help convey SCVFAs to the EBPR bioreactor. The fermenter SRT and sludge inventory are controlled by wasting a portion of the thickened sludge to the sludge handling system and maintaining a target sludge blanket depth. This fermentation concept was first applied at the Kalispell Wastewater Treatment Plant in Montana in the early 1990s (Oldham and Abraham, 1994).

The design of these systems typically accounts for the solids inventory in both the complete-mix tank and the thickener. An option exists to exchange the thickener for a mechanical thickening process such as a centrifuge, rotary drum thickener, or screw press. If this option is implemented, the entire sludge inventory must be contained within the complete-mix tank.

6.1.5 *Unified Fermentation and Thickening Fermenter*

This fermenter configuration consists of two gravity thickeners in series, and was patented by Baur (2002). The first thickener is operated as a fermenter, and the settled solids and supernatant are recombined and directed to the second thickener, where liquid–solids separation occurs. The fermenter SRT is controlled by varying the thickened sludge pumping rate from the bottom of the fermenter. The SCVFA-rich thickener supernatant from the second thickener is conveyed directly to the EBPR bioreactor, while solids are conveyed to the sludge handling system. Elutriation water can be added to the thickener to condition the solids to improve sludge settling and thickening and to help convey SCVFAs to the EBPR bioreactor.

6.1.6 *Volatile Fatty Acid Production*

In a laboratory-scale investigation, Lilley et al. (1990) found VFA production from primary sludge fermentation to involve first-order reaction kinetics. The maximum potential VFA yield at 20 °C was reported to be 0.17 mg SCVFA (as COD)/mg of primary sludge (as COD) applied. Oldham and Abraham (1994) reported on the SCVFA yield from the complete-mix fermenter in Penticton, British Columbia; the single-stage static fermenter in Kelowna, British Columbia; and the two-stage complete-mix fermenter and thickener in Kalispell, Montana. These fermenters added the equivalent of 17, 21, and 58 mg/L of SCVFA, respectively, to the wastewater entering EBPR processes. Practical experience suggests that higher SCVFA yields are observed in fermenters receiving primary sludge from wastewaters with relatively minimal natural fermentation in the collection system.

6.1.7 *Operation of Primary Sludge Fermenters*

The following factors affect the operation of all primary sludge fermenters regardless of configuration. Consequently, these factors should also be considered during design.

6.1.7.1 *Solids Retention Time*

Solids retention time is a key operating parameter for primary sludge fermentation systems because it defines the groups of fermentative organisms that will grow in the system. If the SRT is too short, many of the bacteria that produce the enzymes necessary for hydrolysis and acidogenesis, as well as acetogenesis bacteria, will be washed out of the system and a stable population of acid-formers will not develop. Conversely, if the SRT is too long, either sulfur-reducing bacteria that convert VFAs to CO_2 (rather than to acetate) or a population of methane formers will develop that will consume the VFAs generated and produce methane gas. The growth rate of fermentative bacteria is highly temperature dependent. For this reason, primary sludge fermenters are generally operated at an SRT of 3 to 5 days during the summer months when operating temperatures are above 16 °C, and at an SRT of 4 to 8 days during the winter months when operating temperatures are below 15 °C (WERF, 2011). Given the relatively unique wastewater characteristics and operating conditions found at most facilities, establishment of an optimal SRT is generally based on experience; however, it is typically within the aforementioned limits for various operating temperatures.

Determining the sludge inventory is critical to calculating the SRT of a fermenter system. In systems that use a completely mixed reactor, inventory can be determined from the total volume of the fermenter and suspended solids concentration. However, in a fermenter that uses a gravity thickener, stratification of solids and liquids requires estimation of the sludge blanket depth in the thickener. This measurement can be done directly with sludge blanket probes or sludge blanket instruments. However, use of the latter is difficult because of the low light transmittance of the thickener supernatant. Although other methods using mechanism torque have been used as a surrogate measure of sludge inventory, these measurements can vary seasonally as sludge densification characteristics change in the thickener.

6.1.7.2 Hydraulic Retention Time

Fermenter HRT is a function of tank volume and the primary sludge pumping rate. In general, HRT is not a critical parameter used in the design of primary sludge fermenters, although it is often used to fine tune fermenter operation. Complete-mix fermenters are generally sized with an HRT between 6 and 12 hours while treating between 4 and 8% of incoming facility flow. The primary sludge pumping rate to two-stage, complete-mix, and gravity thickener fermenters is generally lower (i.e., about 2 to 4% of the incoming facility flow) so that the HRT of the complete-mix tank is typically between 12 and 24 hours. Other fermenters are designed to maintain a specific sludge inventory and SRT in the system. In these instances, the fermenter HRT can also be controlled by varying the primary sludge pumping rate and the elutriation water flowrate when added directly to a gravity thickener.

6.1.7.3 pH

The bulk liquid pH is extremely important in anaerobic digestion because methanogenesis proceeds at a high rate when the pH is maintained in the neutral range, with most methane formers inhibited at pH values below 6.3 and above 7.8. However, acid formers are significantly less sensitive to high or low pH values. Hence, acid fermentation typically prevails over methane formation at pH values below 6.5. Primary sludge fermenters tend to operate at ambient pH values between 4.8 and 6.0 because of acid formation reactions that are part of fermentation; raising the fermenter pH value does not improve VFA production. The lower pH is beneficial because it inhibits growth of methane-forming bacteria. Sulfanogenic bacteria are also relatively immune to low pH values and can reduce sulfate to sulfides at pH values below 4.

6.1.7.4 Elutriation Water

In addition to generating VFAs through conversion of the particulate primary solids to soluble organics, these VFAs must be separated from the fermenting solids and conveyed to the EBPR process. Partitioning of the VFAs to the portion of the flow directed to EBPR rather than to sludge disposal is enhanced by introducing elutriation water. In addition, it is useful to provide some level of mixing to enhance diffusion of VFAs from the sludge mass to the liquid fraction.

Use of a water source with a significant level of nitrate has been practiced to control the sulfur-reducing bacteria population. Heterotrophic bacteria, including sulfate-reducing bacteria (SRB), prefer nitrate to sulfates as their terminal electron acceptor. Although SRB that use hydrogen as their electron source are critical to acetogenesis, the proliferation of those bacteria will begin consuming VFAs at a rate in excess of their generation. The addition of some nitrate-rich stream (secondary effluent is most common) can be used to reduce, but not eliminate, SRB dominance, especially at long SRTs. This addition has the secondary benefit of inhibiting the activity of methanogenic bacteria, although intermittent aeration has also been used with some effect.

The key parameter used to control the operation of primary sludge fermenters is the system SRT or sludge inventory. Volatile fatty acid production tends to be low at short SRTs because of the difficulty of establishing a stable population of acid formers that complete the hydrolysis, acidogenesis, and acetogenesis reactions. However, at long-fermenter SRTs, sulfanogenic and methanogenic bacteria will begin to dominate and the generated VFAs will be consumed by methanogenic activity that produces methane gas or sulfanogenic activity that generates hydrogen sulfide.

6.1.8 Return Activated Sludge Fermentation

Effective EBPR can be achieved in processes in which mixed liquor or RAS is fermented to form rbCOD or SCVFAs. Fermentation in anaerobic zones has always been an integral component of EBPR systems. However, configurations have been advanced that depend on fermentation of the secondary sludge, provided that only a portion of the secondary sludge is fermented and that the rbCOD or VFAs formed are made available to PAOs in the mainstream EBPR process. Secondary sludge fermenters typically consist of completely or intermittently mixed reactors that are incorporated to the EBPR process train. These reactors can either be in the main process stream, a sidestream of the process (e.g., on the RAS line), or on a recycle loop that is withdrawn from and returned to the anaerobic zone. Secondary sludge fermentation has been successfully implemented in several EBPR facilities where manipulation of the aeration and mixing systems has created mixed liquor or RAS fermentation zones within the EBPR process train (Barnard et al., 2010).

Clarke (2009) switched off the mixer in one out of two anaerobic zones of a facility with insufficient VFAs in the incoming wastewater and successfully removed phosphorus from about 8 mg/L in the influent to less than 0.5 mg/L through in-basin fermentation. A similar strategy at a Henderson, Nevada, facility, where nitrate added to the sewerage system destroyed the natural VFAs, resulted in effluent orthophosphate concentrations below 0.1 mg/L (Barnard et al., 2010).

Fermentation of a portion of RAS has been used for VFA generation in EBPR processes (WERF, 2011). Although this fermentation option can be used at any EBPR process, it particularly applies to facilities that operate secondary treatment processes without primary clarifiers. There are several variations of the configuration in operation in North Carolina. In a patented sidestream EBPR process using RAS fermentation, a portion of RAS from the secondary clarifier is first directed to a sidestream anaerobic zone. A portion of the RAS from the anaerobic zone is then directed to a sidestream fermentation zone (Lamb, 1994). Effluent from the sidestream fermentation zone is sent back to the anaerobic zone as a VFA source. The sidestream RAS fermentation zone is similar in operation to any anaerobic or anoxic zone in the mainstream EBPR process. As an alternative to the side-stream biological phosphorus removal configuration, the fermentation zone effluent could be sent directly to the anaerobic zone of the mainstream EBPR process.

Although secondary sludge fermentation is a relatively new concept and the biochemical mechanisms involved are not yet well defined, it is regarded as a viable alternative to primary sludge fermentation, primarily because it has few of the operating problems associated with primary sludge fermenters (i.e., odors, corrosion, blockages, abrasion, etc.) (Barnard et al., 2011).

A significant disadvantage of fermenting secondary sludge is the high concentration of ammonia and phosphorus released into the fermentate. Ammonia release is a result of the anaerobic metabolism of protein material in the sludge, while phosphorus is released by PAOs under anaerobic conditions in the absence of SCVFAs. However, practice has shown that the overall benefits of additional SCVFA provided through secondary sludge fermentation can outweigh the negative effect of these additional nutrient loads. The key to successful application of secondary sludge fermentation is to ferment a relatively small portion of the mixed liquor or RAS, thereby fermenting sufficient secondary sludge to generate the VFAs needed to drive EBPR, but not so much that the released masses of nitrogen and phosphorus overwhelm the process (Andreasen et al., 1997; Barnard et al., 2011; Vale et al., 2008; Vollertsen et al., 2006).

Care must be taken to ensure that the VFA-rich fermentate does not come into contact with nitrate-rich mixed liquor because the VFAs generated will be preferentially used for denitrification, thus rendering them unavailable to drive the EBPR process. A key benefit of intermittently mixed or unmixed in-line fermenters is that these reactors allow the mixed liquor to settle, thus increasing the SRT of the fermenting sludge while physically separating it from the nitrate in the supernatant.

6.2 External Carbon Sources

Acetic acid (CH_3COOH) is the most common external carbon source used for EBPR. Acetic acid is a colorless liquid with a strong vinegar odor and is commonly delivered as glacial (approximately 100% solution), 84 or 56% solution. Although not as volatile as methanol, glacial acetic acid has a relatively low flash point (40 °C [104 °F]) and has the added complication of having a 17 °C (62 °F) freezing point. Consequently, continuous heating is required under most climate conditions. However, as acetic acid is diluted with water, the freezing point decreases, and all dilutions below 85% have a freezing point at or below that of water. Similar to methanol, meeting code requirements for a flammable liquid is a vital component of design (unless very dilute solutions are used, nearing the properties of water), and measures must be taken to avoid freezing. Because of flammability concerns, systems designed for use of high concentrations of acetic acid generally have similar requirements to those of methanol systems. Storage tanks, piping, and appurtenances should be constructed of metal materials. Because of the corrosivity of acetic acid, Type 316 stainless steel is typically used. If glacial acetic acid is used in warm climates, it may be necessary to consider an inert gas blanket or floating cover because of the relatively low flash point.

An external carbon feed of pure acetic acid can promote the growth of GAOs, which compete directly with PAOs, thus increasing the chances of EBPR upset. In contrast, a mixture of acetic and propionic acids has been found to be more beneficial to EBPR (Randall and Chapin, 1997).

In addition to acetic and propionic acids, there are other options for carbon source supplementation. Some WRRF operators have looked to industrial waste products as sources of supplemental carbon. In addition, there are a number of possibilities including sugar wastes, molasses, and waste acetic acid solution from pharmaceutical manufacturers. It is important to ensure that these sources are free of contaminants and debris. Finally, the design of chemical feed facilities must comply with all applicable National Fire Protection Association, U.S. Department of Transportation, Occupational Safety and Health Administration, and local code requirements.

7.0 PROCESS MONITORING AND CONTROL

7.1 Dissolved Oxygen

Dissolved oxygen is a measurement of oxygen dissolved in a liquid stream. In the EBPR process, adequate dissolved oxygen should be available for rapid phosphorus uptake in the aerobic zone. If dissolved oxygen is present in the anaerobic zone, however, it could potentially inhibit VFA production and PAO selection. Dissolved oxygen in the aerobic zones should be continuously monitored to ensure sufficient oxygen is provided during all flow and loading conditions. The preferred aeration basin layout for maximum removal efficiency is plug flow. In practice, the aeration basin is often divided into several baffled zones or compartmentalized to approach plug flow conditions. The greater the numbers of zones, the closer the flow regime is to plug flow. However, cost and practical considerations often limit the number of aerated zones.

When dissolved oxygen is not limiting, the phosphorus uptake rate is typically high in the initial oxic zone of the process, as discussed in Chapter 8. Facility operating data have shown that orthophosphate can be reduced to a level as low as 0.1 mg/L when sufficient oxygen is supplied in the first oxic zone of the EBPR process and all other conditions for biological phosphorus removal are met. While it is possible to achieve EBPR at dissolved oxygen concentrations less than 2 mg/L, the initial oxic zone (first 25 to 33%) should be designed to operate at higher than 2.0 mg/L dissolved oxygen to support rapid phosphorus uptake, meet nitrogenous demand, and guard against the growth of low dissolved oxygen filaments such as *Microthrix parvicella*. If the facility has a completely mixed aeration tank, then phosphorus removal efficiency may be increased by partitioning the tank to cost-effectively maintain high dissolved oxygen in the first oxic cell. Because oxygen demand toward the end of a plug flow aeration basin is primarily associated with endogenous respiration, tapering air delivery and reducing the target dissolved oxygen in the subsequent oxic zones will meet the process requirements and reduce the mass of dissolved oxygen in the mixed liquor recycle (MLR) streams. Too much dissolved oxygen introduced by MLR to the anoxic zone can inhibit denitrification. However, it is important to note that the recommended minimum airflow rates for mixing may make it difficult to maintain low dissolved oxygen concentrations at the end of the aeration basin. Dissolved oxygen in the MLR recycle can also be minimized by providing a small deoxygenation zone (nominal HRT of 20 to 30 minutes) before the MLR pumps. The need to mechanically mix the deoxygenation zone should be assessed if problems related to solids settling are experienced.

On the contrary, sufficient dissolved oxygen should be maintained in the bioreactor effluent to avoid problems in final clarifiers such as phosphorus release in the sludge blanket and denitrification (in nitrifying plants). If both objectives (reduced recycle dissolved oxygen load and adequate bioreactor effluent dissolved oxygen) must be met, the bioreactor effluent could be aerated in the final clarifier distribution box or in mixed liquor channels to the clarifiers.

For systems approaching complete-mix conditions (such as an oxidation ditch), it is possible to control the airflow with one dissolved oxygen probe that is strategically located. Plug flow reactors encounter significant oxygen demand gradients. As treatment occurs, oxygen requirements progressively decrease along the length of the bioreactor. As a result, a tapered diffuser arrangement with multiple oxic zones with independent aeration grids may be considered. For these configurations, more sophisticated control systems may be necessary. One approach is to provide a common header system equipped with multiple control valves to regulate the airflow to each grid with dedicated dissolved oxygen probes.

Many facilities use online instrumentation for continuous monitoring of dissolved oxygen in aeration zones. Often, this information is used to control aeration and to reduce energy costs. The dissolved oxygen concentrations in the bioreactor influent, anaerobic zone, internal recycle, and RAS streams typically are not monitored, but should be checked if EBPR is reduced. When instrumentation and control equipment are limited, it is possible to control the aeration system with one strategically placed dissolved oxygen probe in combination with a manually controlled valve to each of the aeration grids to avoid underaeration or overaeration.

7.2 Oxidation–Reduction Potential

Oxidation–reduction potential is a measurement of the oxidation or reduction potential of a liquid. The ORP measures all ionic species in the fluid and cannot be used to make fine adjustments based on subtle changes. Chapter 13 includes a detailed description of process monitoring and control using ORP.

7.3 Online Nutrient Analyzers

Increasingly, sophisticated analyzers are available to measure nutrients directly in the process. Analyzers can be used for monitoring ammonia, nitrate, nitrite, and phosphate through the treatment process. Many of these analyzers are costly to purchase and maintain, but provide invaluable information on dynamic variations in nutrient concentrations through the process and can be used for troubleshooting.

7.4 Solids Retention Time

Solids retention time is an important operational parameter in a biological system because it provides a measure of the time a solid particle (organism) spends in the biological reactor. Controlling the SRT manages the microbial makeup of the mixed liquor. Ideally, the goal is to operate the system close to an SRT that ensures reliable treatment and reduces operating cost.

In an EBPR system, because the emphasis is on the anaerobic storage of the readily available substrate and subsequent consumption of stored carbon, anaerobic and aerobic SRT gain importance to allow just enough time for the PAOs to proliferate and for the EBPR reactions to take place. In addition to SRT, the ratio of feed COD to phosphorus (COD/P) also shapes the microbial composition of the EBPR sludge and effluent phosphorus levels that can be attained. Figure 9.6 illustrates

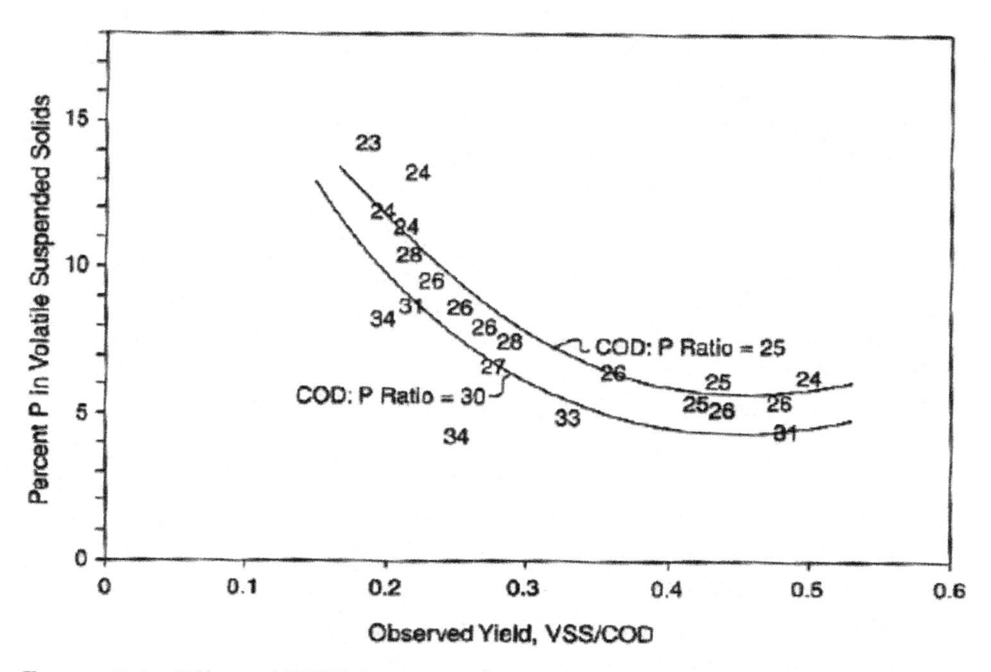

FIGURE 9.6 Effect of EBPR biomass observed yield and SRT on mixed liquor volatile suspended solids phosphorus content (WEF et al., 2005).

the combined effect of the observed yield, as dictated by system SRT, and the feed COD/P on the phosphorus content of the biomass. As the observed yield increases for a constant COD/P value, biomass phosphorus content decreases. In other words, at lower SRT values, less phosphorus is stored in the biomass. Conversely, as the SRT is increased, the sludge will be more robustly enriched by PAOs, as reflected in higher biomass phosphorus values. If, at the same SRT and observed yield value, the feed COD/P is lowered (i.e., increased phosphate or decreased COD in the feed), the biomass phosphorus content will be observed to increase. As feed phosphate increases, more of it can be stored in the biomass, given that the feed COD is sufficient, resulting in an improvement in biomass PAO enrichment. Again, because of the strong effect of feed characteristics and COD/P on the biomass enrichment and microbial composition, results from two different facilities cannot be reliably compared in terms of EBPR performance.

Through years of field and laboratory experience, EBPR systems have been shown to operate at SRTs greater than 3 days. At SRTs between 3 and 4 days, effluent quality declines, and chemical polishing may be needed. At SRTs greater than 4 days and at temperatures greater than 15 °C, nitrification will start to occur and process configurations that include anoxic zones for denitrification of nitrate in system recycles must be used. As the SRT is increased to a level where endogenous reactions become significant (i.e., increased biomass decay), secondary release of phosphorus may lead to decreased performance at given feed VFA and COD values, and phosphorus removal at these higher SRT values can be increased if the feed COD can be increased. Thus, seasonal variations in feed COD and phosphorus can be handled by varying the operational conditions; operators, in turn, can determine the best operational strategy for their influent quality and water resource recovery facility (WRRF) setup.

To explain SRT and EBPR performance, a number of researchers conducted experiments under various operational conditions. McClintock et al. (1991) showed that, at a temperature of 10 °C and an SRT of 5 days, EBPR function of a given activated sludge system would "wash out" before other heterotrophic functions. The washout SRT is a design parameter that defines a critical SRT point, below which no growth of biomass occurs (Grady et al., 2011). Mamais and Jenkins (1992) showed that there is a washout SRT for all temperatures over the range of 10 to 30 °C. Their study clearly indicates that, if the SRT–temperature combination is below a critical value, EBPR ceases before other heterotrophic functions. Lu et al. (2007) demonstrated the effect of SRT on the number of bacterial species present in different biological systems by a comprehensive, simplified microbial model, which could be used in the future for SRT control on biodiversity and treatment performance.

Commercially available programs can automatically control SRT by using sophisticated control algorithms that take into account real-time variations in solids inventory in the aeration tanks and in solids to be wasted. This type of system has been used in various WRRFs such as the San Jose and Santa Clara Water Pollution Control Plant, in the City of Oxnard, California, and in Oakland, California.

In EBPR systems designed for both nitrogen and phosphorus removal, individual zones should be designed to have variable sizing with provisions for mixing and aeration zones that can be expanded or reduced to achieve desired effluent phosphorus concentrations, depending on influent characteristics and operational conditions. At full-scale facilities, VFA uptake is a relatively rapid reaction, requiring an anaerobic zone SRT as low as 0.3 to 0.5 days.

8.0 MINIMIZING FACILITY-WIDE EFFECTS

The biological cycle of anaerobic phosphorus release and aerobic phosphorus uptake in an EBPR facility can significantly affect both upstream and downstream unit operations. These effects can be significant and must be considered during design and operation. Many of the effects are caused by the re-release of phosphorus in anaerobic conditions, such as in thickening or dewatering operations.

Other significant effects may come from filamentous bulking or foaming. Filamentous bulking is rare in biological phosphorus removal facilities because the incorporation of an anaerobic zone for phosphorus release acts like a selector, which will favor predominance of nonbulking organisms. The anaerobic selector does not necessarily inhibit growth of all types of filamentous organisms. Organisms such as Gordona amaerae and Microthrix parvicella are capable of adapting to alternating anaerobic and aerobic cycles. Other low-F/M filaments such as Types 0041 and 0675 may also be able to proliferate in the anaerobic zone. Microthrix parvicella, in particular, are capable of storing polyphosphates and to compete successfully for SCVFAs in anaerobic selectors if wastewater characteristics and environmental conditions are favorable.

8.1 Primary Clarification

In secondary treatment facilities, use of primary clarifiers reduces aeration requirements in the downstream activated sludge tanks. With EBPR processes, excessive organics removal in primary treatment can potentially result in lowering the primary effluent COD/P and can reduce EBPR efficiency. This would only be the case if the anaerobic zone were sufficiently large to affect VFA generation from settleable solids, which typically may not be the case. Another possibility is that primary polymer addition would remove colloidal solids that would otherwise be fermented to SCVFAs in the anaerobic zone. In such instances, primary sludge fermentation could be included to increase the VFA concentration entering the anaerobic zone.

One strategy for thickening waste activated sludge (WAS) is to co-thicken it in primary clarifiers along with primary sludge. In general, this strategy should not be used for EBPR because the combination of primary sludge and WAS will result in significant and rapid secondary phosphorus release. Work by Skalsky and Daigger (1995) showed that combining WAS and primary sludge resulted in more rapid fermentation of the primary sludge than would occur without the addition of WAS. As investigated by Chaparro and Noguera (2003), it may be possible to use this characteristic of co-thickening to affect phosphorus release and create a phosphorus-rich recycle stream for a struvite harvesting process.

Another effect to consider in the primary clarifier is the fate of iron addition to a recycle stream for recycle phosphorus control. If ferric or ferrous chloride is added to precipitate phosphorus from a thickening or dewatering recycle stream, the precipitate formed will settle in the primary clarifiers. The effectiveness of this strategy will be reduced because a pH of approximately 6.5 in the anaerobic environment of the primary clarifier, or in a downstream gravity thickener, will cause a portion of the ferric phosphate precipitated at a higher pH to be reduced to ionized form. This may result in some additional carryover of iron and phosphorus to the secondary process, although it may be insignificant.

8.2 Final Clarification

Separation of biological solids is of primary importance in EBPR. Phosphorus uptake by bacteria will result in biological solids with 2 to 5% phosphorus. The effluent phosphorus concentration is, therefore, highly dependent on the capture of the solids. For example, if the effluent solids concentration is 20 mg/L and the biological solids are 4% phosphorus, the effluent phosphorus concentration will be 0.8 mg/L just from the solids, plus additional soluble phosphorus remaining after treatment. This level of effluent TSS would make it difficult to reliably comply with a 1-mg/L total phosphorus standard in the effluent. Fortunately, the anaerobic selector in an EBPR facility will typically result in good settling sludge that also flocculates well. Effluent TSS concentrations less than 10 mg/L are common in EBPR facilities. With appropriately designed clarifiers, an EBPR facility can produce 1.0 to 0.50 mg/L of total phosphorus in the effluent, and, with filtration, total phosphorus concentrations less than 0.30 mg/L are possible without supplemental chemical addition, provided conditions for efficient EBPR are satisfied. The potential for high flows, development of a deep blanket, and washout of solids must be considered. If the clarifiers do not have sufficient capacity for high-flow conditions, high-effluent phosphorus values may result.

Because mixed liquor solids will generally settle and compact well in an EBPR facility, it is often possible to reduce the return rate and provide a higher degree of thickening in the secondary clarifier. This allows for a lower waste sludge flowrate. Caution must be exercised because a deep sludge blanket may result in secondary release of phosphorus and higher soluble phosphorus levels in the effluent. Improvements to the sludge removal mechanism can be beneficial if the sludge blanket builds up despite good settling and compaction.

8.3 Secondary Release

Bacteria generated in an EBPR process contain two forms of phosphorus. One form is polyphosphate temporarily stored during aerobic respiration, which then allows energy generation by phosphorus release with VFA uptake. The second form is the normal metabolic phosphorus product from microbial growth and reproduction. *Secondary release* is the term for release of temporary phosphorus storage products in anaerobic conditions without the concurrent uptake of VFAs. This release will not be taken up again in the aerobic zone and will result in elevated effluent phosphorus levels. Table 9.4 shows some of the potential causes of secondary phosphorus release.

Secondary release might be caused in the secondary system by long aerobic, anoxic, or anaerobic detention times. Long SRTs in the aerobic system may also result in secondary release. Long SRT operation also requires higher COD/P for the bacteria to take up the available phosphorus. Because of these factors, it is generally best to operate an EBPR system as close to minimum SRTs as possible and to maintain minimum necessary anoxic and anaerobic detention times to meet process goals. If nitrification is required, this may limit the minimum SRT based on minimum winter wastewater temperatures. If effluent phosphorus concentrations rise and jeopardize effluent limitations, it may be necessary to collect samples profiling the secondary process to determine whether the problem may be caused by secondary release.

TABLE 9.4 Location and cause of secondary phosphorus release.

Environment/operation	Reason for secondary phosphorus release
Anaerobic zone	VFA depletion because of oversized anaerobic zone
Anoxic zone	Nitrate depletion because of oversized anoxic zone
Second anoxic zone	Lack of nitrates resulting in anaerobic conditions
Aerobic zone	Long solids retention time leading to cell lysis
Final clarifier	Septic conditions caused by deep sludge blanket and low return rate
Return sludge piping	Septic conditions from long detention time in pipe
Sludge storage	Septic conditions from poorly or unaerated sludge storage
Aerobic digestion	Cell lysis because of long aeration time
Anaerobic digestion	Cell lysis because of long anaerobic detention time

8.4 Recycle Streams

Recycle streams result from sludge thickening and dewatering operations. Thickening of primary sludge may actually benefit the EBPR process because VFAs will be generated in the thickening process. This is similar to inclusion of a dedicated primary sludge fermentation process to enhance EBPR.

Thickening of WAS may have little effect on the EBPR process unless the sludge is held too long, resulting in secondary release. In addition, relatively high levels of recycle phosphorus to the secondary process may result.

Gravity co-thickening of WAS and primary sludge is typically not recommended for an EBPR facility because this will result in the recycle of high concentrations of soluble phosphorus released from WAS. As described for primary clarification, gravity co-thickening of primary sludge and WAS will result in enhanced production of VFAs in the primary sludge and significant release of polyphosphate from WAS.

High concentrations of soluble phosphorus will result from the release of stored polyphosphates and cellular phosphorus in both anaerobic and aerobic digestion. Thickening or dewatering of these streams will result in a stream that is high in soluble phosphorus (200 to 300 mg/L and higher) and high in ammonia-nitrogen (700 to 1000 mg/L and higher). Recycled phosphorus load can represent 5 to 40% of influent total phosphorus load and may alter the COD/P such that successful EBPR to low levels may no longer be possible. Strategies to minimize the effects of digested sludge recycle streams include operating thickening or dewatering continuously to minimize slug loading because of recycled phosphorus and the addition of iron salts to the recycle stream to reduce the soluble phosphorus recycle.

8.5 Bulking and Foam Control

Filamentous bulking and foaming are potential operating problems associated with EBPR facilities. These issues negatively affect effluent quality and create serious housekeeping and odor problems. An in-depth examination of the topic may be found in literature by Eikelboom (2000), Jenkins et al. (2003), and Wenner (1994). Detailed information on microscopic evaluation and identification of activated sludge characteristics, levels, and types of filamentous organisms is presented in literature by Jenkins et al. (2003). The following subsections provide a brief discussion of the causes and potential corrective strategies of bulking and foaming in EBPR systems.

8.5.1 Filamentous Bulking

Excessive filaments often lead to poor settling sludge and high effluent solids. A variety of operating conditions, singly or in combination, can cause growth of filamentous organisms in EBPR systems. These include low dissolved oxygen, low or high F/M, sulfides, and low pH. Filament identification is the first step in resolving the problem. Typically, operational controls focus on removing conditions responsible for bulking or killing filamentous organisms to control their number. Some common strategies include

- Using selectors to provide growth advantage to floc formers,
- Chlorinating RAS,
- Adding nutrients,
- Correcting the dissolved oxygen concentration in the bioreactor, and
- Correcting the pH.

Generally, the anaerobic zone in an EBPR system will select the floc-forming PAOs and minimize settling problems. However, if for any reason VFAs are not totally consumed in the anaerobic selector and enter the aeration tank, septicity-type filaments can grow.

8.5.2 *Filamentous Foaming*

Although *Gordona* (formerly known as *Nocardia*) *amarae* and other nocardiaforms are the most commonly found organisms responsible for filamentous foaming, others (such as *Microthrix parvicella* and Type 1863) also can cause foaming. The surface appearance of tanks with foaming problems may be similar regardless of the organism causing the foaming problem. The presence of some foam in the activated sludge bioreactor is normal. In a well-operated process, 10 to 25% of the bioreactor surface may be covered with a 50- to 80-mm (2- to 3-in.) layer of light tan foam. Under certain operating conditions, foam can become excessive and affect operations.

Three types of problem-causing foams are stiff, white foam; brown and dark tan foam, often incorporating scum; and dark brown or black foam. The most common foaming problem in EBPR systems is brown and dark tan foam associated with nocardiaforms or *Microthrix parvicella.* If this greasy, thick, scummy foam builds up and is conveyed to secondary clarifiers, it will tend to build up behind influent baffles and create additional cleaning requirements. In colder climates and extreme cold conditions, it can build up on scum-removal ramps, hang up the scum-removal skimmers, and cause sludge removal mechanisms to overload. This can result in loss of effluent solids or secondary phosphorus release. It can also plug scum-removal piping. Unmanaged, scum can even overflow aeration tank walls and flow onto walkways and roadways. Waste activated sludge containing foaming organisms can also result in serious foaming problems in anaerobic or aerobic digesters.

Nocardia foaming is typically associated with warmer temperatures, lower pH, grease, oil, fats, and long SRT. *Microthrix parvicella* foaming is more commonly associated with mixed liquor temperatures less than 16 °C and long SRTs.

Microthrix spp. also prefer grease, oils, and fats; these microorganisms are better able to compete for these substrates at colder temperatures.

Because foaming is a surface phenomenon, it typically has a longer SRT than the underlying MLSS. Facilities prone to foaming often receive oil and grease waste from restaurants with poorly performing or missing grease traps, have poor or no primary scum removal or recycle scum, and have bioreactors and final clarifiers that are not properly designed to remove scum and foam. The most effective strategy to deal with foaming is to eliminate conditions that encourage growth and to prevent accumulation within the process. This is not always easy, however, because exact cause-and-effect relationships have not been fully established. The following is a listing of foam and scum control methods that can be implemented during design of EBPR systems:

- Design interzone and intrazone baffles to promote free-flow surface foam and scum;
- Eliminate dead ends, sharp corners, and quiescent zones in channels or bioreactors where there is a potential for foam and scum to accumulate;
- Design secondary sedimentation tank inlet wells and flocculation wells to allow for passage of floating material. Provide effective foam and scum collection mechanisms (such as full-radius collectors);
- Incorporate provisions for selectively wasting foam and scum from the surface of the aeration basin. Collected foam should not be recycled to avoid reseeding the foam-causing organisms. Parker at al. (2003) reported several successful applications of selective wasting in controlling activated sludge foaming. The overall objective of this strategy is to reduce the SRT of the foam-causing organisms while minimizing the effect on process SRT; and
- Consider including selectors; however, selectors have been found to have marginal effectiveness in preventing the predominance of foam-causing organisms.

If foaming occurs, corrective action should be taken quickly. Microscopic analysis of mixed liquor will allow for identification of the predominant causative organisms. This will provide clues as to which operational strategies are likely to be successful. Some examples include the following:

- Operate at the lowest SRT required to reliably meet process goals;
- Avoid opportunities for recycling foam and scum to the mainstream treatment train to prevent recurrence of the problem;
- Apply chlorine (0.5 to 1% solution) spray at localized points of foam and scum collection or accumulation to kill nocardiaforms and to prevent them

from causing problems in either mainstream or sidestream treatment processes. The chlorine dose should be carefully controlled to avoid EBPR inhibition, which can take several days to recover;

- Consider RAS chlorination. However, this has limited effectiveness in controlling foam-causing organisms;
- Add polymers to destroy the hydrophobic properties of the foam and to allow it to mix with the sludge so that it can be removed with waste sludge; and
- Addition of polyaluminum chloride has also been shown to be effective in controlling *Microthrix parvicella* foaming (Melcer et al., 2009).

9.0 REFERENCES

Barnard, J. L. (1974) Cut P and N without Chemicals. *Water Wastes Eng.*, **11**, 33–36.

Barnard, J. L. (1984) Activated Primary Tanks for Phosphate Removal. *Water SA*, **10** (3), 121–126.

Barnard, J.; Houweling, D.; Analla, H.; Stachen, M. (2010) Fermentation of Mixed Liquor for Phosphorus Removal. *Proceedings of the 83rd Annual Water Environment Federation Technical Exhibition and Conference* [CD-ROM]; New Orleans, Louisiana, Oct 2–6; Water Environment Federation: Alexandria, Virginia.

Baur, R. (2002) Unified Fermentation and Thickening Process. U.S. Patent 6,387,264; dated May 14.

Chaparro, S. K.; Noguera, D. R. (2003) Controlling Biosolids Phosphorus Content in Enhanced Biological Phosphorus Removal Reactors. *Water Environ. Res.*, **73** (3), 254–262.

Daigger, G. T.; Littleton, H. X. (2000) Charaterization of Simultaneous Nutrient Removal in Staged, Closed-Loop Bioreactors. *Water Environ. Res.*, **72** (3), 330–339.

Eikelboom, D. K. (2000) *Process Control of Activated Sludge Plants by Microscopic Investigation;* IWA Publishing: London.

Erdal, U. G.; Erdal, Z. K.; Randall, C. W. (2002) The Effect of Temperature on EBPR System Performance and Bacteria Community. *Proceedings of the 75th Annual Water Environment Federation Technical Exposition and Conference* [CD-ROM]; Chicago, Illinois, Sept 28–Oct 2; Water Environment Federation: Alexandria, Virginia.

Grady, C. P. L.; Daigger, G. T.; Love, N. G.; Filipe, C. D. M. (2011) *Biological Wastewater Treatment*, 3rd ed.; CRC Press: Boca Raton, Florida.

Jenkins, D.; Richard, M. G.; Daigger, G. T. (2003) *Manual on the Causes and Control of Activated Sludge Bulking, Foaming and other Solids Separation Problems*, 3rd ed.; IWA Publishing: London.

Jeyanayagam, S. S. (2007) So, You Want to Remove Phosphorus? Part 1: Enhanced Biological Phosphorus Removal. *Buckeye Bull.*, **4** (80) 22–29.

Lilley, I. D.; Wentzel, M. C.; Lowenthal, R. E.; Marais, G. v. R. (1990) *Acid Fermentation of Primary Sludge at 20°C;* Research Report W64; Department of Civil Engineering, University of Cape Town: Cape Town, South Africa.

Lu, T.; Saikaly, P. E.; Oerther, D. B. (2007) Modeling Competition of Aerobic Heterotrophs for Complementary Nutrient in a Biofilm Reactor: Effect of Hydraulic Retention Time on Coexistence. *Water Sci. Technol.*, **55** (8-9), 227–235.

Mamais, D.; Jenkins, D. (1992) The Effects of MCRT and Temperature on Enhanced Biological Phosphorus Removal. *Water Sci. Technol.*, **26** (5-6); 955.

Melcer, H.; Robinson, W.; Jue, P.; Doty, J.; Yanasak, J.; Land, G. (2009) The Application of PAX Compounds for *Microthrix parvicella* Foam Control in BNR Systems. *Proceedings of the 2009 Water Environment Federation Specialty Conference—Nutrient Removal;* Washington, D.C., June 28–July 11; Water Environment Federation: Alexandria, Virginia.

Narayanan, B.; Johnson, B.; Baur, R.; Mengelkoch, M. (2006) Uptake in Biological Phosphorus Removal. *Proceedings of the 79th Annual Water Environment Federation Technical Exhibition and Conference* [CD-ROM]; Dallas, Texas, Oct 21–25; Water Environment Federation: Alexandria, Virginia.

Oldham, W. K.; Abraham, K. (1994) Overview of Full-Scale Fermenter Performance. *Proceedings of the 67th Annual Water Environment Federation Technical Exposition and Conference* [CD-ROM]; Chicago, Illinois, Oct 15–19; Water Environment Federation: Alexandria, Virginia.

Oldham, W. K.; Stevens, G. M. (1984) Initial Operating Experience of a Nutrient Removal Process (Modified Bardenpho) at Kelowna, British Columbia. *Can. J. Civil Eng.*, **11**, 474–479.

Parker, D.; Geary, S.; Jones, G.; McIntyre, L.; Oppenheim, S.; Pedregon, V.; Pope, R.; Richards, T.; Voigt, C.; Volpe, G.; Willis, J.; Witzgall, R. (2003) Making Classifying Selectors Work for Foam Elimination in the Activated-Sludge Process. *Water Environ. Res.*, **75** (1).

Rabinowitz, B.; Koch, F. A.; Vassos, T. D.; Oldham, W. K. (1987) A Novel Design of a Primary Sludge Fermenter for Use with the Enhanced Biological Phosphorus Removal Process. *Proceedings of the IAWQ Specialized Conference on Biological Phosphate Removal from Wastewaters;* Rome, Italy, Oct; International Association on Water Quality: London.

Randall, C. W.; Barnard, J. L.; Stensel, H. D., Eds. (1992) *Design and Retrofit of Wastewater Treatment Plants for Biological Nutrient Removal;* Vol. 5, Water Quality Management Library; Technomic Publishing: Lancaster, Pennsylvania.

Randall, C. W.; Chapin, R. W. (1997) Acetic Acid Inhibition of Biological Phosphorus Removal. *Water Environ. Res.,* **60** (5), 955–960.

Skalsky, D.; Daigger, G. (1995) Wastewater Solids Fermentation for Volatile Acid Production and Volatile Acid Production and Enhanced Biological Phosphorus Removal. *Water Environ. Res.,* **67**, 230.

Stensel, D. (1991) Principles of Biological Phosphorus Removal. In *Phosphorus and Nitrogen Removal from Municipal Wastewater—Principles and Practices;* Sedlak, R., Ed.; Lewis Publishers: Boca Raton, Florida.

Water Environment Federation (2011) *Nutrient Removal;* WEF Manual of Practice No. 34; McGraw-Hill: New York.

Water Environment Federation; American Society of Civil Engineers; Environmental and Water Resources Institute (2005) *Biological Nutrient Removal (BNR) Operation in Wastewater Treatment Plants;* WEF Manual of Practice No. 29; ASCE/EWRI Manuals and Reports on Engineering Practice No. 109; McGraw-Hill: New York.

Water Environment Federation; American Society of Civil Engineers; Environmental and Water Resources Institute (2009) *Design of Municipal Wastewater Treatment Plants,* 5th ed.; Manual of Practice No. 8; ASCE Manuals and Reports on Engineering Practice No. 76; McGraw-Hill: New York.

Water Environment Research Foundation (2011) *Fermenters for Biological Phosphorus Removal Carbon Augmentation;* WERF Nutrient Challenge Project; Water Environment Research Foundation: Alexandria, Virginia.

Wenner, J. (1994) *Activated Sludge Bulking and Foaming Control;* Technomic Publishing: Lancaster, Pennsylvania.

10.0 SUGGESTED READINGS

Reynolds, T. D. (1982) *Unit Operations and Processes in Environmental Engineering;* Brooks/Cole Engineering Division, Wardsworth: Monterey, California.

Thomas, M.; Wright, P.; Blackall, L.; Urbain, V.; Keller, J. (2003) Optimisation of Noosa BNR plant to Improve Performance and Reduce Operating Costs. *Water Sci. Technol.,* **47** (12), 141–148.

Watts, J. B.; Parrott, I.; Shaw, A.; Mason, L. K. (2000) On-Line Monitoring Instrument for the Measurement of Nitrification and Denitrification Rates. *Proceedings of the 73rd Annual Water Environment Federation Technical Exposition and Conference* [CD-ROM]; Anaheim, California, Oct 14–18; Water Environment Federation: Alexandria, Virginia.

Chapter 10

Combined Nitrogen and Phosphorus Removal Processes

Timur Deniz, Ph.D., P.E., BCEE;
Lucas Botero, P.E., BCEE; and
Joel C. Rife, P.E., BCEE

1.0 OVERVIEW

In Chapter 7, the mechanism of enhanced biological phosphorus removal (EBPR) was explained and the possible interference of nitrate in the mechanism for EBPR was discussed. This chapter will focus on biological nutrient removal (BNR) processes that remove nitrogen and phosphorus. While phosphorus removal in a combined nitrogen and phosphorus removal facility can be achieved both chemically and biologically, the biological alternative has a number of significant advantages, such as considerably lower operating costs, less sludge production, and little or no added chemicals in the sludge; however, it requires more skillful operation.

Total nitrogen removal in a combined nitrogen and phosphorus removal BNR facility is most commonly achieved in a two-zone system through nitrification under aerobic conditions and denitrification under anoxic conditions. Because of the possible interference of nitrate on the anaerobic conditions required for the EBPR process, at least three zones or periods in intermittent systems (anaerobic, anoxic, and aerobic) are required to provide the different environmental conditions. Additionally, possible interferences between nitrogen and phosphorus removal processes often mean that additional zones or periods are required (e.g., for the removal of nitrate in return activated sludge [RAS]) (Keller et al., 2001). Lower effluent nitrogen and phosphorus limits, such as those applied to advanced water treatment (AWT), will also require additional zones to reduce effluent total nitrogen and total phosphorus below 3.0 and 1.0 mg/L, respectively. The second anoxic zone (postanoxic) in a five-stage Bardenpho process is an example of this.

Many processes are available for combined nitrogen and phosphorus removal. Key factors for optimizing nitrogen and phosphorus removal in these processes are as follows:

- Complete nitrification should be achieved in aeration zones with as short a solids retention time (SRT) as possible. Longer SRTs will increase microbial decay, thereby increasing nitrogen and phosphorus release;
- Denitrification should be optimized in single-stage anoxic zones and multiple-stage anoxic zones (preanoxic and postanoxic) to achieve lower effluent nitrate levels. Nitrate is often a significant fraction of effluent total nitrogen. Dissolved oxygen must be excluded from anoxic zones. If dissolved oxygen is present, bacteria will preferentially use oxygen as the terminal electron acceptor, thereby reducing the efficiency of denitrification;
- Sufficient nitrate recycle with internal recycle should be provided. Because internal recycle flow is taken from the end of the aeration zone, it will also contain dissolved oxygen in addition to nitrate. Lower residual dissolved oxygen should be maintained at the internal recycle flow withdrawal point rather than the rest of the aeration zones;
- Dissolved oxygen and nitrate return to the anaerobic zones should be minimized to increase EBPR efficiency. Otherwise, ordinary heterotrophic bacteria will outcompete polyphosphate-accumulating organisms (PAOs) for volatile fatty acids (VFAs) using dissolved oxygen and nitrate;
- Secondary phosphorus release should be prevented in anoxic zones with optimized SRT. The optimum operating range can be selected by monitoring nitrate levels or oxidation–reduction potential (ORP) to ensure that secondary phosphorus release does not occur; and
- Highly effective suspended solids removal should be provided for particulate nitrogen and phosphorus removal when low levels of effluent nutrient levels are required.

In this chapter, combined nitrogen and phosphorus removal systems are discussed, along with operational conditions, process control strategies, and troubleshooting, followed by some successful case studies of combined nitrogen and phosphorus removal facilities.

2.0 COMMONLY USED PROCESS CONFIGURATIONS

2.1 The Anaerobic–Anoxic–Oxic Process

The three-stage anaerobic–anoxic–oxic (A2/O) process consists of a modified Ludzack–Ettinger (MLE) process for nitrogen removal, with an anaerobic zone in front for phosphorus removal (Figure 10.1). The process is called *A2/O* because it consists of anaerobic–anoxic–oxic (aerobic) tanks in series. The RAS is recycled to the anaerobic zone; the anaerobic zone is followed by an anoxic zone for nitrogen removal through denitrification. Internal recycle is recycled from the aerobic zone to the anoxic zone at a rate of 100 to 400% of influent. Nitrogen removal is limited by the internal recycle flow. Because nitrate-nitrogen concentrations between 5 and 10 mg/L may still be present in RAS, the phosphorus removal capability of A2/O is reduced unless there is a surplus of VFA that can denitrify the nitrate with a sufficient amount remaining to activate the phosphorus removal mechanism. This process has largely been replaced with other configurations that allow for denitrification of nitrate in RAS. This is a relatively simple process to operate. In addition, the process has a small footprint. Existing A2/O BNR facilities may be retrofitted with denitrifying tertiary filters to achieve AWT limits.

2.2 Five-Stage Bardenpho

Barnard (1976) proposed a number of flow sheets for combined nitrogen and phosphorus removal. These all consisted of adding an anaerobic zone ahead of either a high-rate non-nitrifying aeration zone (later called *anaerobic–oxic* [A/O]), an MLE process (later called *A2/O*), or a four-stage Bardenpho process, which was then referred to as the *modified* or *five-stage Bardenpho process*. The goal was to ensure that nitrate in RAS could be reduced to avoid interfering with EBPR. The five-stage Bardenpho process is illustrated in Figure 10.2. All recycle rates are

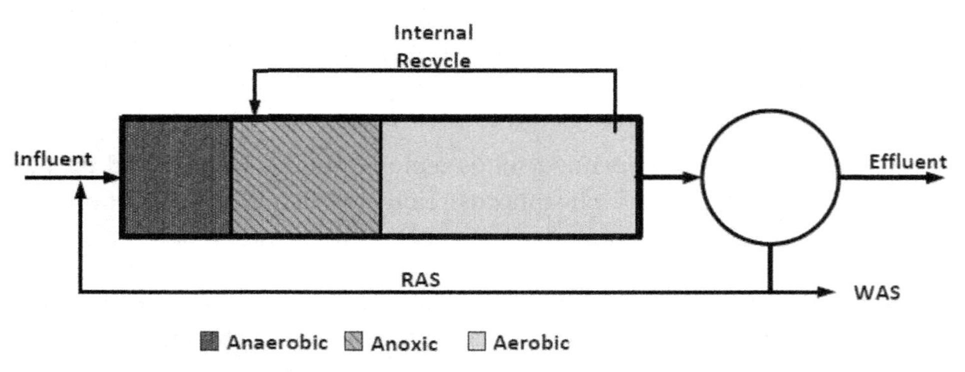

FIGURE 10.1 Three-stage A2/O process.

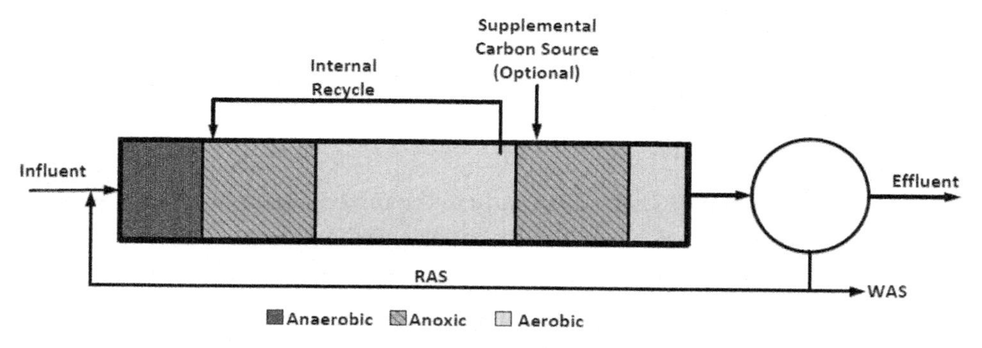

FIGURE 10.2 Five-stage Bardenpho process.

defined in terms of the average influent flowrate. The four-stage Bardenpho process described in Chapter 6 may reduce effluent nitrate to less than 2 mg/L when ratios of influent carbon to nitrogen are favorable and sufficient SRT is provided. When the ratio of influent carbon to nitrogen is lower, an external carbon source may be needed to reduce effluent nitrate to less than 2 mg/L. There will thus be little nitrate in RAS to interfere with the mechanism for EBPR.

In the five-stage Bardenpho process, influent wastewater combines with RAS from secondary clarifiers in an anaerobic zone. In this zone, phosphorus is released and VFAs are stored by PAOs in the absence of dissolved oxygen and nitrate. The flow then enters the first anoxic zone, where it combines with the internal recycle stream containing nitrate from nitrified mixed liquor from the end of the first aerobic zone. In this anoxic zone, nitrate is reduced to nitrogen gas while the bacteria are using readily biodegradable substrate. The flow then enters the first aerobic zone, where ammonia is oxidized to nitrate and the released orthophosphate and more orthophosphate present in mixed liquor is taken up by PAOs. Mixed liquor is typically recycled at a rate between 100 to 400% of influent flow from the end of the first aerobic zone to the head end of the first anoxic zone.

The remainder of the mixed liquor enters a second anoxic zone to reduce (or polish) the effluent nitrate concentration from the first aerobic zone. The second anoxic zones can be as large as the first anoxic zone, although less nitrate removal occurs because denitrification occurs via endogenous respiration and degradation of slowly biodegradable substrate. The second anoxic zone should be sized to optimize denitrification and prevent secondary phosphorus release, which can occur after all of the nitrate is denitrified. The mixed liquor then enters a reaeration zone, where nitrogen gas is stripped from the bulk liquid, and then passes to the secondary clarifiers. Phosphorus is removed from the process in accordance with the theory described in Chapter 7, along with the sludge wasted from the secondary clarifier underflow as waste activated sludge (WAS). Even though it helps to take up phosphorus and to prevent the occurrence of anaerobic conditions and the associated release of phosphorus from sludge to the final effluent, the main purpose of the final aerobic zone is to help release nitrogen gas from the preceding denitrification process; hence, it is typically much smaller than the preceding anoxic and aerobic zones. Other purposes of the reaeration zone are to

- Raise dissolved oxygen sufficiently in the mixed liquor to prevent secondary phosphorus release in the secondary clarifier as a result of low nitrate concentrations in mixed liquor entering the secondary clarifiers;
- Provide nitrification of ammonia released in the second anoxic zone when underloaded with respect to nitrate, resulting in anaerobic conditions; and
- Provide oxidation of excess methanol to prevent high biochemical oxygen demand (BOD) in the effluent.

The intent of the original five-stage Bardenpho process was to reduce nitrogen to low levels. At low levels of the effluent total nitrogen concentration, coupled with some unavoidable denitrification in the sludge blanket, there is no danger of recycling high levels of nitrate to the anaerobic zone through RAS. Today, five-stage Bardenpho processes are commonly used in facilities that require low total nitrogen levels coupled with low-effluent total phosphorus limits such as 1.0 mg/L.

Five-stage Bardenpho processes are highly efficient for nitrogen removal, but less efficient for phosphorus removal. Two-stage anoxic zones (i.e., preanoxic and postanoxic) are the main reason for efficient nitrogen removal. On the other hand, the process requires a long SRT, which diminishes the efficiency of phosphorus removal. When it is not possible to lower total nitrogen and total phosphorus to the effluent limits (i.e., AWT limits of 3.0 and 1.0 mg/L, respectively), metal salt addition for chemical phosphorus removal and external carbon source addition to the second anoxic zone can be implemented.

2.3 University of Cape Town, Modified University of Cape Town, and Virginia Initiative

The purpose of these configurations is to minimize nitrate input to the anaerobic zone, particularly when the effluent nitrate is high. The University of Cape Town (UCT) process consists of anaerobic, anoxic, and aerobic zones, with RAS directed to the anoxic zone for achieving denitrification (Ekama and Marais, 1984) (Figure 10.3). The influent wastewater flows directly to the anaerobic zone, providing a source of VFAs to PAOs. Nitrified mixed liquor from the aerobic zone is pumped to the anoxic zone at rates of 200 to 400% of influent flow to increase nitrogen removal through denitrification. Denitrified effluent from the end of the anoxic zone is recycled to the anaerobic zone at a rate of 200% of influent flow to provide the mixed liquor recycle to the anaerobic zone. The mixed liquor suspended solids (MLSS) concentration in the anaerobic zone will

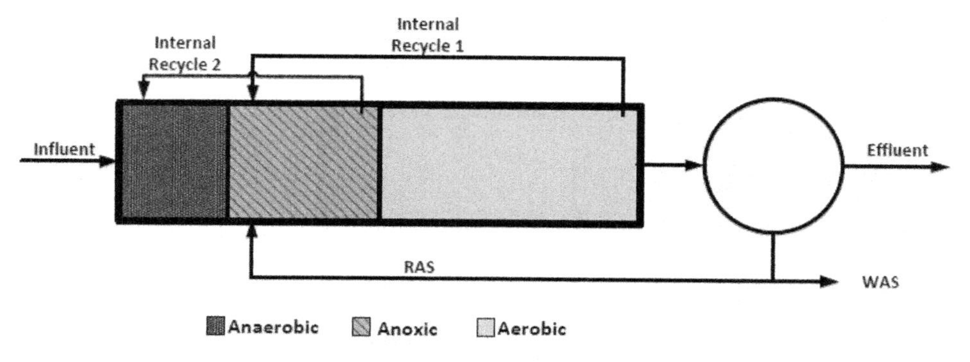

FIGURE 10.3 The UCT and VIP configurations.

be lower than the anoxic and aeration zones and, therefore, a longer hydraulic retention time (HRT) should be provided in the anaerobic zones.

The UCT process can provide more efficient EBPR than A2/O, although both processes have similar nitrogen removal efficiencies because they use single-stage denitrification zones. On the other hand, larger anaerobic zones and higher pumping requirements are the main disadvantages of the UCT process.

The Virginia Initiative Process (VIP) is a high-rate version of the UCT process (Daigger et al., 1988). All zones are staged, consisting of at least two cells in series. The idea behind these processes is that, when there is insufficient carbon in the influent for both nitrogen and phosphorus removal, preference could be given to phosphorus removal. For example, if the effluent and RAS nitrate is high, the nitrate in the RAS could be denitrified in the anoxic zone before it is recycled back to the anaerobic zone. If the anoxic zone cannot remove all the nitrate, the mixed liquor recycle (MLR) can be reduced to ensure that the nitrate at the end of the anoxic zone will be low, thus ensuring that nitrate will not be recycled to the anaerobic zone. This approach led to the development of the modified UCT (MUCT) process. The MLR flowrate from the aeration to the anoxic zone could be reduced to ensure that all nitrate is removed before passing the mixed liquor to the anaerobic zone. This may then result in higher effluent nitrate. If there is a need for further reduction in nitrate, a post-attached-growth denitrification system, such as denitrifying sand filters or a moving bed biofilm reactor, could be used with external carbon source addition.

In the MUCT process, the anoxic zone is divided into two cells (Figure 10.4). The first anoxic zone receives the effluent of the anaerobic zone and RAS. The denitrified mixed liquor is then recycled to the anaerobic zone. The first anoxic cell is, therefore, required to reduce only the nitrate in RAS, and can be better controlled. The second anoxic cell receives internal recycle from the aerobic zone, where bulk denitrification occurs. Accurate control of the internal recycle is not as vital. Facility influent mixed with denitrified anoxic recycle from the first anoxic cell is routed into the anaerobic zone, where multistaged compartments are used

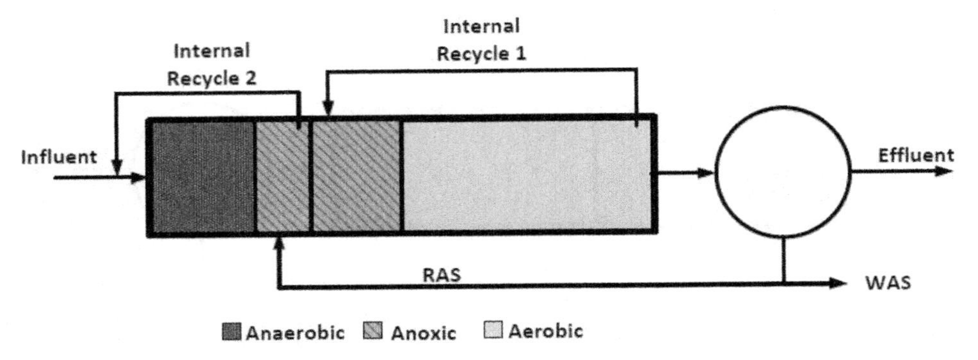

FIGURE 10.4 The MUCT process configuration.

to minimize the effects of nitrate recycle to the anaerobic zone. Flow from the anaerobic zone then enters the first anoxic cell and then the second anoxic cell. From there, it flows to the aerobic zone, where nitrification occurs. The nitrified aerobic mixed liquor is recycled back to the second anoxic cell for denitrification. It is not vital that MLR to the second anoxic cell be accurately controlled because it will not affect the return of nitrate to the anaerobic zone. This configuration is highly effective and, therefore, a large number of MUCT facilities are in operation, giving good performance of removal of phosphorus to low levels. The only disadvantage of the process is that the mixed liquor concentration in the anaerobic tank is lower than that of other processes, requiring larger anaerobic zone volume for the same anaerobic SRT.

2.4 Johannesburg and Modified Johannesburg

The City of Johannesburg, South Africa, solved their EBPR problem of nitrate in RAS by providing an endogenous pre-denitrification zone, as illustrated in Figure 10.5 (Pitman, 1992). The MLSS concentration in the preanoxic zone (equal to RAS from the secondary clarifier underflow) is typically approximately twice that of the MLSS concentration in subsequent zones, which results in endogenous denitrification taking place at a much higher rate, reducing the nitrate to low levels. This is known as the *Johannesburg* (JHB) *process*. The JHB process was later improved by adding a recycle stream from the end of the anaerobic zone to the preanoxic zone to use readily biodegradable compounds not taken up by PAOs for denitrification (the modified JHB process).

This recycle stream is typically set at 10% of influent flow, resulting in an inexpensive way of providing more carbon for denitrification before the anaerobic zone. Because of the high concentration of solids in the pre-denitrification zone, the volume of this zone may be as little as 25% of the first anoxic zone of the MUCT process. Refer to *Design of Municipal Wastewater Treatment Plants* (WEF, 2010) for more information about the MUCT process. Figure 10.5 shows a flow schematic for the modified JHB.

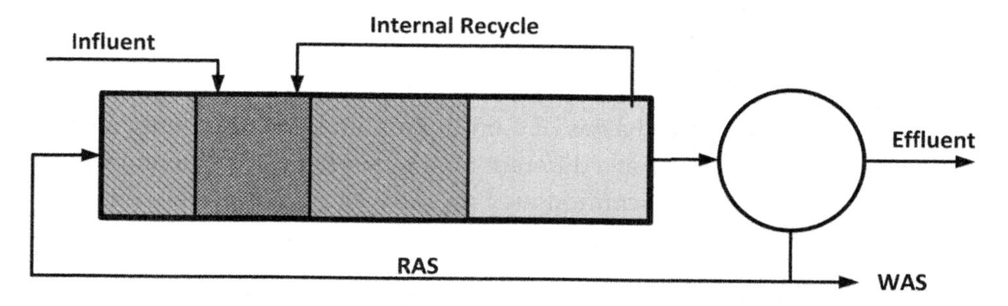

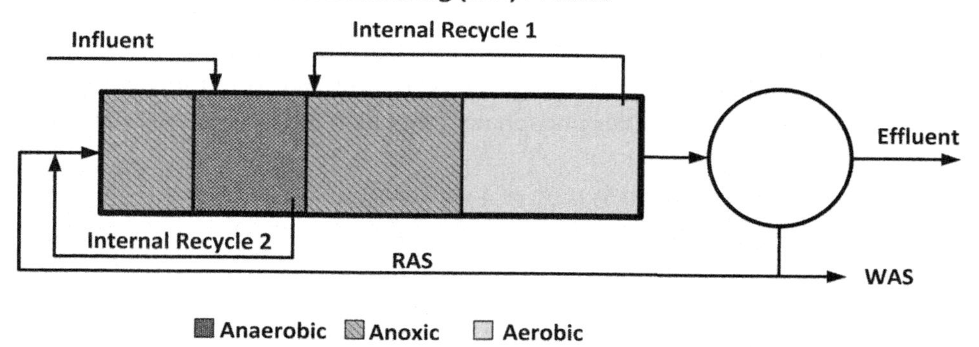

FIGURE 10.5 The JHB and modified JHB process configurations.

2.5 Process Configurations with Cyclic Aeration

Cyclic aeration refers to alternating cycles of "air on" and "air off" in the bioreactor. Nitrification and denitrification occurs in the same tank. During extended air-off cycles, EBPR can occur in cyclic aeration systems. There are several types of alternating aeration processes currently available. These include

- Sequencing batch reactors (SBRs),
- Phased isolation oxidation ditch, and
- Sequential aeration.

Sequencing batch reactors for phosphorus removal consist of a large batch reactor with a reseeding anaerobic zone. A typical sequence for phosphorus removal incorporates a number of stages, namely the following:

- An anoxic idle period, during which nitrate in the sludge is reduced;
- Feed to an anaerobic zone, with recycle of sludge to this zone from the anoxic stage;

- Aeration of the main reactor while still feeding to the anaerobic zone;
- Settling; and
- Decanting from the main tank.

Figure 10.6 shows two basins of a multiple-SBR configuration. Each environmental condition occurs at a different time rather than in a different volume compared to conventional, continuous-flow, activated sludge processes. After decanting, the sludge is again kept under anoxic conditions to reduce nitrate. Mixed liquor is pumped from under the decanters to the anaerobic zone, where it is contacted with the feed that contains VFAs and phosphorus is released. The mixed liquor then overflows into the batch reactor during a mixing and no-aeration (anoxic) period, when denitrification takes place. This is followed by the aeration stage, during which nitrification and phosphorus uptake takes place. Endogenous denitrification takes place during settling and decant. It is essential that the remainder of the nitrate be removed during the settling and decant phases to achieve better phosphate removal in the next cycle (Kazmi and Furumai, 2000).

A phased isolation ditch system is a continuous-flow activated sludge process in which the main treatment phases of the process are isolated into separate oxidation ditches. Process conditions within the oxidation ditches will alternate or phase between oxic, anoxic, and/or settling. Phased isolation ditches can be designed to operate with or without external clarifiers. The process can be designed to incorporate an anaerobic selector for EBPR. These processes include double ditch (D-Ditch), triple ditch (T-Ditch), BioDenitro, and BioDenipho.

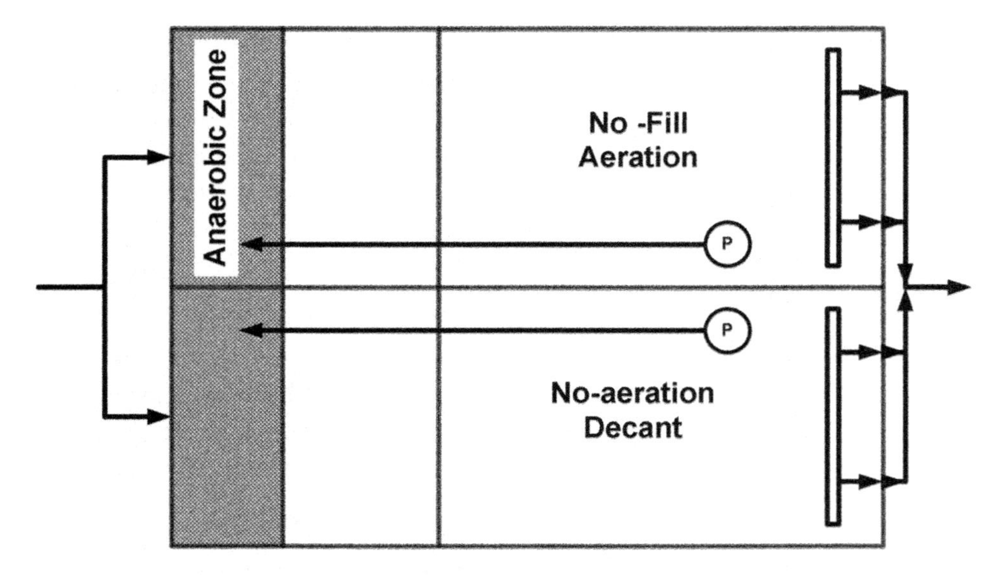

FIGURE 10.6 Biological nutrient removal SBR configuration (WEF, 2005).

The sequential aeration process uses a single alternating aeration tank (first-stage aeration) that feeds two parallel alternating aeration tanks (second-stage aeration). A patented in-line clarifier is used with this process. The clarifier is submerged in the mixed liquor using RAS air lifts, allowing for continuous flow through the process. When EBPR is desired, an anaerobic zone is incorporated before the first-stage aeration tank. One difference between sequential aeration and other alternating aeration processes is that mixing is not provided during the air-off cycles, with no apparent effect on treatment.

Alternating aeration processes can have lower operating costs than continuously aerated processes because, when properly designed and operated, only enough air is applied to satisfy the organic and nitrogenous oxygen demands. Air-on and air-off periods will be unique to each facility and are determined by operational experience. A typical cycle is 4 hours, with 2 hours of aeration and 2 hours of no aeration. During periods of no aeration, residual dissolved oxygen will allow the process to continue to nitrify, particularly if fine-bubble aeration is used. Determining aerobic SRT in alternating aeration systems is, therefore, based on system response and the type of aeration diffusers used. Total system SRT tends to be longer in alternating aeration systems to make up for perceived inefficiencies associated with times when conditions are not fully aerobic or anoxic. As more knowledge is gained about simultaneous nitrification and denitrification (SND), particularly organisms capable of nitrification at low dissolved oxygen, it may be possible to reduce the total system SRT in alternating aeration systems.

2.6 Process Configurations with Oxidation-Ditch Systems

Oxidation-ditch systems are common, with more than 10,000 installations in the United States and abroad, because of their relatively simple and reliable operation and cost effectiveness (*Design of Municipal Wastewater Treatment Plants* [WEF et al., 2009]). Oxidation-ditch operation is based on the use of mechanical surface aerators, mechanical aerators, brush aerators, and slow-speed mixers to provide flow of mixed liquor around channels in a racetrack configuration. Diffused aeration with slow-speed submersible mixers is also used to provide more flexibility in controlling aeration and mixing separately. Oxidation ditches are commonly designed as extended aeration systems with HRTs of 24 hours and SRTs in excess of 15 days; however, there is a tendency to reduce the design SRT of these processes to make them more competitive with other processes.

Oxidation ditches that use point-source aerators, such as surface aerators and brush aerators, push the mixed liquor around the racetrack channel, passing it through aerated and unaerated zones, resulting in various degrees of nitrogen removal by SND. A higher degree of SND is achieved in oxidation ditches as high-to-low dissolved oxygen zones are created. Theoretically, all that is needed for EBPR is to add an anaerobic zone in front and perhaps a reaeration zone after the channel, as was suggested by Barnard (1976).

For phosphorus removal, an anaerobic zone is added in front of the anoxic zones to result in a three-stage A2/O process, with the channel as the aeration zone. Enhanced biological phosphorus removal to reduce effluent phosphorus well below 1 mg/L also requires that the SRT be kept as short as possible. At longer SRTs (in excess of 15 days), there is more endogenous breakdown of sludge; therefore, there is less sludge production and the phosphorus content of the sludge must be higher for the same degree of removal. In addition, some secondary release of phosphorus by PAOs may occur when all the carbon has been removed and the nitrate-nitrogen is less than 2 mg/L.

2.7 Process Configurations with Membrane Bioreactors

Biological nutrient removal configurations for membrane bioreactor (MBR) systems are generally the same as those for conventional systems. The one significant difference is related to the RAS rate. When designed for typical MLSS values of 8000 mg/L in the bioreactor, a RAS rate of 400% of the influent flow is required to keep the solids concentration below the required 10 000 mg/L in the membrane tank. Because this is a typical mixed liquor recirculation rate for a BNR configuration for total nitrogen removal, the RAS stream of the MBR can also provide the necessary mixed liquor recirculation for nitrogen removal in a BNR process.

The RAS of an MBR can also contain high concentrations of dissolved oxygen (typically, 2 to 6 mg/L); as such, this dissolved oxygen will reduce the soluble BOD available for the BNR processes. For this reason, a common practice is to return the RAS stream of an MBR to the aeration tank and to use separate MLR pumps for return of the mixed liquor to the anaerobic and/or anoxic zones, similar to conventional facilities. This also lowers the aeration requirement in the bioreactor by using dissolved oxygen in RAS for this purpose. What is lost with this approach is the more concentrated solids in the membrane tank and RAS, resulting in a higher solids concentration in the BNR zones and increasing nutrient removal efficiency.

As an alternative design, the RAS stream can be returned to a deoxygenation tank. Effluent with low dissolved oxygen from the deoxygenation tank enters the anaerobic tank, which also receives the raw influent wastewater. This alternative design allows higher concentrations of mixed liquor in BNR basins with lower pumping requirements to be maintained. A variation to achieve EBPR is incorporation of the anaerobic zone downstream of the anoxic zone, allowing for a single RAS recycle. While the availability of VFAs in the raw wastewater is lost in this configuration, the high MLSS concentration in MBRs enhances the production of VFAs in the anaerobic zone. Figure 10.7 shows the simplest configuration of a BNR MBR that is typically appropriate for meeting limits of 8 mg/L of total nitrogen and 1.0 mg/L of total phosphorus (assuming no recycle of nitrogen and phosphorus from sidestreams). Figure 10.7 also shows the more complex BNR MBR configuration capable of achieving slightly lower total phosphorus limits,

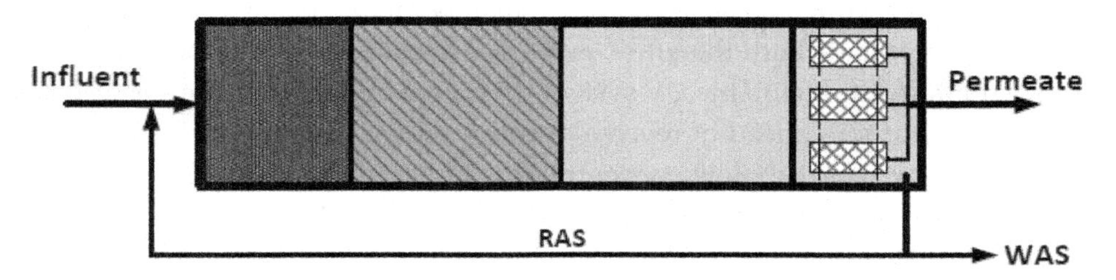

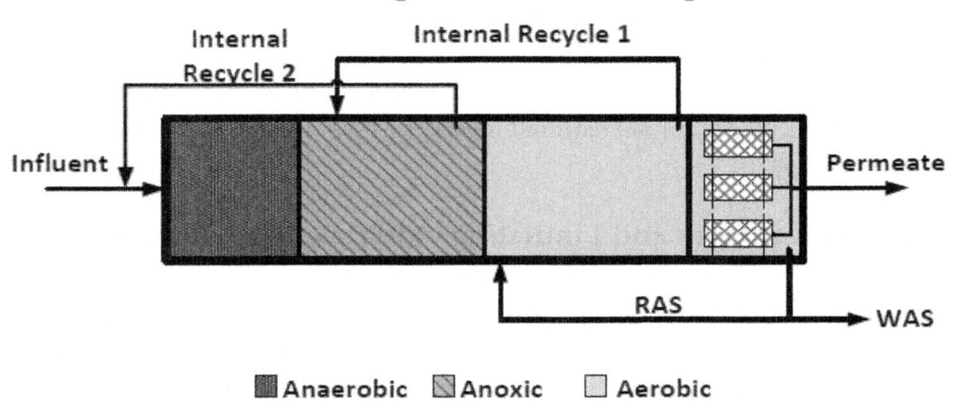

FIGURE 10.7 Biological nutrient removal MBR process configurations.

although achieving equivalent total nitrogen to the simpler configuration requires high recirculation rates.

As a result of high dissolved oxygen in the RAS stream, the MBR has gained a reputation for inefficient EBPR and, as a result, chemical phosphorus removal is more common in MBR facilities. In fact, the high RAS solids concentration can also enhance EBPR and, as manufacturers work to continually lower dissolved oxygen in the membrane tank, EBPR will become just as appropriate in MBRs as it is in conventional facilities. Even at higher RAS dissolved oxygen, EBPR can be effective, depending on the ratio of influent COD to total phosphorus, emphasizing the need for a reasonable modeling to properly assess the appropriate method of phosphorus removal in MBRs.

Chemical phosphorus removal can be effective in MBRs. The total retention of solids in the bioreactor typically results in lower dosages of the coagulant chemical required per gram of phosphorus removed than conventional facilities. Both aluminum and iron coagulants are used. Although iron can result in staining of the membranes, performance is not affected. Ultraviolet disinfection is frequently used following MBRs. It is important to remember that the membranes will pass soluble compounds and, therefore, excessive iron doses should be avoided to prevent interference with the UV system. Membrane bioreactors are also used as a pretreatment step ahead of reverse osmosis membranes; reverse osmosis can be affected by relatively low concentrations of aluminum. It is also important to avoid excessive chemical dosages in MBRs that are achieving some degree of EBPR because the excessive coagulant combined with the high RAS rate of an MBR will result in suppression of PAO growth.

Because MBR effluent is essentially free of solids, low concentrations of total phosphorus can be achieved using MBRs (i.e., even lower than conventional activated sludge treatment in combination with conventional filters). Because of their high degree of filtration, MBRs can achieve effluent total nitrogen concentrations equivalent to those achievable with denitrification filters without the use of external carbon sources. Second-stage anoxic zones following the aerated portion of the bioreactor are required to achieve these low-effluent total nitrogen concentrations.

2.8 Advantages and Limitations of Process Configurations

The various aforementioned process configurations can be categorized as staged processes, cyclic aeration processes, oxidation ditches, and MBRs. Each process is capable of achieving low-effluent nitrogen and phosphorus if designed and operated correctly.

2.8.1 Staged Processes

With the exception of the five-stage Bardenpho configuration, which is specifically designed for lower total nitrogen effluent limits, variations in staged processes have been found to have only minor differences in process performance. This is because the beneficial effect of higher solids concentration in RAS tends to negate the detrimental effects of RAS nitrate on the anaerobic zone. The greatest advantage of using staged processes is that the anaerobic, anoxic, and oxic volumes and recycles can be well defined and independently manipulated, allowing a tailoring of the process to fit the specific influent characteristics and effluent goals. Disadvantages of staged processes are complexity and pumping energies required for internal recycles.

2.8.2 Cyclic Aeration Processes

Cyclic aeration processes offer the advantage of being able to perform all nutrient removal processes in a simpler tank configuration. In some SBR configurations and phased isolation ditches without clarifiers, no recycles are used and the same biomass is sequentially exposed to anoxic, anaerobic, and oxic conditions. Because the anaerobic cycle must follow the anoxic cycle, EBPR is not as efficient in true cyclic aeration processes as in staged processes. This is the reason for the separate-stage anaerobic zone shown in Figure 10.6, where nitrification in the settled sludge is used to create a first-stage anaerobic zone.

For phased isolation ditches with clarifiers and sequential aeration, RAS is used to create an anaerobic zone in the same way as staged processes; it is only the denitrification process that differs. The biggest operational advantage of cyclic aeration processes is that, when operated correctly, only enough aeration is provided to achieve nitrification, thereby reducing aeration costs. The denitrification efficiency of cyclic aeration processes has typically been considered inferior to staged processes because of a less concentrated food source for the denitrifiers; however, with increasing evidence of efficiencies associated with SND, cyclic aeration processes may eventually overcome this perceived disadvantage.

2.8.3 *Oxidation Ditches*

Ditches are one of the most commonly used processes for BNR for small- to medium-sized facilities, primarily because of operational simplicity and resistance to upsets caused by the long SRT typically used. This makes space requirements greater than staged and cyclic aeration facilities. The oxygen-transfer efficiency of surface aeration equipment is considerably less than the diffused aeration equipment typically used in staged and cyclic aeration processes. When phosphorus removal is required, the design SRT for ditches should be decreased to enhance EBPR, making it necessary to supplement surface aerators with diffused aeration or to switch to diffused aeration entirely. The high recirculation rate of ditches can make it difficult to establish distinct anoxic zones and, for this reason, there is a tendency to use SND in ditch processes. The exception to this is the concentric-ring ditch design that is designed for isolated anaerobic, anoxic, and oxic zones. Because of the multiple zones and high recirculation rates, properly operated ditches are capable of achieving lower nitrogen and phosphorus concentrations than staged processes.

2.8.4 *Membrane Bioreactors*

The basic definition of an MBR is the substitution of membranes for final clarifiers. Membrane bioreactors can be configured with any of the aforementioned BNR configurations, even oxidation ditches, and have two key operational advantages: (1) the ability to use the membrane tank as a mixed liquor concentrator and (2) a high degree of filtration provided by the membranes. The main bioreactors of MBRs have traditionally been designed to operate at a high MLSS concentration of 8000 mg/L, making it necessary to use a RAS rate of 4 times the influent flow. This RAS rate can be detrimental to the anaerobic zone, reducing EBPR efficiency in MBR facilities. If the MBR bioreactor is designed and/or operated at lower MLSS, this allows the RAS rate to be reduced, increasing the efficiency of EBPR. Separate anoxic recycle may be required in this instance to meet effluent total nitrogen limits; however, the need for separate recycles must be weighed against the advantage of the high mixed liquor concentration occurring in the membrane tank. When MBRs are operated at high MLSS, the efficiency of second-stage anoxic zones is increased because of the higher concentrations of endogenous carbon. Combining this with the high degree of filtration provided by the membranes, it is possible for an MBR in a Bardenpho configuration to be capable of achieving the lowest nitrogen and phosphorus concentrations of all the BNR processes. The biggest disadvantage of MBRs is their high operational costs associated with aeration of the membrane tank, although recent advances in the technology are reducing this disadvantage.

3.0 PROCESS VARIABLES

3.1 Influent Characteristics

Combined nitrogen and phosphorus removal processes that mainly rely on biochemical reactions (excluding chemical phosphorus removal) are particularly sensitive to certain influent wastewater characteristics.

3.1.1 *Influent Flow Variability*

A key aspect of most BNR processes is adequate control of SRT. To a certain extent, this is affected by HRT. When the influent flow variability to these processes is large, then the HRT of each process could be significantly affected to the point that either the influent flow does not have enough contact time with active biomass (which, for example, can cause ammonia breakthrough) or it has too much contact time, which could lead to nitrate depletion and/or secondary phosphorus release. Operationally, the best strategy to manage flow variability is to increase or decrease recycle rates to maintain proper HRTs or SRTs. Influent flow equalization and/or additional chemical feed systems (carbon or metal salts) can also be used to achieve consistent nitrogen and phosphorus removal when large diurnal influent flow and load variations are experienced.

3.1.2 *Influent Substrate Requirements*

Combined nitrogen and phosphorus removal processes have specific substrate requirements to ensure sufficient levels of nutrient removal. There are several ratios commonly used in the industry for measuring the adequacy of combined nutrient removal processes. In general, higher influent carbon-to-nitrogen and carbon-to-phosphorus ratios will make the wastewater more amenable to BNR (Grady et al., 2011).

3.1.2.1 *Biological Phosphorus Removal Substrate Requirements*

Table 9.2 showed the minimum recommended substrate requirements for EBPR. Typically, design engineers use the ratio of VFA to phosphorus as the most commonly used parameter for verifying EBPR. When there is a sufficient supply of VFAs in influent flow (or from a supplemental VFA process such as a fermenter), the biology for phosphorus removal is forgiving. However, in many facilities, the VFA supply is marginally sufficient or may have to be augmented. Figure 10.8 illustrates the performance of the Blue River Main Wastewater Treatment Plant in Johnson County, Kansas, as it relates to VFAs and total phosphorus in the effluent. Operational experience in this facility has validated the influent correlation between available VFAs and total phosphorus removal in EBPR systems (Nolkemper, 2012).

Insufficient levels of VFAs present in the influent of the facility should be evaluated. For example, chemical addition to the collection system for corrosion and odor control can inhibit and hinder the formation of VFAs (Kobylinski et al., 2008). In addition, smaller collection systems in cold climates may also generate lower VFAs.

Factors such as temperature, pH, SRT, dissolved oxygen, and inhibitors influence the extent of nitrification. Ratios of total BOD or total chemical oxygen demand to total Kjeldahl nitrogen (TKN) are principal factors controlling the extent of denitrification. The ratio of readily biodegradable COD (rbCOD) to phosphorus into biological reactors controls the reliability of phosphorus removal because a significant portion of rbCOD will be converted to VFAs in the anaerobic

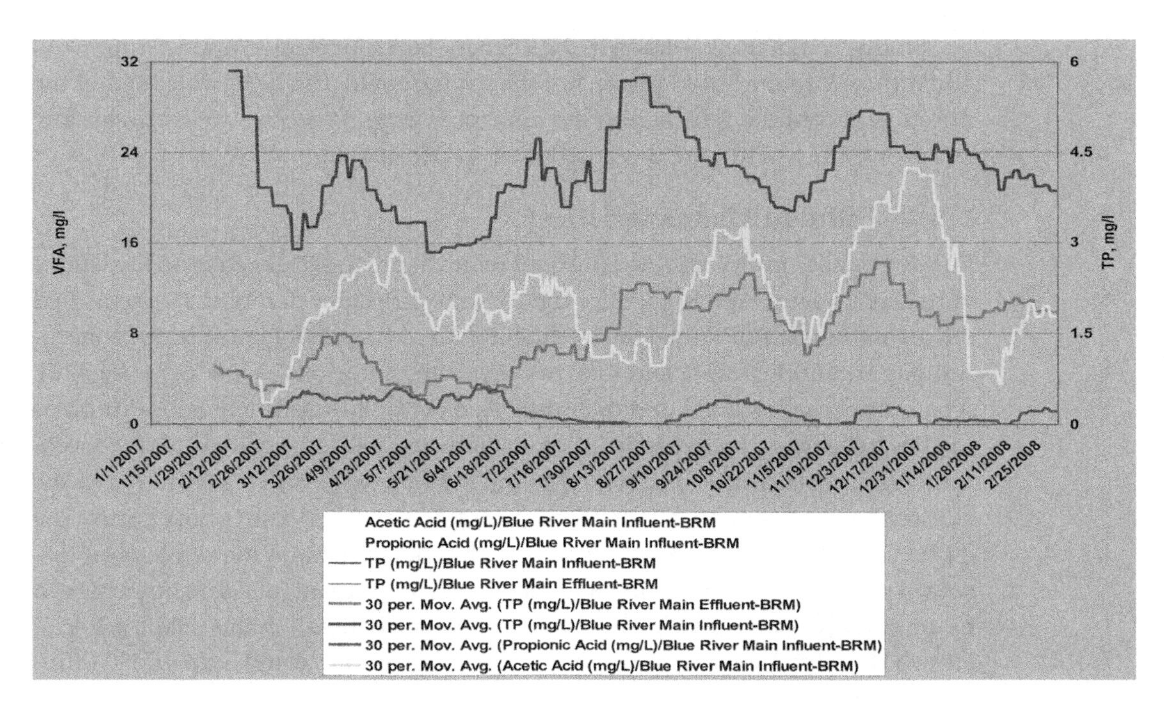

FIGURE 10.8 Typical Influence of VFAs in the influent vs total phosphorus in the effluent of EBPR processes (Nolkemper, 2012).

zone. The ratio of COD to TKN will determine the process configuration for achieving good phosphorus removal, nitrification, and/or denitrification.

3.1.2.2 Biological Nutrient Removal Process Substrate Requirements

Similar to EBPR processes, BNR processes also have minimum substrate requirements to make the processes viable. Most wastewaters have a ratio of COD to TKN on the order of 8 to 10:1 after primary sedimentation. If this ratio is lower, all aeration of the influent should be avoided so that the carbon is not oxidized by influent oxygen. This ratio may also be influenced by pretreatment processes. In addition, there are instances in which the groundwater contains nitrate that infiltrates the sewers or nitrate is added from an industrial source. This will have a negative effect on the EBPR process. Phosphorus uptake is less of a problem at high ratios of COD to TKN when there is sufficient organic matter for both EBPR and denitrification. Therefore, denitrification does not interfere with EBPR. At lower ratios of COD to TKN, greater reliance on endogenous respiration is necessary to optimize use of the available carbon. This could take the form of a second anoxic zone or a preanoxic zone to reduce nitrate through endogenous respiration before entering the anaerobic zone. Where the organic carbon content of the influent is not sufficient and good nitrogen and phosphorus removal is required, primary sludge may be fermented and the supernatant directed to the anaerobic zone.

Sludge could also be stored in the primary tanks for fermentation. Some VFAs will diffuse into the liquid phase, but the sludge could also be dewatered and the filtrate and centrate returned to the anaerobic zone. Processes to deal with low ratios of COD to TKN are discussed later in this chapter.

3.2 Effluent Characteristics

Effluent quality from combined nutrient removal processes depends on the degree of nutrient treatment designed for the combined system. Combined systems that rely primarily on EBPR may have effluent with average total phosphorus concentrations around 1.0 mg/L and total nitrogen concentrations as low as 3.0 mg N/L when properly designed and operated. Achieving these effluent concentrations can be challenging, particularly in colder climates. Total suspended solids concentration in the effluent is also linked to the total phosphorus concentration in the effluent because roughly 6% of TSS is typically associated with phosphorus. The effluent total nitrogen concentration is highly dependent on the recalcitrant dissolved organic nitrogen portion that remains in the effluent, which is not removed by commonly used conventional BNR processes discussed in this chapter. Recalcitrant dissolved organic nitrogen is discussed in detail in Chapter 11. Effluent that undergoes chemical phosphorus removal (as standalone treatment for phosphorus or as a polishing step) can achieve total phosphorus concentrations below 0.1 mg/L with a tertiary, highly efficient solids separation step. However, regardless of the amount of chemical dosed, there will be a recalcitrant phosphorus fraction that will not react with metal salts and may constitute a significant portion of the effluent total phosphorus. Recalcitrant dissolved organic nitrogen and recalcitrant total phosphorus vary within each sewershed and should be carefully determined if low nutrient levels are required in the effluent.

3.3 Temperature

Temperature effects on nitrification and denitrification are described in Chapters 3 and 5. There are several trends regarding the effects of temperature in EBPR processes. Some researchers indicate that higher temperatures affect EBPR because of the shift from PAOs to glycogen-accumulating organisms (GAOs) (*Nutrient Removal* [WEF, 2010]), whereas other researchers have found that these processes are relatively insensitive to temperature changes compared to other biological processes. Sell (1981) concluded that the responsible bacteria are *Psychrophiles*, which grow better at lower temperatures than typical activated sludge BOD-removing bacteria that have their maximum growth rate at 20 °C. Laboratory-scale studies found that there was no difference in EBPR efficiency at temperatures of 15 and 10 °C. However, full- and pilot-scale studies have also shown that EBPR can be affected by low temperatures. Fermentation in the collection system will decrease with a decrease in temperature. Therefore, during cold weather, there may be insufficient VFAs in influent, which will result in less phosphorus removal.

A facility in Grimstad, Norway, is removing phosphorus consistently, even though the mixed liquor temperature is approximately 5 °C for about 3 months of the year. However, VFAs are supplied by an efficient fermenter. There is some evidence that GAOs may successfully compete with PAOs at temperatures approaching 30 °C. Phosphorus removal in the VIP facility at Hampton Roads, Virginia, becomes unstable during the summer when the temperature of the mixed liquor exceeds 30 °C (WERF, 2005). In practical terms, if the temperature of EBPRs is within the 5 to 20 °C range, experience has shown little effect on performance.

3.4 Solids and Hydraulic Retention Times

Solids retention time is the most important operational parameter in any BNR facility. When implementing nutrient removal, it is important to make a distinction between aerobic SRT and total SRT. Aerobic SRT will determine the ability of the facility to nitrify and is the basis for sizing the aerobic portion of the bioreactor. Total system SRT for BNR facilities is defined as the total mass of solids in the entire bioreactor (aerobic plus anaerobic plus anoxic portions) divided by the sum of the total solids in WAS and the total solids in the effluent. Although total system SRT is occasionally discussed for comparative purposes, it is, in fact, of limited usefulness and can lead to erroneous conclusions. This is because, while the required aerobic SRT is a well-defined term based on bioreactor temperature, total system SRT can vary greatly depending on the required sizes of the anaerobic and anoxic zones, as dictated by the ratios of influent COD to total nitrogen to total phosphorus and required effluent total nitrogen and total phosphorus. Therefore, it is important to understand clearly which SRT is being used to control nitrogen and phosphorus removal processes.

3.5 Aeration and Dissolved Oxygen

As discussed in Chapters 3 and 5, a dissolved oxygen concentration of more than 2 mg/L is required for optimal nitrifier growth. In contrast, denitrification requires the absence of dissolved oxygen.

The effect of dissolved oxygen on the uptake of phosphorus has not been studied as extensively. Polyphosphorus-accumulating organisms need a positive dissolved oxygen concentration for phosphorus uptake. In addition, some of the PAOs are also capable of using nitrate for phosphorus uptake.

Reactions in the aerobic zone determine system phosphorus removal. There must be sufficient dissolved oxygen for PAOs to completely metabolize stored organics to gain the energy for phosphorus uptake. The dissolved oxygen concentration should not limit the rate of oxygen transfer to the cells. There should also be enough dissolved oxygen or nitrate in bioreactor effluent to prevent phosphorus release in the secondary clarifier; however, the amount contained in the RAS should be near zero. High SRT oxidation-ditch systems have shown that EBPR performance is determined by the mass of oxygen transferred rather than the concentration maintained at any point in the tank.

The dissolved oxygen concentration will vary from high to low, but the gradient will not be constant because of daily variations in load. It is difficult to control the dissolved oxygen gradient with a probe in a fixed position. Instruments that determine the oxygen-uptake rate, ORP, or some other surrogate parameter such as effluent turbidity or alkalinity have been used instead of dissolved oxygen monitoring for aeration control purposes. A simple way of controlling oxygen input is to use online ORP meters controlled by a supervisory control and data acquisition (SCADA) system. The ORP increases with increasing dissolved oxygen. The aeration rate is then reduced, and the decline in ORP is monitored. When the mixed liquor runs out of nitrate, there is a discernable bend in the downward line as a result of the sharp drop in redox potential, which is an indication that oxygen supply must be resumed.

There is also evidence that phosphorus removal in the aerobic zone does not begin until all of the VFAs have been removed from solution (Pattarkine, 1991). Therefore, if the VFAs are not completely assimilated in the anaerobic zone, the size of the aerobic zone should be enlarged to ensure completeness of the phosphorus removal reaction in addition to soluble substrate removal. The breakthrough of readily available organics to the aerobic zone is also likely to stimulate filamentous growth, particularly if the aerobic zone is completely mixed rather than plug flow.

3.6 Alkalinity and pH

The pH in EBPR facilities typically varies from 6.6 to 7.4; however, indications are that there is a decline in efficiency of both nitrification and phosphorus removal when the pH drops to below approximately 6.9. There is evidence that GAOs may also compete well at pH values below 7, reducing the VFAs available for PAOs (Filipe, 1999).

Higher pH inputs on BNR processes have not been studied adequately. The effect of various operations on alkalinity in mixed liquor is shown in Table 10.1. Hydrolysis of organic nitrogen will supply 3.6 g of alkalinity/g of nitrogen hydrolyzed. Nitrification consumes 7.14 g of alkalinity for every 1 g of ammonia-nitrogen converted to nitrate-nitrogen. Approximately 3.5 g is regained for every 1 g of nitrate-nitrogen reduced to nitrogen gas by denitrification. When adding chemicals for phosphorus removal, more alkalinity is consumed. This may affect nitrification and require the addition of buffering chemicals to prevent the pH from dropping too far.

The most probable cause of lowering alkalinity and pH is when a facility is fully nitrifying without denitrification while treating a low-alkalinity wastewater. The pH can generally be adjusted back toward the optimum range by including denitrification in the single sludge system. There are, however, instances in which alkalinity must be added, not only for phosphorus uptake, but also because of the sensitivity of nitrifying organisms to low pH values. Low pH can be a problem when treating industrial wastewaters containing high concentrations of VFAs, such as acetic acid. Pully (1991) found that the anaerobic zones of both two- and three-stage EBPR systems were only capable of adjusting the pH of a wastewater high in acetic acid from 4.5 to 5.5, even though COD removal was complete in the aerobic zone where the pH was 7.5. Consequently, EBPR could not be established without pH adjustment of the wastewater, even though neutralization was not necessary for COD removal.

TABLE 10.1 Alkalinity consumed or produced by certain processes (WEF, 2005).

Process	Alkalinity change, mg/L	Per mg/L of
Ammonification of organic nitrogen	+3.6	Organic nitrogen hydrolyzed
Nitrification	−7.1	Ammonia-nitrogen oxidized
Denitrification	+3.6	Nitrate-nitrogen reduced
Phosphorus removal	−5.6*	Aluminum added
Phosphorus removal	−2.7*	Iron added

*For reducing phosphorus to 1 mg/L. For lower effluent phosphorus values, proportionally more chemicals will be required.

3.7 Return Activated Sludge

Return activated sludge brings settled solids from secondary clarifiers to process zones, thereby maintaining proper MLSS concentration and SRT in these zones. In BNR facilities, one way to determine the optimum RAS flowrate is to adjust the RAS flowrate to maintain about 30 cm of sludge blanket in the secondary clarifiers. Another way to determine optimum RAS flow is based on a mass balance around a secondary clarifier using influent flow, MLSS, and TSS concentration of the RAS. The optimum RAS flowrate can be determined using the following equation, which is simplified by assuming that the effluent TSS is negligible:

$$Q_{RAS} = (Q_{INF} \times MLSS)/(TSS_{RAS} - MLSS) \quad (10.1)$$

Where

Q_{RAS} = RAS flowrate (m^3/d),
Q_{INF} = facility influent flowrate (m^3/d),
MLSS = mixed liquor suspended solids concentration (g/m^3), and
TSS_{RAS} = TSS concentration in RAS (g/m^3).

Although optimum RAS flowrate can be calculated from eq 10.1, the actual RAS flowrate required to maintain a 30-cm sludge blanket will change depending on the actual conditions and sludge settleability. The optimum RAS flowrate should be determined as a fraction of the influent flowrate and the RAS pump rate should be modulated based on this ratio. To keep the sludge blanket variations to an acceptable range, Q_{RAS}/Q_{INF} may be adjusted based on the observed sludge blanket.

Applying rates lower or higher than the optimum RAS flowrates will affect BNR facility performance. Lower RAS flow will improve EBPR efficiency because the mass of nitrate load to the anoxic and anaerobic zone is reduced and more VFAs will be available for PAOs. On the other hand, lower RAS flow increases the sludge blanket and solids inventory in the secondary clarifier, perhaps allowing denitrification to occur via endogenous respiration. Nitrogen gas bubbles formed as result of denitrification in clarifiers can cause rising sludge and increased effluent TSS. In addition, deeper blankets also increase residence time of PAOs in the secondary clarifiers, perhaps resulting in secondary release of phosphorus.

Higher RAS rates, which can be as much as 75 to 100% of influent flow, may improve nitrogen removal as more nitrate is returned to the anoxic or anaerobic zones. However, EBPR will be impaired because denitrifiers will use more of the available VFAs.

In MBRs, RAS flow is returned from the membrane basins; the RAS will have high dissolved oxygen because of the scour air. When the oxygen-rich mixed liquor is returned to the anaerobic or anoxic zones, more VFAs will be used aerobically, impairing the denitrification and EBPR efficiency. Minimum possible RAS flow should be used in MBR facilities to maintain sufficient MLSS distribution between the membrane and process basins. Even then, the oxygen load in RAS can impair the BNR efficiency of the facility. For this reason, modern MBR designs use membrane tank aeration systems that reduce the dissolved oxygen in the RAS significantly.

3.8 Internal Recycle

There are a few significant types of internal recycle streams used in combined BNR processes. The first type of internal recycle returns nitrified mixed liquor from the aeration zone to the anoxic zone. Nitrified mixed liquor is high in nitrate and, therefore, higher internal recycle flowrate will increase the nitrate load to the anoxic basin, increasing denitrification efficiency.

Lower internal recycle flow (as low as 200% of influent flowrate) may be applied for low-efficiency denitrification processes consisting of single-stage anoxic zones, such as the A2/O process. On the other hand, much higher internal recycle flowrates (as high as 400% of the influent flowrate) may be applied for high-efficiency nitrogen removal processes consisting of preanoxic and postanoxic zones, such as the five-stage Bardenpho process. The internal recycle flowrate should be determined with respect to the residual nitrate level at the end of the anoxic zones. If the process consists of at least two or three anoxic zones in series, the internal recycle flowrate should be optimized to drive nitrate down to low levels in the last anoxic zone. Lower internal recycle flowrates may drive nitrate concentration down to low levels in most anoxic zones. This can lead to secondary release of phosphorus.

Because internal recycle is returned from the aeration zone, it may also have high dissolved oxygen that will reduce denitrification in anoxic zones. Therefore, if possible, low dissolved oxygen should be maintained at the point of internal recycle removal. However, the minimum dissolved oxygen that can be maintained is determined by the minimum airflow requirement for mixing at that point in the aeration zone. The internal recycle flowrate can be increased up to 400% of the influent flow as long as dissolved oxygen in internal recycle is below 1.0 mg/L (*Nutrient Removal* [WEF, 2010]).

Another type of internal recycle is used for returning denitrified mixed liquor from the end of the anoxic zone to the anaerobic zone, such as in the UCT process (Figure 10.3). The purpose of this internal recycle flow is to minimize nitrate return to the anaerobic zone. This internal recycle flow can be 100 to 300% of the influent flow.

3.9 Sidestreams

Sidestream management and treatment is discussed in detail in Chapter 12. Sidestreams can affect combined nitrogen and phosphorus removal processes just like any other BNR process. Sidestreams containing nitrogen and phosphorus are typically returned to process basins after getting mixed with facility influent.

Sidestreams may contain ammonia, nitrate, and phosphorus depending on the sludge treatment and handling processes that are used.

As previously discussed, commonly used measures of influent amenability to BNR are based on the amount of readily biodegradable substrate available to sustain EBPR and denitrification. This determination should be made on the influent to the BNR bioreactor and should include all significant recycle loads. Primarily domestic municipal wastewater typically contains sufficient carbon substrate for BNR if excess BOD removal does not occur in the primary clarifiers. Recycle streams, which are characteristically low in BOD and relatively high in nitrogen and phosphorus, can depress the ratios of BOD to total phosphorus and BOD to TKN in the bioreactor influent. In addition, if solids capture is not optimized in EBPR sludge thickening and dewatering operations, significant phosphorus-rich solids would be returned to the primary clarifier and mixed with substrate-rich primary sludge. This can potentially result in secondary phosphorus release if anaerobic conditions prevail in the sludge blanket. This released phosphorus will lower the bioreactor ratio of influent BOD to total phosphorus.

3.10 Operation of Other Unit Treatment Processes for Nutrient Removal

3.10.1 Primary Clarifiers

The influent ratios of BOD to TKN and BOD to total phosphorus may be reduced substantially when primary clarifiers are used. The majority of influent TKN in facility influent is typically in the form of ammonia. Ammonia and orthophosphate are soluble species and pass through primary clarifiers. On the other hand, a substantial amount of influent BOD is associated with influent TSS, and primary clarification can commonly remove 25 to 40% of influent BOD. Therefore, primary clarifier effluent may have lower ratios of BOD to TKN and BOD to total phosphorus. Enhanced biological phosphorus removal is not negatively affected by primary clarifiers because VFAs also pass through primary clarifiers. However, denitrification will be reduced as less BOD is available. It is important to recognize that although removing more influent solids at the primary clarifiers may reduce loading to the BNR facility, it will not be beneficial for denitrification.

Maintaining shallow blankets in primary clarifiers can reduce hydrolysis of organic matter, which is the mechanism that converts particulate nitrogen and phosphorus to soluble forms. Therefore, soluble nutrient loading to process basins will increase. It is also important to maintain a reasonable blanket depth in primary clarifiers to prevent low concentrations of primary sludge.

3.10.2 Tertiary Filters

Tertiary filters are used for suspended solids removal from effluent. Effluent suspended solids contain particulate nitrogen and phosphorus at varying fractions depending on the upstream biosolids and chemical treatment process. Suspended solids removal with tertiary filters is more critical when low-effluent nitrogen and phosphorus levels are required.

Deep-bed filters combine suspended solids removal and additional nitrogen removal with external carbon source addition. Deep-bed denitrifying filters provide good suspended solids removal while reducing effluent nitrate to a 1.0- to 2.0-mg/L level.

4.0 OTHER CONSIDERATIONS

4.1 Competition for Available Substrate in Combined Biological Nutrient Removal Systems

Operators of combined BNR facilities must be aware of the competition for soluble COD between PAOs and denitrifying organisms. For example, excessive nitrate in the RAS return to the anaerobic zone results in soluble COD taken up by denitrifying organisms in the anaerobic zone, thereby reducing EBPR efficiency. For facilities with anaerobic zones ahead of anoxic zones, the uptake of soluble COD by PAOs in the anaerobic zone reduces denitrification efficiency in the anoxic zone, possibly requiring the use of supplemental carbon to meet low nitrogen limits.

To quantify potential effects for equal anoxic volumes in an MLE process (refer to Chapter 6) treating conventional domestic wastewater, the presence of an anaerobic zone ahead of the anoxic zone will result in an effluent nitrate concentration of 6 to 7 mg/L; without the anaerobic zone, an effluent nitrate concentration of 3 to 4 mg/L may be achievable.

4.2 Secondary Release

Other competing reactions when both anoxic and anaerobic zones are present are related to the secondary release of phosphorus that has initially been taken up by the PAO. When there is a sufficient supply of VFAs in the feed or from a fermenter, EBPR is forgiving. However, in many facilities, the VFA supply is just sufficient or may have to be augmented. When secondary release of phosphorus takes place, more VFAs are needed to take up the excess phosphorus that is released. Secondary release takes place in any zone where there are unaerated conditions without a supply of VFAs or nitrate.

Secondary release can take place in anaerobic zones that are too large. Figure 10.9 shows the effect of insufficient VFAs on phosphorus release in anaerobic zones. The top curve represents combined primary and secondary release. The primary release stops when VFAs are all absorbed and there is only secondary release. Therefore, the zone should be just large enough for completion of the primary release to minimize the secondary release. Adjusting the RAS recycle rate or the amount of flow going to the anaerobic zone can be used for optimization.

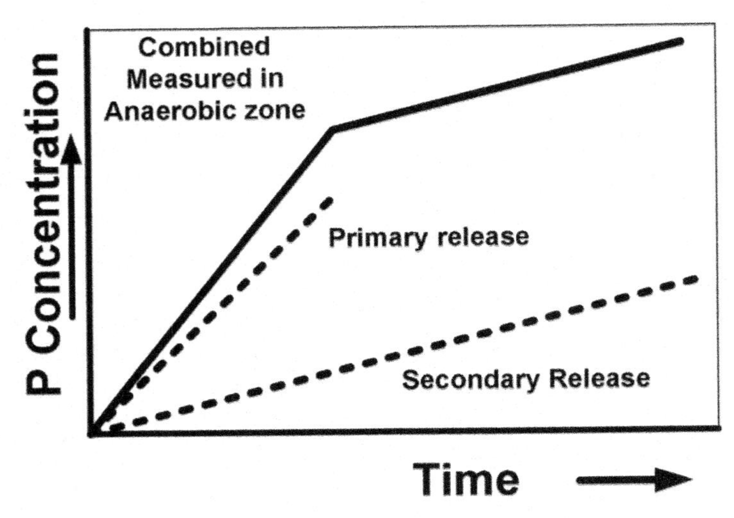

FIGURE 10.9 Primary and secondary release in the anaerobic zone (WEF, 2005).

Secondary release can also take place in the anoxic zones when all the nitrate is denitrified to nitrogen gas. Therefore, if the first or second anoxic zones are too large and run out of nitrate, secondary release will take place. As long as nitrate is present, PAOs, which can use nitrate as the electron acceptor, will take up any released phosphorus. Without nitrate and oxygen, there is no means of taking up phosphorus that is released by other organisms, and too much secondary release takes place. This can be verified by measuring soluble phosphorus profiles through the anoxic zone.

Figure 10.10 presents an example profile of a facility where, as a result of the low MLR rate, secondary release took place in the anoxic zone. In Figure 10.10, the concentration of phosphorus in the anoxic zone is the same as that in the anaerobic zone of a three-stage A2/O facility. The recycle of mixed liquor from the aeration to the anoxic zone should have diluted the phosphorus in the anoxic zone; however, because there is no reduction, there must have been secondary release. This could possibly be overcome by increasing the MLR rate, reducing the retention time in the anoxic zone, or by recycling more nitrate and dissolved oxygen into this zone. This will stop excessive secondary release and reduce the mass of released phosphorus that must be taken up in the aeration basin. In this particular instance, the facility manager confirmed that only one out of three recycle pumps were operational.

Secondary release can also take place in the second anoxic zone of a five-stage Bardenpho facility, as illustrated in Figure 10.11. In this instance, SND in the aeration basin, using draught tubes with surface aerators, was so efficient that there was no nitrate in the mixed liquor to the second anoxic zone. Phosphorus was released and could not be taken up again with aeration alone. If VFAs were

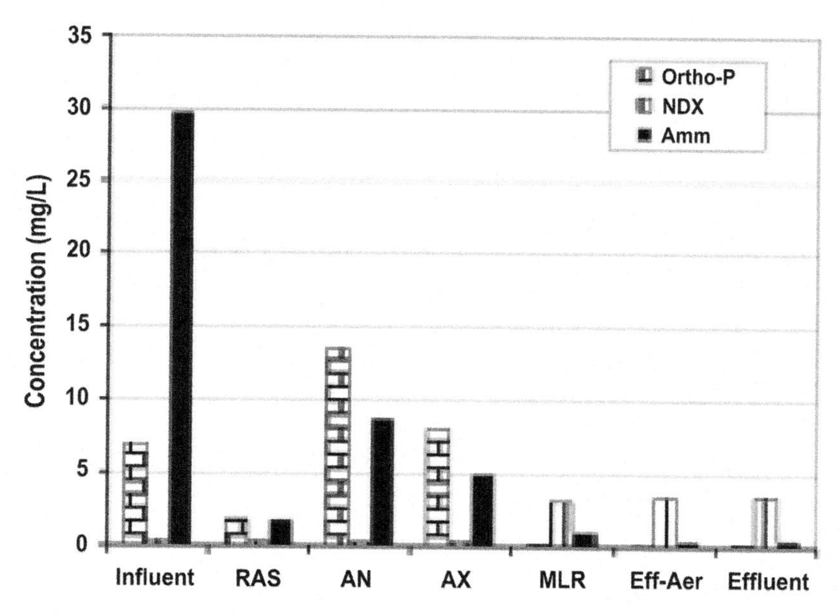

FIGURE 10.10 Profile of soluble phosphorus through Rooiwal, South Africa (WEF, 2005).

added to the second anoxic zone, phosphorus would be taken up in the reaeration zone. In this instance, the solution was to aerate the second anoxic zone and, as a result, no further release took place.

Phosphorus may also be released when there is a low nitrate concentration in the mixed liquor flow to final clarifiers. When a deep sludge blanket develops in the final clarifiers, nitrate will be denitrified. When the nitrate is low, phosphorus

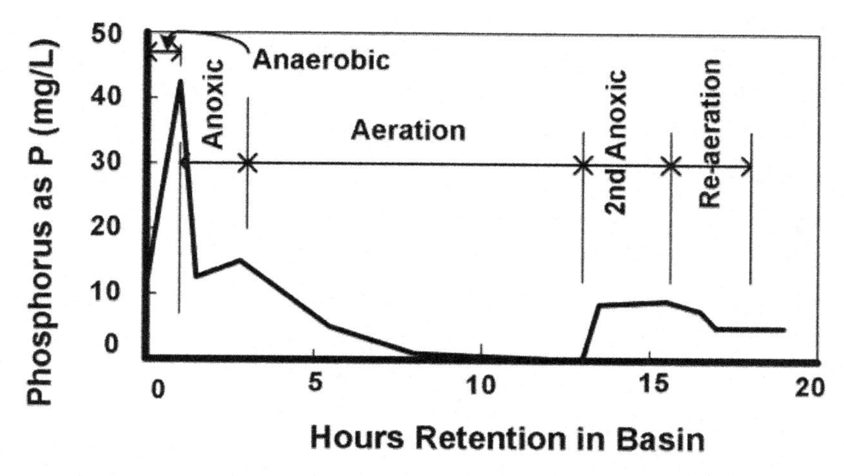

FIGURE 10.11 Secondary release of phosphorus in the second anoxic zone (WEF, 2005).

will be released. This release may not affect effluent phosphorus directly, but may return a large portion of the released phosphorus back to the anaerobic zone, where there may not be enough VFAs for uptake of this additional released phosphorus. At one facility with a relaxed requirement for nitrogen removal but a total phosphorus limit of less than 0.15 mg/L, the operator tried to reduce nitrate in RAS by reducing the recycle rate. A sludge blanket formed in the clarifiers and 10 mg/L of phosphorus was released in the RAS. Increasing the RAS rate reduced the secondary release of phosphorus, resulting in effluent soluble phosphorus dropping from 1.5 to 0.08 mg/L. Therefore, when the facility is operated for low-effluent nitrate in addition to low phosphorus, the sludge blanket should not be deep.

4.3 Considerations for Low-Effluent Nitrogen and Phosphorus Limits

For high levels of nitrogen and phosphorus removal in combined systems, the five-stage Bardenpho process typically is required followed by metal salt addition with filtration. Alternatively, processes that do not remove nitrate completely, such as the UCT and Westbank processes, could be used with attached-growth denitrification as a tertiary process.

A 2011 Water Environment Research Foundation (WERF) report concluded that, even though several flow sheets for combined nutrient removal statistically perform well, their effluent quality with respect to total nitrogen and total phosphorus varies somewhat depending on factors such as geographic location of the facility, climate, and other factors. The following are other important findings of this report:

- Separate-stage nitrogen removal facilities outperform combined nitrogen removal facilities because of a higher degree of denitrification control possible with a separate-stage process regarding total nitrogen;
- Bardenpho (four- or five-stage) facilities come close to meeting the monthly total nitrogen goal of 3.0 mg/L 95% of the time;
- Single-stage chemical addition processes for total phosphorus removal outperformed multiple-stage processes, although often at the expense of higher chemical dosages; and
- Tertiary chemical addition and effective filtration (gravity, media, or membrane) is required to achieve low phosphorus. Facilities with some form of tertiary chemical addition, clarification, and filtration slightly outperform those that have only effluent filters.

When a final effluent total nitrogen of less than 3 mg/L is required, the effluent nitrate must not exceed 1.5 mg/L to allow for some ammonia and recalcitrant soluble organic nitrogen that cannot be removed.

Thus far, most instances of consistently good removals of both nitrogen and phosphorus have been performed by five-stage Bardenpho-type facilities having point-source aerators because of the effects of SND. The Kelowna Wastewater Treatment Plant in British Columbia, Canada, is an example of such a facility. Aeration is supplied by a turbine aeration system, which allows for turning down the dissolved oxygen without loss of mixing. This, in turn, allows for a considerable amount of SND in the aeration basin. The facility was able to achieve an effluent of less than 1 mg/L ammonia-nitrogen and nitrate-nitrogen, which was maintained over months. Figure 10.12 presents representative data regarding total nitrogen performance for different BNR systems.

Table 10.2 summarizes the effects of factors identified in nutrient removal process selection. The primary factor affecting process selection is the degree of nitrification and/or nitrogen removal desired for the process. If nitrification only or only a moderate degree of nitrogen removal (effluent total nitrogen of 6 to 12 mg/L) is desired, then either the A2/O, UCT, or VIP process would typically be selected. Selection among these three options depends on projected wastewater characteristics (ratio of BOD to total phosphorus) and the degree of phosphorus removal capability required. The phosphorus removal capability of the A2/O process is generally lower than that of the UCT or VIP process; however, when a preanoxic zone is added, it is equal or better. The Bardenpho process is selected when extensive nitrogen removal (e.g., effluent values of approximately 3 mg/L)

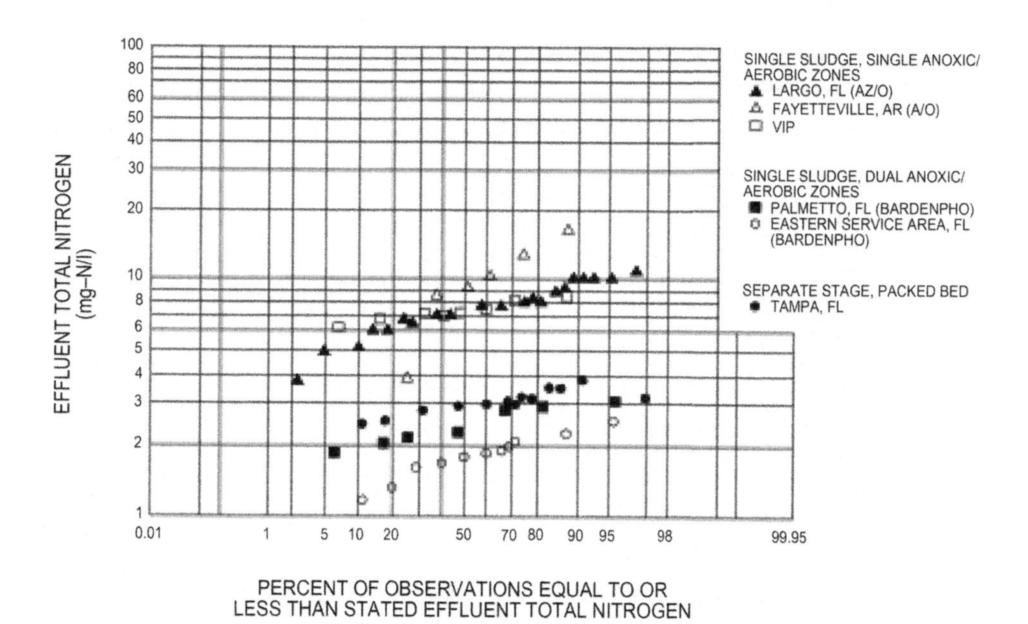

FIGURE 10.12 Removal of nitrogen in BNR facilities in the United States (WEF, 2005).

TABLE 10.2 Nitrogen and phosphorus removal process selection (WEF, 2005).

Process	Nitrogen removal[a]	Sensitiviy to TBOD/ total phosphorus[b]
A2/O	6 to 8 mg/L	High
JHB, MJHB	6 to 8 mg/L	Moderate
UCT/VIP	6 to 12 mg/L	Moderate
Modified UCT	6 to 12 mg/L	Moderate
Bardenpho	3 mg/L	High

[a]Approximate effluent concentrations.
[b]All processes could benefit from using fermenters.

is desired or a tertiary denitrification attached growth system could be added to any of the other processes.

While exceptions exist to the generalizations presented in Table 10.2, they represent a good starting point for preliminary biological process selection (Sedlak, 1991). The 2011 WERF report also found the following factors responsible for increasing the variability of effluent total nitrogen and total phosphorus concentrations:

- Infrequent toxic event upsets;
- Unexpected interruptions in chemical supply. The majority of facilities in the survey use chemicals for either nitrogen or phosphorus removal;
- Facility upgrading projects and the effects of construction on effluent reliability;
- Peak flow events were the most difficult operating issues along with seasonal variations in flows and loads;
- Biological treatment capacity issues affected performance during more stressed periods;
- Sludge supernatant recycle streams containing ammonia;
- Chemical feed control issues for phosphorus removal; and
- Fermenter control issues were the most difficult aspect of operations in facilities that rely solely on EBPR.

4.4 Effect of Chemical Phosphorus Removal on Biological Nutrient Removal Systems

The addition of chemicals to different points in the BNR system and how this affects nitrogen and phosphorus removal are discussed in the following subsections.

4.4.1 *Primary Clarifiers*

In addition to precipitation of phosphorus compounds, metal salt addition upstream of primary clarifiers enhances suspended solids and BOD removal in primary clarifiers as a result of coagulation of suspended organic matter. Therefore, primary sludge will contain a greater amount of organic matter because it is captured with inorganic flocs. Removal of organic material in the primary clarifier reduces loading to secondary treatment facilities, resulting in capital and operation and maintenance cost savings for secondary treatment. However, removal of BOD and phosphorus from wastewaters could cause a low ratio of BOD to TKN, with many effects on the process including reduced nitrogen removal in the BNR process, phosphorus deficiency to the biological process if too much phosphorus is removed by chemical precipitation, and loss of alkalinity needed in the biological process. Chemical enhanced primary clarifiers will not remove soluble COD, and some rbCOD could be generated by fermentation of some of the primary sludge for denitrification and phosphorus removal.

4.4.2 *Secondary Clarifiers*

Addition of metal salts upstream of the secondary clarifiers provides a high level of phosphorus removal when used in conjunction with EBPR. At this point in the treatment process, phosphorus is typically in the orthophosphate form, which can be precipitated with metal salts. The metal salt and phosphorus precipitate can be removed with the flocculent biomass in the secondary clarifier. Chemical addition could also enhance nitrogen removal because colloidal nitrogen is flocculated as a result of chemical addition, thereby reducing total nitrogen in the effluent. The addition of chemicals to secondary clarifiers increases the amount of inerts carried in the mixed liquor because RAS returns inorganic precipitates to the aeration basin. This effectively means that the nitrification capacity of the aeration basin is reduced because of the lower ratio of mixed liquor volatile suspended solids to MLSS. This phenomenon is rarely considered when designing or operating a combined nutrient removal reactor and, depending on the amount of salts added to the clarifier, the effect on the nitrification system could be severe. This effect should be studied during the design phase because it may restrict the ability of a secondary clarifier to accept metal salts and subsequent provisions for polishing to lower total phosphorus levels may be required. To minimize the effects on the nitrification system, a two-stage chemical addition approach may be better suited for combined nutrient removal facilities performing chemical phosphorus removal in secondary or tertiary systems.

4.5 External Carbon Source Addition

In combined nitrogen and phosphorus removal processes, external carbon sources may be added to anaerobic and anoxic basins to enhance EBPR and denitrification, respectively. Use of an external carbon source for denitrification is a more common practice. Methanol is the most commonly used external carbon source by far because it provides lower sludge yield and higher denitrification efficiency indicated by relatively low ratios of gram of methanol per gram of nitrate and lower costs (Gu and Onnis-Hayden, 2010). On the other hand, the main concern regarding methanol is the safety issue related to storage and transportation because methanol is a highly flammable chemical. In addition, specialized methanol-oxidizing bacteria must be enriched in the mixed liquor and, therefore, methanol cannot be added on an "as needed basis".

Carbon sources can be directly added to the preanoxic basins of BNR processes with only preanoxic basins (i.e., A2/O) to lower effluent total nitrogen to the 5- to 8-mg/L range. The level of effluent that can be achieved is limited by internal recycle flow in this instance. The A2/O process can be coupled with denitrifying filters with carbon source addition to reduce effluent total nitrogen to 3.0 mg/L or lower.

Carbon sources can also be added to the postanoxic basins of BNR processes (i.e., five-stage Bardenpho) with both preanoxic and postanoxic basins. These BNR processes can lower effluent total nitrogen levels to 3.0 mg/L with carbon source addition to the postanoxic basins. The carbon source dose should be optimized to achieve target effluent total nitrogen.

Carbon source addition to anoxic basins rather than to denitrification filters can reduce nitrate load in the RAS, thereby improving EBPR as well. On the other hand, carbon source addition to the filters can be controlled better, thereby allowing better nitrate removal than carbon source addition to the anoxic basins.

Depending on the BNR configuration and influent VFA levels, moderate effluent total phosphorus levels of 0.5 to 1.0 mg/L may be achieved. If there is insufficient VFA present in the influent, external VFA sources may be added to anaerobic basins. Fermentation of primary solids is also commonly used at BNR facilities in Western Canada, the United States, and Europe to generate additional VFAs, thereby improving nutrient removal efficiency (Oleszkiewicz and Barnard, 2006).

Research shows that a mixture of acetic and propionic acids are optimal for improving EBPR. Full-scale observed supplemental VFA requirements are typically on the order of 5 to 10 mg/mg of phosphorus removed. The most common choice of VFA added for EBPR is acetic acid. Although the acetic acid unit cost may be high compared to the addition of metal salts (i.e., alum) for the removal of phosphorus by chemical precipitation, the cost of chemical sludge treatment and disposal should also be considered when making a decision. In some instances, use of metal salts for phosphorus removal may be needed when low-effluent phosphorus levels (i.e., 0.3 mg/L or lower) are required.

5.0 REFERENCES

Barnard, J. L. (1976) A Review of Biological Phosphorous Removal in Activated Sludge Process. *Water SA,* **2** (3), 136–144.

Daigger, G. T.; Waltrip, G. D.; Roman, E. D.; Morales, L. M. (1988) Enhanced Secondary Treatment Incorporating Biological Nutrient Removal. *J.—Water Pollut. Control Fed.,* **60**, 1833.

Ekama, G. A.; Marais, G. v. R. (1984) Biological Nitrogen Removal. In *Theory, Design and Operation of Nutrient Removal Activated Sludge Processes;* Water Research Commission: Pretoria, South Africa.

Filipe, C. D. M. (1999) Competition between Phosphate and Glycogen Accumulating Bacteria: Stoichiometry, Kinetics and the Effects of pH. Ph.D. Dissertation, Clemson University, Clemson, South Carolina.

Grady, C. P. L.; Daigger, G. T.; Love, N. G.; Filipe, C. D. M. (2011) *Biological Wastewater Treatment,* 3rd ed.; IWA Publishing and CRC Press: London and Boca, Raton, Florida.

Gu, A.; Onnis-Hayden, A. (2010) *Protocol to Evaluate Alternative External Carbon Source for Denitrification at Full-Scale Wastewater Treatment Plants;* Water Environment Research Foundation: Alexandria, Virginia.

Kazmi, A. A.; Furumai, H. (2000) Filed Investigations on Reactive Settling in an Intermittent Aeration Sequencing Batch Reactor Activated Sludge Process. *Water Sci. Technol.,* **41** (1), 127–135.

Keller, J.; Watts, S.; Battye-Smith, W.; Chong, R. (2001) Full Scale Demonstration of Biological Nutrient Removal in a Single Tank SBR Process. *Water Sci. Technol.,* **43** (3), 355–362.

Kobylinski, E.; Durme, G. V.; Barnard, J; Massart, N.; Koh, S. (2008) How Biological Phosphorus Removal Is Inhibited by Collection System Corrosion and Odor Control Practices. *Proceedings of the 81st Annual Water Environment Federation Technical Exhibition and Conference* [CD-ROM]; Chicago, Illinois, Oct 18–22; Water Environment Federation: Alexandria, Virginia.

Nolkemper, D., Johnson County Wastewater, Overland Park, Kansas (2012) Personal communication regarding sidestreams.

Oleszkiewicz, J. A.; Barnard, J. L. (2006) Nutrient Removal in North America and European Union: A Review. *Water Qual. Res. J. Can.,* **41** (4), 449–462.

Pitman, A. R. (1992) Process Design of New BNR Extensions to Northern Works, Johannesburg. Paper presented at the Biological Nutrient Removal Conference; Leeds University: Leeds, U.K.

Pully, T. (1991) Investigation of Anaerobic-Aerobic Activated Sludge Treatment of a High-Strength Cellulose Acetate Manufacturing Wastewater. Unpublished research; Department of Civil Engineering, Virginia Polytechnic Institute and State University: Blacksburg, Virginia.

Sedlak, R. (1991) *Phosphorus and Nitrogen Removal from Municipal Wastewater,* 2nd ed.; Lewis Publishers: Boca Raton, Florida.

Sell, R. L. (1981) Low Temperature Biological Phosphorus Removal. *Proceedings of the 54th Annual Water Environment Federation Technical Exposition and Conference;* Detroit, Michigan, Oct 4–9; Water Pollution Control Federation: Washington, D.C.

Water Environment Federation; American Society of Civil Engineers; Environmental and Water Resources Institute (2005) *Biological Nutrient Removal (BNR) Operation in Wastewater Treatment Plants;* WEF Manual of Practice No. 29; ASCE/EWRI Manuals and Reports on Engineering Practice No. 109; McGraw-Hill: New York.

Water Environment Federation; American Society of Civil Engineers; Environmental and Water Resources Institute (2009) *Design of Municipal Wastewater Treatment Plants,* 5th ed.; Manual of Practice No. 8; ASCE Manuals and Reports on Engineering Practice No. 76; McGraw-Hill: New York.

Water Environment Federation (2010) *Nutrient Removal;* Manual of Practice No. 34; McGraw-Hill: New York.

Water Environment Research Foundation (2005) *Factors Influencing the Reliability of Enhanced Biological Phosphorus Removal;* Project NO. 01-CTS-3; Water Environment Research Foundation: Alexandria, Virginia.

Water Environment Research Foundation (2011) *Nutrient Management Volume II: Removal Technology Performance & Reliability;* Water Environment Research Foundation: Alexandria, Virginia.

Chapter 11

Optimization of Nutrient Removal Systems

William C. McConnell, P.E.; Jason Beck, P.E.; Timur Deniz, Ph.D., P.E., BCEE; and Joel C. Rife, P.E., BCEE

1.0 INTRODUCTION

Operators are under constant pressure to optimize the performance of the nutrient removal process in terms of effluent quality and/or reducing ongoing operational costs associated with power and chemical consumption. The complexity of nutrient removal processes and the many interrelated and often competing operational parameters dictate that a deliberate, orderly process occur to achieve process optimization.

2.0 PHASE 1 PROCESS EVALUATION

The first step in optimizing process performance is to evaluate the current operation and set process improvement goals. This first phase of process evaluation may include review of facility influent and effluent discharge data, evaluation of each unit process performance, and a review of operations efficiency in terms of chemical dosages, electricity consumption, and sludge production. Optimization goals may include improving effluent performance and/or reducing the cost of operation. Based on this initial evaluation, it is likely that additional data will be required to meet optimization goals.

2.1 Data Review

Review of facility data and process optimization should be an ongoing task. Interpreting data requires technical knowledge, experience, and a strong familiarity with the facility. Depending on the specific facility and the particular goals of process optimization, there are numerous areas in which to focus attention. The factors that are discussed in Sections 2.1.1 through 2.1.4 of this chapter were identified as those that will typically result in the highest effect on a facility's effluent performance and operating efficiency.

2.1.1 Nutrient Loading and Loading Variability

Nutrient mass loads that are received by a water resource recovery facility (WRRF) are products of the flowrate and pollutant concentration. Loading rates may vary considerably during a typical day and can result in a biological nutrient removal (BNR) process experiencing loads both significantly higher and lower than the average design. Figure 2.2 in Chapter 2 illustrates a typical diurnal variation of flow and load. This variation can be more pronounced in smaller collection systems, where there is less storage capacity and transit time to dampen the effects. Typical effects of high nutrient loading may include deteriorated effluent quality (i.e., higher effluent concentrations of effluent total nitrogen and total phosphorus), an inability to maintain target dissolved oxygen concentrations, and reduced sludge settleability. A common effect of a lowly loaded biological process is higher dissolved oxygen concentrations and the potential for interference with denitrification and enhanced biological phosphorus removal (EBPR).

Table 11.1 presents typical peaking factors for nutrient loading in raw wastewater (WEF et al., 2009b). The table illustrates typical peaking factor relationships, how they vary by size of the facility, and whether the collection system is separately sewered or combined. Using Table 11.1 as a guide, for a facility with an average flow of 10 mL/d (2.6 mgd) and an average annual biochemical oxygen demand (BOD) load of 2500 kg/d (5400 lb/d), the maximum monthly BOD load would be approximately 3250 kg/d (7150 lb/d) ($2500 \times -0.05 \ln(10) + 1.44$). However, there can be considerable variability in these ratios, depending on the characteristics of the collection system.

2.1.2 *Influent Fractions and Constituent Ratios*

Commonly used measures of influent amenability to BNR are based on the amount of readily biodegradable substrate available to sustain EBPR and denitrification. As a first approximation, ratios of minimum 5-day BOD (BOD_5) to total Kjeldahl nitrogen (TKN) and BOD_5 to total phosphorus of 3 to 1 and 25 to 1, respectively, may be used to assess the site-specific availability of an adequate carbon source for BNR. This determination should be made on the influent to the BNR bioreactor and should include all significant sidestream loads. Municipal wastewater that is primarily domestic in origin typically contains sufficient carbon substrate for BNR if excessive BOD removal does not occur in the primary clarifiers. Sidestreams, which are characteristically low in BOD and relatively high in nitrogen and phosphorus, can depress the ratios of BOD_5 to total phosphorus and BOD_5 to TKN in bioreactor influent. It is important to note that the aforementioned minimum ratios are a first approximation of nutrient removal capability. Even if these ratios indicate that adequate readily biodegradable substrate is available, other influent characteristics and operating factors could compromise the ability of the system to achieve reliable BNR. For a detailed discussion of this topic, refer to other chapters in this manual.

2.1.3 *Effluent Quality and Fractions*

Biological nutrient removal facilities are designed and operated to achieve effluent criteria for total nitrogen and/or total phosphorus. As performance requirements become increasingly stringent and required effluent concentrations of total nitrogen and total phosphorus are reduced to levels that challenge the capabilities of feasible treatment processes, the fractions of effluent total nitrogen and total phosphorus become an important aspect in the evaluation of process performance. The parameter of concern for compliance with total nitrogen limits is refractory dissolved organic nitrogen (rDON) and, for a low total phosphorus limit, the parameter of concern is recalcitrant dissolved organic phosphorus (rDOP). Refractory dissolved organic nitrogen passes through BNR processes and typically ranges from 0.5 to 2.0 mg/L in BNR process effluent. Likewise, rDOP passes

TABLE 11.1 Design peaking factor summary (WEF, 2009b).

	Flow	BOD	TSS	TKN	Ammonia-nitrogen	Phosphorus
All data sets	60	57	54	16	33	16
Separate collection systems						
Number of valid data sets	40	33	32	6	15	6
Peaking factors*						
Minimum daily	$0.023\text{Ln}(x) + 0.67$	$0.032\text{Ln}(x) + 0.45$	$0.03\text{Ln}(x) + 0.46$	0.74	0.70; or $0.048\text{Ln}(x) + 0.55$	0.72
Maximum monthly	$-0.033\text{Ln}(x) + 1.38$	$-0.050\text{Ln}(x) + 1.44$	$-0.04\text{Ln}(x) + 1.43$	1.13	1.22; or $-0.074\text{Ln}(x) + 1.45$	1.14
Maximum daily	$-0.027\text{Ln}(x) + 1.47$	$-0.051\text{Ln}(x) + 1.68$	$-0.08\text{Ln}(x) + 1.91$	1.27	1.34; or $-0.08\text{Ln}(x) + 1.59$	1.26
Combined collection systems						
Number of valid data sets	20	19	18	2	11	3
Peaking factors*						
Minimum daily	0.68	0.60	0.53	0.67	0.66	0.73
Maximum monthly	1.32	1.26	1.31	1.24	1.21	1.20
Maximum daily	1.62	1.61	1.88	1.40	1.39	1.36

*Average values or equation of logarithmic regression lines shown in Figures 3.1, 3.2, 3.3, and 3.4, where x = flow in ML/d.

through both EBPR processes and chemical phosphorus removal and may typically comprise about 0.05 mg/L in BNR process effluent. These parameters are discussed in detail in Section 3.0.

2.1.4 *In-Facility Sidestreams*

Internal facility sidestreams, such as those from solids dewatering processes, supernatant from sludge digestion processes, thickening systems, and filter backwashes, can have a significant effect on the nutrient removal process and can introduce tremendous variability in treatment facility flows and loads. Chapter 12 describes alternatives for addressing these effects.

Sidestreams containing nitrogen and phosphorus are typically blended with facility influent and recycled through the bioreactor. The sidestream loads may be intermittent and highly variable. Biological nutrient removal process reactions are sensitive to influent load variations. While the average sidestream loading may or may not be significant, the short-term peak loads imposed by return streams can overwhelm the BNR system. For example, a peak load of ammonia introduced back to the head of a treatment process may exceed the nitrification capacity of the biological process, thereby resulting in a spike of ammonia in the effluent.

Sidestream flows also have the potential to upset the carbon-to-nitrogen ratio (as measured by the BOD_5-to-TKN ratio) or the carbon-to-phosphorus ratio (as measured by the BOD_5-to-total phosphorus ratio) described earlier in this chapter.

Facilities that remove phosphorus biologically need to be careful in their handling of waste sludge, particularly waste activated sludge. If the sludge is held in an anaerobic condition, such as in a holding tank, gravity thickener, or anaerobic digester, then sidestream flows from these processes can have high concentrations of phosphorus. These heavy loads can overwhelm the ability of the process to remove phosphorus biologically, especially if there is not enough readily degradable carbon available.

Facilities that incinerate their sludge may be introducing cyanide from the scrubber blow-down water to their process through sidestream flows. Depending on the concentration, the cyanide may be toxic or at least inhibitory to nitrifying organisms.

2.2 Establishing Optimization Goals and Metrics

Successful optimization of BNR process operation first requires setting optimization goals and a system by which optimization will be measured.

2.2.1 *Consistent Effluent Quality*

Biological nutrient removal optimization is perhaps most often targeted to improve process effluent quality to comply with regulatory requirements. The performance goals and means of measuring progress can, therefore, be easily established.

2.2.2 *Chemical Doses*

Operation of nutrient removal processes typically requires dosing of chemicals to increase available alkalinity, adjust pH, provide supplemental carbon to improve denitrification and/or biological phosphorus removal, precipitate reactive phosphorus, and improve solids separation. The dose required for these chemicals is dependent on effluent quality requirements, although there may be opportunities to reduce dose by modifying operations or dosing locations or using alternate chemicals.

2.2.3 *Energy Use*

Biological nutrient removal processes are typically energy intensive. Therefore, finding a means to reduce power consumption should be a high priority. Setting goals for energy consumption by process optimization should follow a step-by-step procedure as follows: (1) assess current energy use at the facility, setting a baseline from which to measure progress; (2) take an inventory of facility assets, equipment, and systems to identify areas of significant energy consumption; and (3) prioritize areas of potential optimization. U.S. Environmental Protection Agency's *Energy Management Guidebook for Wastewater and Water Utilities* (U.S. EPA, 2008) contains detailed guidance procedures for energy management.

2.2.4 *Sludge Production*

Sludge produced through operation of a BNR process is primarily a function of influent load, which is beyond an operator's control. However, the net solids yield is also affected by the operating solids retention time (SRT), the process configuration, and chemical addition.

2.2.5 *Operation and Maintenance Requirements*

Efficient operation of a BNR process requires qualified, diligent attention by operators and maintenance staff. The level of effort and associated costs may be a significant drain on resources, and optimization may focus on ways of reducing the labor and resources required to operate the process.

2.3 Identification of Additional Data Needs

Once the goals of the BNR optimization process are established, it will be necessary to collect and review relevant information such as facility historical performance data, maintenance records, energy use data, and operations budgets. Typically, available facility data do not provide sufficient information on which to complete an evaluation (e.g., facility data may include influent BOD, but no carbonaceous oxygen demand [COD] data when needed). In this situation, it will be necessary to develop a supplemental process sampling and/or monitoring plan to collect the necessary data. Section 3.0 provides detailed guidance on preparing such a plan and describes key parameters that should be considered.

3.0 PROCESS SAMPLING, TESTING, AND MONITORING

3.1 Sampling and Testing

Sampling and testing includes determining sampling locations, frequency, techniques, and appropriate tests that will allow evaluation of process performance and identify operational improvements needed to meet optimization objectives.

3.1.1 Sampling Plan

Every WRRF is unique and, therefore, will require a site-specific sampling plan to monitor performance. Operations personnel and facility administrators should work together to develop a sampling plan that allows for a complete understanding of key operating parameters without creating unneeded effort and excessive expense. During this exercise, everyone involved should ask the following questions:

- What are our regulatory reporting requirements?
- What reactions occur at each point in the process and how do they affect the facility's ability to meet regulatory requirements?
- What environmental conditions are needed for the reactions to occur? and
- What parameters must be sampled to evaluate if these reactions are occurring?

When creating a sampling plan, it is helpful to start with a copy of the facility's flow schematic(s) to illustrate the reactions in each process and various sample locations of interest at the facility. Figure 11.1 presents a sample flow schematic of a typical BNR facility. It is important to record the general reactions that occur in each process and the required environmental conditions to understand the parameters of interest. Operators should identify the location and type of samples that are required for permit compliance and the location of key parameters that indicate reactor performance for a specific step in a treatment process. Operators should then build a table summarizing the data that should be collected for each of the sample locations identified on the schematic. Each sample location and type is chosen to provide specific information or an indication of the process and reaction that is meant to occur. Table 11.2 contains an example of a sampling plan that could be prepared for the facility characterized in Figure 11.1.

3.1.2 Sample Handling and Collection

Sample handling should occur in accordance with regulatory requirements that govern the facility and/or as suggested by a generally accepted resource, such as 40 CFR 136 (U.S. EPA, 2005) or *Standard Methods* (APHA et al., 2012). Once a facility's staff has determined the information needed and marked it on the sampling

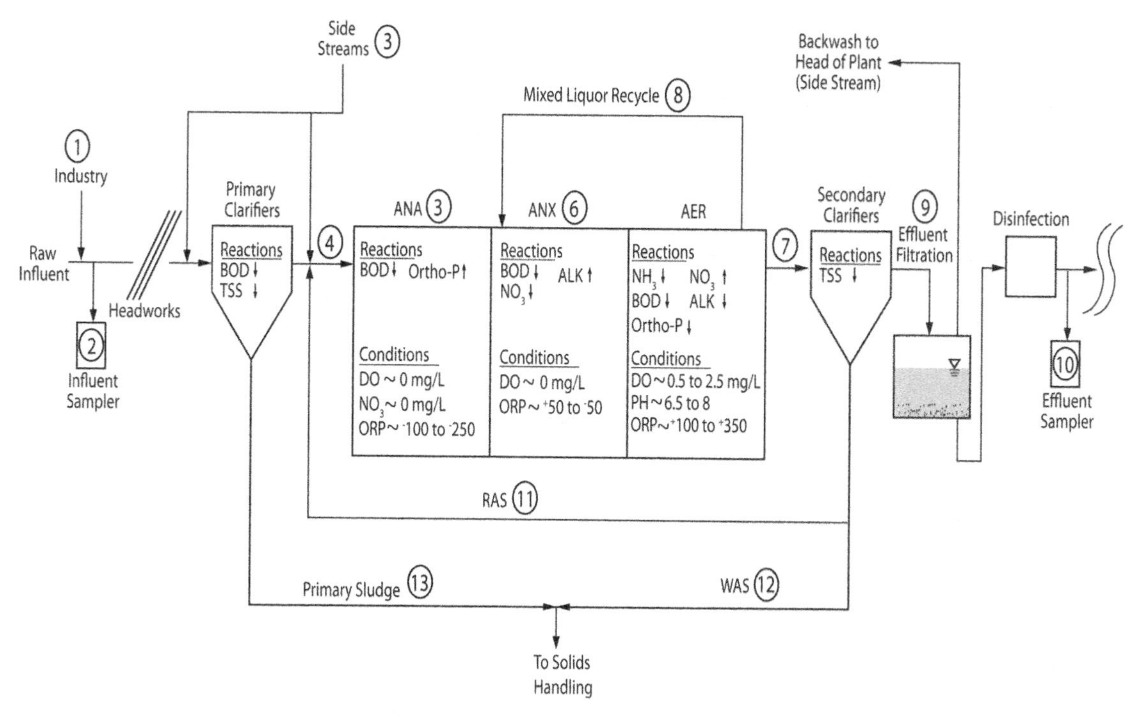

FIGURE 11.1 Flow schematic and sampling plan example.

plan, a decision must be made regarding how the raw data will be collected. The raw data will generally be collected in one of the following three forms:

- In situ sampling—typically, samples are measured with probes to achieve a real-time indication of the reactor environment and performance. Examples include dissolved oxygen, temperature, oxidation–reduction potential (ORP), ammonia, nitrate, and pH;
- Grab sampling—a sample taken as a "snapshot" in time, used to measure real-time reactor performance. Grab samples differ from in situ sampling in that they are typically used for parameters that require laboratory analysis; and
- Composite sampling—small volumes of sample taken throughout the day at the same location and combined to create a single sample to represent an average for that day. Samples can be collected based on a time interval (e.g., 5 mL of sample per hour) or flow paced (e.g., 5 mL of sample per 4000 L of flow).

Chapter 15 provides additional information on sample collection and storage.

3.1.3 *Sample Locations*

After a sample location is selected, it is important that all personnel collect the sample at precisely the same place and in the same manner. The sample location

TABLE 11.2 Sampling plan example of a typical BNR facility.

Sample location	Sample description	Flow	Temperature	pH	ORP[a]	BOD or COD	TSS[b]	TKN	NH_3[c]	NO_3[d]	Total phosphorus	Orthophosphorus	Dissolved oxygen	Alkalinity
	Sample type (grab [G]/ composite [C])	—	G	G	G	C	C	C	C	C	C	C	G	C
1	Industry effluent	Daily				M/W/F[e]		M/W/F			M/W/F			
2	Raw influent	Daily	Daily	Daily		Daily	Daily	Daily	Daily	Daily	Daily	Daily	Daily	Daily
3	Sidestreams													
	a. Filtrate	Daily				M/W/F	M/W/F		M/W/F			M/W/F		
	b. Thickening decant	Daily				M/W/F	M/W/F		M/W/F			M/W/F		
4	Primary effluent			Daily		Daily	Daily		Daily			M/W/F		
5	Anaerobic basin effluent (supernatant, other than TSS)			Daily	Daily	Weekly	Weekly			M/W/F	Daily	Daily	Daily	
6	Anoxic basin effluent (supernatant, other than TSS)			Daily	Daily	Weekly			M/W/F			M/W/F	Daily	M/W/F
7	Aerobic basin effluent (supernatant, other than TSS)			Daily	Daily	Daily	Daily	Daily	Daily	M/W/F	M/W/F	Daily	Daily	M/W/F
8	Mixed liquor recycle	Daily												
9	Secondary effluent						Daily				M/W/F			
10	Final effluent	Daily				Weekly	Weekly	Weekly	Weekly	Weekly	Weekly	Weekly	Weekly	Weekly
11	RAS	Daily												
12	WAS	Daily						M/W/F			M/W/F			
13	Primary sludge	Daily						M/W/F			M/W/F			

[a]ORP = oxidation reduction potential.
[b]TSS = total suspended solids.
[c]NH3 = ammonia.
[d]NO3 = nitrate.
[e]M/W/F = Monday, Wednesday, and Friday.
[f]RAS = return activated sludge.
[g]WAS = waste activated sludge.

should be clearly marked. Sample locations should be selected that allow representative samples to be collected that provide the desired information related to process performance. Additional samples should be collected if there is a possible change in conditions as the flow passes through the tanks. The following are characteristics of a good sample location:

- The contents are mixed well enough at that location to collect a representative sample and
- There is accessibility, allowing safe collection of the sample.

3.2 Online Process Monitoring and Automation

Use of online monitoring and automation has been more prevalent in recent years because of low effluent nutrient requirements and an increasing focus on energy and chemical costs at many facilities. Advancements in online analyzers allow operators to monitor virtually every step in a BNR process in real time. When considering online instrumentation, the benefits of each parameter and monitoring capability must be compared to the relative need based on regulatory requirements and the capital and maintenance costs of the instrumentation. The following subsections briefly discuss key parameters to consider when using online analyzers for process monitoring and control. Online analyzers are discussed in further detail in Chapters 13 and 14.

3.2.1 *Monitoring Plan*

Similar to the sampling plan discussed in Section 3.1.1, a process flow schematic should be the starting point of any online monitoring plan. It is important to understand the function of each unit process to select the appropriate parameter and location to monitor.

3.2.2 *Monitoring Locations*

Table 11.3 lists several common parameters and corresponding locations that can be monitored continuously as an indication of treatment performance.

3.2.3 *Use of Monitoring for Automation*

Data from online monitors can also be used to control chemical feed rates, internal recycle flowrates, influent loadings, and blower output, among other operations. Examples of some common process monitoring and corresponding control parameters are given in Table 11.4.

3.3 Parameters

This section describes many parameters and how they affect the operation of nutrient removal processes. Refer to Chapter 15 for a description of the procedures necessary to analyze these parameters.

TABLE 11.3 Potential monitoring locations of common online analyzers and process indication.*

Parameter	Monitoring location	Purpose and process indication
Ammonia (NH_3)	• Raw influent • Aerobic zone effluent • Final effluent	Monitor influent ammonia loading, nitrification performance, and total nitrogen in the final effluent.
Alkalinity	• Raw influent • Aerobic zone influent (upstream of the aerobic zone) • Aerobic zone effluent	Monitor alkalinity consumption and recovery to ensure sufficient alkalinity levels required for nitrification.
pH	• Aerobic zone effluent	Indication that pH level is sufficient for nitrification
Dissolved oxygen	• Various locations in the aerobic zone	Indication of sufficient dissolved oxygen for nitrification and biophosphorus removal in upstream portions of the aerobic zone. Monitor for low enough dissolved oxygen in aerobic zone effluent and internal recycle to minimize oxygenation of preanoxic zones.
Orthophosphorus (PO_4^{-3})	• Raw influent • Aerobic zone effluent • Final effluent	Monitor biological phosphorus removal.
Oxidation–reduction potential	• Anaerobic, anoxic, and aerobic zones	Monitor the environmental conditions in each zone. For example, ORP levels can indicate high levels of nitrate in an anaerobic zone, high dissolved oxygen in anaerobic or anoxic zones, or other environmental factors that reduce process performance.
Nitrate (NO_3^-)	• Anoxic zone effluent • Aerobic zone effluent • Secondary effluent • Filtered effluent (denitrification filters)	Indication of nitrification and denitrification performance.
Total suspended solids	• Various locations in BNR basins • Secondary effluent • Final effluent	Monitor MLSS concentrations. Monitor filter performance.

*Parameters given are relevant to a BNR treatment process to remove both total nitrogen and total phosphorus. Some parameters may not be applicable depending on the treatment process used and permit limits at each facility. Instrument reliability can vary depending on the application, wastewater characterization, and environment. All online instruments should be tested in the field during a trial period before purchase and installation at full scale. Sample preparation and location must be closely coordinated with vendor recommendations.

3.3.1 *Flow*

Flow is an important wastewater parameter that quantifies the bulk of fluid movement in facility streams. Flow of streams must be measured, along with concentrations, to determine mass loading rates. Essential flow measurements include process influent, effluent, return activated sludge (RAS), waste activated sludge (WAS), and internal recycles. Flow measurements for many other streams, such as sidestreams, are also necessary depending on the facility configuration and unit treatment processes included.

TABLE 11.4 Commonly monitored parameters used to control unit processes in a WRRF.*

Parameter	Monitoring location	Process and equipment controlled
Facility flow	• Influent (upstream of sidestream addition)	• Coagulant for phosphorus removal • RAS and internal recycle flowrates • Influent flow and load pacing • Disinfection chemical dosing
Alkalinity	• Raw influent • Aerobic zone influent (upstream of the aerobic zone) • Aerobic zone effluent	• Alkalinity addition
Dissolved oxygen	• Various locations in the aerobic zone	• Aeration blowers' airflow and energy consumption
Orthophosphorus (PO_4)	• Raw influent • Aerobic zone effluent • Final effluent	• Coagulant feed rate for phosphorus removal • Supplemental carbon feed for biophosphorus removal
Nitrate	• Anoxic zone effluent • Aerobic zone effluent • Secondary effluent • Filtered effluent (denitrification filters)	• Carbon feed for denitrification

*Use and effectiveness of the instruments for process control are dependent on site-specific conditions including permit limits, chemical and energy costs, and the BNR process being implemented.

The two main types of flow measurement are open channel and closed pipe. One of these types may be preferred for flow measurements for a certain location and purpose. Open-channel flow measurement consists of a weir or flume measuring flow in open channels or partially full pipes. Parshall flumes are examples of this type of flow-measurement device. Closed-pipe flow measurement devices include magnetic, mechanical, ultrasonic, and mass flow meters.

3.3.2 *Mixed Liquor Suspended Solids, Mixed Liquor Volatile Suspended Solids, Return Activated Sludge, and Waste Activated Sludge*

The total suspended solids (TSS) concentration in mixed liquor, RAS, and WAS is measured for process monitoring and control. Waste activated sludge concentration, along with mixed liquor suspended solids (MLSS) concentration, is typically measured daily to maintain the desired SRT, which is an important process parameter. The volatile fraction of the mixed liquor is measured with mixed liquor volatile suspended solids (MLVSS) measurements. All of the samples should be collected from a well-mixed, representative location. Ideally, samples to determine MLSS, MLVSS, RAS, and WAS should be collected a minimum of several times a day. These samples can be tested individually, or, if the facility is stable and does not have significant variation in incoming flow, the samples can be composited and then tested. Daily composite samples are ideal because the facility's daily variation in flow and load may be missed with infrequent sample collection.

The WAS test is run on a sample taken from the WAS pipe (or channel or box) in a location that is well mixed and representative. When the WAS is pumped from a single hopper, composite sampling may be necessary over the duration of the WAS pumping cycle to accurately determine the mass of solids wasted out of the process. Selection of sample type and timing for WAS should consider the facility's wasting strategy (i.e., continuous or intermittent).

The MLSS concentration is also used to calculate the sludge volume index (SVI) in conjunction with settleability test results (refer to Section 3.3.3). The MLVSS result divided by the MLSS result provides the volatile fraction. The volatile fraction is a rough indicator of active biomass in suspended growth activated sludge processes. In typical BNR facilities, the volatile fraction is approximately 70 to 80%, depending on the relative concentration of inert material in the process influent and the SRT. Longer SRTs will result in a lower volatile fraction because volatile material is reduced in the mixed liquor through oxidation of organic material in the biomass (endogenous respiration). Changes to the volatile fraction of MLSS can be an indicator of process problems and should be considered.

The MLSS and WAS solids testing results (with WAS flow) are typically used in conjunction with the volumes of anaerobic, anoxic, and aerobic zones in the treatment process to calculate each phase and total process SRT. Aerobic SRT is important for nitrification because the nitrifiers are obligate aerobic organisms that grow slower than other organisms used in BNR (refer to Chapter 3). The MLSS and RAS results can also be used to evaluate whether the RAS rate is appropriate in relation to process influent flow. The required RAS flowrate is typically selected based on maintaining a target ratio with the influent flowrate, and is the result of controlling sludge blanket depth (0.6 to 0.9 m) and settled mixed liquor detention time in the secondary clarifiers.

3.3.3 *Settleability and Sludge Volume Index*

A settleability measurement is used to indicate the settling and compaction characteristics of mixed liquor in suspended growth treatment processes at the instantaneous MLSS concentration. Settleability is reported as milliliters per liter, which represents the volume corresponding to the top of the settled solids interface from a 1-L sample of mixed liquor after 30 minutes of settling. The SVI "normalizes" the settleability using the MLSS concentration and is reported as milliliters of settled volume per gram of MLSS. The SVI is generally interpreted as follows:

- An SVI less than 80 indicates excellent settling and compacting characteristics,
- An SVI from 80 to 150 indicates moderate settling and compacting characteristics, and
- An SVI greater than 150 indicates poor settling and compacting characteristics.

A "good" settling sludge is not necessarily the same as a "fast" settling sludge; nonetheless, the settleability and SVI tests verify that the BNR facility has stable operation and may not experience problems with solids carryover at the clarifiers under high-flow conditions. It should be noted that clarifier capacity and performance depend on many factors. The SVI test serves as an indicator of how the settleability of the mixed liquor may change over time. The SVI should then be trended against other treatment indices to evaluate how these parameters affect settleability.

Solids settling and compacting characteristics can vary during the day as a result of diurnal variations in organic or nutrient loadings. Samples can be pulled manually once (or more, if necessary) per 8-hour shift, tested for settleability and MLSS concentration, and the SVI can be calculated.

To determine unstirred SVI, a sample of mixed liquor is pulled from a well-mixed representative location of the aeration basin upstream of the clarifiers. This sample is gently mixed and placed in a 1-L graduated cylinder with a stirring mechanism (or a settleometer, calibrated beaker, or an unstirred 1-L graduated cylinder, the latter of which is less desirable). The settled volume is recorded approximately every 5 minutes up to 30 minutes as milliliters per liter. The 30-minute reading is generally used for settleability and SVI. One of the reasons for recording the settled volume every 5 minutes is that the mixed liquor in some nitrifying facilities denitrifies, and solids float up before the 30-minute period is complete.

3.3.4 *pH*

pH is a measurement of the acid, neutral, or basic condition of wastewater. A pH of 7 is neutral, while a pH below 7 is acidic and a pH above 7 is basic. The pH of incoming wastewater is a function of the drinking water supply; acids or bases added through household, commercial, or industrial activity; residence time in the collection system; and buffering capacity (refer to Section 3.3.5). The pH can be changed through chemical addition, byproducts of biological activity, and interaction of the wastewater with air. The pH should be measured in wastewater on at least a daily basis; in addition, many facilities may find it advantageous to install a permanent in situ pH analyzer to provide continuous indication of potential upset conditions or unusual wastewater discharges.

The pH in the BNR process should be within a narrow range for stable operation. If the pH is too low or varies significantly, the nutrient removal activity can be slowed to where nutrient limits are not met. pH can be used with alkalinity to determine if wastewater characteristics and/or the treatment process are responsible for a pH drop and whether further pretreatment by industrial contributor(s), process modification, or alkalinity addition is needed.

The optimum pH range for nitrifying bacteria is pH 7.5 to 8.6, while an acceptable pH range is 7.0 to 9.0. Nitrifying bacteria can acclimate to a pH in the 6.5 to 7.0 range, with little decrease in activity, if the pH does not vary significantly. The optimum pH for denitrification is between 6.5 and 8.5 (WEF et al., 2009).

For EBPR, the pH should be above 6.5. Although the optimum pH for chemical removal of phosphorus is 5.3 for ferric or 6.3 for alum, chemical removal of phosphorus will occur at higher pH values, with increased chemical dosage. The pH of the wastewater is rarely adjusted to accomplish chemical removal of phosphorus, unless lime is used.

Although pH is generally not used for control purposes, it can be a quick indicator of a significant industrial constituent in the wastewater entering a BNR facility. Typically, pH in BNR facility effluent represents a compliance parameter. In larger facilities, pH is monitored continuously in the facility effluent; some facilities also monitor pH in the BNR process.

3.3.5 *Alkalinity*

Alkalinity is a measurement of the buffering ability of wastewater and is a rough indicator of carbonate and bicarbonate concentration in the wastewater. Alkalinity can result from hydroxides, carbonates, bicarbonates, and, less commonly, phosphates, silicates, borates, and similar compounds in the wastewater. With sufficient alkalinity, water can take in acid (or base) with no change or only a small change in pH. An alkalinity over 50 mg/L is necessary to prevent pH from dropping sharply when small amounts of acid are introduced to the water by either chemical addition or biological activity. Wastewater alkalinity results from alkalinity in the water supply plus alkalinity added during domestic use (e.g., cleaners and detergents). A typical alkalinity value for raw wastewater is 150 mg/L, although it can range from 60 to 250 mg/L, depending on regional alkalinity in the drinking water and the amount of infiltration and inflow entering the collection system. Samples for alkalinity testing are typically taken from the raw wastewater entering a BNR facility.

Alkalinity is most important during the nitrification process, which reduces alkalinity through the metabolism of inorganic carbon by nitrifying bacteria and produces acid when ammonia is oxidized. The nitrification reaction predicts that, for each gram of ammonia-nitrogen oxidized to nitrate-nitrogen, 7.08 g of alkalinity as calcium carbonate ($CaCO_3$) is destroyed. When too much alkalinity is destroyed, the pH can decrease to a level outside the narrow pH range for good activity of the nitrifying bacteria.

The denitrification process replenishes alkalinity by destroying acid and producing carbon dioxide. The denitrification reaction predicts that, for each gram of nitrate-nitrogen reduced to nitrogen gas, 3.5 g of alkalinity (as $CaCO_3$) is produced. Where alkalinity in the raw wastewater is low, the denitrification step can be performed before, or simultaneously with, nitrification to reduce net alkalinity destruction.

Although alkalinity is not used for process control in most BNR facilities, alkalinity addition is occasionally necessary in situations with low background alkalinity concentrations and/or significant nitrification needs. It is tested when pH problems are experienced, or initially when a BNR process is being designed, to identify whether alkalinity needs to be added. Alkalinity is sometimes used for process control in high-rate BNR facilities that are trying to minimize nitrification by operating with low SRT while removing phosphorus only; in this instance, the process can be controlled by preventing a decrease in alkalinity that signals the start of nitrification.

3.3.6 *Temperature*

The relevant temperature of wastewater in suspended growth systems is that at which the microorganisms function and oxygen transfer is occurring. In some biofilm systems, temperature is a function of both wastewater temperature and air temperature. Wastewater temperature typically ranges from 3 to 27 °C in the United States and as high as 30 to 35 °C in areas of Africa and the Middle East (Metcalf and Eddy, 2003). Wastewater temperature arises from the temperature of the drinking water supply; warm water from households, commercial, and industrial water uses; and soil temperature. In addition, biological activity and aeration adds heat, and heat is lost from surfaces and turbulent areas open to the atmosphere.

Temperature affects the growth rate and metabolism of microorganisms. Nitrification and denitrification rates are affected by temperature, and nitrifying bacteria are particularly sensitive to temperature variations. Low temperatures reduce the biochemical reaction kinetic rates and longer SRT is needed to maintain adequate population of slow-growing microorganisms (e.g., nitrifying bacteria) in BNR facilities (refer to Chapter 3). Temperature also affects oxygen transfer into the wastewater because warmer temperatures reduce the solubility of oxygen in water and, consequently, the oxygen-transfer rate. At the same time, when the temperature is high, biological activity is higher, thereby increasing oxygen demand for treatment.

3.3.7 *Dissolved Oxygen*

In water, the dissolved oxygen concentration is limited by the maximum equilibrium concentration, or saturation concentration, which depends on the gas used (air or pure oxygen), water temperature, dissolved solids in the water, and elevation (atmospheric pressure). The dissolved oxygen saturation concentration decreases with increasing temperature, increasing dissolved solids concentration, and decreasing atmospheric pressure. Oxygen-uptake rates are higher with higher temperature as a result of increased microbiological activity. The critical time for maintaining adequate dissolved oxygen is during high-temperature periods at peak loading. It should be noted that dissolved oxygen, as measured in an activated sludge system, provides the dissolved oxygen concentration in the bulk solution and may not be directly representative of dissolved oxygen concentrations within the floc. The dissolved oxygen within the floc will generally be lower than that measured in the bulk solution.

Dissolved oxygen is essential for aerobic microorganisms that are responsible for BOD removal, nitrification, and EBPR. In aeration zones, dissolved oxygen concentrations must be sufficient to meet oxygen requirements of the reactions taking place and should be high enough to achieve the necessary removal rates. Biochemical oxygen demand removal can occur with dissolved oxygen concentrations of 0.5 mg/L or less; however, low dissolved oxygen conditions can lead to filamentous bulking. In nitrification processes, a minimum dissolved oxygen concentration of approximately 2.0 mg/L is typically adequate, with the nitrification rate decreasing with lower dissolved oxygen concentrations. Operation at lower dissolved oxygen concentrations is increasingly being considered as an energy-savings measure because oxygen transfer is more efficient at lower dissolved oxygen concentrations. Dissolved oxygen should not be present in any significant quantity in anoxic zones because denitrification will be inhibited. Similarly, dissolved oxygen should not be present in the anaerobic (fermentation) zone because volatile fatty acid (VFA) uptake and subsequent biological phosphorus removal will be inhibited.

Dissolved oxygen in aeration zones should be monitored a minimum of several times per day and, preferably continuously, if instrumentation is used. Dissolved oxygen is commonly used to control aeration to enhance BNR and reduce energy costs. Although dissolved oxygen concentrations in anoxic zones, anaerobic zones, and internal recycle or RAS streams are typically not monitored, they should be checked if an upset condition is experienced or as part of a process optimization effort. Chapter 13 contains a detailed discussion of process control using dissolved oxygen.

3.3.8 *Oxidation–Reduction Potential*

Oxidation–reduction potential is a measurement of the oxidative state of a liquid. Dissolved oxygen, as described in the preceding section, is simply the concentration of dissolved oxygen in the mixed liquor. Oxidation–reduction potential, on the other hand, indicates if aerobic, anoxic, or anaerobic biochemical reactions are occurring in the mixed liquor. The highest positive ORP values indicate the most oxidative conditions (i.e., aerobic), while the highest negative ORP values indicate the most reductive conditions (i.e., anaerobic) in mixed liquor. Chapter 13 contains a detailed description of the meaning, interpretation, and process control using ORP.

3.3.9 *Ammonia and Total Kjeldahl Nitrogen*

Total Kjeldahl nitrogen is the sum of organic nitrogen and ammonia-nitrogen. The TKN is measured in the influent to determine the nitrogen load to the BNR facility, as municipal wastewater typically does not contain nitrate-nitrogen unless there is an industrial source. A large fraction (60 to 75%) of the TKN is ammonia-nitrogen. The rest of the influent nitrogen is organic nitrogen that is associated with soluble and particulate organic matter in municipal influent.

Some of the organic nitrogen in the influent is hydrolyzed by microorganisms to ammonia-nitrogen at the treatment facility. Nitrifying organisms oxidize ammonia to nitrate under aerobic conditions. Ammonia is also incorporated to new cell growth by the biomass. The oxidation of each milligram of ammonia-nitrogen requires 4.6 mg of oxygen. The nitrogen uptake in cell growth results in approximately 0.08 to 0.12 mg N/mg volatile suspended solids (VSS) in MLSS. Considering that BNR facilities have an MLVSS-to-MLSS ratio of 0.70 to 0.80, the nitrogen content of WAS is estimated to be in the range of 5 to 10%.

Ammonia-nitrogen and TKN should be measured in the influent wastewater before the return point of sidestreams, influent to the BNR process, and BNR effluent. The TKN in facility influent is typically run at the same frequency as compliance monitoring in facilities that have a total nitrogen limit; ammonia-nitrogen in facility influent can also be run at the same time. The TKN and ammonia-nitrogen should also be periodically checked in sidestreams from solids handling processes to evaluate their effect on BNR process performance. If the BNR facility uses anaerobic digestion, the sidestreams may contain substantial ammonia-nitrogen.

Ammonia-nitrogen is sometimes monitored at the end of aerobic zones with nitrite-nitrogen and/or nitrate-nitrogen to evaluate nitrification performance. These measurements can also be used to control the aeration system. Aeration can be reduced to levels such that the ammonia-nitrogen setpoint at the end of the aeration basins is not exceeded. This type of control can be implemented with automation and real-time measurements using online ammonia-nitrogen analyzers and probes.

3.3.10 *Nitrate-Nitrogen*

Influent wastewater typically has little or no nitrate-nitrogen, although, in some instances, nitrate-nitrogen can enter the collection system with industrial discharges or by chemical addition for hydrogen sulfide control. Nitrate-nitrogen can be generated through complete nitrification in the aeration basins. In addition, nitrate-nitrogen can be measured in facility influent, aeration zones, clarifier effluent, and final effluent after disinfection.

Nitrate-nitrogen in treatment facility effluent can be limited by total nitrogen effluent limits or nitrate-nitrogen limits for groundwater discharge. The nitrate-nitrogen profile through the BNR process indicates performance of nitrification and denitrification and can be used to control aeration or internal recycle rates and to make decisions regarding SRT.

3.3.11 Nitrite-Nitrogen

In wastewater, nitrite-nitrogen can enter the collection system with industrial discharges and can be generated through incomplete nitrification at the treatment facility. Insufficient aerobic SRT at low temperatures, low dissolved oxygen, pH, and inhibition can result in incomplete nitrification, thereby resulting in nitrite-nitrogen. Nitrite-nitrogen can be measured in facility influent, aeration zones, clarifier effluent, and final effluent after disinfection.

Nitrite-nitrogen in wastewater effluent can have a toxic effect on micro-organisms, macroinvertebrates, and fish in surface waters. In addition, nitrite-nitrogen, if present in high concentrations, can increase chlorine demand. The presence of significant concentrations of nitrite-nitrogen can be used to indicate the need for additional aeration in aerobic zone(s).

3.3.12 Refractory Dissolved Organic Nitrogen

Refractory dissolved organic nitrogen is the fraction of influent nitrogen load that is resistant to chemical or biological treatment (Stensel et al., 2008). Compounds that contribute to rDON are largely unknown. Organic nitrogen levels in facility effluent depend on industrial wastewater contributions, organic nitrogen present in the drinking water, and microbial decay that occurs during biological treatment. Organic nitrogen resistant to further biological degradation can represent a significant fraction of the effluent total nitrogen when low effluent total nitrogen limits are required (Bott and Parker, 2011).

Refractory dissolved organic nitrogen becomes particularly important when the facility is required to achieve low levels of nitrogen. Levels of rDON in municipal WRRF effluents can vary between 0.5 to 2.8 mg/L, as shown by a survey of 30 municipal WRRF effluents conducted by Pagilla (2007). Dissolved organic nitrogen levels in facility effluent are often used as indicators of rDON. Dissolved organic nitrogen is estimated by subtracting effluent ammonia from effluent soluble TKN.

3.3.13 Total Phosphorus

Phosphorus is a macronutrient needed by all living things to reproduce and grow cell mass. Phosphorus is generally regarded as the limiting factor for algae growth in freshwater streams and lakes. For this reason, many WRRFs have phosphorus effluent limits that attempt to reduce the total phosphorus load to receiving water. When algae are produced in streams and lakes, they eventually die and are decomposed by organisms, resulting in a reduction of dissolved oxygen (eutrophication). In addition to the potential immediate effects of low dissolved oxygen, decomposed algae settle to the bottom of the body of water and contribute to the process of filling the stream or lake.

Total phosphorus should be monitored as required by the facility permit and as often as is practical from industry at the influent, from any sidestream, and at the effluent for process control. It is important that enough data are collected when the facility is running well to form a reliable baseline of information.

Because phosphorus is needed for reproduction and cell growth, there is a need for influent phosphorus to biological wastewater treatment systems. Activated sludge phosphorus content is generally 0.025 to 0.035 mg P/mg VSS without EBPR. Phosphorus content of activated sludge can be substantially larger when EBPR is achieved. Most municipal WRRFs will have more than enough phosphorus to satisfy cell growth, and the remainder needs to be removed through treatment enhancements.

Phosphorus can occur in many forms and typically changes form through biological or chemical wastewater treatment. Among the forms of phosphorus, some will be dissolved and some will be part of the suspended solids. The reactive soluble fraction is generally in the orthophosphorus state (refer to Section 3.3.14); although these two forms of phosphorus (reactive soluble phosphorus and orthophosphorus) are not the same, they are often used interchangeably.

3.3.14 *Orthophosphorus*

Orthophosphorus is a soluble form of inorganic phosphorus that is common in treated BNR process effluent. Most of the particulate phosphorus will be contained within the activated sludge system and, therefore, should not be discharged from the facility in the effluent, with the exception of phosphorus included in effluent suspended solids. For this reason, the majority of effluent phosphorus in a facility with low effluent TSS will be in the orthophosphorus form.

Test kits are available to allow for quick determination of orthophosphorus, making this an attractive process control tool to some operators. It is important to note that orthophosphorus data generated by a test kit typically do not provide a reportable value and should be considered an estimate because the kits are generally not as precise as laboratory versions. In an effluent with low suspended solids, many operators test their effluent quality frequently with orthophosphorus and routinely compare it to their total phosphorus data. These comparisons allow operators to determine how reliable their orthophosphorus data are as an indicator of total phosphorus.

Orthophosphorus can also be used within EBPR processes to quickly monitor various stages of the system for phosphorus release in the anaerobic zone and subsequent luxury uptake in the aerobic zone. Solids should be removed by allowing the samples to settle or by filtering them; in addition, they should be analyzed as quickly as possible.

3.3.15 *Recalcitrant Dissolved Organic Phosphorus*

Total phosphorus consists of particulate phosphorus and soluble phosphorus. Particulate phosphorus must be hydrolyzed and converted to reactive orthophosphorus form to be removed by EBPR or by chemical precipitation with metal salts. Similar to rDON, there is also a refractory fraction of phosphorus, which is known as *rDOP*. Compounds that contribute to rDOP are largely unknown.

Recalcitrant dissolved organic phosphorus becomes a greater concern when WRRFs are required to reduce effluent total phosphorus to low levels (i.e., below 0.10 mg/L). Advanced treatment technologies may be required to achieve low levels of effluent total phosphorus because biological and chemical treatment will not be effective for rDOP. The rDOP levels in WRRF effluent may be in the range of 0.05 mg/L or lower. Analysis and quantification of rDOP is also an issue because of such low levels of effluent total phosphorus.

3.3.16 *Carbonaceous Oxygen Demand*

Carbonaceous oxygen demand is a measure of the carbonaceous (organic) matter in municipal wastewater. At approximately 3 hours, COD analysis is much faster than the BOD_5 test. Variations of COD analysis have been developed in an attempt to better characterize the effect of COD on biological systems by isolating certain fractions of COD that are directly used by polyphosphate-accumulating organisms (PAOs). Figure 11.2 shows these influent COD subdivisions, as presented by Ekama et al. (1984).

The influent total COD (Sti) is first subdivided into biodegradable and un-biodegradable fractions. The un-biodegradable soluble fraction will typically pass through the treatment system ($Susi$) in the facility effluent while the un-biodegradable particulate COD fraction (S_{upi}) becomes part of the MLSS and is

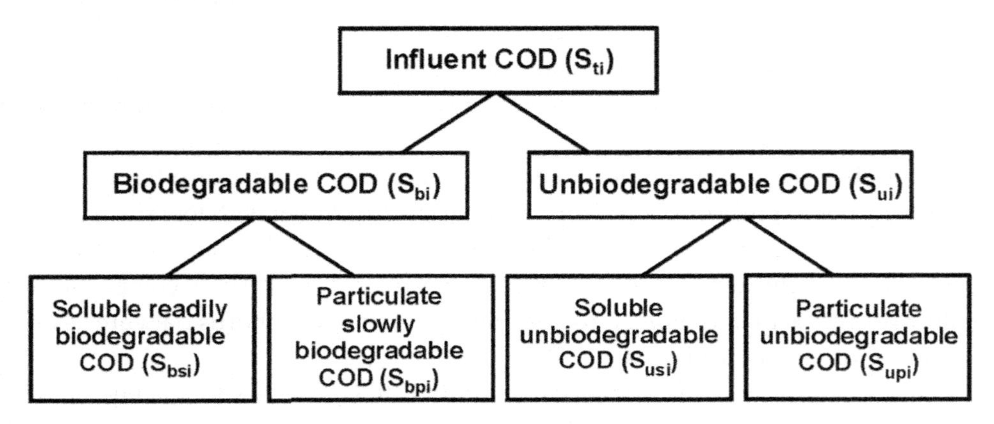

FIGURE 11.2 Division of total influent COD in municipal wastewater.

eventually wasted from the system in WAS. The biodegradable fraction is broken into readily biodegradable (*Sbsi*) and slowly biodegradable (*Sbpi*). The *Sbsi* fraction is the key component when dealing with any BNR system evaluation.

Samples for total COD should be collected from facility influent, just before the influent reaches the BNR process, and effluent from the BNR process. Other locations may be sampled to quickly evaluate organic contributions to loading on the BNR process. Samples for soluble COD should be collected at locations just before the influent reaches the BNR process (*Sbsi*) and the effluent from the BNR process (*Susi*).

In the anaerobic zone, typically only the *Sbsi* component is susceptible to fermentation to form VFAs within the short retention time (1 to 2 hours). Ekama et al. (1984) found that phosphorus release in the anaerobic zone increased as the influent *Sbsi* increased. They concluded that a *Sbsi* concentration of at least 25 mg/L was necessary surrounding the PAO in the anaerobic zone for phosphorus release to occur.

Other research indicates that a *Sbsi* influent concentration of 60 mg/L is needed for reliable EBPR to occur. Every facility has its own unique wastewater characteristics and *Sbsi* concentrations that correspond to good EBPR performance. Operations staff can develop a database to determine if changes in treatment efficiencies are related to changes in influent wastewater characteristics, such as *Sbsi*. These data can be trended against effluent quality and other operating parameters to better understand conditions that affect treatment.

A facility can elect to use total COD in their sampling protocol; however, variations in the amount of inorganic oxygen demand may reduce the value of this data. The *Sbsi* analysis may be the preferred parameter for tracking feed to the EBPR process.

The COD or *Sbsi* data are most useful when they are collected regularly so that results obtained during good treatment can be compared to results obtained during poor treatment. Although sample frequency will likely depend on the resources of each facility, it is recommended to occur at least weekly to monthly. Additional sampling should be conducted when treatment is poor and during seasonal changes.

The analysis involves chemical oxidation of the sample followed by either a titrimetric or color reading of the result to determine the result expressed as milligrams per liter of oxygen. Because the COD process oxidizes both the inorganic and organic components of the sample, the COD results will always be greater than the BOD results for the same sample. The BOD results are typically between 50 and 60% of the COD results, although each wastewater will have its own ratio.

Analysis of *Sbsi* requires two samples of COD to be run: one on the influent to the BNR process and one from the effluent. Both samples receive chemical flocculation and filtration before COD analysis is conducted. *Sbsi* is estimated by subtracting the effluent truly soluble value from the influent truly soluble COD.

Particulate, colloidal, and soluble are the three main COD fractions. The colloidal fraction can pass through a 1.2-μ glass fiber filter. Therefore, COD filtered through a 1.2-μ glass fiber filter represents the sum of colloidal and soluble COD. Particulate COD is estimated by subtracting colloidal and soluble COD from the total COD measured in the unfiltered sample. To measure truly soluble COD, colloidal and soluble fractions must be determined. The flocculated and filtered COD (ffCOD) method is used to determine truly soluble COD. The ffCOD method provides removal of colloidal and particulate COD. The method described by Mamais et al. (1993) consists of

- One milliliter of 100-g/L zinc sulfate (flocculating agent) solution is added to 100 mL of sample,
- The sample is vigorously mixed for 1 minute,
- The sample pH is adjusted to 10.5 using 6 M of sodium hydroxide solution,
- The sample is rested for suspended solids to settle, and
- The supernatant is run through a 0.45-μ membrane filter and the filtrate is analyzed for COD.

Once ffCOD is determined, colloidal COD is estimated by subtracting ffCOD from filtered COD. It is important to determine these two fractions of COD because colloidal COD is slowly degradable and believed to be removed from wastewater by being incorporated to mixed liquor. Soluble COD has biodegradable and un-biodegradable fractions. The un-biodegradable fraction is refractory and will appear in facility effluent along with the small amount generated in the biological treatment as a result of bacterial decay. The biodegradable fraction of soluble COD consists of VFA and non-VFA.

3.3.17 *Volatile Fatty Acids*

The performance of an EBPR process will vary with the specific VFAs available in the anaerobic zone. Based on past research, COD consumption during EBPR has been estimated to be 50 to 60 mg/L COD per milligram per liter of phosphorus removed from municipal wastewater. According to Abu-ghararah and Randall (1991), acetic acid (HAc) is the most efficient VFA for enhanced BPR. Fortunately, acetic acid is typically the primary VFA formed from wastewater fermentation, with propionic acid as the secondary VFA. Sufficient VFA production can occur in wastewater collection systems, especially in long force mains or gravity sewers with long detention times in warmer climates. Several facilities observed improved reliability of EBPR with the addition of a VFA-laden waste stream, such as a long force main serving an area several miles from the publicly owned treatment works. Fermentation of wastewater or primary sludge involves the reduction of proteins and carbohydrates into, most commonly, acetates, propionates, and butyrates.

Samples should be collected from the raw wastewater influent. For facilities with primary clarifiers, the sample should be primary effluent; for facilities without primary clarifiers, the sample should be the effluent from preliminary treatment. Samples should also be collected on the elutriate from prefermentation facilities to determine the performance of these units in generating additional VFAs that are directed to the EBPR anaerobic zone(s).

Enhanced biological phosphorus removal requires the presence of VFAs in the anaerobic zone of any BNR wastewater treatment system. When the influent wastewater is weak and not septic, VFA production can potentially be accomplished in the anaerobic zone of the BNR process or outside the BNR system in a separate process called *prefermentation.* A facility can elect to continue monitoring VFAs as a process control parameter. This may prove especially useful at BPR facilities that do not have an abundance of VFAs in the influent to the BPR tanks. These facilities can potentially be optimized by tracking the abundance of VFAs and adjusting their processes accordingly.

Samples should be taken during times of adequate BPR to obtain a baseline to compare to periods of poor performance. Typically, approximately 40 to 50 mg/L of HAc as COD is desired in the influent to the anaerobic zone; however, this can vary by facility. If the anaerobic zone is larger (greater than 1.5 hours of detention time), then VFA production can potentially be generated within the zone.

An alternative method to determine sufficient fermentation of influent wastewater is to compare the ratio of COD to total phosphorus to the ratio of BOD_5 to total phosphorus. If the ratio of COD to total phosphorus is considerably higher than 40 to 1 and the ratio of BOD_5 to total phosphorus is considerably lower than 20 to 1, then the wastewater has likely not undergone sufficient fermentation (Randall et al., 1992).

3.3.18 Soluble Biochemical Oxygen Demand

Soluble BOD (SBOD) is the BOD of the filtered sample. The soluble fraction of BOD that is loaded to the anaerobic zone of a BPR facility may be more accurate in predicting the response of PAOs than total BOD. This is because PAOs are able to use only the soluble fraction of BOD in gaining their competitive advantage while in the anaerobic zone. Although SBOD should not replace BOD_5 for calculating oxygen demand exerted on the biological system, it may be useful in predicting phosphorus removal success.

The BPR process is dependent on having a readily degradable food source available to the PAOs to remove phosphorus. The greater the ratio of BOD to total phosphorus or SBOD to soluble phosphorus, the better the chance that the system will be capable of adequately removing enough phosphorus to achieve a high-quality effluent. Review of this data over time will allow a facility to determine if variations in effluent quality are the result of changes in the amount of food available to the system. It should be noted that the ratio of SBOD to soluble phosphorus offers the advantage of being in the correct form to predict how the system will react and is likely a better control parameter than the ratio of BOD to total phosphorus; however, it may not be as good an indicator as Sbsi or short-chain VFAs. The ratio of SBOD to soluble phosphorus may be the best indicator for a facility that does not have the equipment and staff needed to run COD or VFA analysis. The key disadvantage of using BOD or SBOD is that it takes 5 days to get the results; therefore, it is not a good predictor of the immediate future, but instead could be used to identify long-term trends. Designers may use ratios of 15 to 1 (SBOD to soluble phosphorus) or greater and, possibly, a ratio of 20 to 1 (BOD to total phosphorus) as a threshold to determine if a facility is a good candidate for BPR. If the ratios being measured at a given facility are at or greater than those values, it is unlikely that the facility is BOD limited.

3.3.19 Nitrification Rate

The purpose of the nitrification test is to initially determine a baseline indicator value of nitrification rate for a specific facility and to determine the performance of the facility relative to that baseline under the various operating conditions encountered. Samples for the nitrification test should be collected from the influent to the BNR process before the introduction of RAS and from RAS.

The performance of the nitrification process depends on the concentration of nitrifiers in the system, the ammonia concentration, and environmental factors such as wastewater temperature, dissolved oxygen concentration, pH and alkalinity, and the presence or absence of inhibitory compounds in the influent. The nitrification test provides adequate supply of ammonia, dissolved oxygen, and alkalinity. Influent pH, influent wastewater temperature, and biomass concentration (MLSS and MLVSS) are recorded for the test, and the biomass is presumably acclimated to the influent and contains a sufficient proportion of nitrifiers to achieve nitrification.

Establishing a baseline indication of nitrification performance is useful at the various temperatures encountered and concentrations of MLSS and MLVSS controlled by facility operations. Periodic use of the test reinforces the baseline indicator values of nitrification rate and will indicate if there is some factor, such as wastewater temperature, inhibitory compounds, or nitrifier population, that is adversely affecting the nitrification process if poor process performance occurs. If the nitrification test shows that adequate nitrification can occur but the that facility performance is lacking, then facility operators should look for conditions related to insufficient dissolved oxygen, inadequate detention time or inadequate mixing, insufficient alkalinity and lowered pH in the process, or intermittent slugs of inhibitory compounds. If the nitrification test shows poor nitrification compared to the baseline, then facility operators should look for lowered SRT, insufficient dissolved oxygen, high organic loading, low wastewater temperature, and the presence of inhibitory compounds.

The test should be performed regularly to maintain baseline information and to assist in judging the effectiveness of operational adjustments necessary to maintain nitrification during seasonal changes in wastewater temperature. The test also can be performed in multiple reactors several times a day if slugs of inhibitory compounds are suspected based on observations of full-scale nitrification performance.

There are three basic tests that determine nitrification maximum growth rate. These are the low food-to-microorganism (F/M) sequencing batch reactor method, the washout method, and the high F/M method, as described in the Water Environment Research Foundation's *Methods for Wastewater Characterization in Activated Sludge Modeling* (WERF, 2003).

3.3.20 *Microbiological Activity*

Monitoring microbiological activity involves the use of a sensor (probe) to detect biological activity by monitoring the strength of a fluorescence signal reflecting from the biomass in an activated sludge system. This, in theory, allows more reliable monitoring and control of the biological state or activity in aeration tanks at low dissolved oxygen concentrations than a dissolved oxygen probe alone could accomplish.

The fluorescence signal is dependent on the concentration of a coenzyme nicotinamide adenine dinucleotide (NADH) contained within the cells of the biomass in the aeration tanks. The fluorescence signal reflected from the biomass originates from the probe, which emits a UV light with a wavelength of 340 nm. The NADH has a unique property in that it fluoresces back at 460 nm when struck with a UV light of 340 nm. The level of NADH contained in each cell of the biomass changes with the metabolic state of the cells in the system. Under anaerobic conditions (no or low dissolved oxygen concentrations), the concentration of NADH is high, corresponding to a strong signal; conversely, when aerobic conditions prevail (higher dissolved oxygen concentrations), the NADH reading is low. Under anoxic conditions, the signal is somewhere in between.

Monitoring microbiological activity can be useful in optimizing the operation of an activated sludge system, operating in a simultaneous nitrification and denitrification mode, detecting influent toxicity or inhibition, detecting a potential slug loading into the facility, and for use with a dissolved oxygen probe for blower aeration automated control.

4.0 PHASE 2 PROCESS EVALUATION—DATA ANALYSIS AND INTERPRETATION

4.1 Nitrification

The parameters critical to nitrification are related to operating conditions that favor or indicate the activity of nitrifying organisms. These conditions include the chemical environment, SRT, and performance indicators.

4.1.1 *Environmental Conditions*

Chemical environment conditions most affecting nitrification are

- Alkalinity and pH,
- Temperature,
- Dissolved oxygen, and
- Inhibitory compounds.

The pH in the bioreactor can greatly affect nitrification. Low pH can occur when the alkalinity concentration is insufficient to buffer the loss of alkalinity resulting from nitrification and tends to inhibit nitrification. The pH and alkalinity of the incoming wastewater affect the pH and alkalinity in the aeration basin, where nitrification occurs. An important factor to consider regarding the effects of pH on the nitrification process is the degree of acclimation that the particular system has achieved. Wide swings in pH have been demonstrated to be detrimental to nitrification performance, although acclimation generally allows satisfactorily performance with consistent pH control within the range of 6.5 to 8.0 standard pH units. It is generally recommended that sufficient alkalinity be present through the reactors by maintaining a minimum effluent alkalinity of at least 50 and, preferably, 100 mg/L. Low pH inhibition of nitrification is most common in the following instances:

- Nitrification-only (no denitrification) facilities with low influent alkalinity and
- Continuously aerated aerobic digesters operated at SRT exceeding 15 days.

The growth rates of *Nitrosomonas* and *Nitrobacter* are particularly sensitive to the liquid temperature in which they live. Nitrification occurs in wastewater temperatures from 4 to 45 °C (39 to 113 °F), with an optimum growth rate occurring in the temperature range of 35 to 42 °C (95 to 108 °F) (U.S. EPA, 1993). However, most WRRFs operate with a liquid temperature between 10 and 25 °C (50 and 77 °F). It is generally recognized that the nitrification rate doubles for every 8 to 10 °C rise in temperature. The effect of temperature on the nitrification rate is shown in Figure 3.2 in Chapter 3.

Nitrifiers are only able to function under aerobic conditions. Consequently, the dissolved oxygen concentration in the bulk liquid can have a significant effect on the nitrifier growth rate. The value at which the dissolved oxygen concentration reduces the rate of nitrification varies on a site-specific basis depending on temperature, organic loading rate, SRT, and diffusional limitations. It has generally been accepted that a bioreactor dissolved oxygen concentration of greater than 2 mg/L is required to ensure full nitrification. However, with greater emphasis on the reduction of energy use in WRRFs, many facilities are capable of achieving full nitrification while operating at dissolved oxygen levels lower than 2 mg/L.

4.1.2 *Solids Retention Time*

Nitrifying bacteria include species that do not grow as quickly as bacteria that oxidize organic material (COD or BOD). A sufficiently long aerobic SRT is needed for a stable, adequate nitrifier population. Nitrification tends to be an "all or nothing" process, meaning that effluent ammonia concentrations will quickly rise once favorable conditions for nitrification are not met. *Washout SRT* is a term applied to the SRT at which no nitrifying organisms are present and no nitrification is occurring. For a more detailed discussion of this topic, refer to Section 2.0 of Chapter 3.

The target operating aerobic SRT for a BNR facility is determined by applying a peaking factor to the washout SRT, with a peaking factor of 2 to 3 being most common. Highly variable influent wastewater characteristics; intermittent internal facility recycles; stringent permit limits; intermittent, manually controlled operation; and smaller process tanks warrant greater peaking factor values. Highly variable influent ammonia-nitrogen loading may also suggest that swing zones be used to quickly increase aerobic SRT when needed (refer to Chapter 6). Consistent influent and internal facility recycles; less frequent (i.e., moving annual average) permit requirements; well-automated processes; and facilities with larger, more forgiving process tanks or equalization allow lower peaking factor values.

4.1.3 *Performance Indicators*

The profile of nitrogen species ammonia, organic nitrogen, nitrite, and nitrate throughout the treatment process indicates nitrification performance. Raw wastewater will typically have about 80% ammonia and 20% organic nitrogen. It is typically assumed that all but 1 to 2 mg/L of organic nitrogen will be converted to ammonia in the activated sludge process. There is typically little to no nitrate or nitrite in raw wastewater. Ammonia is converted to nitrate in the nitrification process in the aeration basins and nitrate is converted to nitrogen gas in the denitrification process in the anoxic basins.

Nitrite typically does not accumulate in large concentrations in biological treatment systems under stable operation because the maximum growth rate of *Nitrobacter* is significantly higher than that of *Nitrosomonas*. As a result, the growth rate of *Nitrosomonas* generally controls the overall rate of nitrification. However, facilities that transition into and out of nitrification may experience a condition where the growth rate of *Nitrobacter* is not able to keep up with that of *Nitrosomonas*, and effluent nitrite concentrations increase. This condition is known as "nitrite lock" and results in a significant increase in effluent chlorine demand (5.1 mg chlorine/mg nitrite). In situations where effluent permits contain both seasonal nitrification and seasonal disinfection requirements (using chlorine), it is advisable to establish complete nitrification before the start of chlorination.

4.2 Denitrification

Parameters critical to denitrification are related to operating conditions that favor or indicate the activity of denitrifying organisms (heterotrophic organisms in an anoxic environment). These conditions include the chemical environment, SRT, and performance indicators.

4.2.1 *Environmental Conditions*

The environment for denitrification is anoxic, that is, it lacks molecular oxygen (i.e., dissolved oxygen). The important parameters are pH, temperature, dissolved oxygen, ORP, nitrate-nitrogen, and COD or BOD. While optimum denitrification occurs within a narrow pH range, denitrification is significant unless pH is lower than pH 6 or higher than pH 8. Denitrification can be affected by low temperature as a result of lowered activity. Because the denitrifiers are capable of using both dissolved oxygen and nitrate as the electron acceptor and the use of dissolved oxygen requires less energy than nitrate, denitrification is inhibited by the presence of dissolved oxygen. The organic concentration, reported as COD or BOD, needs to be adequate for the desired amount of denitrification to occur. The minimum ratio of BOD to TKN for good denitrification is 4.5 to 1, with 6 to 1 being preferred.

4.2.2 *Solids Retention Time*

Denitrifying bacteria include many species, such as those that can use dissolved oxygen or nitrate for metabolism. Therefore, anoxic SRT need not be equal to aerobic SRT for nitrification.

4.2.3 *Performance Indicators*

The nitrate profile throughout the treatment process will indicate denitrification performance. Using the MLE process described in Chapter 6 as an example, a properly operated MLE process should have little to no nitrate in the anoxic reactor effluent because a fraction of the nitrate will not be recycled from the aerobic reactor to the anoxic reactor.

4.3 Enhanced Biological Phosphorus Removal

Parameters critical to EBPR are related to operating conditions that favor or indicate the activity of PAO organisms. These conditions include the chemical environment, SRT, and performance indicators.

4.3.1 *Environmental Conditions*

The environment for EBPR is anaerobic. The important parameters are temperature, dissolved oxygen, ORP, nitrate-nitrogen, and the VFA fraction of COD or BOD. Operation of EBPR is affected by temperature because the ability of PAOs, glycogen-accumulating organisms, and other heterotrophic microorganisms to out-compete each other for organics varies with temperature (refer to Chapter 7 for more information). Dissolved oxygen and nitrate input to anaerobic basins must be minimized to optimize the uptake of VFAs by PAOs. Recycle streams are the main sources of dissolved oxygen (internal recycles) and nitrate-nitrogen (RAS) input to anaerobic basins. Sufficient VFA should be available for the desired level of EBPR.

Oxidation–reduction potential can be used as a quick indicator of anaerobic basin conditions. For example, low ORP is an indicator of conditions that favor EBPR and higher ORP would be a reason to look at other indicators, such as dissolved oxygen, nitrate-nitrogen, or toxicity. In the anaerobic stage, ORP can also go too low, resulting in reduction of sulfate to sulfide with associated odors and bulking in the system. Chapter 13 contains a detailed description of operations based on ORP.

4.3.2 *Solids Retention Time*

Aerobic SRT and total system SRT affect EBPR rather than an "EBPR" SRT. Anaerobic SRT of 1 to 2 days is sufficient for EBPR. A long SRT results in lower mass removed in WAS, which lowers the amount of total phosphorus removed through biomass. A long SRT will also lower the EBPR rate as a result of more endogenous activity of the biomass and less storage capability within PAOs. Long anoxic SRT, particularly in postanoxic basins, in five-stage Bardenpho or similar process configurations, can cause secondary release of phosphorus by PAOs when nitrate-nitrogen levels are low. Contact time under anaerobic conditions is important to produce VFAs, which is the preferred food source of EBPR organisms. This contact time is determined by the flowrate entering the process and the anaerobic volume that is provided. In some facilities, VFAs are formed in the collection system and no contact time or only a short contact time is needed to produce adequate VFAs for EBPR.

4.3.3 *Performance Indicators*

The orthophosphorus profile throughout the treatment process will indicate EBPR performance. Higher phosphorus release in the anaerobic basin is a good indicator, while low phosphorus release may indicate low VFA in the facility influent or nitrate or dissolved oxygen input to the anaerobic basin. Phosphorus concentration in the aeration basin indicates the efficiency of phosphorus uptake by PAOs. Increased phosphorus levels in anoxic basins and secondary clarifier effluent indicate secondary phosphorus release, which may occur when dissolved oxygen and nitrate are absent and bacterial decay is occurring.

4.4 Chemical Phosphorus Removal

The basic principle of chemical phosphorus removal is precipitation followed by solids removal. The critical parameters for chemical phosphorus removal are physical and chemical and are related to the type and dose of chemical being added and to sufficient solids removal or filtration. Parameters to monitor include

- Total phosphorus and orthophosphorus influent design loadings and target removal rates;
- Chemical injection location, dispersion, and mixing;
- Chemical dosing and usage;
- Alkalinity and pH control; and
- Sedimentation and filtration performance.

Metal salts containing aluminum, iron, or calcium are commonly used to form insoluble precipitates with orthophosphate and other reactive phosphorus and to turn soluble phosphorus into a solid for removal. Metals salts used for chemical phosphorus removal consume alkalinity and can lower pH. It is important to monitor alkalinity consumption and influent alkalinity to determine if supplemental alkalinity is required.

After phosphorus is removed from the liquid stream, it is important to monitor the phosphorus mass flow in other areas in the facility because sidestreams containing high concentrations of phosphorus that return to the facility can be detrimental to the treatment process. Chemically bound phosphorus is removed from the liquid phase in the following three primary forms:

- Primary sludge,
- Secondary sludge, and
- Filter backwash.

Sidestream flows and phosphorus loadings from the handling of the aforementioned processes need to be properly managed and equalized so that high loads of phosphorus are slowly added back to the secondary process.

4.4.1 *Environmental Conditions*

Environmental conditions for chemical phosphorus removal are related to the physical process configuration and include four key parameters: injection dose, injection location and mixing, alkalinity, and solids removal performance. The chemical dose is unique to each chemical and is dependent on the target total phosphorus to be removed. After the dose has been selected, the chemical injection point should be placed far enough upstream of the sedimentation (primary or secondary clarifiers) and/or effluent filtration process to allow enough contact time for the chemical to react with the reactive phosphorus. The final condition requires successful performance of the solids removal or filtration process to remove the phosphorus that has been chemically bound as a solid.

4.4.2 *Performance Indicators*

The main performance indicator of chemical phosphorus removal is simply a measure of whether the phosphorus removed from the treatment process meets the permit objective. Alkalinity consumption should also be monitored in relation to the permit and process requirements of the treatment facility. Chemical dosage of both the metal salt and alkalinity supplementation should be closely monitored to minimize operation costs.

5.0 USE OF PROCESS MODELING FOR OPTIMIZATION

5.1 Introduction

Operators are now using models and simulators to understand nutrient removal processes, evaluate new designs, and to optimize and troubleshoot. Models and simulators are becoming easier to understand and more user friendly. This chapter briefly describes what they are and how they can be used. More detailed information on this topic can be found in *An Introduction to Process Modeling for Designers* (WEF, 2009).

5.2 History and Development of Models for Biological Nutrient Removal

Activated sludge systems become more complex as their function is expanded from carbonaceous removal alone to include nitrification, denitrification, and biological phosphorus removal. Typically, a nutrient removal system involves multiple reactors, some that are aerated and some that are not, and the internal circulation of mixed liquor between reactors. The number of biological reactions and the number of compounds involved in the process also increase correspondingly. This is because a nitrification and denitrification, biological phosphorus removal process involves three separate groups of microorganisms: polyphosphorus heterotrophs, non-polyphosphorus heterotrophs, and nitrifying autotrophs. The microorganisms operate on a large number of chemical compounds in three distinct environmental regimes: aerobic, anoxic (where nitrate is present but there is no dissolved oxygen), and anaerobic (where there is neither nitrate nor dissolved oxygen). These features make for complex behavior, which has increased the level of difficulty in design, operation, and control.

Given this complexity, the performance of any proposed system design can be determined only by experimentation (i.e., in pilot facilities) or by a mathematical model that simulates the behavior accurately. Comprehensive experimentation to evaluate the influence of a wide range of parameters is costly and time consuming. Increasingly, mathematical modeling of system behavior is being used as a tool to facilitate design and evaluate operation.

5.3 Description of Available Process Simulators

Simulators are computer- or spreadsheet-based programs used to solve the various established equations that have been developed to simulate the response of a given activated sludge system to changes in various parameters. Typically, this is done in two steps. First, the reactor configuration and the flow scheme must be specified, including the reactor sizes, influent characteristics, recycle flowrates, wastage rate, and other specifics. After this information is fixed, it is possible to perform mass balances over each reactor (or reactor zone) for each model. These mass balances constitute the state equations that relate the dependent variables (compound concentrations) to the independent variables, such as reactor volume. The mass balances form a set of simultaneous equations, which, when solved, characterize the system behavior. The simultaneous solution provides values of the concentrations in different reactors and time. In this way, the change in concentrations throughout the system is related to the input and output and conversion processes occurring within the system.

Simulating the response of a nutrient removal system based on a compressive biological model is mathematically complex and is typically achieved with a computer program. A simulation program is useful for the following reasons:

- System analysis and optimization—if a system model provides accurate predictions of response behavior, then these predictions can be compared to observed responses in analyzing the operation of existing systems. Any discrepancies can be useful in identifying problems in operation. An accurate model can also be used to optimize performance of existing systems. Various operating strategies can be proposed and tested rapidly, without having to resort to potentially difficult practical evaluation; and
- System design—a simulation program does not design a system directly. However, a simulation program can be a useful tool for the design engineer to evaluate proposed system designs or improvements rapidly. In addition, a dynamic model can provide valuable design information that has often only been available through empirical estimates. For example, a parameter such as peak oxygen demand can be obtained directly from the simulation program run under time-varying input patterns. This means that peak aeration capacity can be quantified accurately.

It is important to emphasize that models are not perfect. This is particularly true for nutrient removal systems, in which there are many unresolved technical aspects that are subject to ongoing research. Perhaps the biggest limitation is the expense and effort required for comprehensive influent wastewater characterization. Furthermore, the nitrifier growth rate is a parameter with high variability and, therefore, using the default value may not give accurate model predictions. The user must understand the kinetics of the process and how the various coefficients relate to the prediction of BNR performance.

5.4 Use of Simulators to Optimize Facility Operations

Ease of simulator use is dependent on program configurations. Although some are relatively easy, they require training to be used effectively. Model use also requires an understanding of the theory of nutrient removal, including kinetic and stoichiometric relationships. Some terms may be difficult for operators to understand, such as *half-saturation constant* and *specific growth rates*; however, even without fully comprehending their meanings, operators can use simulators to optimize facility performance. Developers of these software models typically include extensive training as part of the purchase agreement.

Developing data for use with these models may be expensive, time consuming, and somewhat complicated. The facility laboratory or a contract laboratory must determine both soluble and particulate components of COD and TKN. For COD, the soluble and particulate portions are further characterized to determine biodegradable and non-biodegradable fractions.

The total organic nitrogen fraction of TKN is further characterized as soluble and particulate biodegradable and soluble and particulate non-biodegradable organic nitrogen. Procedures for these tests are available in the literature and through each software provider. A review of these procedures helps evaluate whether to make these determinations in house or if outsourcing is more appropriate.

As mentioned previously, engineers and operators use simulators to assist in the design of nutrient removal systems. Another important function of the simulator is to help troubleshoot the process. The following are some troubleshooting examples:

- Models can be used to evaluate changing operational parameters such as RAS rate, mixed liquor recycle rate, dissolved oxygen concentrations in the bioreactor, and distribution of influent and recycles at various points within the bioreactor;
- Influent wastewater characteristics often change from those used during design, as industry leaves or a new industry comes to town, for example. An operator can input the new characteristics to the simulator to determine how those changed characteristics have affected predicted effluent quality. Through the simulator, the operator can decide whether to change operational parameters, to take tanks offline or to add tanks (if no remedies exist within the existing facilities), or to request a capital expenditure for improvements;
- The simulator can be used to understand the effect of taking tanks out of service for maintenance. An operator can evaluate what time of year is the best for preventive maintenance by entering in various temperatures to determine which temperatures favor the available volume. The operator can also determine if methanol addition would attenuate the effects of the reduced volume;

- The potential for an existing facility to meet stricter effluent discharge requirements can be evaluated by operating at a higher mixed liquor suspended solids concentration or a longer SRT. It is important to note that models do not predict sludge settling characteristics and how they are affected as environmental conditions change. Thus, model results alone may not result in meaningful effluent quality predictions unless sludge settleability is also addressed;
- Simulators give tools to the operator to understand what the design engineer is proposing, to help meet permits under all conditions, improve the economics of operating a BNR facility, and to understand exactly what is happening within these processes. Models are not easy to use and require not only training, but also knowledge of the kinetics of the biological system. However, these simulators are improving continuously. Indeed, modeling and simulation are assuming a prominent role in nutrient removal system design and will be important for operation and control.

6.0 TROUBLESHOOTING

Refer to Appendix A for an optimization and troubleshooting guide.

7.0 REFERENCES

Abu-ghararah, Z. H.; Randall, C. W. (1991) The Effect of Organic Compounds on Biological Phosphorus Removal. *Water Sci. Technol.*, **23**, 585–94.

American Public Health Association; American Water Works Association; Water Environment Federation (2012) *Standard Methods for the Examination of Water and Wastewater*, 22nd ed.; American Public Health Association: Washington, D.C.

Bott, C.; Parker, D. S. (2011) *Nutrient Management Volume II: Removal Technology Performance and Reliability;* Water Environment Research Foundation: Alexandria, Virginia.

Ekama, G. A.; Marais, G. v. R.; Siebritz, I. P.; Pitman, A. R.; Keay, G. F. P.; Buchan, L.; Gerber, A.; Smollen, M. (1984) *Theory, Design and Operation of Nutrient Removal Activated Sludge Process;* Water Research Commission: Pretoria, South Africa.

Mamais, D.; Jenkins, D.; Pitt, P. (1993) A Rapid Physical-Chemical Method for the Determination of Readily Biodegradable Soluble COD in Municipal Wastewater. *Water Res.*, **27**, 195–197.

Metcalf and Eddy, Inc. (2003) *Wastewater Engineering Treatment and Reuse,* 4th ed.; Metcalf and Eddy: New York.

Pagilla, K. (2007) Organic Nitrogen in Wastewater Treatment Plant Effluents. Presented at the Water Environment Research Foundation and Chesapeake Bay

Science and Technology Advisory Committee Workshop; Baltimore, Maryland, Sept 27–28; p 97; Chesapeake Research Consortium, Inc.: Edgewater, Maryland.

Randall, C. W.; Barnard, J. L.; Stensel, H. D. (1992) *Design and Retrofit of Wastewater Treatment Plants for Biological Nutrient Removal;* Technomic Publishing: Lancaster, Pennsylvania.

Stensel, H. D.; Bronk, D.; Khan, E.; Love, N.; Makinia, J.; Neethling, J. B.; Pelligrin, M.; Pitt, P.; Sedlak, D.; Sharp, R.. (2008) *Dissolved Organic Nitrogen (DON) in Biological Nutrient Removal Wastewater Treatment Processes;* Water Environment Research Foundation: Alexandria, Virginia.

U.S. Environmental Protection Agency (1993) *Nitrogen Control Manual;* EPA-625/R-93-010; U.S. Environmental Protection Agency: Washington, D.C.

U.S. Environmental Protection Agency (2008) *Energy Management Guidebook for Wastewater and Water Utilities;* U.S. Environmental Protection Agency: Washington, D.C.

U.S. Environmental Protection Agency (2005) *Guidelines Establishing Test Procedures for the Analysis of Pollutants Code of Federal Regulations;* Part 136, Title 40; U.S. Environmental Protection Agency: Washington, D.C.

Water Environment Federation (2009) *An Introduction to Process Modeling for Designers;* Manual of Practice No. 31; Water Environment Federation: Alexandria, Virginia.

Water Environment Federation; American Society of Civil Engineers; Environmental and Water Resources Institute (2009) *Design of Municipal Wastewater Treatment Plants,* 5th ed.; Manual of Practice No. 8; ASCE Manuals and Reports on Engineering Practice No. 76; McGraw-Hill: New York.

Water Environment Research Foundation (2003) *Methods for Wastewater Characterization in Activated Sludge Modeling;* Water Environment Research Foundation: Alexandria, Virginia.

Chapter 12

Recycle Streams Management

Joseph A. Husband, P.E., BCEE, and
Scott Phipps, P.E.

1.0 INTRODUCTION

This section addresses concerns associated with recycle streams and provides potential management strategies for the recycle of solids, nutrients, and other constituents from solids management processes. These recycle streams can represent a significant nutrient load on a water resource recovery facility's (WRRF's) nutrient removal process and must be managed to ensure compliance with effluent quality goals and to improve treatment efficiency.

In this chapter, *recycle streams* refer to flows typically generated from solids processing units such as thickening, stabilization, and dewatering units. *Sidestream flows* refer to the effluent of add-on treatment processes that modify the quality of the recycle stream, whether via physical and chemical processes, biological processes, or both.

2.0 ISSUES AND CONCERNS

2.1 Introduction

Key parameters in recycle streams that can affect biological nutrient removal (BNR) treatment processes are solids, nutrients, pH, and alkalinity. Internal facility recycles such as those from solids dewatering processes, supernatant from sludge digestion processes, and overflow and underflow from thickening systems can introduce tremendous variability in treatment facility solids and nutrient loadings on the BNR treatment process.

2.2 Suspended Solids

Recycle solids loads from solids thickening and solids dewatering processes can significantly increase solids loading returned to the BNR treatment process. If discharged to the BNR treatment process (either directly discharged or indirectly discharged at facilities without primary clarification), there can be a reduction in the solids retention time (SRT) within the biological treatment process, thus reducing the microorganism population and treatment effectiveness. Accumulation of inert solids from recycle streams limits the maximum SRT achievable for a desired mixed liquor suspended solids (MLSS) concentration. Lower SRT can compromise nitrification, particularly during the winter because of colder wastewater temperatures. The effect of these internally recycled solids, particularly for facilities without primary clarification, can have a significant effect on the BNR treatment process.

2.3 Nutrients

2.3.1 *Nitrogen*

2.3.1.1 *Recycle of Ammonia*

Anaerobic processes, including fermentation and thermal stabilization processes that cause cell rupturing, will result in significant ammonia concentrations in the associated recycle streams (thickening and dewatering). The following are concerns of higher ammonia loading to the biological treatment process:

- Adds increased aeration demands, especially if dewatering occurs during the daytime when diurnal influent ammonia loads are also highest;
- Increase in the ammonia concentration being discharged to the BNR treatment process that can translate into higher effluent nitrogen concentrations;
- Increased short-term (slug) peak nitrogen loading caused by intermittent dewatering operation; and
- Reduces the ratio of carbon to nitrogen, which stresses the denitrification process in the biological treatment process.

2.3.1.2 *Recycle of Oxidized Nitrogen*

Aerobic digesters that are operated without significant anoxic (nonaerated) periods for partial denitrification can result in a significant load of oxidized nitrogen discharged to the BNR treatment process. The elevated oxidized nitrogen loading recycled to the biological treatment process can

- Reduce the ratio of carbon to nitrogen, which may negatively affect the denitrification capacity of the BNR treatment process;
- Increase the effluent oxidized nitrogen concentration because of overloading of the anoxic zones implemented for denitrification in the BNR treatment process; and
- Increase the oxidation–reduction potential of anaerobic zones (less negative) because of higher nitrate concentration.

2.3.2 *Phosphorus*

Phosphorus recycled from anaerobic digestion processes can significantly affect the performance and operation of biological and chemical phosphorus removal processes at BNR treatment facilities. Concerns with recycled phosphorus are

- Slug loading of phosphorus because of intermittent dewatering operation,
- Decrease in the ratio of carbon to phosphorus for the enhanced biological phosphorus removal (EPBR) process,

- Increase in metal salt demand for chemical phosphorus removal, and
- Increased potential for struvite formation.

Struvite ($MgNH_4PO_4$) formation can occur because all of the required constituents (ammonia, phosphorus, and magnesium) are present following anaerobic digestion. Struvite formation occurs when the pH increases in the digester and/or subsequent piping and process handling equipment because of the release of carbon dioxide. While struvite is the most common chemical precipitate formed in the digester environment, other compounds such as bushite ($CaHPO_4$ $2H_2O$) and vivanite [$Fe_2(PO_4)$ $8H_2O$] may also form if favorable conditions are encountered. These favorable conditions include high calcium or reduced iron concentration.

3.0 RECYCLE STREAM CHARACTERISTICS

3.1 Potential Concentrations from Various Solids Management Processes

Recycle streams from BNR solids processes typically contain biochemical oxygen demand (BOD), total suspended solids (TSS), nitrogen, and phosphorus. The characteristics of recycle streams can vary significantly depending on the biological treatment process that is implemented. In general, recycle streams of concern are from thickening, solids stabilization, and dewatering activities. The characteristics of these recycle streams cannot be generalized because of the wide variability exhibited within an individual treatment facility and the strong influence of many site-specific factors, including the following:

- Facility influent characteristics;
- Mainstream unit processes that are implemented and corresponding performance;
- Solids stream unit processes that are implemented and corresponding performance;
- Nitrogen and phosphorus content of the solids;
- Operating schedule for solids handling processes, particularly dewatering; and
- Extent of other unintentional internal removal mechanisms (struvite formation).

The recycled BOD and TSS loads resulting from most conventional sludge processing operations are modest and generally do not pose any treatment challenges. Table 12.1 provides typical BOD and TSS concentration ranges for sludge processing recycle streams.

TABLE 12.1 Biochemical oxygen demand and TSS levels in sludge processing sidestreams (adapted from U.S. EPA, 1987).

Source	Recycle stream	BOD_5, mg/L	TSS, mg FL
Thickening			
Gravity thickening	Supernatant	100 to 1200	200 to 2500
DAF	Subnatant	50 to 1200	100 to 2500
Centrifuge	Centrate	170 to 3000	500 to 3000
Stabilization			
Aerobic digestion	Decant	100 to 2000	100 to 10 000
Anaerobic digestion	Supernatant	100 to 2000	100 to 10 000
Composting (static pile)	Leachate	2000	500
Wet air oxidation	Decant liquor	3000 to 15 000	100 to 10 000
Incineration	Scrubber water	30 to 80	600 to 10 000
Dewatering			
Belt filter press	Filtrate	50 to 500	100 to 2000
Centrifuge	Centrate	100 to 2000	200 to 20 000
Sludge drying beds	Underdrain	20 to 500	20 to 500

Characteristics of recycle streams can vary significantly depending on the treatment process. Therefore, facility operators should review their facility operation and develop real-time data. Additional care must be taken if and when future modifications to the treatment process are being considered. For example, measuring existing phosphorus recycle concentrations before EBPR will not reflect the additional phosphorus that will be in the recycle stream following implementation of EBPR. The following guidelines can be used to estimate recycle nitrogen and phosphorus loads:

- Developing a mass balance of the wastewater treatment facility (liquid and solids) is critical to understanding the magnitude of the flows and loads contained in the recycle streams. Even small flows with high concentrations of a particular wastewater constituent can have a significant effect on the biological treatment process. Additional sampling of internal process and recycle streams may be required for a period of time to develop a complete mass balance. This effort is critical to developing an understanding of the recycle streams and in the formulation of an effective strategy to manage the recycle streams;
- Mass balances should be developed for average conditions with adjustments to assess the effect of operating schedules and peak conditions. Both

present and projected mass balances should be performed and examined. It is important to note that a mass balance provides a "snapshot" of operating conditions and, therefore, cannot be used to predict the dynamic behavior of the biological treatment system and the effects of the recycle streams;

- Nitrogen release during digestion is directly related to volatile solids reduction. Nitrogen that is released during the digestion process is equivalent to approximately 8 to 10% of the volatile suspended solids (VSS) removed via the digestion process; and
- Phosphorus release from EBPR sludges represents approximately 4 to 8% of VSS removed during digestion, whereas the potential phosphorus release from non-EBPR sludges represents approximately 1 to 3% of the digester influent VSS removed during the digestion process.

3.1.1 *Thickening*

In general, primary and waste activated sludge (WAS) should be gravity thickened separately to minimize biological activity in the sludge thickening process to reduce the effect of nutrient recycle. Mechanical thickening, such as via gravity belt or dissolved air flotation, may be a viable thickening approach that may minimize biological activity when operated continuously and immediately after mixing. Sludge thickening processes can significantly affect the biological treatment process with elevated solids recycle loads because of poor capture efficiency. The recycle solids can increase the influent inert solids when directed to the BNR treatment process. These inert solids decrease the active fraction at a given MLSS concentration, which can negatively affect treatment performance. Recycle of microorganisms from biological sludge thickening processes can cause treatment process upsets, which depend on the discharge location and inert solids load of the recycle stream.

Nutrient release in thickening processes is typically not significant when primary and WAS sludges are thickened separately while minimizing the anaerobic detention time. For EBPR facilities, co-mixing of primary and WAS must be avoided to prevent the release of phosphorus from WAS. Anaerobic conditions in the thickening device can be similar to those in the anaerobic zone of an EBPR treatment process, which causes a significant release of phosphorus under anaerobic conditions.

3.1.2 *Stabilization*

3.1.2.1 *Aerobic Digestion*

Aerobic digestion processes have an extremely long SRT (typically greater than 40 days) to achieve biosolids designation as per 40 CFR Part 503. This is more than sufficient time for complete nitrification of the ammonia released via the breakdown of organic nitrogen in the digestion process. Therefore, aerobic digestion processes that are operated continuously and fully aerobic can produce a significant loading of oxidized nitrogen (nitrite and/or nitrate). This oxidized nitrogen load will be present in the aerobic digester decant recycle stream or subsequent dewatering recycle stream.

As mentioned previously, these oxidized nitrogen recycle streams can negatively affect the anaerobic zones required for EBPR or overwhelm the anoxic zones used for denitrification. Some facilities with aerobic digestion processes have incorporated unaerated (anoxic) periods to reduce oxidized nitrogen recycle loadings and electrical power requirements associated with the aerobic digestion process. Additionally, this anoxic period allows potential alkalinity recovery because of denitrification of the oxidized nitrogen in the aerobic digestion process.

Another strategy is to recycle the elevated oxidized nitrogen recycle streams to the preliminary treatment processes. This provides odor control benefits because it reduces hydrogen sulfide production and can cause sulfide oxidation through autotrophic denitrification.

3.1.2.2 *Anaerobic Digestion*

Anaerobic digestion, through fermentation processes, will release high concentrations of ammonia because of the breakdown of organic nitrogen. Typically, the ammonia concentrations associated with anaerobic digestion process recycle streams will be greater than 85% of the total nitrogen present. High levels of phosphorus may also be released, particularly at EBPR facilities. These elevated phosphorus concentrations are present in EBPR because WAS contains high levels of stored polyphosphates (4 to 10%, by dry weight). A portion of the stored polyphosphates will be released from WAS when it is subjected to anaerobic conditions in the digestion process. Ultimately, the actual ammonia and phosphorus concentrations in the anaerobic digester will depend on influent solids characteristics and the volatile solids destruction performance of the anaerobic digestion process.

3.1.3 *Dewatering*

Dewatering of unstabilized sludges will have insignificant nutrient recycles, with the possible exception of EPBR WAS that have been subjected to anaerobic conditions in which significant phosphorus release can occur. Dewatering recycle streams from stabilized sludges (both aerobic and anaerobic digestion processes) represents more of a concern because of the potential significant nutrient concentrations associated with these recycle streams. The following two dewatering process examples reflect those dewatering processes without (or with minimal) dilution water vs dewatering processes with significant dilution water.

3.1.3.1 *Centrifuges*

Centrifuges can dewater influent feed solids to between approximately 20 to 35% total solids. The soluble nutrient concentration of the centrate is equivalent to the stabilization process soluble nutrient concentration because makeup water is not used in the dewatering centrifuge operation. Accordingly, the centrate will be similar in terms of soluble constituent characteristics of the upstream stabilization process, but the particulate constituent characteristics will vary from the upstream stabilization process.

3.1.3.2 Presses (With and Without Spray Water in Filtrate)

Presses, such as belt filter presses and screw presses, can have recycle streams (called *filtrate*) that are diluted by spray water that is used to clean the cloth or other screening material used for dewatering the stabilized solids. A number of belt filter presses now have the ability to isolate the filtrate discharged from the gravity and belt pressure zone from the belt cleaning operation. However, many treatment facilities direct all of this liquid (filtrate and spray water) to the internal drain system as a combined recycle stream. In many applications, the spray water can be equal in volume to the filtrate, thus diluting and cooling the filtrate recycle stream. As will be discussed in Section 4.0, add-on sidestream treatment processes can take advantage of the higher temperature and concentration of the filtrate, suggesting that separation of the makeup water from the filtrate is desirable.

3.2 Intermittent Versus Continuous Recycle Issues

Recycle streams containing released nitrogen and phosphorus are typically blended with the facility influent and recycled through the biological treatment process. These recycle streams are not always continuous and, in many facilities, occur intermittently. For example, sludge dewatering operations conducted 5 days per week, 8 hours per day, will increase the recycle nutrient loading rate to the biological treatment process by a factor of 4 to 5 vs a 7-day-per-week, 24-hour-per-day operation. Accordingly, significant nutrient concentration increases can occur during this intermittent discharge, which challenges the ability of the BNR treatment process to achieve effluent quality goals. Additionally, the time of day and discharge location of the recycle loadings to the BNR treatment process can significantly affect loading on the BNR treatment process. Figure 12.1 shows an example of three different modes of recycle loadings discharging directly to the BNR treatment process. The dashed line shows a fully equalized recycle loading (24-hour-per-day recycle of dewatering recycle stream) and the solid line shows the effect of dewatering 5 days per week between 0800 and 1600 hours; the dotted line shows the loading on the BNR treatment process when dewatering is performed 5 days per week between 2300 and 0700 hours. In this example, dewatering overnight provides the most equalized ammonia loading on the BNR treatment process. It also represents a cost savings (electricity) because this facility uses centrifuges for dewatering and thus saved significant costs by operating during nonpeak periods of electrical demand. However, there are two other considerations. The first consideration is that the carbon available in the

FIGURE 12.1 Centrate equalization.

raw wastewater will be lower during morning hours; thus, the ratio of carbon to nitrogen may not be favorable for good denitrification. The second consideration is if the centrate for the overnight dewatering operation is sent to the facility headworks, which will significantly increase the early morning recycle loads. For example, in a facility with primary clarifiers, the concentration in the primary clarifiers can increase significantly and be "washed out" during the early morning hours. This can cause a significant increase in ammonia loading to the BNR process.

This example illustrates that careful examination of each facility's operation and the potential effect of different dewatering schedules and operating schemes can be beneficial to the BNR treatment process. Monitoring of recycle loads and daily diurnal loadings of facility influent can provide valuable information on how to optimize recycle management. Similarly, facilities that accept significant septic tank loads and other high-nutrient wastes should consider the effect of this operation on recycle loads to the biological treatment process.

4.0 SIDESTREAM TREATMENT OPTIONS

This section provides a general overview of sidestream treatment processes and the significant benefits and tradeoffs of the different sidestream processes. There are numerous management and treatment options available to control the effect of recycle loading on the biological treatment process. Figure 12.2 shows various options available for managing nitrogen and phosphorus in dewatering recycle streams at the time of publication of this manual of practice. The process configuration that works best for a particular facility depends on multiple factors, including influent loading, effluent quality requirements, existing infrastructure and facilities, and the economics of treatment. Recycle streams discussed in this section are those for stabilization processes that generate significant nutrients (whether ammonia, phosphorus, or both) and are typically associated with anaerobic digestion processes. The key factor in sidestream treatment processes is taking advantage of the high loading and low volume of the recycle stream. The type and degree of sidestream treatment is facility specific and must always be weighed against alternatives. The remainder of this section describes principles behind the various sidestream technologies.

Although a detailed discussion of each process cannot be provided in this manual, a general description of these treatment options is included. Further details can be obtained from *Design of Municipal Wastewater Treatment Plants* (WEF et al., 2009), *Nutrient Removal* (WEF et al., 2010), and vendor and technology-owner Web sites. These technologies (some of which are proprietary) can be as simple as equalization tanks and as complex as physicochemical treatment. The applicability and feasibility of these processes for a particular facility depend on

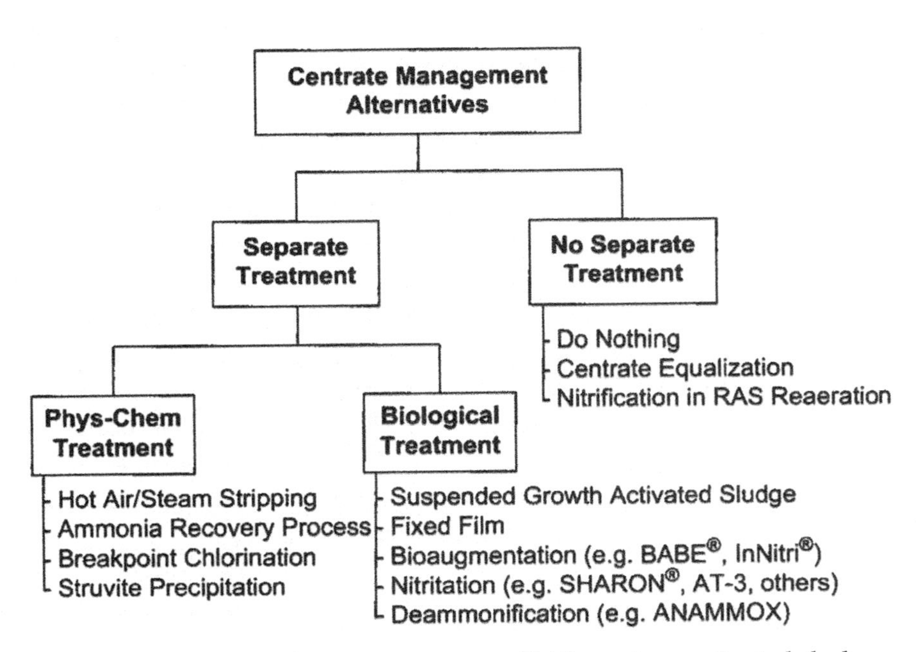

FIGURE 12.2 Overview of centrate processes (RAS = return activated sludge; and BABE = bioaugmentation batch-enhanced) (Constantine et al., 2005).

numerous factors such as size of facility, staffing numbers and capability, space availability, and effluent quality requirements.

4.1 Alternatives

4.1.1 *Recycle Equalization*

Recycle equalization is one of the first tools for mitigating the effect of recycle streams on the BNR treatment process. As shown in Figure 12.1, equalizing the dewatering recycle loading on the BNR treatment process may be achieved by modifying dewatering operations; however, this may not be possible because of labor availability. However, the incorporation of an equalization tank to hold the dewatering recycle stream volume is one of the simplest operating strategies that permits a controlled discharge back to the BNR treatment process at the most opportune time. A well-mixed tank (mechanical or hydraulic) will not only ensure a homogeneous mixture, but can reduce the potential for stripping of odors.

4.1.2 *Recycle Equalization and Semitreatment*

4.1.2.1 Solids Removal

The equalization tank should be designed to allow solids removal because solids in the dewatering recycle stream typically settle very rapidly because of the polymer present. This equalization tank can be mixed, but incorporating an operational strategy to operate unmixed for a portion of the day will allow the solids to settle in the equalization tank, thereby providing for solids removal. This mode of equalization tank operation is designed to reduce the solids recycled to the BNR treatment process.

4.1.2.2 Partial Nitrification

Some facilities have incorporated aeration of equalization tanks to allow for some nitrification to occur. Partial nitrification can occur in the equalization tank, with the maximum amount of nitrification being approximately 50% of the recycle stream ammonia load because of alkalinity limitations. However, aeration of the equalization tank can promote struvite formation because of the stripping of carbon dioxide from the liquid, thus causing an increase in pH and precipitation of struvite. Therefore, extensive coating of concrete wall, aeration diffusers, and other equipment has been observed.

4.2 Ammonia Removal (Ammonia Oxidation Only)

For facilities with ammonia and/or total nitrogen limits, the additional ammonia in the recycle stream will reduce the overall facility capacity (vs no recycle loading). Having a higher ammonia concentration in the BNR influent will require additional oxygen and time to achieve complete nitrification and, for total nitrogen removal processes, additional nitrate that must be removed. For those facilities that may be borderline in meeting effluent discharge limits, use of a sidestream nitrification process, particularly warm anaerobic digester dewatering liquor, represents a cost-effective approach to managing the higher ammonia loading.

The concept for a separate nitrification process is to nitrify the warmer recycle stream and discharge the oxidized nitrogen to the main facility. One example is the centrate treatment process at the 91st Avenue Water Pollution Control Plant in Phoenix, Arizona (Coughenhour II et al., 2010). The main facility BNR process is required to reduce a total nitrogen influent (with centrate) of 35 mg/L down to less than 8 mg/L because of the high concentration of centrate. Pilot studies were conducted that demonstrated that a separate complete-mix activated sludge facility with a detention time of approximately 1.8 days could oxidize more than 95% ammonia with the addition of alkalinity. Actual operation has demonstrated that nitrification to the alkalinity limit of the centrate was possible (no alkalinity addition), even with a basin pH of less than 6. In addition to reducing the ammonia loading on the biological treatment process, the discharge of oxidized nitrogen to the headworks reduced the cost of odor control chemicals (from $460 per day to $150 per day) because of the oxidation of influent hydrogen sulfide.

Another benefit of separate nitrification is the use of WAS from the centrate nitrification process as a seed source for nitrifiers to the nitrification reactors of the main facility; this process is referred to as *bioaugmentation*. A patented process by M2T Technologies, Lotepro Environmental Systems & Services, called In-Nitri® (Kos et al., 2000), uses WAS from the separate nitrification reactor to supplement the activated sludge system of the main facility with nitrifiers, which allows facilities to operate at lower SRTs while maintaining nitrification. Although the viability of nitrifiers grown in the warm recycle stream in colder wastewater (main facility BNR) has been questioned, research on laboratory-scale systems by the University of Manitoba, Winnipeg, Manitoba, Canada, indicated a slight decrease in nitrification vs theory when taking temperature correction into account.

There are some sidestream processes that send a portion of the return activated sludge (RAS) from the main biological treatment process to the sidestream process (e.g., BABE®). This offers the advantage of using nitrifiers already acclimated to conditions within the biological treatment process to continue to grow using ammonia in the recycle stream.

Another process that has gained widespread use (but may not be strictly considered a sidestream process) is performed using RAS from the main biological treatment process within the existing infrastructure of the main facility. Essentially, a portion or all of the RAS from the main facility biological treatment process is mixed with the recycle stream and aerated in the first pass of a step-feed and contact stabilization activated sludge facility. The advantages of this process are the relatively high concentration of MLSS (RAS concentration) and ammonia to increase nitrification population (bioaugmentation) under similar environmental conditions as those treating raw wastewater. In addition to the higher concentration of microorganisms and food (ammonia), the detention time in this reactor can be significantly longer than the remainder of the biological reactor, thus increasing the nitrification population and oxidation of ammonia.

This process of RAS reaeration and centrate treatment, known as BioAugmentation Reaeration, uses the first zone of the biological treatment tank to treat the high-strength ammonia recycle stream with recycled RAS sludge, while main process influent is introduced to the downstream zone (Parker and Wanner, 2007). Other variations include the common centrate and RAS reaeration basin process (CaRBB) (Luna et al., 2010) developed at the Robert W. Hite Treatment Facility in Denver, Colorado. The centrate generated in the combined solids handling facility is distributed equally to the head of each sludge reaeration tank (CaRRB unit) along with a portion of RAS. The CaRRB effluent (partially nitrified centrate plus RAS) is discharged into the upstream end of the main influent channel to be mixed with primary effluent. The portion of RAS not routed through CaRRB is sent to the head of the main influent channel. It should be noted that some facilities have installed a new, separate RAS reaeration tank before the main activated sludge tank to take advantage of this process.

4.3 Nitrogen Removal

There are numerous sidestream biological treatment processes available to remove nitrogen. Typically, these processes take advantage of the warm temperature and high ammonia concentration of the recycle stream to more efficiently remove nitrogen vs performing the same function in the main biological treatment process. Using a conventional approach, sidestream nitrification and denitrification (using an external carbon source) can achieve nitrogen removal in a smaller volume than returning the recycle stream to the main biological treatment process. However, this sidestream treatment scheme typically requires increased energy and external carbon requirements. The following subsections describe treatment processes that improve upon this "traditional" sidestream nitrogen removal process.

4.3.1 *Nitritation and Denitritation*

One approach is to stop nitrification at nitrite by promoting an environment that encourages ammonia-oxidizing bacteria (AOB) to produce nitrite (nitritation), but attempts to minimize the population of nitrite-oxidizing bacteria (NOB) that convert nitrite to nitrate. Together with denitritation (i.e., conversion of nitrite to nitrogen gas), this process can reduce the aeration requirement by approximately 25% and the carbon demand for denitrification by approximately 40% because denitrification is only from nitrite (NO_2) and not nitrate (NO_3). The primary operational strategy is to maintain a minimal SRT and a high temperature (25 to 35 °C). At this temperature, AOB, which have a faster growth rate than NOB, will survive in the system while NOB will be washed out of the system. Because denitritation can occur either via recycle or within the same reactor, a majority of the alkalinity demand (80 to 90%) for nitrification can be satisfied with little supplemental alkalinity addition. One of the earliest documented examples of the nitritation and denitritation process is the SHARON® (Single Reactor High Activity Ammonia Removal Over Nitrite) process developed by the Delft University of Technology, Delft, Netherlands, and Grontmij. Other nitritation and denitritation processes rely on controlling dissolved oxygen and pH conditions to promote AOB and to discourage NOB.

4.3.2 *Nitritation and Anammox Process*

Nitritation and anaerobic ammonium oxidation (Anammox) is a relatively new treatment process that offers significant advantages over other sidestream BNR processes. Anammox is achieved by a highly specialized group of bacteria belonging to the *planctomycete* group, which remove nitrogen by nitrifying ammonia to nitrogen gas using nitrite as an electron acceptor instead of oxygen while producing alkalinity (Strous et al., 1998; 1999), as follows:

$$\begin{aligned} &NH_3 + 1.32\ NO_2^- + 0.066\ HCO_3^- + 0.13\ H^+ \Rightarrow \\ &1.02\ N + 0.26\ NO_3^- + 0.066\ CH_2O_{0.5}N_{0.15}\ 2.03\ H_2O \end{aligned} \qquad (12.1)$$

Microorganisms responsible for this reaction occur in nature, where ammonia and low levels of nitrite coexist under anaerobic conditions. This process offers the following benefits over conventional nitrification and denitrification and nitritation and denitritation processes:

- Only approximately one-half of the ammonia needs to be converted to nitrite, thus reducing the energy required for aeration, and
- The resulting nitrite and remaining ammonia are converted to nitrogen gas, thus eliminating the need for an external carbon source for denitrification.

Challenges associated with this process include the following:

- Anammox bacteria have a slow growth rate, with a doubling of population in the range of 6 to 12 days. Startup of an Anammox process without seed sludge requires from 100 to 180 days. Accordingly, these microorganisms must be handled with care; and
- Anammox bacteria are reversibly inhibited by dissolved oxygen concentrations as low as 0.32 mg/L. In addition, there is wide acceptance in the literature that exposure to elevated nitrite is inhibitory and possibly toxic at concentrations as low as 100 mg/L as nitrite-nitrogen, whereas more recent studies indicate that this toxic limit is as high as 350 mg/L as nitrite-nitrogen (Rosenthal et al., 2009).

Several process configurations that combine nitritation and Anammox exist and can be categorized as having either "separate" or "integrated" nitritation and Anammox steps (Wett, 2005). In the first step of the "separate" process, a nitritation reactor is typically operated without addition of supplemental alkalinity, which results in conversion of approximately half of the ammonia in the dewatering centrate to nitrite. The effluent from the nitritation reactor, which typically is composed of approximately equal parts of ammonia and nitrite, is fed into the Anammox reactor, which is operated at relatively high temperatures (25 to 40 °C). The Anammox bacteria present in this second reactor convert ammonia and nitrite to nitrogen gas, typically without carbon addition.

In an "integrated" nitritation and Anammox process scheme, a sequencing batch reactor (SBR) is used with intermittent aeration to avoid a buildup of nitrite to toxic levels. In Europe, the DEMON process (SBR) has been used at more than 20 WRRFs to remove more than 80% of the total nitrogen from filtrate or centrate from solids dewatering. A pH-based control system determines the length of aeration intervals depending on the production of hydrogen ions (pH depression) or nitrite (Wett, 2005).

A newer, "integrated" nitritation and Anammox process system is one that includes the use of attached growth systems in moving bed reactors. The biological processes take place inside the biofilm developed on the media, which are typically plastic carriers. The outer layer of biomass is aerobic, providing the location for AOB to produce nitrite. Below this aerobic layer is the anoxic layer, where the Anammox bacteria use the nitrite produced by AOB to nitrify excess ammonia. Claimed advantages of this option include the ability to operate with a higher dissolved oxygen concentration within the reactor, potentially protecting the Anammox bacteria located under the aerobic film on the media. Further oxidation of nitrite to nitrate by NOB in the aerobic zone of the biofilm should be avoided to maximize the amount of nitrite available for the Anammox bacteria.

4.3.3 *Physicochemical Nitrogen Removal*

Ammonia stripping can be achieved by raising the pH of the recycle stream so that the aqueous ammonia exists in equilibrium with its gaseous counterpart in accordance with Henry's Law. When the pH exceeds 9.5, un-ionized ammonia prevails and can be stripped from the recycle stream. This requires a large volume of air to achieve high ammonia removal. In addition, the exhaust stream should be captured to eliminate odor problems. This requires condensing the exhaust gas and capturing the ammonia. Stripping via pH can cause scaling problems and is subject to freezing during cold weather.

A second option is to steam strip the ammonia from the recycle stream. Studies conducted by the New York City Department of Environmental Protection (Gopalakrishnan et al., 2000) indicated that steam stripping is more cost effective than hot air stripping and can achieve 70 to 90% removal of ammonia from the recycle stream. The adjustment of pH (typically 7.0 to 7.5) is not required in a stream stripper, thereby reducing scaling issues. The stripped ammonia associated with exhaust gas is sent to a distillation column, where the exhaust gas is cooled and a pure ammonia solution is formed. Clogging did occur in heat exchangers as a result of organic material, including hair, calcium, and magnesium in the recycle centrate stream. Pretreatment requirements for steam stripping include fine screening and suspended solids settling to prevent clogging issues.

A third option is to use a vacuum distillation process to remove most of the ammonia in the recycle stream. A proprietary vacuum distillation system (RCAST®) that uses a combination of physical separation methods has reportedly demonstrated the ability to remove ammonia from municipal centrate and filtrate (Orentlicher et al., 2009). This process includes pretreatment of the centrate (i.e., removal of hair and other material interfering with the mechanical process) and conversion of aqueous soluble ammonium ions to ammonia gas by elevation of pH before introduction to the RCAST® system. The RCAST® system liberates soluble ammonia gas from the recycle stream by increasing the temperature and lowering the operating pressure of the recycle stream. The liberated gaseous ammonia is entrained in a sulfuric acid solution and converted to highly soluble and stable ammonia sulfate.

4.4 Phosphorus Removal

Recycle streams from anaerobic digestion processes and dewatering can contain a significant amount of soluble phosphorus because of the hydrolysis of microorganisms and solids. Typically, the phosphorus concentration in microorganisms is approximately 2% on a dry-weight basis to satisfy metabolic requirements and is organically bound in the microorganisms. Elevated phosphorus concentrations between 4 and 10% on a dry-weight basis can result from EBPR treatment processes.

4.4.1 *Chemical Precipitation*

Highly concentrated orthophosphate recycle streams can be managed using metal salts to precipitate the phosphorus. This process is described in detail in Chapter 8. The advantage of adding metal salts to the more concentrated recycle stream is that the ratio of metal to phosphorus is closer to theoretical demand because of the higher concentration of orthophosphorus, lower pH (which must be managed for subsequent treatment), and lower competing reactions with other constituents. While use of metal salts is a method to precipitate and capture the recycle stream phosphorus, this practice requires a significant amount of chemicals and increased dewatering volume because of the increased chemical solids generated.

4.4.2 *Struvite Precipitation*

Struvite (magnesium ammonium phosphate [$MgNH_4PO_4$]) is typically a nuisance for WRRFs that operate anaerobic digestion because it precipitates on piping and pumps from the anaerobic digester effluent through the dewatering facilities. Once formed, it is an extremely hard substance that can cause piping to clog without an easy removal method. Often, facilities will need to replace piping (vs the expense of cleaning the pipe). However, struvite is also an excellent fertilizer with the ability to release nutrients over an extended period of time, reducing costs to the agricultural user while also reducing nutrient runoff from farmland. Extensive work on physicochemical treatment processes was carried out in the late 1980s and early 1990s with processes to induce struvite precipitation.

Struvite precipitation had limited applicability because orthophosphate and magnesium often are present in relatively low levels in anaerobically digested dewatering recycle streams. Typically, these concentrations are approximately 10% of the mass loading of the ammonia concentration. The stoichiometric requirements for the precipitation of 1 g of NH_4-N as struvite are 1.7 g/L of Mg^{2+} and 2.2 g/L of PO_4-P. For facilities that have an EBPR biological treatment process, the orthophosphate concentration of the dewatering recycle stream is typically much higher. Magnesium addition is also typically required because magnesium is the limiting element in the formation of struvite. Magnesium is typically added in the form of magnesium chloride or as magnesium hydroxide to maximize the capture of phosphorus. In addition, the ammonia concentration remaining after struvite formation will remain high, thereby requiring additional management.

There are a number of struvite precipitation processes that have been commercially developed. In North America, a process using a fluidized bed reactor for struvite precipitation (marketed by Ostara under the commercial name Pearl®) has been demonstrated to be technically and commercially viable at multiple full-scale facilities (Baur, 2011). A recirculation flow suspends the struvite particles in a reactor. The particles are segregated by size by the upflow velocity. The heaviest, large particles are in the lowest, smaller section, where the velocity is the greatest, the magnesium, ammonia, and phosphorus concentrations are the highest, and precipitation of struvite on existing particles occurs. The desired particle size is taken from the reactor, dried, and can be sold as a fertilizer supplement.

5.0 REFERENCES

Baur, R. (2011) Results of the First Year of Operation of North America's First Full Scale Nutrient Removal Facility. *Proceedings of the WEF/IWA Specialty Conference—Nutrient Management and Recovery;* Miami, Florida, Jan 9–12; Water Environment Federation: Alexandria, Virginia.

Constantine, T.; Murthy, S.; Bailey, W.; Benson, L.; Sadick, T.; Daigger, G. T. (2005) Alternatives for Treating High Nitrogen Liquor from Advanced Anaerobic Digestion at the Blue Plains AWTP. *Proceedings of the 78th Annual Water Environment Federation Technical Exposition and Conference* [CD-ROM]; Washington, D.C., Oct 30–Nov 2; Water Environment Federation: Alexandria, Virginia.

Coughenour II, J. R.; Walz, T.; Blatchford, G.; Phillips, J.; Husband, J. (2010) Innovative Approach to Centrate Nitrification Accomplishes Multiple Goals—Nitrogen Removal, Odor Control, and Potential for $40 Million Savings. *Proceedings of the 83rd Annual Water Environment Federation Technical Exhibition and Conference* [CD-ROM]; New Orleans, Louisiana, Oct 2–6; Water Environment Federation: Alexandria, Virginia.

Gopalakrishnan, K.; Anderson, J.; Carrio, L.; Abraham, K.; Stinson, B. (2000) Design and Operational Considerations for Ammonia Removal from Centrate by Steam Stripping. *Proceedings of the 73rd Annual Water Environment Federation Technical Exposition and Conference* [CD-ROM]; Anaheim, California, Oct 14–18; Water Environment Federation: Alexandria, Virginia.

Kos, P.; Head, M. A.; Oleszkiewicz, J.; Warakomski, A. (2000) Demonstration of Low Temperature Nitrification with a Short SRT. *Proceedings of the 73rd Annual Water Environment Federation Technical Exposition and Conference* [CD-ROM]; Anaheim, California, Oct 14–18; Water Environment Federation: Alexandria, Virginia.

Luna, B.; Narayanan, B.; Rogowski, S.; Walker, S. (2010) Metro's CaRRB Diet—Centrate Treatment Process Tackles Big Challenges in Small Package. *Proceedings of the 83rd Annual Water Environment Federation Technical Exhibition and Conference* [CD-ROM]; New Orleans, Louisiana, Oct 2–6; Water Environment Federation: Alexandria, Virginia.

Orentlicher, M.; Ginzburg, H.; Grey, G. (2009) Greenhouse Gas and Energy Savings from Integration of Centrate Ammonia Reduction with BNR Operation: Simulation of a New York City WPCP. *Proceedings of the 82nd Annual Water Environment Federation Technical Exhibition and Conference* [CD-ROM]; Orlando, Florida, Oct 10–14; Water Environment Federation: Alexandria, Virginia.

Parker, D.; Wanner, J. (2007) Review of Methods for Improving Nitrification through Bioaugmentation. *Proceedings of the 80th Annual Water Environment Federation Technical Exhibition and Conference* [CD-ROM]; San Diego, California; Oct 13–17; Water Environment Federation: Alexandria, Virginia.

Rosenthal, A.; Ramalingam, K.; Park, H.; Deur, A.; Beckmann, K.; Chandran, K.; Fillos, J. (2009) Anammox Studies Using New York City Centrate to Correlate Performance, Population Dynamics and Impact of Toxins. *Proceedings of the 82nd Annual Water Environment Federation Technical Exhibition and Conference* [CD-ROM]; Orlando, Florida, Oct 10–14; Water Environment Federation: Alexandria, Virginia.

Strous, M.; Heijnen, J. J.; Kuenen, J. G.; Jetten, M. S. M. (1998) The Sequencing Batch Reactor as a Powerful Tool for the Study of Slowly Growing Anaerobic Ammonium-Oxidizing Microorganisms. *Appl. Microbiol. Biotechnol.*, **50**, 589–596.

Strous, M.; Fuerst, J. A.; Kramer, E. H. M.; Logemann, S.; Muyzer, G.; van de Pas-Schoonen, K. T.; Webb, R.; Kuenen, J. G.; Jetten, M. S. M. (1999) Missing Lithotroph Identified as New Planctomycete. *Nature*, **400**, 446–449.

U.S. Environmental Protection Agency (1987) Sidestreams in Wastewater Treatment Plants; U.S. EPA Design Information Report; *J.—Water Pollut. Control Fed.*, **59**, 54–59.

Water Environment Federation (2010) *Nutrient Removal*; Manual of Practice No. 34; McGraw-Hill: New York.

Water Environment Federation; American Society of Civil Engineers; Environmental and Water Resources Institute (2009) *Design of Municipal Wastewater Treatment Plants*, 5th ed.; Manual of Practice No. 8; ASCE Manuals and Reports on Engineering Practice No. 76; McGraw-Hill: New York.

Wett, B. (2005) Solved Scaling Problems for Implementing Deammonification of Rejection Water. *Proceedings of the IWA Specialized Conference on Nutrient Management*; Krakow, Poland, Sept 19–21; International Water Association: London.

Chapter 13

Process Control Using Oxidation–Reduction Potential and Dissolved Oxygen

Sara Arabi, Ph.D., P.Eng.; Nancy Afonso; George Nakhla, Ph.D., P.E., P.Eng.; and Alvin Pilobello

1.0 INTRODUCTION

Biological nutrient removal (BNR) systems can benefit significantly from using oxidation–reduction potential (ORP) and dissolved oxygen as process control parameters. These benefits include reductions in cost and improved reliability. This chapter provides an overview of process monitoring, optimization, and controls using ORP and dissolved oxygen for BNR processes. Information on ORP and dissolved oxygen instruments, including general considerations and concerns, is also provided.

2.0 OXIDATION–REDUCTION POTENTIAL

Oxidation–reduction potential is a measurement of the oxidation or reduction potential of a liquid (i.e., measurement of the ability of a solution to receive or donate electrons). The ORP value, expressed in millivolts, is measured using a sensor that measures electrical charges from the ions; these charges produce either negative or positive voltage readings on the meter. Positive ORP value indicates the ability to accept electrons and an oxidative environment, whereas a negative ORP value indicates the ability to donate electrons and a reductive environment. A solution that is neither oxidizing nor reducing has an ORP value near zero. Other water conditions that affect ORP readings include pH and temperature; therefore, care should be taken when comparing ORP values.

Measurement of ORP is useful in wastewater treatment because the various oxidative and reductive reactions can be measured in a spectrum of strengths, from the highest positive values (most oxidative) to the highest negative values (most reductive). Oxidation–reduction potential can be used to indicate what type of biochemical activity (i.e., aerobic, anoxic, or anaerobic) is occurring in a BNR facility. Important biological activities (oxidation–reduction reactions) in wastewater treatment systems include nitrification, denitrification, enhanced biological phosphorus removal (EBPR), and removal of carbonaceous biochemical oxygen demand (BOD). In a biological solution, such as an activated sludge reactor, an aerobic environment will result in an ORP reading higher than would be measured in an anoxic environment and an anoxic environment would result in a higher ORP reading than would be measured in an anaerobic environment. By monitoring the ORP value of wastewater, an operator can determine which biological reaction is occurring and if operational conditions should be changed to promote or prevent that reaction. Oxidation–reduction potential in BNR systems can be manipulated by aeration, internal recirculation, and return activated sludge (RAS) flowrates.

Because certain wastewater characteristics, the types of ORP sensors used (silver vs platinum), and reference electrodes affect ORP readings, typical values cannot be assigned for aerobic, anoxic, and anaerobic conditions. It should be noted that the absolute value of an ORP instrument may not be sufficiently reliable to make process changes in some conditions; however, the trend in ORP readings changes can indicate changes in a wastewater treatment process. Therefore, ORP readings should also be used with readings from dissolved oxygen and pH meters to verify the stage of the process in which the ORP reading is made. Oxidation–reduction potential measurement is dependent on the reference electrode in the ORP sensor (probe). Different manufacturers use different reference electrodes with specific reference potentials. To compare ORP values for two instruments, manufacturers should have the same reference electrode or one measurement should be compensated for the difference in reference potential. The ORP value for various standard solutions is reported in *Standard Methods for the Examination of Water and Wastewater* (APHA et al., 1998). The particular reference electrode should be noted along with ORP results.

3.0 NITRIFICATION–DENITRIFICATION CONTROL USING OXIDATION–REDUCTION POTENTIAL

3.1 Nitrification Control and Monitoring Using Oxidation–Reduction Potential

Oxidation–reduction potential can be used as a quick indicator of aeration basin conditions; high ORP value is an indicator of conditions that favor nitrification and low ORP value is a symptom that other process indicators should be looked at, such as dissolved oxygen, ratios of carbonaceous oxygen demand (COD) to total Kjeldahl nitrogen (TKN) or BOD to TKN, or toxicity. Autotrophic nitrification requires aerobic ORP greater than +50 mV (Zehnder and Stumm, 1988). During nitrification, the oxidation of ionized ammonia (NH_4) to nitrate (NO_3) is performed by nitrifying bacteria when the ORP of the wastewater is +100 to +350 mV (Grissop, 2010). Lower ORP could indicate an inactive sludge in the aerobic zone even though dissolved oxygen levels could be around 2.0 mg/L or higher. Occasional ORP measurement by a portable ORP meter is sufficient to ensure a positive millivolt potential.

3.2 Denitrification Control and Monitoring Using Oxidation–Reduction Potential

Denitrification occurs at ORPs in the range of −100 mV (Zehnder and Stumm, 1988) to +25 mV (Zhao et al., 1999). Dabkowski (2008a) identified ORP ranges in the anoxic tanks for five of the most common combined BNR processes (i.e., five-stage Bardenpho, Anaerobic, anoxic, oxic [A2/O], University of Cape Town [UCT], Virginia Initiative Process [VIP], and modified Johannesburg [JHB]; see Chapter 10 for detailed process descriptions). The ORP in the anoxic reactors was in the range of −100 to +100 mV for all five BNR processes. It should be noted that the ORP values suggested for denitrification are general recommendations that may differ with changing operating conditions and facility design. The suggested ranges are starting points; therefore, the optimal range of ORP values for a specific location should be based on additional criteria such as sludge settleability and effluent total nitrogen concentrations. One of the most valuable uses for an online ORP sensor is trending the process to see any sudden changes in ORP value that indicate the environment is no longer anoxic.

In the five-stage Bardenpho process, the majority of denitrification typically occurs in the first anoxic tank. Factors affecting ORP of this zone include the amount of recycled aerobic mixed liquor and the amount of anaerobic effluent from the first zone. The ORP in this zone should be between −100 and 100 mV (provided that the dissolved oxygen is zero), indicating a fairly neutral solution for denitrification to take place. The ORP in the second anoxic zone is especially important if a carbon source is being added to further denitrify the wastewater. While carbon addition is best determined by pilot studies, it can be controlled through ORP measurement combined with COD, total inorganic nitrogen, and volatile total suspended solids testing (Dabkowski, 2008b).

Similar to the five-stage Bardenpho process, the first and most important location for ORP measurement for the A2/O process is in the anoxic zone, where the denitrification is taking place. Factors affecting ORP of this zone are also the same as the Bardenpho process, that is, the amount of recycled aerobic mixed liquor and the amount of anaerobic effluent from the first zone along with the biological state of the bacteria from each zone. Like most anoxic zones, the ORP should be between −100 and 100 mV.

For the UCT and VIP processes, ORP measurements and installations are the same as those for the A2/O system. The first and most important location for ORP measurement is in the anoxic zone, where denitrification occurs. The difference is the factors affecting the ORP of this zone, that is, the amount of recycled aerobic mixed liquor, the amount of nitrate in RAS, the amount of anaerobic effluent from the first zone, and the biological state of bacteria from each zone. Again, like most anoxic zones, the ORP should be between −100 and 100 mV. The difference between the standard UCT and VIP processes and the modified UCT (MUCT) process is that, in the MUCT process, an additional anoxic zone is created solely for denitrifying RAS before it is returned to the anaerobic zone. This way, it is free from the influence of the recycled aerobic zone nitrate and further increases flexibility in the "phosphorus-first" philosophy. An additional ORP sensor installed in the first anoxic zone of the MUCT would help ensure an environment conducive to denitrification of the RAS. In this first anoxic zone, ORP values should be between −100 and 100 mV, with the most likely scenario being between −100 and 0 mV.

The modified JHB process allows for quick denitrification of RAS through an initial preanoxic zone, which typically has twice the solids concentration as the mixed liquor in the remaining zones. The best location for the first ORP sensor is in the preanoxic zone. Premature fouling of the sensor could be a concern because of the high level of suspended solids in this preanoxic zone. In addition, with the addition of the anaerobic recycle stream, it is possible for this zone to become inactive if the detention time increased significantly. Controlling the ORP in this zone between −100 and 100 mV is critical to proper denitrification and seeding of the remaining process. If this zone is mixed using on–off aeration, ORP control becomes even more important to ensure that it does not become aerobic. Another ORP sensor can be installed in the second anoxic zone. This second anoxic zone is responsible for the majority of denitrification that takes place and, therefore, it is necessary to ensure a fairly neutral environment. As with other anoxic zones, ORP values should be between −100 and 100 mV. Oxidation–reduction potential in this zone will ensure that an appropriate amount of aerobic effluent is recycled for the biomass present.

4.0 ENHANCED BIOLOGICAL PHOSPHORUS REMOVAL CONTROL USING OXIDATION–REDUCTION POTENTIAL

In the two-phase phosphorus removal process (phosphorus release and uptake), an ORP probe is considered a reliable indicator to determine aerobic and anaerobic conditions. Sudden changes in the ORP value, for example, could help indicate whether the process is no longer aerobic. It is important to note that the ORP values mentioned in this section are general recommendations and may differ with changing operating conditions and facility design. During the biological phosphorus release stage (the anaerobic zone), an ORP range of −100 to −225 mV is typical. The phosphorus uptake occurs in the aerobic tank. Studies by de la Menardiere et al. (1991) and Lee et al. (2001) have indicated that ORP-reading-controlled aeration in the aeration tank can yield more efficient removal of phosphorus than fixed-timing aeration control.

In the five-stage Bardenpho process, an ORP sensor is installed in the initial anaerobic zone. This basin requires a strongly negative ORP value to break down the polyphosphates. The ORP values in this zone are less than −100 mV to ensure adequate electron donors for the respiration of polyphosphate-accumulating organisms (PAOs) (Dabkowski, 2008a). In certain applications where volatile fatty acids (VFAs) are limited in the raw wastewater, the initial anaerobic zone can be used to ferment the wastewater and create VFAs necessary for EBPR. In these applications, the ORP value can drop to −350 mV as fermentation takes place, with nominal values averaging −250 mV (Dabkowski, 2008a). Mixers in the anaerobic zone can be turned on and off to facilitate this fermentation, controlled by time or ORP values. Similar to the five-stage Bardenpho process, the anaerobic zone is one of the locations for ORP measurement for the A2/O, UCT, and VIP processes. In this zone, the breakdown of polyphosphates and the release of orthophosphates occur by PAOs, which require a strongly reducing environment. The ORP values should be less than −100 mV in the anaerobic zone for these processes (Dabkowski, 2008a). In the modified JHB process, to ensure a proper anaerobic and reducing environment for PAOs, an ORP sensor should be installed in the anaerobic zone. The ORP values in the anaerobic zones should be less than −150 mV (Dabkowski, 2008a).

5.0 OXIDATION–REDUCTION POTENTIAL METERS

5.1 Principles of Operation

An ORP sensor is comprised of a measurement half cell and a reference half cell, which allow for potentiometric measurement. The measurement half cell consists of a noble metal electrode (typically platinum for most wastewater applications), which acts as an electron donor, or electron acceptor, depending on the ORP of the sample. Gold must be used with solutions containing copper, lead, or zinc because these elements will not work with platinum. The other half cell is the reference electrode, typically a silver–silver chloride (Ag–AgCl) electrode, which acts as the stable output comparison. The two half-cell potentials are required to complete a circuit (Liptak, 2003).

5.2 Types of Meters (Field, Laboratory, and Continuous Monitoring)

Field measurements using portable ORP meters and online continuous monitoring systems can be used for measurements of ORP. In general, online process ORP sensors will give more precise data and real-time values and will allow operator trending of the data. Although portable meters are less sensitive to all potentials, they are good for spot checks. Portable meter readings, which are ±10% compared to online measurements, are considered acceptable. Many online process systems can be configured to be "portable" data-logging systems (Dabkowski, 2008b).

5.3 Positioning of Oxidation–Reduction Potential Probes (Installation)

Similar to any measurement device, the ORP probe should be installed in an area of the tank that is representative of the process within that tank. Some probe manufacturers recommend immersing the ORP sensor to one-quarter of the depth of the tank. Care must be taken to mount the ORP sensor such that the electrode end is pointed down, ±15 deg from horizontal, to prevent an air bubble in the reference side from causing an air gap between the fill solution and the junction (Dabkowski, 2008a).

Proper in-line mixing or tank agitation may be necessary to ensure that the sensor is measuring a representative sample. Where the sensor is mounted in a sidestream, the sample line length should be as short as possible to minimize response time and to keep the sensor "in sync" with the process.

5.4 Concerns with Oxidation–Reduction Potential Probes

The specific construction and design of the ORP probe itself can influence the absolute measurement value (in millivolts) or response to the same sample, as provided by the electrode. For example, the differences in purity and state of the platinum electrode surface found between manufacturers can contribute to variability in measurements. An uneven or rough surface contributes to higher absorption of oxygen on the surface, leading to higher formation of an oxidized layer on the platinum surface, which affects measurements. Therefore, before comparing the absolute ORP measurement values between two different probes, the construction type of the probe and their known ORP measurement profiles for similar samples should be taken into consideration.

5.5 Drifting and Fouling

As with any probe that conducts readings between a reference electrode and the measuring electrode, fouling of the ORP probe will occur. The rate of fouling is likely to be faster in process areas where the mixed liquor has high levels of suspended solids. As per *Standard Methods* (APHA et al., 1998), any interferences that contaminate the ORP electrode's surface, salt bridge, or electrode can lead to drift, poor response, or incorrect readings. As such, effective and regular cleaning of the ORP probe should be undertaken, as explained in Section 5.7.

5.6 Accuracy and Repeatability

As with any control system that depends on a measurement device or probe, the probe should ideally be able to provide readings that are accurate and repeatable. However, because of fluctuating characteristics of wastewater samples and the fouling that it causes for most measurement devices, accuracy and reproducibility becomes challenging. It is more important to have a reliable, low-maintenance probe that can produce highly reproducible results than a highly accurate probe that requires high maintenance to maintain reproducibility.

The ORP sensor should be calibrated on a regular basis. As a general rule, ORP sensors should be calibrated every few months or as necessary. The proper standard solution for calibration will depend on the manufacturer's recommendation for specific probes. It is recommended to calibrate the sensors every 3 months using 200- or 600-mV calibration standards (Dabkowski, 2008b), unless otherwise recommended in the sensor's instruction manuals. Ideally, the probe should be calibrated using solutions that span the range of the measurements expected to be seen for the specific BNR process characteristics.

The ORP sensor should not be exposed to air for extended periods of time because this can cause it to dry out. Similarly, it also should not be allowed to freeze. Damage to the ORP sensor because of drying or freezing can be irreversible. Proper storage of the ORP probe, when not in use, must be practiced per the manufacturer's instructions; generally, this involves the sealed immersion of the ORP probe in a cap filled with neutral solution media to avoid drying.

5.7 Cleaning and Maintenance Requirements

Specific instructions on care and maintenance of ORP probes are typically found in the manufacturer's operating manual. Because of the spatial variations of wastewater characteristics in the BNR process, frequent cleaning of the sensor's noble metal electrode may be required for specific locations. For example, the high level of suspended solids in the preanoxic zone may contribute to higher cleaning regiments than other parts of the BNR process. The specific time period between cleanings (days, weeks, etc.) can only be determined by operating experience because it is affected by the characteristics of the process wastewater. Automatic cleaning systems using air or water jets are available to reduce cleaning frequency in difficult applications. Probes should be inspected regularly and cleaned as necessary or as determined by operational experience. In addition, it is important to keep the sensors clean.

6.0 DISSOLVED OXYGEN

Dissolved oxygen is a measurement of the oxygen dissolved in a liquid stream. The quantity of dissolved oxygen in water is typically expressed in milligrams per liter. In water, the dissolved oxygen concentration is limited by the maximum equilibrium concentration, or saturation concentration, which depends on the gas used (air or pure oxygen), water temperature, dissolved solids in the water, and elevation (atmospheric pressure). The dissolved oxygen saturation concentration decreases with increasing temperature, increasing dissolved solids concentration, and decreasing atmospheric pressure. It is important to note that, in activated sludge systems, dissolved oxygen measurements provide the dissolved oxygen concentration in the bulk solution and may not be directly representative of dissolved oxygen concentrations within the floc. The dissolved oxygen within the floc will generally be lower than that measured in the bulk solution.

7.0 PROCESS CONTROL AND MONITORING USING DISSOLVED OXYGEN

Dissolved oxygen should not be present in any significant quantity in anoxic zones because denitrification will be inhibited. Similarly, dissolved oxygen should not be present in anaerobic (fermentation) zones because VFA uptake and subsequent biological phosphorus removal will be inhibited. Dissolved oxygen in aeration zones should be monitored a minimum of several times per day, or, continuously if instrumentation is used. Often, dissolved oxygen is used to control aeration to enhance BNR and to reduce energy costs. Dissolved oxygen concentrations in anoxic zones, anaerobic zones, and internal recycle or RAS streams are typically not monitored, but can be checked if an upset condition is experienced.

7.1 Nitrification–Denitrification Control and Monitoring Using Dissolved Oxygen

Dissolved oxygen is an important operating parameter affecting both nitrification and denitrification kinetics. This section discusses the control of both processes using dissolved oxygen.

7.1.1 Nitrification

In wastewater systems designed for carbonaceous BOD removal, a minimum average aeration tank dissolved oxygen concentration of 0.5 mg/L is generally acceptable under peak loading conditions, and 2.0 mg/L is typical for average conditions unless low dissolved oxygen filamentous microorganisms begin to predominate. For nitrifying systems, a minimum average aeration tank dissolved oxygen of 2.0 mg/L under all conditions has historically been the general practice, although nitrification in suspended growth systems has occurred at dissolved oxygen concentrations of 0.7 to 1.0 mg/L.

7.1.2 *Denitrification*

Denitrification is negatively affected by increasing dissolved oxygen, which reduces denitrification rates by two mechanisms: (1) dissolved oxygen inhibition and (2) aerobic consumption of readily biodegradable COD (rbCOD), leading to a reduction in the biomass specific denitrification rate. The dissolved oxygen inhibition coefficient (K_{O2}), with the typical value of 0.2 mg/L, is actually site specific depending on the floc size. At a dissolved oxygen concentration of 0.2 mg/L, the denitrification rate may be only 10 to 50% of the maximum denitrification rate (Metcalf and Eddy, 2003). Aeration tank dissolved oxygen recycled back to the anoxic tank will reduce the rbCOD available for denitrification.

7.2 Enhanced Biological Phosphorus Removal Control and Monitoring Using Dissolved Oxygen

The control of dissolved oxygen entering the anaerobic zone is important. Recycle streams with significant concentrations of dissolved oxygen (and nitrate) can have an adverse effect on process performance (Metcalf and Eddy, 2003). Strict anaerobic conditions must be maintained to provide the PAOs the first opportunity to take up the substrate. This means that the anaerobic zone should be protected from dissolved oxygen and nitrate sources, which eliminate anaerobic conditions and place the PAOs at a competitive disadvantage to other heterotrophs. Screw pumps and free fall over weirs may introduce dissolved oxygen to the influent. Likewise, internal mixed liquor recycles used in total nitrogen removal processes may be a significant source of dissolved oxygen and nitrate, and the return sludge in nitrifying systems can also recycle nitrate (Jeyanayagam, 2005). To improve BNR efficiency, it is important to reduce the amount of nitrate and/or dissolved oxygen entering the anaerobic zone.

7.2.1 *Phosphorus Release*

Under anaerobic conditions, PAOs release phosphorous in the form of orthophosphates and store short-chain fatty acids as intracellular products such as polyhydroxyalkanoates and polyhydroxybutyrate. Theoretically, the ratio of VFA uptake (as acetate acid [HAc]) to phosphorus release is 1.4 to 2.0 mg HAc/mg P (Schuler and Jenkins, 2003). A widely accepted ratio of VFA to phosphorus removed is 7 to 10 mg HAc/mg P removed (Metcalf and Eddy, 2003). Meanwhile, the recirculation of dissolved oxygen and nitrate to the anaerobic tank depletes rbCOD at 2.3 g rbCOD/g O_2 and 6.6 g rbCOD/g NO_3-N (Metcalf and Eddy, 2003). Accordingly, dissolved oxygen recirculation likely sequesters approximately 2.5 mg HAc/mg O_2 and reduces phosphorus release by 1.08 to 1.5 mg P/mg O_2. Similarly, nitrate recirculation reduces phosphorus release by as much as 3.1 to 4.3 mg P/mg NO_3-N. As is apparent from the aforementioned ratios, on a mass basis, nitrate is more detrimental to phosphorus release (and subsequent phosphorus uptake) than dissolved oxygen, which has spurred development of processes such as UCT and MUCT for low-carbon wastewater.

7.2.2 *Phosphorus Uptake*

Polyphosphate-accumulating organisms are heterotrophic bacteria that predominantly use oxygen as an electron acceptor, although a small fraction (~15 to 20%) can use both oxygen and nitrate (Hu et al., 2002). Generally, denitrifying PAOs (DPAOs) have a much lower specific denitrification rate than ordinary heterotrophic organisms (OHOs) and, hence, can only exist in systems that have excess nitrate that cannot be used by OHOs. According to *Activated Sludge Model No. 2* (Henze et al., 1995), the dissolved oxygen half-saturation concentration for a PAO is 0.2 mg O_2/L and, hence, phosphorus uptake under aerobic conditions is typically insensitive to dissolved oxygen at a concentration above 0.5 mg O_2/L. During full-scale comparative testing of A2/O and UCT at the 341-ML/d (90-mgd) City of Las Vegas Water Pollution Control Facility (Las Vegas, Nevada), Gu et al. (2006) observed that aerobic phosphorus uptake was insensitive to dissolved oxygen concentrations in the range of 0.5 to 3.5 mg O_2/L.

The effect of dissolved oxygen and nitrate on phosphorus release and phosphorus uptake strongly depends on the availability of rbCOD in the bulk solution. You et al. (2004) found that, in the absence of rbCOD, both DPAO and non-DPAO could uptake phosphates. However, when rbCOD increased, phosphorus release was much more pronounced under aerobic conditions than under anoxic conditions.

8.0 DISSOLVED OXYGEN INSTRUMENTS

Dissolved oxygen is typically measured using the membrane electrode Method 4500-OG in *Standard Methods* (APHA et al., 1998). The equipment includes a probe with an oxygen-permeable, membrane-covered, electrode-sensing element with a thermistor for temperature reading and a meter that provides temperature compensation of the sensing element's signal that indicates the dissolved oxygen concentration.

Sensor types, which are the online dissolved oxygen meter principles of operation, are described in Sections 8.1 through 8.5. Recommended calibration, maintenance, and installation practices are provided in this section (ITA, 2003) along with information from existing technical publications, report journals, manufacturer literature, and practical experience.

8.1 Dissolved Oxygen Measurement Systems

The laboratory method for determining dissolved oxygen, such as azide modification of the iodometric titrimetric method (i.e., *Standard Method* 4500-O C; APHA et al. [1998]), can be used. However, this method is time consuming and impractical for most BNR facilities because of the number of sample points and frequency of sampling required. There are generally three types of technologies to measure dissolved oxygen. These are Clark (polarographic), galvanic, and optical (fluorescence and luminescence). Galvanic and polarographic measurements have been accepted or approved by the U.S. Environmental Protection Agency, while acceptability of the optical method varies by region (Dabkowski, 2008b).

Polarographic and galvanic electrochemical sensors are used in dissolved oxygen meters. Electrochemical sensors consist of a probe that can be directly immersed into the process solution and uses the following six basic components to measure dissolved oxygen: measuring cathode, reference anode, potassium chloride (KCl) or potassium hydroxide (KOH) electrolyte solution, oxygen porous membrane, electronics module, and temperature sensor (thermistor). Electrochemical online dissolved oxygen measurement uses an electronics module and a special measuring electrode sensor that outputs current to the electronics module in proportion to the concentration of dissolved oxygen in the process stream.

The polarographic sensor uses a thermistor, a KCl electrolyte, a membrane, a silver anode, and a gold or platinum cathode. The thermistor is a temperature-sensitive resistor and is used to compensate the sensor electronics for varying process conditions and to ensure accurate dissolved oxygen readings. The KCl electrolyte provides a path for the current that is generated by the reaction of the oxygen molecules on the cathode to flow. The membrane is placed over the tip of the sensor and serves to contain the electrolyte and provide even diffusion of the oxygen molecules to the cathode. A polarizing voltage from the sensor electronics is applied to the anode and cathode. When oxygen molecules diffuse across the membrane and come into contact with the cathode, an electrical current flows from the cathode to the anode in proportion to the amount of dissolved oxygen in the process. The sensor electronics sense the flow of the electrical current and produce a direct output reading of dissolved oxygen expressed in percent saturation or milligrams per liter.

The galvanic sensor uses the same components; however, KOH electrolyte is used instead of KCl. Typically, the anode is lead and the cathode is silver. When oxygen molecules diffuse across the membrane, they react with the surface of the cathode and an electrical current is produced without the application of polarizing voltage to the electrodes. Again, the sensor electronics compensate for temperature and produce a direct reading output.

All of the aforementioned system components (i.e., the anode, cathode, electrolyte, and the temperature sensor) are contained inside a single dissolved oxygen probe. The transmitter, which is separated from the dissolved oxygen probe, is typically remotely mounted and can be connected to a control and automation system. The combination of the dissolved oxygen probe and the transmitter comprises the dissolved oxygen sensor. The dissolved oxygen sensor uses sensitive input electronics and a microprocessor to process all of the input and output signals.

The fluorescence sensor consists of a probe that is immersed in the process solution and uses the following components: light emitter, oxygen-permeable fluorescing sensor element, light detector, temperature sensor, and electronics module for signal processing and output. The fluorescence sensor uses an electronics module and a measuring fluorescence sensor that outputs a signal to the electronics module in proportion to the concentration of dissolved oxygen in the process stream. The fluorescence sensor uses a temperature sensor to compensate the sensor electronics for varying process conditions and to ensure accurate dissolved oxygen readings. The sensing element is constructed of a fluorescing dye that has been immobilized or embedded in an oxygen-permeable structural material that varies from one manufacturer to another. The immobilizing material protects the fluorescing dye from any direct liquid contact, but allows oxygen to pass freely and affect the fluorescing dye.

The emitter transmits blue light of a specific wavelength. In the presence of this blue light, the dye will fluoresce, returning a pink light with a different wavelength. The light detector is configured to respond to this fluorescing pink light. The output of the light detector with the temperature sensor output is processed by the sensor electronics and produces a direct output reading of dissolved oxygen expressed in milligrams per liter. Manufacturers of these systems use a variety of techniques to convert the detected signal to oxygen concentration, relating either amplitude, time-based measurements, or some combination of the two.

This luminescence-based sensor procedure measures the light-emission characteristics from a luminescence-based reaction that takes place at the sensor–water interface. A light-emitting diode (LED) provides incident light required to excite the luminophore substrate. In the presence of dissolved oxygen, the reaction is suppressed. The resulting dynamic lifetime of the excited luminophore is evaluated and equated to dissolved oxygen concentration. Luminescent dissolved oxygen technology has been proven to reduce maintenance requirements and energy costs while also being reliable and accurate (Dabkowski, 2009).

8.2 Installation Practices

Most dissolved oxygen instruments can be installed using standard fittings available from the manufacturer. If the application requires custom installation, care should be taken so that the dissolved oxygen probe and electronics module are mounted to allow easy access for calibration and maintenance. If necessary, short sample lines should be used, bypass piping for process streams that cannot be shutdown should be included, and the electronics module should be mounted near the dissolved oxygen probe for easy calibration adjustments. In-line dissolved oxygen sensors require a minimum process flow velocity at the sensor.

This is because in-line dissolved oxygen measurement sample lines can be long, allowing for dissolved oxygen to deplete during transit to the probe. This creates a discrepancy when comparing dissolved oxygen concentrations to immersed dissolved oxygen probe readings.

8.3 Accuracy and Repeatability

Although the accuracy and repeatability of membrane-type sensors will vary by manufacturer, typical specifications are

- Accuracy—0.10% of span and
- Sensitivity—0.05% of span.

For florescent sensors, the accuracy should be

- Less than 1 mg/L ± 0.1 mg/L,
- Greater than 1 mg/L ± 0.2 mg/L,
- Repeatability of 0.05 mg/L, and
- Resolution of dissolved oxygen at 0.01 mg/L or 0.01% saturation.

In addition, sensitivity should be ±0.05% of the span.

8.4 Maintenance Practices

Manufacturer recommendations for maintenance practices should be followed because each type of probe has unique requirements. Generally, dissolved oxygen sensors require minimal maintenance; however, maintenance frequency depends on sensor design. Cleaning the dissolved oxygen sensor is the most important maintenance procedure. Membranes are subject to biofouling and puncture. Algae, grease, or sludge can coat all dissolved oxygen electrode sensors, interfering with the ability to measure accurately. Periodic manual cleaning to remove grease or sludge buildup will lessen the interference. Typically, the sensor can be hand wiped as needed. Weekly cleaning frequency is sometimes adequate; however, automatic cleaning using air or water blasts should be considered for automatic dissolved oxygen control applications.

8.5 Calibration Practices

Manufacturer recommendations for calibration practices should be followed. Calibration frequency of a dissolved oxygen meter will depend on the type of measuring electrode sensor used, the process that is measured, and the accuracy required. Typically, dissolved oxygen meters should be calibrated quarterly.

Dissolved oxygen meters require a minimum of one-point calibration. An air calibration can be performed with saturated air comprised of 20.9% oxygen. If the range of the dissolved oxygen meter is 0 to 10 mg/L, then the meter is switched to air calibration mode and the dissolved oxygen sensor is suspended just above the water level in a tank or bucket. Care should be taken to ensure that the temperature of the sensor is allowed to equalize with the existing air temperature (about 15 to 20 minutes). Next, the meter air calibration function is started. Not all dissolved oxygen meters can be calibrated in saturated air. Some meters must be calibrated against a known reference such as a portable dissolved oxygen meter measurement.

9.0 REFERENCES

American Public Health Association; American Water Works Association; Water Environment Federation (1998) *Standard Methods for the Examination of Water and Wastewater;* American Public Health Association: Washington, D.C.

Dabkowski, B. (2008a) Applying Oxidation–Reduction Potential Sensors in Biological Nutrient Removal Systems. *Proceedings of the 81st Annual Water Environment Federation Technical Exhibition and Conference* [CD-ROM]; Chicago, Illinois, Oct 18–22; Water Environment Federation: Alexandria, Virginia.

Dabkowski, B. (2008b) ORP in BNR Systems: A Magic Wand? *Proceedings of the British Colombia Water and Waste Association (BCWWA) Annual Conference;* Whistler, British Columbia, Canada, April 26–30; British Colombia Water and Waste Association: Burnaby, British Colombia, Canada.

Dabkowski, B. (2009) New Sensor Technology Optimizes DO Control. Water Online, August 19. http://www.wateronline.com/doc/New-Sensor-Technology-Optimizes-DO-Control-0001 (accessed June 2013).

de la Menardiere; Charpentier, J.; Vachon, A.; Martin, G. (1991) ORP as a Control Parameter in a Single Sludge Biological Nitrogen and Phosphorus Removal Activated Sludge System. *Water SA,* **17** (2), 123–132.

Grissop, G. (2010) Biological Nutrient Removal Operation. Paper presented at the Shenandoah Valley Pure H_2O Forum, Wastewater Treatment Plant Network; December 1. http://www.acsawater.com/sites/default/files/websitefiles/SVWWTPN/BNR%20Operations.pdf (accessed June 2013).

Gu, A. Z.; Hughes, T.; Fisher, D.; Swartzlander, D.; Dacko, B.; Ellis, W. G.; He, S.; McMahon, K. D.; Neethling, J. B.; Wei, H. P.; Chapman, M. (2006) The Devil Is in the Details: Full-Scale Optimization of the EBPR Process at the City of Las Vegas WPCF. *Proceedings of the 79th Annual Water Environment Federation Technical Exhibition and Conference* [CD-ROM]; Dallas, Texas, Oct 21–25; Water Environment Federation: Alexandria, Virginia.

Henze, M.; Gujer, W.; Mino, T.; Matsuo, T.; Wentzel, M. C.; Marais, G. v. R. (1995) *Activated Sludge Model No. 2;* IAWQ Scientific and Technical Report No. 3; IWA Publishing: London.

Hu, Z. R.; Wentzel, M. C.; Ekama, G. A. (2002) Anoxic Growth of Phosphate-Accumulating Organisms (PAOs) in Biological Nutrient Removal Activated Sludge Systems. *Water Res.*, **36**, 4927–4937.

Instrumentation Testing Association (1993) Online Dissolved Oxygen Analyzers: Maintenance Benchmarking Study, Henderson, Nevada; Instrumentation Testing Association: Pensacola, Florida.

Jeyanayagam, S. (2005) True Confessions of the Biological Nutrient Removal Process. *Florida Water Resour. J.*, January, 37–46.

Lee, D. S.; Joen, C. O.; Park, J. M. (2001) Biological Nitrogen Removal with Enhanced Phosphate Uptake in a Sequencing Batch Reactor Using Single Sludge System. *Water Res.*, **35** (16), 3968–3976.

Liptak, B. G. (2003) *Process Measurement and Analysis, Volume I, Instrument Engineers' Handbook*, 4th ed.; CRC Press: Boca Raton, Florida.

Metcalf and Eddy, Inc. (2003) *Wastewater Engineering, Treatment and Reuse*, 4th ed.; Tchobanoglous, G.; Burton, F. L.; Stensel, H. D., Eds.; McGraw-Hill: New York; pp 801–804.

Schuler, A. J.; Jenkins, D. (2003) Enhanced Biological Phosphorus Removal from Wastewater by Biomass with Different Phosphorus Contents, Part I and II. *Water Environ. Res.*, **7** (16), 485–511.

You, S. J.; Shen, Y. J.; Ouyang, C. F.; Hsu, C. L. (2004) The Effect of Residual Chemical Oxygen Demand of Anoxic and Aerobic Phosphate Uptake and Release with Various Intracellular Polymer Levels. *Water Environ. Res.*, **76**, 149–154.

Zehnder, A. J. B.; Stumm, W. (1988) Geochemistry and Biochemistry of Anaerobic Habitats. In *Anaerobic Microbiology*; Wiley & Sons: New York.

Zhao, H. W.; Mavinic, D. S.; Oldhan, W. K.; Koch, I. A. (1999) Controlling Factors for Simultaneous Nitrification and Denitrification in a Two Stage Intermittent Aeration Process Treating Domestic Sewage. *Water Res.*, **23** (4), 961–970.

10.0 SUGGESTED READINGS

Goronszy, M. C.; Bian, Y.; Konicki, D.; Jogan, M.; Engle, R. (1992) Oxidation–Reduction Potential for Nitrogen and Phosphorus Removal in a Fed-batch Reactor. *Proceedings of 65th Annual Water Environment Federation Technical Exposition and Conference*; New Orleans, Louisiana, Sept 20–24; Water Environment Federation: Alexandria, Virginia.

Jee, H. S.; Mano, T.; Nishio, N.; Nagai, S. (1998) Influence of Redox Potential on Methanation of Methanol by *Methanosarcina barkeri* in Eh-Stat Batch Cultures. *J. Ferment Technol.*, **66**, 123–126.

Law, I. B. et al. (1987) Influence of Readily Biodegradable COD Fraction on the Degree of Biological Phosphorous Removal. *Proceedings of the Australian Water and Wastewater Association 12th Federal Convention;* Adelaide, Australia, March 23–27.

Wentzel, M. C.; Ekama, G. A.; Dold, P. A.; Marais, G. v. R. (1990) Biological Excess Phosphorus Removal—Steady State Process Design. *Water SA,* **16** (1), 29–48.

Chapter 14

Process Control, Instrumentation, and Automation

Marie S. Burbano, Ph.D., P.E., BCEE;
Georgine Grissop, P.E., BCEE;
Jane Saulnier, P.E.; William Schilling, P.E.;
and James Taylor, P.E.

1.0 INTRODUCTION

Biological nutrient removal (BNR) processes often require more sophisticated operations than conventional processes that do not involve nutrient removal; in addition, capital and operational costs of a BNR facility may be higher than a conventional activated sludge facility. Therefore, BNR system operation can significantly benefit from using online analyzers for both monitoring and automatic control. This chapter provides an overview of online analyzers and sensors, process parameter optimization, automatic control, and requirements for control systems.

2.0 PROCESS CONTROL PARAMETERS

2.1 General Considerations

External conditions, such as flowrate and ammonia load variations, vary significantly at wastewater facilities. For example, the ratio of maximum to minimum process oxygen demand, consisting of demand for both biochemical oxygen demand (BOD) and ammonia removal, typically varies from 1.5 to 1 to 3 to 1 between peak and off-peak hours. For small treatment facilities, this ratio is sometimes as high as 10 to 1, depending on the amount of infiltration, inflow, and other extraneous sources to the collection system. To maintain optimum process performance, it is necessary to compensate for changing ammonia and carbonaceous oxygen demand (COD) load by changing airflow supply, waste flow, internal recycling flow, and other control parameters. This can be accomplished through the implementation of control schemes discussed in the previous chapters.

2.2 Selecting Optimum Setpoints

Selecting the optimum setpoint for each of the control approaches discussed in the previous chapters is a complicated task. Table 14.1 presents considerations that must be taken into account when selecting the setpoint for each parameter.

2.3 Basic Automatic Control

Optimum performance of BNR systems can be achieved only when the solids retention time is maintained at an optimum value. Below this value, a system can experience low dissolved oxygen filamentous bulking, poor ammonia removal and nitrite breakthrough, dispersed growth of biomass, and an overloaded thickening facility. At an SRT above the optimum value, a system may experience a low food-to-microorganism ratio (F/M), filamentous bulking and foaming, increased oxygen demand, and increased clarifier loading.

Wasting mixed liquor is the simplest method to implement SRT because it does not require measurement of the waste activated sludge solids concentration. In addition, such wasting can help control foam. However, this simplicity comes

TABLE 14.1 Considerations for selecting setpoints.

Process	Parameter	Potential problem is that value is: Lower than optimum	Potential problem is that value is: Higher than optimum
Primary treatment	Sludge depth	Reduction of VFA production and, as a result, reduction of influent denitrification potential; reduction of efficiency of subsequent sludge treatment processes as a result of excess water in thin sludge discharge from primary clarifiers.	Increased effluent TSS, soluble and particulate BOD, and hydrogen sulfide (H_2S) generation; phosphate release in case of co-thickening of primary and BNR wasted sludge; increased floating solids resulting from gasification; increased mechanical stress on sludge and scum collection mechanism.
Biological nutrient removal process	Solids retention time	Chlorination problems resulting from the presence of nitrite in the BNR effluent; inadequate removal of phosphate, ammonia, and nitrate; deterioration of waste sludge thickening as a result of thin sludge discharged from secondary clarifiers.	Low F/M ratio of foaming and bulking; increased clarifier solids loading; phosphate release; increased energy demand to sustain endogenous respiration.
Biological nutrient removal process	Dissolved oxygen concentration	Chlorination problems resulting from the presence of nitrite in the BNR effluent; inadequate removal of ammonia; foaming problems caused by *Microthrix;* gasification in the clarifiers resulting from the presence of nitrogen (N_2).	Inhibition of denitrification; high energy cost resulting from excessive airflow supply; breakup of floc and, as a result, increased effluent TSS.
Biological nutrient removal process	Internal recycle	Reduction of the amount of nitrate to be denitrified.	Decreased denitrification rate resulting from increased oxygen concentration in the anoxic compartment; increased pumping cost.

at a price. As a result of the low solids concentration, the volume of excess sludge may be several times larger than the volume of sludge to be wasted from the return activated sludge (RAS) line. As a result, the pumping cost of the wasted sludge and recycled water from the thickening facility is also higher. Pipes, pumps, and thickening facilities may need to be upsized to accommodate an increased flow, which is why wastage from the RAS line alone, or in combination with wasting foam from mixed liquor, is more widespread. At the same time, SRT control for this popular wasting method is somewhat more challenging than wasting methods previously mentioned in this manual.

2.4 Nitrification

Nitrification performance is affected by a complex relationship between several key parameters, including SRT, F/M, dissolved oxygen concentration in the aeration basin, mixed liquor volatile suspended solids (MLVSS), hydraulic retention time, temperature, and available alkalinity, as described in detail in Chapter 3. Each parameter has an optimum value for achieving the desired treatment level. When one parameter is outside of its optimum range, other parameters can often be adjusted to compensate. Some parameters, such as available alkalinity, cannot be compensated through manipulation of other parameters, and would need outside correction such as adding a chemical to meet the process requirement.

2.5 Denitrification

Denitrification is affected by the F/M, dissolved oxygen concentration, temperature, MLVSS, and hydraulic detention time, as described in detail in Chapter 5. In some nitrified wastewaters, there is not an adequate source of carbon for denitrification to occur. In these situations, a supplemental source of food (carbon) is required. Most commonly, the carbon source is methanol; however, other chemicals that are less hazardous to handle have been introduced to the marketplace. Some facilities that do not use a supplemental carbon source purposely reduce their F/M to allow more naturally occurring carbon to be available for denitrification. This approach is particularly well suited to facilities using a chemically enhanced primary process and those using two stages of treatment, such as a trickling filter followed by a conventional aeration basin.

The denitirification process requires good mixing to ensure that any supplemental chemical is well dispersed, biomass remains in suspension, and anoxic conditions are maintained (the minimum dissolved oxygen available can quickly be depleted in a settled sludge environment, leading to anaerobic conditions). The mixing process allows surface agitation to provide for the minimum dissolved oxygen required for an anoxic environment.

2.6 Enhanced Biological Phosphorus Removal

Available phosphorous is typically only partially removed in conventional biological treatment processes. The mass fraction of phosphorous in MLVSS is typically between 1.5 and 2.5%. Generally, this represents less than half of the available phosphorous. However, implementing an enhanced biological phosphorus removal (EBPR) process, as described in Chapters 7 and 9, can increase this mass fraction to between 6 and 10%.

2.7 Chemical Phosphorus Removal

Chemical phosphorous removal, as described in detail in Chapter 8, is often used when the required effluent level is below 1 mg/L or to supplement or back up

EBPR where lower regulatory limits are required. The two most commonly used chemicals are ferric chloride and aluminum salts, which are both well-known and effective coagulants. When added to advance treated wastewater upstream of polishing filters, effluent levels of 0.1 mg/L are typically achievable.

3.0 ONLINE AND IN-LINE INSTRUMENTATION

Instruments to measure flow, temperature, pH, total suspended solids (TSS), COD, dissolved oxygen, ORP, ammonia-ammonium, nitrate-nitrite, and phosphorus-orthophosphate are commonly used at BNR facilities for monitoring, control, and automation. The principles of operation, accuracy and repeatability, installation, maintenance requirements, and specific applications for each of these instruments are discussed in this chapter, with the exception of dissolved oxygen and ORP, which are discussed in Chapter 13. Installation and maintenance of flow meters are discussed in Section 3.1. Installation and maintenance of all other instruments is discussed in Sections 3.2 through 3.8. For more detailed information on these topics, refer to *Automation of Wastewater Treatment Facilities* (WEF, 2013).

3.1 Flow

There are many types of flow meters used at water resource recovery facilities (WRRFs), including flumes; weirs; and electromagnetic, ultrasonic, differential-pressure, mechanical, variable-area, and mass flow meters.

3.1.1 Principle of Operation

Electromagnetic flow meters operate based on the principle that the voltage induced by a conductive fluid moving through a magnetic field is proportional to the velocity of the fluid. The flowrate is then calculated based on the velocity of the fluid and the size of the conduit carrying the fluid.

There are two types of ultrasonic flow meters: transit time and Doppler. Transit-time ultrasonic flow meters measure fluid velocity by calculating the difference in travel time for one sonic pulse to go a specific distance downstream and another sonic pulse to go the same distance upstream. In a Doppler ultrasonic flow meter, a signal of a specific frequency is sent into the fluid. The suspended solids of gas bubbles in the fluid then reflect the signal back to a transducer. If the fluid is moving, the reflected signal will be of a different frequency than the original signal. The magnitude in shift in frequency between the original signal and the reflected signal is proportional to flowrate.

Weirs and flumes can be used to measure flow when combined with a level-sensing device. Fluid flows through the weir or flume and the water depth is measured at a specific point. The flow is then determined using characteristic head-to-flow relationships that have been developed for specific weir and flume geometries.

Differential-pressure flow meters operate based on the principle that when a fluid passes through a pipe, it will cause a pressure drop that is proportional to the square of the flowrate. Pitot tubes and Venturi meters are two types of differential-pressure flow meters.

Mechanical flow meters operate based on the principle that the movement of an object that has been moved by a fluid can be correlated to the flowrate of the fluid. Turbines and propellers are examples of mechanical flow meters.

Variable-area flow meters typically consist of a glass flow tube and a stainless steel float. Fluid passes vertically through the tube. The level of the stainless steel float rises in proportion to the flowrate through the tube.

Mass flow meters directly measure the mass flowrate of a material. Coriolis meters and thermal-dispersion flow meters are examples of mass flow meters. A Coriolis meter measures mass flow by inducing a vibration and comparing it to the measured vibration of the material flowing through the meter. Thermal-dispersion mass meters use heat transfer and the known characteristics of the fluid being measured to determine the mass of material flowing through the meter per unit of time.

3.1.2 *Accuracy and Repeatability*

Each style of meter provides a different level of accuracy. Most applications do not require an extreme amount of precision in flow measurement. Typically, 0.5% is sufficient for most wastewater applications. If the meter is being used for control, repeatability is typically more important than accuracy.

3.1.3 *Installation*

To obtain the most accurate readings, electromagnetic flow meters should be installed in a portion of a pipe that has a low amount of turbulence. A manufacturer typically specifies the minimum distance that the flow meter should be installed from any fittings or valves. The distance varies based on the diameter of the pipe and whether the fittings or valves are located upstream or downstream of the instrument. Flow meters should be installed on the discharge side of pumps and on the upstream side of throttling valves. Pipe fittings should be kept at the minimum specified distances on the upstream and downstream side of the meter. Flow meters that require a full-flowing pipe should be installed in pipes where the flow is going up or upward. Horizontally mounted flow meters need air-bleed valves in many applications. All flow meters should be installed in areas where there is enough space to access the meter or to remove the meter body for calibration or maintenance. Providing bypass around the flow meter is critical in applications where the process cannot be shut down.

3.1.4 *Maintenance Requirements*

Maintenance requirements vary based on the type of flow meter and the manufacturer. The manufacturer should be consulted for the required maintenance of any flow meter under consideration. Maintenance requirements can include periodic inspection, periodic cleaning of devices such as orifice plates, cleaning of pipes where grease has accumulated, altering the pipe diameter, replacing moving parts that wear, and calibration checks to maintain accuracy.

3.2 Temperature

3.2.1 *Principle of Operation*

Thermal couples and thermal bulbs, resistance temperature detectors (RTDs), and thermistors are the three main types of temperature sensors that are commonly used. Thermocouples operate based on the principle that heating a metallic junction composed of two dissimilar metals will cause a flow of current, also known as the *Seebeck effect*. Resistance temperature detectors consist of a wire made of a pure metal, typically platinum, coiled around a ceramic or glass core. The RTD measures the temperature by correlating the resistance in the metal coil to temperature. Thermistors are semiconductors whose resistivity changes in response to temperature change. The resistance of most thermistors decreases with increasing temperature.

3.2.2 *Accuracy and Repeatability*

Temperature sensors typically have a resolution of 0.1 °C and an accuracy of about 0.5 °C.

3.3 pH Measurement

3.3.1 *Principle of Operation*

The heart of a pH sensor is the glass membrane. An electrical potential, varying with pH, is generated across the membrane. The difference between this potential and a reference electrode is measured and amplified by an electronic signal conditioner. The complete electric circuit includes the glass electrode wire, glass membrane, process fluid, reference electrode fill solution, and reference electrode wire.

A pH electrode assembly, or *sensor*, as it is sometimes called, consists of the following two primary parts:

- Measuring electrode—the measuring electrode is sometimes called the *glass electrode*, and is also referred to as a *membrane* or *active electrode*; and
- Reference electrode—the reference electrode is also referred to as a *standard electrode*.

The measuring and reference electrodes can be in one of the following two forms: two physically separate electrodes, known as an *electrode pair*, or joined together in a single glass body assembly, known as a *combination electrode*. The combination pH electrode is the most widely applied.

3.3.2 *Accuracy and Repeatability*

Manufacturer claims for pH meter accuracy range from ±0.02 to ±0.2 pH units. This represents the combined accuracy of the electrodes and the signal conditioner or transmitter. Without temperature compensation, an additional error of 0.002 pH units per 1 °C difference from the calibration temperature can be expected. The repeatability of pH meter measurements varies by manufacturer from 0.02 to 0.04 pH units. Stability (drift) is an important performance parameter that indicates how often meters must be recalibrated. Manufacturer claims for stability vary from 0.002 to 0.2 pH units of drift per week. With flow-through probe mounts, the velocity of the sample can cause a shift (0.2 to 0.3 pH) in measured values. High velocity creates non-uniformity around the glass, displacing some of the charges and affecting the pH measurement.

Methods of reporting performance specifications vary among manufacturers. Adjustment of the method of reporting performance specifications to equal units of measure shows that there is large variance in the accuracy and stability claimed by different manufacturers. Typically, good pH meters achieve the following performance standards in WRRFs: accuracy, ±0.1 pH units; repeatability, ±0.03 pH units; and stability, ±0.02 pH units per week.

3.3.3 *Applications*

In wastewater treatment, pH sensors are used to monitor facility conditions and biological treatment process conditions and to control acid or base additions for pH adjustment. Regulatory agencies require measurement of facility influent and effluent pH to extrapolate overall facility conditions.

It may also be necessary to monitor the pH of specific industrial discharges to give advance warning of possible toxic conditions. While the activated sludge and most other biological processes can tolerate a pH variance of 5 to 9, some, such as anaerobic digestion, are pH sensitive. Normal monitoring of facility influent and primary effluent or mixed liquor suspended solids (MLSS) (if applicable) is typically sufficient to detect impending toxic conditions.

An online pH monitor can provide feedback for control of other processes requiring pH adjustment. For example, pH adjustment may be required to neutralize low-pH industrial wastes, enhance phosphorus removal by alum addition, or adjust pH to optimum ranges for nitrification and denitrification.

3.4 Total Suspended Solids

3.4.1 *Principle of Operation*

Currently, most TSS instruments use near infrared technology. However, instruments that use microwave, nuclear, or ultrasonic technologies are also available.

There are two main types of analyzers using near infrared technology: transmittance and reflectance. Transmittance analyzers are available in both two-beam and four-beam configurations. The two main types of reflectance analyzers are those that measure beams backscattered at 90 deg from the source and those that measure forward-scattered beams. There are some reflectance analyzers that have multiple detectors. For example, one analyzer has detectors positioned 90 and 140 deg from the light source and another analyzer uses as many as six channels of multi-angle measurement. Depending on the application, one method might perform better than another; therefore, facility operators may need to be diligent to identify the proper sensor selection.

The most important feature for any technology is color compensation or color independent analysis. Blackish, highly concentrated sludge is more difficult to measure than light-colored sludge with a lower suspended solids concentration. Multichannel analyzers may provide some advantages for these applications.

Another important feature is a method for sensor cleaning. The following automatic cleaning methods are available: water or air purging, ultrasonic, and wiper cleaning. Some manufacturers claim that their products are not susceptible to active biofouling and, as a result, do not require a self-cleaning system. According to information from these manufacturers, infrequent manual cleaning is still required. However, it is always advisable to consider self-cleaning systems, even for those analyzers for which cleaning is not specified by the manufacturer.

3.4.2 *Accuracy and Repeatability*

The accuracy of a suspended solids analyzer is typically 5% of reading, and several ranges of operation are available. The repeatability of solids analyzers is typically ±1% of reading or ±0.1 g/L, whichever is greater. It is important to note that different sensors and different manufacturers have different accuracy, precision, and error specifications, and investigation of these parameters is necessary to find the best TSS sensor for the application.

3.4.3 *Applications*

Suspended solids analyzers are used to control sludge wasting to maintain a desired SRT and MLSS.

3.5 Carbonaceous Oxygen Demand

The laboratory method used for direct measurement of COD, which involves wet chemistry, can be completed in a few hours. The method uses hazardous chemicals (potassium dichromate, silver sulfate catalyst, and mercuric sulfate). Therefore, the laboratory COD test has limited use in online analysis. Online wet chemical COD analyzers are available, but these are not commonly used. In addition, online methods for measuring COD are available, but these methods are surrogate methods in which COD is estimated using a correlation with the laboratory method. The three main methods of online measurement of COD are

UV absorbance, thermal combustion, and electrochemical oxidation. These methods are discussed in the following subsections.

3.5.1 *Principle of Operation*

In the UV absorbance method of measuring COD, a spectrophotometer measures the intensity and wavelength of UV light that passes through a sample. The intensity and wavelength of light passing through the sample is compared to the intensity and wavelength of light being shined on the sample to determine the intensity and wavelength of light absorbed by the sample. The intensity and wavelength of UV light absorbed by the sample is then correlated to the COD concentration in the sample.

In thermal combustion, a sample is pumped from a process area and injected to a furnace that is operated at temperatures up to 1200 °C. Once the sample enters the furnace, it is completely oxidized. The gas from the furnace goes through a cooler and an absorption column, then into a gas analyzer. The gas analyzer determines the amount of COD in the sample.

In the electrochemical oxidation method, an electric current is passed through an electrode producing hydroxyl (OH) radicals. The sample is brought into contact with the OH radicals, where it is oxidized and the OH radicals are converted into OH ions. This process causes a change in the electrode current. The COD concentration is correlated to the change in current.

3.5.2 *Accuracy and Repeatability*

Although the accuracy of COD analyzers varies with the type of instrument used, most instruments can achieve accuracies of ±3%.

3.6 Nutrient Analyzers

Ammonia-ammonium, nitrate-nitrite, and phosphorus-orthophosphate analyzers are discussed in detail in Sections 3.6.1 through 3.6.4. Regarding the accuracy of nutrient analyzers, it is important to note that consistency is often more important than accuracy. Analyzer results are often discredited because they do not match laboratory tests. However, full-scale experience has shown that the information they provide is just as valuable as long as they are consistent because operators will learn how to interpret the readings.

3.6.1 *Ammonia and Ammonium*

Ammonia (NH_3) is a gas which, when dissolved in wastewater, exists in equilibrium with ammonium (NH_4). The equilibrium between the two molecules is dependent on pH and temperature. When applicable, most WRRFs have a permit limit expressed in terms of ammonia. However, depending on the measuring principle used, instrument manufacturers may express results as either ammonia or ammonium. To avoid confusion, most results are reported in terms of nitrogen (ammonia-nitrogen [NH_3-N] or ammonium-nitrogen [NH_4-N]).

Instruments for measuring ammonia and ammonium are discussed in the following subsections.

3.6.1.1 *Principle of Operation*

The three principal methods for measuring ammonia-ammonium are ion-selective electrode (ISE), gas-sensitive electrode (GSE), and colorimetry. In principle, ammonium ISEs are similar to a pH electrode. However, instead of a glass bulb, the sensors have a specialized membrane that allows ammonium to pass through the membrane into the cell. Ammonium is destroyed in the cell through a chemical reaction. The voltage created by the reaction is proportional to the ammonium concentration in the sample. Ammonium ISEs are the most common method used for online measurement of ammonium.

An ammonia GSE is, effectively, a pH electrode behind a gas-permeable membrane. Sodium hydroxide is added to the sample to raise the pH above 11 and to drive all the ammonium to ammonia. The ammonia gas is allowed to pass across the GSE membrane and then redissolves in an electrolyte surrounding a pH electrode. The ammonia gas increases the pH of the electrolyte. The analyzer then converts this change in pH to an ammonia reading. A GSE is not suitable for online measurement because the sample has to be withdrawn from the tank and taken to the laboratory.

Ammonia can be measured using the colorimetric method. In the colorimetric method, a chemical reaction causes a color change in the sample. The color change is proportional to the ammonia concentration in the sample. There are three main colorimetric methods: Indophenol blue method; Monochloramine-F method; and a third colorimetric method in which ammonium in the sample is converted to ammonia gas, stripped from the sample, and then redissolved in a liquid pH indicator solution. The change in color of the indicator solution is converted to an ammonia reading. Colorimetric methods are not commonly used at wastewater facilities.

3.6.1.2 *Accuracy and Repeatability*

Ammonia ISEs measure between 0.2 and 1000 mg/L NH_4-N, with an error of 5% of the measured value, or 0.2 mg/L NH_4-N, whichever is greater. Gas-sensitive electrodes are more precise, measuring between 0.02 and 1000 mg/L NH_4-N with an error of 2% of the measured value or 0.02 mg/L NH_4-N, whichever is greater.

3.6.1.3 *Applications*

Applications of ammonia and ammonium analyzers are discussed in Section 3.6.4.

3.6.2 *Nitrate and Nitrite*

3.6.2.1 *Principle of Operation*

Online measurement of nitrate (NO_3) in wastewater is most commonly achieved using UV absorbance. Nitrate and nitrite in water absorb UV light between 215 and 240 nm. Online instruments direct UV light through the sample and measure the amount of light that passes through to a detector. The absorbance is calculated and converted to a nitrite-nitrate value. The result is provided in milligrams per liter of NO_x-N (where $NO_x = NO_3 + NO_2$) because both nitrate and nitrite absorb light in that spectrum. In wastewater facilities, the concentration of NO_2 is generally insignificant and considered part of the error of the measurement. In some instruments, light of a different wavelength (a reference wavelength) is also passed through the sample to allow for compensation for interference resulting from solids and organic material that also absorb UV light.

Another method of measuring nitrate uses ISE. The ISE works in the same manner as was described for ammonium. This technology is not frequently applied because of the convenience of the direct UV absorbance method.

Instruments that measure nitrite using UV absorbance have recently been developed. These instruments measure the UV absorbance of a sample and then separate the absorbance of nitrite from that of other compounds that absorb UV light by either using a proprietary algorithm or comparing the spectra of the sample to a library of known spectra. Nitrite can also be measured using online colorimetric methods.

3.6.2.2 *Accuracy and Repeatability*

Ultraviolet nitrate probes typically provide a lower detection limit of 0.05 to 0.1 mg/L NO_x-N. The measurement range may extend as high as 50 mg/L NO_x-N, although this may require a longer path length. Claimed accuracy varies from 2 to 5% of the reading, and repeatability is in the same range.

3.6.2.3 *Applications*

Applications of nitrate and nitrite analyzers are discussed in Section 3.6.4.

3.6.3 *Phosphorus and Orthophosphate*

Total phosphorus consists of orthophosphate, organic phosphate, and polyphosphate. There are instruments available that measure total phosphorus. Most total phosphorus in wastewater is typically present as orthophosphate (PO_4). Therefore, most instruments measure orthophosphate. Results are typically reported in terms of orthophosphate as phosphorus (PO_4-P).

3.6.3.1 *Principle of Operation*

All online orthophosphate instruments use colorimetry. The most commonly used methods are the following: the molybdovanadate method, also referred to as the *yellow method*, and the ascorbic acid method, also referred to as the *blue method* (or *molybdenum blue method*). In practice, the yellow method is most widely applied because of the relative simplicity of the instrument and related reagents. In instances where low detection levels are required (<0.1 mg/L PO_4-P), the blue method is used.

Online total phosphorus instruments first convert polyphosphate and organic phosphorus compounds to orthophosphate. Orthophosphate is then measured colorimetrically. The conversion step requires high pressures and temperatures; it is this step that makes online total phosphorus instruments relatively expensive and maintenance intensive.

3.6.3.2 *Accuracy and Repeatability*

Most orthophosphate analyzers that use the yellow method are capable of measuring over a range from 0.05 mg/L to as high as 50 mg/L PO_4-P. The measurement error is approximately 2% or 0.05 mg/L, whichever is higher. Analyzers that use the blue method can measure down to 2 mg/L PO_4-P. The measurement error is typically ±4% of measured value or ±0.02 mg/L PO_4-P, whichever is greater. Total phosphorus analyzers are available with measuring ranges from as low as 0.3 mg/L to as high as 100 000 mg/L. The measurement error is typically 3% of the measured value.

3.6.3.3 *Applications*

Most discharge permits for phosphorus in the United States are written in terms of total phosphorus. Although online total phosphorus instruments are available, they are relatively expensive and require a high level of maintenance. As a result, orthophosphate instruments are more widely accepted, particularly for process monitoring and control, and, in many instances, for effluent monitoring. Applications of phosphorus and orthophosphate analyzers are discussed further in Section 3.6.4.

3.6.4 *Nutrient Analyzer Applications*

Online nutrient analyzers discussed in Sections 3.6.1 through 3.6.3 are used in the following applications:

- Influent monitoring for feed-forward control of chemical dosing and mass-load balancing;
- Mass-load equalization;
- Aeration control;
- Recirculation control;
- At the exit of an anaerobic zone to ensure adequate release of phosphorus from phosphate-accumulating organisms; it can also be used to limit excessive phosphate release by controlling recirculation (containing nitrate) to the anaerobic zone;
- At the exit of the aeration basin to ensure adequate uptake of phosphorus; and
- Effluent monitoring (facility effluent, clarifier effluent, etc.) for feedback control of a chemical dose.

3.7 Instrument Installation

There are two methods to install the instruments discussed in Sections 3.1 through 3.6: in-tank or open-channel installation and free-standing installation. With the in-tank or open-channel approach, the analyzer is installed directly to the process tank or channel. This provides a faster response time and alleviates the need to transport sample to the analyzer with a pump. The analyzer should be installed in a well-mixed zone to provide a representative sample of the process. If the probe is installed in an open channel, it should be located in a free-flowing zone.

The main disadvantage of the in-tank or open-channel approach is the inconvenience of having to remove the analyzer for reagent replacement and maintenance. This task may become overwhelming if analyzers require frequent maintenance.

Under the free-standing approach, the instrument is installed in a sampling station located away from the process tank or channel. Sample must be delivered to the analyzer by a pump and is often required to be filtered. The design of sample pumping, piping, and filtration has critical implications on the quality of data being produced by the overall system. An important factor to consider is the detention time in the piping, which should be kept to a minimum. One approach to minimize detention time is to deliver a high-volume sample stream to the analyzers. Only a small portion of the sample stream will be transferred to the analyzer and the rest of the stream will be wasted. Another important consideration is potential buildup of biomass in the piping. Biomass can react with wastewater and alter the characteristics of the sample stream before it reaches the analyzer. Therefore, it is necessary to provide some means of periodically flushing the line (often with a hypochlorite solution) to strip biomass from the pipe walls. Installation site lists, which are typically provided by an analyzer manufacturer, should be checked to learn more details about successful design of a sample delivery and preparation system.

Free-standing analyzers are generally wall mounted and sometimes require a shelter. When this approach is implemented, all analyzers and other high-maintenance instruments are often located in the same area to provide easier service. Buffer solutions needed for standardization and other analytical instrument reagents can also be conveniently stored in the analyzer area.

3.8 Instrument Maintenance

Most instrument manufacturers provide a variety of features to reduce maintenance, such as automatic cleaning and calibration and the use of peristaltic pumps and pinch valves (so the sample stays inside tubing and does not contact any moving parts). Nevertheless, some of the more sophisticated instruments may require extensive routine maintenance.

The following maintenance steps and frequencies are common to most online analyzers:

- Routine replacement of reagents typically occurs monthly or quarterly, but can vary according to the measurement, calibration, and cleaning intervals chosen;
- Cleaning or replacement of electrodes (for pH probes and ISE and GSE instruments) on a monthly basis. Automatic sensor cleaning systems are available on some instruments;
- Replacement of sample tubing every 3 to 6 months;
- Regular inspection (and cleaning, if necessary) of the sample filtration system;
- Regular cleaning of optical windows (for spectrometer instruments). Although this procedure is sometimes performed manually, there are instruments that have automatic cleaning systems; and
- Regular calibration (weekly calibration is a reasonable frequency); calibration should be based on laboratory analysis of the same grab sample processed in triplicate.

With the increased complexity and sophistication of control system hardware, it has become increasingly important to train and retrain control system staff to ensure that systems are operated correctly and that instruments and network systems are functioning properly. Many facilities that are upgrading from conventional treatment to BNR systems are not staffed to handle the new instrumentation that goes along with the process conversion. Facilities that are upgrading to BNR should familiarize instrumentation and control staff with the new process by involving them in construction and startup of the process.

4.0 FACILITY CONTROL AND SUPERVISORY CONTROL AND DATA ACQUISITION SYSTEMS FOR NUTRIENT REMOVAL PROCESSES

An important component of the successful operation of a BNR facility is the control system used to monitor process data and adjust the operation of process equipment. Essentially, the purpose of a BNR facility control system is to provide timely information and, where possible, to make process adjustments automatically to keep the facility operating safely within the desired parameters so that facility staff can perform their jobs properly and efficiently.

The control system, as a whole, is referred to by different names, including *plant control system* (PCS), *supervisory control and data acquisition* (SCADA), and *distributed control system* (DCS), with some seeing a distinction between these terms and others using them interchangeably. Traditionally, *SCADA* has referred to systems that are used to monitor and send supervisory control commands to remote locations, such as pump stations, while *DCS* and *PCS* are used for systems that monitor and control processes within a facility. In addition, *DCS* generally refers to proprietary industrial control system packages, while *SCADA* and *PCS* refer to a collection of components that are based on widely accepted hardware and software and that use open technology to integrate the components to a unified control system. In the water and wastewater industry, *SCADA* is most often used to refer to either the control system as a whole or to just the server-based data collection and operator-interface components.

Regardless of the terminology or which approach is taken (i.e., DCS or SCADA), all systems will perform the following core functions:

- Process data acquisition and process equipment adjustments (i.e., input and output [I/O]);
- Basic process control (regulatory control);
- Alarm generation and management;
- Supervisory, or advanced, control;
- Providing a user interface to the process;
- Storage and retrieval of historical data for display, analysis, and reporting; and
- Control of user access to facility areas or control system functions (i.e., security).

Table 14.2 lists some of the common terminology and definitions used for control systems. Some of these terms are used differently by different users; therefore, operators should obtain a clear definition of how these terms are being used in discussions about the control system for a complete understanding of how an approach will affect the way a process is monitored and controlled.

The components of a facility control system can be divided into three broad categories: the control system hardware and software that comprise the "infrastructure" of the system, the custom control algorithms that interface to field devices and control the process, and the user interface systems that provide a human-friendly window to the process. In addition to these three fundamental categories, additional features such as system redundancy; application revision control; interfaces to other systems such as change management systems and learning management systems, libraries of wastewater graphic symbols, and Web-based user interfaces; and advanced historical data collection and analysis can often be purchased to expand system functionality with "off-the-shelf"

TABLE 14.2 Common terminology.

Term	Explanation
Client	Workstation or application that receives its data and/or files from a "server".
Data I/O server	A server application that is tasked with handling inputs and outputs to and from the SCADA system. The application may run on a separate server computer in some systems or it may be bundled with other SCADA server applications.
Distributed control system	Originally referred to as *system architecture,* where I/O and process controllers are distributed throughout the facility instead of centrally located, but is now also used to describe proprietary control systems where all of the parts of the system are designed and supplied by the same manufacturer to work as a system instead of as stand-alone components that can work within other systems.
Developer license	A software license that allows programming and configuration of control system applications. The developer licenses are typically more expensive than "runtime" licenses.
Domain name system server	A server application that assigns and manages Internet Protocol addresses on a network and from which user groups can be managed.
Driver	A software application that allows a computer to use a device or allows communication between a SCADA computer and other networked devices such as programmable logic controller.
Engineering workstation	A workstation from which programming and configuration can take place. In some systems, especially in DCSs, it may be a computer configured with special software and interfaces not available on a typical personal computer.
Fiber optic	A networking physical media that allows data to pass through it via pulses of light.
Firewall	A hardware or software application that limits access to a network, computer, or application. Often, both are used in SCADA systems to provide a broader range of protection.
Historical data	A measurement or event that has occurred in the past.
Historical database server	A server application that collects and stores time-based data from the SCADA system and makes it available to clients for trending and analysis.
Human–machine interface	Provides human-friendly interface to the process, most often through graphical displays and through management of alarms and display of historical information as time-based trends.
Operator interface terminal	This is often meant to refer to a graphical user interface with limited functionality (such as limited alarming and trending), intended only to provide a user with convenient local access to the facility control system information for a particular area of the facility.

(continued)

TABLE 14.2 Common terminology (*Continued*).

Term	Explanation
Object Linking and Embedding for Process Control	A software standard that is based on the open OPC Foundation standards for process control devices to communicate with each other.
Programmable logic controller	Programmable logic controller.
Protocol	The method that data on a network are structured and interpreted by the computers using the data. Can view this as the "syntax" that a computer network uses. Examples are Ethernet, RS-232, Modbus, and so on.
Protocol converter	Either a software application or a specialized industrial module that converts one protocol to another.
Real-time data	A measurement or event that is being updated from the process controller as it changes.
Runtime license	A software license that allows a software application program to execute, but does not allow the application to be modified.
Supervisory control and data acquisition	Supervisory control and data acquisition.
Scripting	Using an HMI application's built-in high-level programming language to perform functions that do not have a built-in software method already defined.
Server	Either a class of computer or a software application. A server as a class of computer is designed with more computer resources. A server application provides data, files, and other computing resources to client applications.
Source code	The programming statements or configuration written using a particular programming application. The source code is then converted to a program (a process called *compiling*) that can be executed by a computer or controller. The compiled program is not modifiable, so modifications to a program must be done through the source code.
Terminal server	A server application that allows multiple client computers to use applications on the server.
Thick client	A client computer that runs client application software. This is typically a workstation that has peripherals and adequate memory and drive space to run applications and to allow additional software to be installed.
Thin client	A client device that runs only a Web browser and communication software required to communicate on a network and interface with a user through a monitor, keyboard, and mouse. It connects with a terminal or Web server that gives the user a window into the server's applications, but the client itself runs only limited software and, therefore, needs few computer resources. Thin client devices are typically small and inexpensive.
Web server	A software application that stores a variety of types of files such as documents, text, and images and makes them available to Web browsers.

software. The remainder of this section will discuss each of these categories as they apply to BNR facilities.

4.1 System Infrastructure and Components

One of the advantages of modern facility control systems is their flexibility and scalability, which allows them to be affordable for different sizes of BNR systems and owner organizations. Figures 14.1 and 14.2 show typical SCADA infrastructures for a smaller and larger facility or organization, respectively.

4.1.1 *Packaged Equipment Control Systems*

Some BNR process equipment, such as process air blowers and facility water filter systems, are typically supplied with an integral control system. Depending on the complexity of the equipment, the controls may be programmable-processor-based

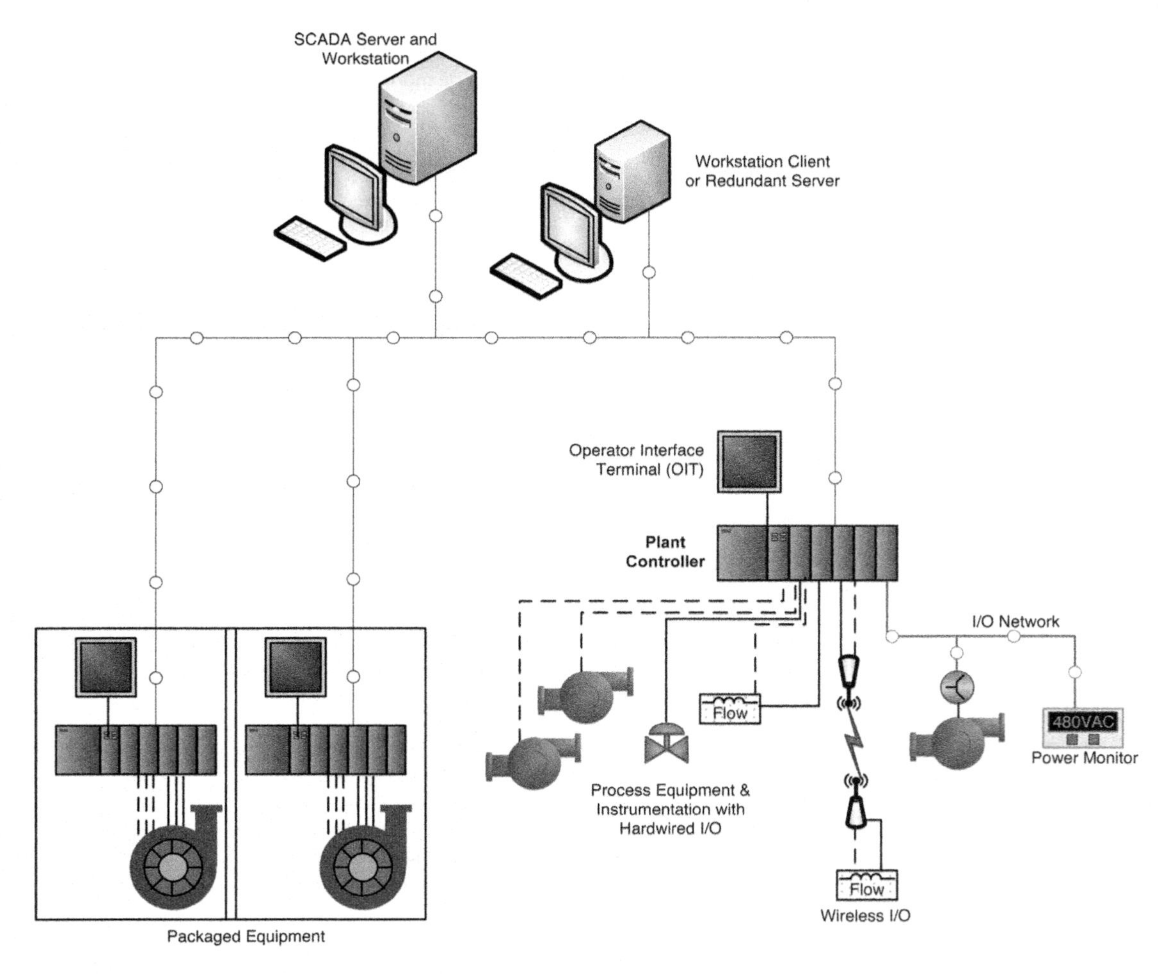

FIGURE 14.1 Small-system SCADA.

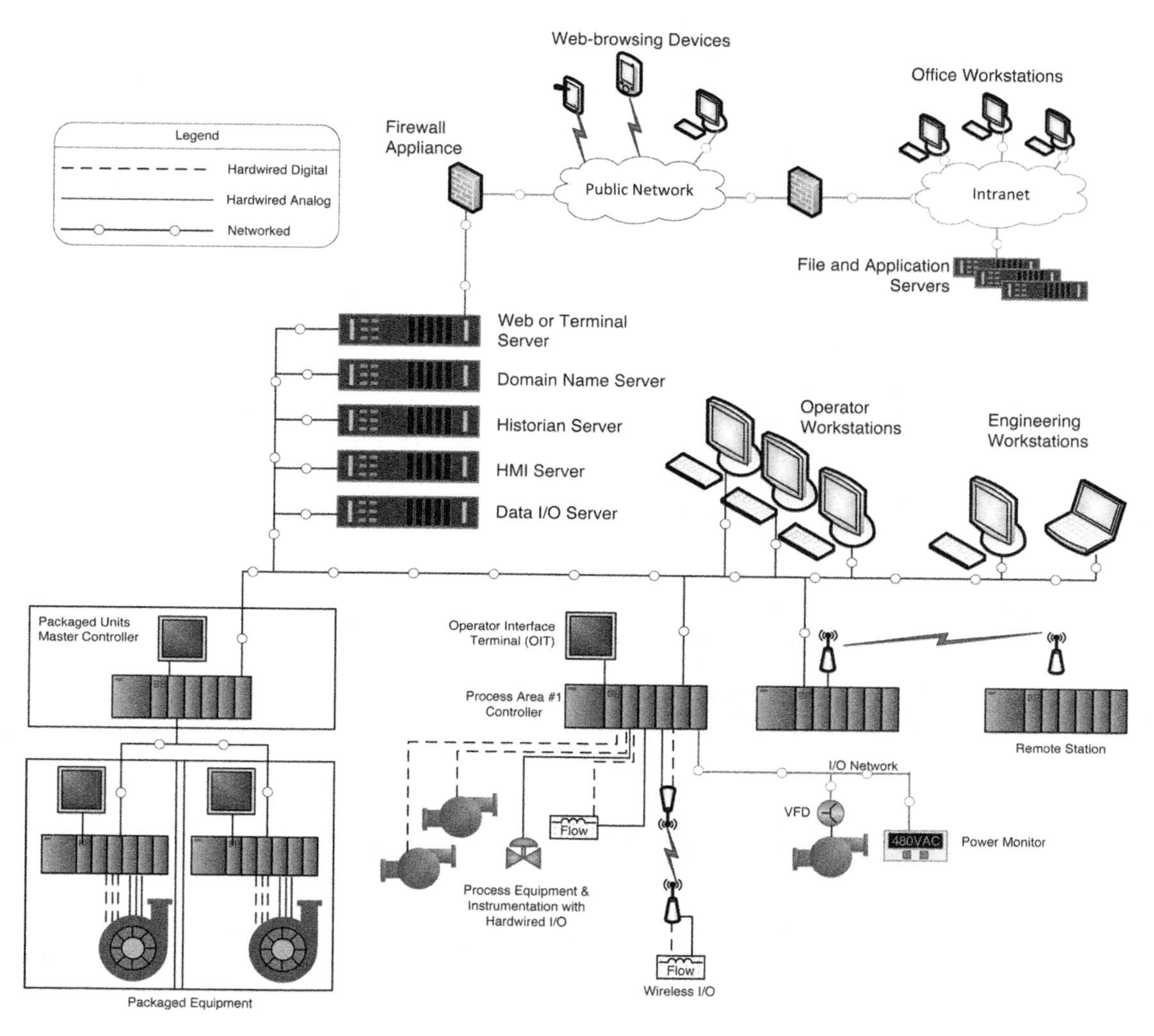

FIGURE 14.2 Larger-system SCADA.

and include a local operator interface and a communication port to the facility control system. In other instances, the control package may be simple and provide only a few hardwired signals to interface to the facility control system. For more complex equipment to be networked into the facility control system, it is preferable to have the controller and operator interface within the same family of facility controllers and user interfaces to reduce capital and maintenance costs. While a few vendor-supplied control systems are viewed as proprietary by the equipment supplier, most will provide application programs for maintenance and troubleshooting purposes. Care should be taken to ensure that any passwords for the system are made available upon equipment turnover and that any required software licenses are registered to the owner.

4.1.2 *Input and Output Subsystems*

There are many different ways of getting information to and from process field instrumentation and devices. The most common way is to "hardwire" individual instrument I/Os to the process controller panel; however, I/O networks can also be used where an instrument or device has more detailed information to be monitored (such as variable-frequency drives, power monitors, analyzers, and even valve controllers). Deciding whether to use network-based I/O depends on how much information from field devices is available and useful to operations or maintenance and how critical the information is to the control of the process. Because hardwiring is simple and easy to troubleshoot, critical signals from a control device might be hardwired (e.g., the run status, run command, speed setpoint, and actual speed for a variable-frequency drive), while other information (e.g., frequency, amperes, and fault conditions) may be monitored via the I/O network. The advantage that networked I/O has over hardwired I/O is that more information can be monitored with less wiring, which, therefore, can reduce installation, operating, and maintenance costs.

4.1.3 *Process Controllers*

Process controllers are basically specialized industrial computers that run user programs to condition I/O for digital processing and run the algorithms that allow humans to monitor and control the process. It is often useful to dedicate one or more controllers to a specific process area, and especially to different lines of the same process area; this way, if one processor fails, the rest of the facility may continue to operate.

4.1.4 *Human–Machine Interface Systems*

There are several methods of providing access for operating personnel to monitor and control the process, and it is common to use several methods within the same facility. Most equipment and many instruments provide field-mounted, operator-accessible switches, pushbuttons, and indicators for convenient maintenance and testing. In addition, some equipment and process areas may have a local control panel that will provide manual operation of equipment such as start and stop commands and, sometimes, basic automatic operation such as on–off control based on a field input (such as a level switch) and automatic switching of duty and standby equipment.

While some equipment will only be operated through these local interfaces (such as skimmers), most will be operated remotely through the SCADA system. As shown in Figures 14.1 and 14.2, operator interfaces can be run on a combination of servers and workstations or, for small systems, on one high-end workstation. Many human–machine interface (HMI) suppliers also have terminal or Web-server applications available so that system displays can be viewed from thin clients (i.e., computers that depend on another computer, or server, to fulfill some of their computational roles) or other devices such as tablets and smartphones that have Web browsers.

4.1.5 *Control Network*

The control network is the means through which process controllers and HMI computers communicate with each other; therefore, it is a critical part of a facility control system. The control network type is specified by the media (i.e., physical means that are used for data to travel through), topology (i.e., how the network is arranged), and protocol (i.e., the "syntax" used for devices to talk to each other). In addition to media that comprise the network, various types of connecting devices (such as switches), media converters, and protocol converters can be part of the network.

4.1.6 *Alarm and Event Notification*

Some systems use an alarm dialer to e-mail and/or call a telephone number when certain critical alarms or events take place. There are two basic approaches to using an alarm dialer: a hardware-based dialer that either connects directly to the area control processor or has hardwired inputs or a software-based dialer application that is installed on a SCADA server or client. The hardware-based dialer is simple and provides the advantage of being active even if the network is down. However, the software-based approach allows more flexibility and, often, a greater number of alarms. Some owners have used a hardware-based approach with software as backup, or vice versa.

4.1.7 *System Software and Licenses*

The system hardware components described in the previous sections require software applications to configure, program, and run, as intended. There are two categories of software: system applications and user applications or configurations. System applications include programming and runtime software, operating systems, device drivers, and any other software required to allow the device to operate. The user applications are the programs or configurations written to perform the specific operations required to monitor and control the process. Many system applications require software licenses that are purchased from the supplier either one time or are renewable over some time period, typically annually. In addition to license fees, some software providers offer a support agreement that allows access to online resources, technical assistance, and software-version upgrades.

4.2 Process Control Schemes and Algorithms

Control of the process is carried out at different levels between the field device and through to the programmable process controllers and HMIs. It is important to have a basic understanding of the layers of control and the typical algorithms (in programming, an approach to performing a specific task) so that appropriate action can be taken under different operating conditions such as startups, shutdowns, maintenance, normal operations, and process upsets. In most instances, the operator should be allowed to choose the level of automatic operation, except for bypassing of safety or equipment protection interlocks. With some equipment, however, it may be undesirable to allow the operator take direct control of a device (such as when two valves must operate together). Table 14.3 summarizes common process control schemes and algorithms used in BNR facilities. The table is arranged from the lowest level of operation (closest to the field device) to the highest level of operation (advanced and supervisory control).

4.3 Human–Machine Interface Configuration

Given that the HMI is the primary method to show a large amount of complex process information and to enter commands and setpoints for process controller algorithms, good planning and thorough user reviews of database tags, graphical displays, alarms, historical and real-time trends, and user access to facility areas and levels of control as they are being developed or modified should be exercised. The time invested in planning and reviews will more than pay for itself in terms of effective operation of the facility, lowered training costs, faster process and system troubleshooting, and easier maintenance and modifications.

4.3.1 *User Displays*

User displays are typically developed hierarchically based on the level of detail required for different users and different operating conditions. To create easy-to-use and intuitive displays, a standard set of colors, symbols, fonts, screen navigation order and methods, information to display, nomenclature, menus, and display layouts should be developed and maintained.

4.3.2 *Alarm Management*

Proper alarm management is critical to effectively operating a BNR system and should, therefore, be planned during system development and made part of facility control system standards. Methods such as time delays and deadbands should be programmed into process controllers to prevent nuisance alarms. Alarming should be disabled for informational-type tags and alarms should be classified into two or three criticality categories so that the most critical can be displayed first in alarm summaries and can be assigned a different audible alarm tone. In addition to assigning alarms to alarm criticality categories, alarms may be assigned to one or more process areas so that they can be displayed in different alarm summaries or so that only authorized users can acknowledge or clear alarms for different process or system areas.

TABLE 14.3 Process control algorithms.

Control approach	Function
Emergency stop	Unconditionally shuts off power to the control circuit, immediately stopping the equipment being controlled. This must be hardwired directly to the equipment control circuit and the switch that activates it must be easily accessible to the operator near the rotating equipment it is meant to stop.
Interlocks	A condition that will stop operation of a control element or equipment. Safety and equipment protection interlocks are often hardwired into the control circuit. The software program may include additional interlocks.
Permissives	A condition that will prevent operation of a control element or equipment, but will not shut it down. Safety and equipment protection permissives are often hardwired into the control circuit. The software program may include additional permissives.
Proportional–integral–derivative loop control	Controls a measured value (the process variable) by adjusting the output to a final control element (such as a valve or variable-frequency drive speed), to maintain the process variable at the desired level (the setpoint). The operator typically enters the setpoint.
Cascade	This control mode automatically adjusts the setpoint of a secondary controller to maintain the desired process variable of a different process variable. When in cascade mode, the operator cannot adjust the setpoint of the secondary controller.
Ratio	This control mode adjusts the setpoint of a secondary controller to maintain a desired ratio of the secondary controller's process variable to the ratio controller's process variable.
Trim	An adjustment made to a control output based on the measurement from another process variable. This is typically used to provide feedback control to a flow-pacing approach.
Bias	A positive or negative adjustment for one or more devices being controlled by a common control output. The operator typically enters a bias, but this may be determined by the logic.
Duty and standby	This logic will automatically start the standby equipment when the duty equipment has failed. The operator typically designates which equipment will be assigned as duty, but the logic may also be set up to automatically rotate the duty equipment based on some condition such as runtime.
Lead-lag	This logic will automatically start or stop the next piece of equipment in series when it is needed to meet a certain setpoint.
Function curve	A function curve may be used to adjust setpoints to secondary controllers based on a function of some value.
Most open valve	This is used when a set of valves is used in combination to maintain a process variable setpoint, but the position of the valve that is most open is maintained at a desired setpoint so that the valves will remain in a controllable range.

4.3.3 *Security and User Access*

To prevent inadvertent modification of setpoints, control parameters, or alarm limits or operation of equipment by unauthorized users, a user access configuration can be implemented within the HMI. Some possible user types include "administrator", "engineer", "manager", "supervisor", "senior operator", "operator", "maintenance", and "guest". In addition to the user types, process or control system areas can be created to allow access to only certain areas of the facility for the same type of user. For example, if a user category of "senior operator" is defined such that these operators are allowed to make setpoint and alarm limit changes, one senior operator may be allowed access to primary treatment and another may be allowed access to secondary treatment.

Full-system security is not limited to preventing access to different types of operation in different process areas. Care must be taken to physically protect the system from environmental damage, unauthorized access, and introduction of harmful software. Locking servers in a secure, environmentally controlled room; using thin clients for operator workstations; disabling workstation peripherals such as external drives and universal serial bus ports or locking the workstations in a cabinet; using monitor, keyboard, and mouse extenders; and keeping remote access to the control network protected through firewalls can reduce the risk of damage to the control system.

4.4 Additional System Features

Features of the facility control system described in the previous sections can optionally be expanded to provide additional system reliability through redundancy, more flexible and robust management of historical data, the ability to share and retrieve data from other types of systems, software-version backups and management, compliance and operations reports, and commercial advanced control applications developed specifically for wastewater treatment. While custom features can often be programmed into the system using a scripting or programming language, this should be kept to a minimum to prevent problems during troubleshooting, maintenance, and system upgrades.

4.5 Maintenance Requirements

The facility control system requires regular maintenance to keep performance at the required level. Process alarms should be checked daily and corrected so that no more than a few alarms for temporary conditions are active. Prompt attention to instrument and equipment malfunctions, loop-tuning issues, and control system malfunctions will allow advanced control features to remain on more often, thereby improving the effectiveness of BNR processes. A planned schedule for system software upgrades to address issues with the system or to implement security and feature upgrades needs to be established and maintained. The status of software licenses needs to be monitored and managed so that licenses will not expire and disable required applications. In addition to these routine maintenance requirements, modifications to the system such as the addition of new process areas or equipment need to be reviewed for compliance with system standards and integration to the existing display hierarchy so that the system will continue to be intuitive and user friendly.

4.6 Conclusion

The facility control system is an essential part of BNR and, if well planned and maintained, can greatly improve the effectiveness of the process. Because people that will make the process control successful are key users of the system, it is essential that they understand how to use the control system and to spot and correct problems so that the advanced control strategies designed to optimize control of the process can be kept online.

5.0 OPTIMIZATION AND ADVANCED CONTROLS

5.1 Diurnal Load Variations

The effects of diurnal variations in flow and loadings do not allow facilities to achieve true steady-state operating characteristics. For certain process technologies, this can be particularly challenging. A batch process like sequencing batch reactors (SBRs) (see Chapter 6), repeating several times a day, may experience highly variable loads over the fixed process cycle time. In these instances, without sophisticated control systems to manage the aeration blowers for dissolved oxygen control and allowing for adequate oxygen depletion time in transitioning to anoxic conditions, it is likely that either incomplete nitrification will occur because of oxygen limitations or incomplete denitrification will result from a lack of available carbon or excessive residual dissolved oxygen levels.

Conventional aeration basins provide a naturally occurring buffering of flow and load variations. It is also easier to maintain targeted dissolved oxygen levels at certain points in the treatment process. However, separate clarifiers, waste and return MLSS pumping systems, and other factors generally result in more costly construction to achieve high-quality effluent characteristics.

5.2 Respirometry

Respirometry is the continuous measurement of the specific oxygen uptake rate (SOUR) of microorganisms in the activated sludge process as they consume organic matter. Respirometry has been widely used to characterize waste and as a method of determining kinetics. Development of respirometric control has been made possible by the use of an online respirometer. This instrument has been used to conduct surveys of WRRFS, to develop diurnal load profiles, and to optimize the operation of many facilities. Several control approaches have been suggested, and they are dependent on the type of activated sludge facility that is to be controlled. Oxidation ditches, completely mixed tanks, SBRs, and plug flow reactors are all suitable candidates for respirometric control.

Respirometric control was shown to be the most effective approach for improving nitrogen removal in an oxidation ditch in Beemster, Holland (Draaijer et al., 1997). In this application, an in situ respirometer was used to measure the oxygen load present in the ditch and to simply control the number of aerators in operation to match demand.

Respirometric control is an obvious choice for SBRs, and some promising investigations have been carried out in this field by Cohen et al. (2003) and Yoong et al. (2000). The simplest method of respirometric control, proposed by Shaw and Watts (2002), is to discontinue aeration once the respirometric measurement indicates that the biomass is endogenous and allow an idle phase that makes use of endogenous denitrification to improve overall nitrogen removal (Shaw and Watts, 2002). Some SBR facilities have implemented a dual-process cycle approach in which two aerated and anoxic cycles are used per batch to allow for initial nitrogen removal, followed by denitrification, and then repeated a second time within the overall treatment cycle. This allows for more readily available carbon during the initial denitrification step.

Plug flow reactors provide the greatest challenge to implementation of respirometric control, although they were one of the first activated sludge processes considered for advanced control by the foremost proponent of respirometric control, Dr. John Watts (Watts and Garber, 1993). Watts and Garber proposed that anoxic and aerobic swing zones could be used with respirometric measurements to adjust the overall aerobic volume to match influent load. Rapid respirogram generation is required to facilitate dynamic control; alternatively, respirometry can be used to build a picture of typical load variations, and this can be used to adjust anoxic and aerobic volumes.

Regardless of which BNR process is used, respirometric control can be used to improve the performance of the facility. There is great potential to provide better nitrogen removal, to increase nitrifier viability by reducing the amount of time that the biomass is endogenous, and to reduce energy costs by matching aeration to the nitrogenous oxygen demand in the facility.

5.3 Intermittent Aeration

Intermittent aeration in a continuous-flow system (as opposed to SBRs) has been proposed as an effective way to achieve nitrification and denitrification in a single reactor. In this process, as described in Chapter 6, the biomass in the main reactor (or reactors) is subjected to a cycling of aerobic and then anoxic conditions as the aeration supply is switched on and off to that particular basin. Because the reactor repeatedly changes from an aerobic state to an anoxic state and back again, an obvious measurement choice for control is ORP (refer to Chapter 15). Several examples of ORP control systems exist, including those described by Caulet et al. (1997), Charpentier et al. (1998), and Mori et al. (1997). In these systems, ORP is used to determine when denitrification is complete under anoxic conditions and when nitrification is complete under aerobic conditions. Carucci et al. (1997) also recognized the potential for using ORP to control intermittent aeration, but contended that pH was a more suitable control parameter based on their experiments at the Cisterna di Latina Wastewater Treatment Plant (Cisterna di Latina, Italy). This approach is now being used in SBRs that, by design, are intermittent aeration applications.

The other main approach to controlling intermittent aeration is to simply use dissolved oxygen and timers to achieve the required treatment. In their publication, *Introduction of Japanese Advanced Environmental Equipment*, the Japan Society of Industrial Machinery Manufacturers (2001) describes a simple control scheme that runs the aeration system to low dissolved oxygen (0.2 mg/L) for a set period to provide mixing, but no nitrification, and then increases the dissolved oxygen to 2.5 mg/L to provide fully aerobic conditions for nitrification.

5.4 Cooperation in the Field of Scientific and Technical Research Model for Control Strategy Development

Much of the current development of control algorithms uses process models. To standardize the way in which models are used to develop control approaches, in 2001, the benchmark group of the European Cooperation in the Field of Scientific and Technical Research (COST) Action 624 published guidelines for several benchmark models that can be used to test control strategies (Copp, 2001). This work is now continued under the International Water Association Task Group on Benchmarking of Control Strategies for WRRFs.

5.5 Model-Based Process Control

Some facilities are working to combine the use of modern process simulation models to be a predictive tool for active process control. This can be approached in two ways. The first is to use historical trend data and to input that information as a range of conditions that the treatment facility sees on a daily, weekly, or seasonal basis and to project the best combination of process control parameters to use based on the details of the information and the goal of the effort.

The second approach is to take online information, such as MLSS, dissolved oxygen, ammonia, and nitrate from a field instrument and feed that information in real time to a simulation model. This allows for fine tuning both the model and process performance simultaneously. Although this approach is more labor and capital intensive, it may make sense for large facilities to meet stringent regulatory compliance standards.

5.6 Supplemental Carbon, Chemical Phosphorus Removal, and Ammonia-Based Aeration Control

Additional optimization opportunities include control approaches for supplemental carbon addition, chemical phosphorus removal, and ammonia-based aeration control. Carbon supplementation using a source such as methanol, acetate, glycerol, or molasses to support denitrification in anoxic reactors or denitrification filters may be controlled based on flow and effluent nitrate analyzer trim. Overfeeding chemicals is a waste and will increase the BOD of the process effluent stream. Underfeeding reduces the amount of nitrate removed and may cause an elevation of nitrite levels. Initial dosage may be calculated based on the carbon source, but actual dosage will be determined in the field. Once an optimal dosage is determined, the chemical feed systems are typically flow paced to increase or decrease with changes in flow. More advanced control could be achieved with nitrate analyzers, although this is not typical because flow control is generally sufficient. Additional information on this topic can be found in Chapters 5 and 6.

Chemical phosphorus removal is typically controlled based on flow pacing of chemical addition. The optimal ratio of chemical to phosphorus is determined in the field and can be fine tuned periodically as conditions change. This assumes that the influent phosphorus and the phosphorus needed for biological growth are consistent over a period of time and, therefore, flow pacing is appropriate. A more detailed discussion on chemical phosphorus removal is provided in Chapter 8.

Ammonia-based aeration control is the control of aeration based on ammonia concentration, rather than dissolved oxygen. There are two main types of ammonia-based control: feedback control and feed-forward control. In feedback control, aeration is adjusted to maintain an ammonia setpoint in the bioreactor. In feed-foward control, aeration is adjusted based on monitoring the influent ammonia concentration.

5.7 Conclusion

It is important to remember that each treatment facility has a specific performance objective to meet and that they seek to exceed, and that an approach that is appropriate for one facility may not be suitable for another. All of the factors discussed in this manual, whether related to facility size, pollutant loadings, process technology, variations in flows and loadings, or availability of capital and operating funds, must be considered in terms of what is the most effective and efficient way to achieve treatment objectives of the facility.

6.0 REFERENCES

Carucci, A.; Rolle, E.; Smurra, P. (1997) Experiences of On-Line Control at a Wastewater Treatment Plant for Nitrogen Removal. *IAWQ Specialist Group on Instru-*

mentation, Control and Automation 7th International Workshop, Brighton, United Kingdom, July; International Association on Water Quality: London.

Caulet, P.; Lefevre, F.; Bujon, B.; Reau, P.; Philippe, J. P.; Audic, J. M. (1997) Automated Aeration Management in Waste Water Treatment: Interest of the Application to Serial Basins Configuration. *IAWQ Specialist Group on Instrumentation, Control and Automation 7th International Workshop;* Brighton, United Kingdom, July; International Association on Water Quality: London.

Charpentier, J.; Martin, G.; Wacheux, H.; Gilles, P. (1998) ORP Regulation and Activated Sludge: 15 Years of Experience. *Water Sci. Technol.,* **38** (3), 197–208.

Cohen, A.; Hegg, D.; de Michele, M.; Song, Q.; Kasbov, N. (2003) An Intelligent Controller for Automated Operation of Sequencing Batch Reactors. *Water Sci. Technol.,* **47** (12), 57–63.

Copp, J. (2001) *The COST Simulation Benchmark-Description and Simulator Manual.* COST (European Cooperation in the Field of Scientific and Technical Research): Brussels, Belgium.

Draaiger, H.; Bunnen, A. H. M.; van Dijk, J. W. (1997) Full-Scale Respirometric Control of an Oxidation Ditch. *IAWQ Specialist Group on Instrumentation, Control and Automation 7th International Workshop;* Brighton, United Kingdom, July; International Association on Water Quality: London.

Japan Society of Industrial Machinery Manufacturers (2001) *Introduction of Japanese Advanced Environmental Equipment;* Japan Society of Industrial Machinery Manufacturers: Osaka, Japan.

Mori, Y.; Sasaki, K.; Yamamoto, Y.; Tsumura, K.; Ouchi, S. (1997) Countermeasures for Hydraulic Load Variation in Intermittently Aerated 2-Tank Activated Sludge Process for Simultaneous Removal of Nitrogen and Phosphorus. *IAWQ Specialist Group on Instrumentation, Control and Automation 7th International Workshop;* Brighton, United Kingdom, July; International Association on Water Quality: London.

Shaw, A.; Watts, J. (2002) The Use of Respirometry for the Control of Sequencing Batch Reactors—Principles and Practical Application. *Proceedings of the 75th Annual Water Environment Federation Technical Exhibition and Conference* [CD-ROM], Chicago, Illinois, Sept 28–Oct 2; Water Environment Federation: Alexandria, Virginia.

Water Environment Federation (2013) *Automation of Wastewater Treatment Facilities;* Manual of Practice No. 21; Water Environment Federation: Alexandria, Virginia.

Watts, J.; Garber, W. (1993) On-Line Respirometry: A Powerful Tool for ASP Operation and Design. *6th IAWQ Specialized Conference, Sensors in Waste Water Technology;* Hamilton, Canada; June 17–23; International Water Association: London.

Yoong, E. T.; Lant, P. A.; Greenfield, P. F. (2000) In Situ Respirometry in an SBR Treating Wastewater with High Phenol Concentrations. *Water Res.*, **34** (1), 239–245.

7.0 SUGGESTED READINGS

Geary, S. (2004) Selection of Automation Equipment. *Proceedings of the 77th Annual Water Environment Federation Technical Exhibition and Conference* [CD-ROM]; New Orleans, Louisiana, Oct 2–6; Water Environment Federation: Alexandria, Virginia.

Guisasola, A.; Marcelino, M.; Lemaire, R.; Baeza, J. A.; Yuan, Z. (2010) Modeling and Simulation Revealing Mechanisms Likely Responsible for Achieving the Nitrite Pathway through Aeration Control. *Water Sci. Technol.*, **61** (6), 1459–1465.

Ingildsen, P.; Rosen, C.; Gernaey, K. V.; Nielsen, M. K.; Guildal, T.; Jacobsen, B. N. (2006) Modelling and Control Strategy Testing of Biological and Chemical Phosphorus Removal at Avedore WWTP. *Water Sci. Technol.*, **53** (4/5), 105–113.

Jahan, K.; Hoque, S.; Ahmed, T.; Türkdoan, L.; Arslankaya, E.; Patarkine, V. (2011) Activated Sludge and Other Suspended Culture Processes. *Water Environ. Res.*, **83,** 1092–1149.

Joh Kang, S.; Olmstead, K. P.; Takacs, K. M.; Collins, J.; Wheeler, J.; Zharaddine, P. (2009) Energy Sustainability and Nutrient Removal from Municipal Wastewater. *Proceedings of the 82nd Annual Water Environment Federation Technical Exhibition and Conference* [CD-ROM]; Orlando, Florida, Oct 10–14; Water Environment Federation: Alexandria, Virginia.

Kestel, S.; Duffy, G. J.; Gregory, J.; Gray, M.; Stahl, T.; Lee, G. (2009) DO Control Based on On-Line Ammonia Measurement. *Proceedings of the 82nd Annual Water Environment Federation Technical Exhibition and Conference* [CD-ROM]; Orlando, Florida, Oct 10–14; Water Environment Federation: Alexandria, Virginia.

Kim, Y. M.; Cho, H. U.; Lee, D. S.; Park, D.; Park, J. M. (2011) Influence of Operational Parameters on Nitrogen Removal Efficiency and Microbial Communities in a Full-Scale Activated Sludge Process. *Water Res.*, **45** (17), 5785–5795.

Kiser, P., DeSantis, G.; et al. (2005) Monitoring Treatment Plants Automatically: Automated Monitoring Makes Wastewater Treatment More Efficient. *Public Works*, **136** (5), 84–86.

Kobylinski, E.; Shaw, A.; Steichen, M.; Analla, H. (2004) The Basics of Automated Activated Sludge Process Control: Keep It Simple. *Proceedings of the 77th Annual Water Environment Federation Technical Exhibition and Conference* [CD-ROM]; New Orleans, Louisiana, Oct 2–6; Water Environment Federation: Alexandria, Virginia.

Littleton, H. X.; Daigger, G. T.; Amad, S.; Strom, P. F. (2009) Develop Control Strategy to Maximize Nitrogen Removal and Minimize Operation Cost in Wastewater Treatment by Online Analyzer. *Proceedings of the 82nd Annual Water Environment Federation Technical Exhibition and Conference* [CD-ROM]; Orlando, Florida, Oct 10–14; Water Environment Federation: Alexandria, Virginia.

Liu, W.; Lee, G. J. F.; Goodley, J. J. (2003) Using Online Ammonia and Nitrate Instruments to Control Modified Ludzack–Ettinger (MLE) Process. *Proceedings of the 76th Annual Water Environment Federation Technical Exhibition and Conference* [CD-ROM]; Los Angeles, California, Oct 11–15; Water Environment Federation: Alexandria, Virginia.

Olsson, G. (2006) Instrumentation, Control and Automation in the Water Industry—State-of-the-Art and New Challenges. *Water Sci. Technol.*, **53** (4/5), 1–16.

Olsson, G. (2011) Automation Development in Water and Wastewater Systems. *Environ. Eng. Res.*, **16** (4), 197–200.

Powell, R.; Thompson, R. (2004) South West Water (Cornborough WWTP) Automatic Sludge Age Control Benefits. *Proceedings of the 77th Annual Water Environment Federation Technical Exhibition and Conference* [CD-ROM]; New Orleans, Louisiana, Oct 2–6; Water Environment Federation: Alexandria, Virginia.

Rosen, C.; Jeppsson, U.; Vanrolleghem, P. A. (2004) Towards a Common Benchmark for Long-Term Process Control and Monitoring Performance Evaluation. *Water Sci. Technol.*, **50** (11), 41–49.

Schmit, C. G.; Jahan, K.; Debik, E.; Pattarkine, V. (2010) Activated Sludge and Other Aerobic Suspended Culture Processes. *Water Environ. Res.*, **82,** 1073–1123.

Ma, T.; Zhao, C.; Peng, Y.; Liu, X.; Zhou, L. (2009) Applying Real-Time Control for Realization and Stabilization of Shortcut Nitrification-Denitrification in Domestic Water Treatment. *Water Sci. Technol.*, **59** (4), 787–796.

Turney, H. G.; Wells, C. D. (2004) The Use of Automated Oxygen Uptake Rate Measurements to Improve the Reliability of a Nitrification/Denitrification Activated Sludge Treatment Plant. *Proceedings of the 77th Annual Water Environment Federation Technical Exhibition and Conference* [CD-ROM]; New Orleans, Louisiana, Oct 2–6; Water Environment Federation: Alexandria, Virginia.

Vrecko, D.; Hvala, N.; Stare, A.; Burica, O. (2006) Improvement of Ammonia Removal in Activated Sludge Process with Feedforward–Feedback Aeration Controllers. *Water Sci. Technol.*, **53** (4–5), 125–132.

Chapter 15

Laboratory Analyses

Marie S. Burbano, Ph.D., P.E., BCEE;
Tania Datta, Ph.D.;
Katherine Bell, Ph.D., P.E., BCEE;
Cory Lancaster; Huijie Lu; and
Jun Meng, M.S., E.I.T

1.0 INTRODUCTION

Facility operators with supervisory experience who are responsible for overall facility performance and compliance need valid and legally defensible data to be developed by the facility's internal analytical laboratory. Therefore, from managers who are trained to have extensive knowledge of federal, state, and local wastewater treatment and discharge regulations to laboratory technicians who have the specific education, experience, and training to competently generate valid and legally defensible analytical data, all staff at a water resource recovery facility (WRRF) should understand the fundamental concepts underlying the analytical procedures performed to meet these objectives. This chapter provides background information on the parameters that are encountered in wastewater treatment for process control and regulatory compliance. In addition, a brief introduction to the importance of wastewater microbiology is provided, including a description of recent advances in this field as nutrient removal is becoming a higher priority for many facilities. More detailed information on this topic can be found in *Basic Laboratory Procedures for the Operator-Analyst* (WEF, 2012).

1.1 Regulatory Compliance

Satisfying the requirements of the National Pollution Discharge Elimination System (NPDES) permit or the requirements of a specific re-use application for treated effluent, which will vary from state to state (U.S. EPA, 2004), is possible only with accurate and defendable laboratory results. The analyses necessary for compliance require that quality assurance, which is critical to producing sound and defensible data, is necessary to provide the empirical evidence upon which compliance is determined. Therefore, in addition to the description of the parameters that are necessary for regulatory compliance, a discussion is provided in this chapter on considerations for sample collection, the appropriate use of sample statistics, reporting, and common pitfalls in sample collection.

1.2 Process Control

A WRRF must operate properly to consistently achieve regulatory compliance. However, with increasingly stringent limits on nutrient discharges, operation of facilities is becoming more complex. Gathering information about the performance of a nutrient removal facility is the key to optimizing its operations.

In many instances, the analyses that are required for permit compliance may not be comprehensive enough for process control. Therefore, many facilities will collect analytical data that are not associated with compliance monitoring or will implement real-time online monitoring to aid in developing the necessary information for process control. Permit language must be carefully checked for sampling and reporting requirements because some permits may require sample data to be reported to a regulatory agency, even if the sampling was not specified in the permit. For example, biological phosphorus removal may require the addition of supplemental carbon source if adequate concentrations of volatile fatty acids (VFAs) are not available. Thus, some facilities monitor this parameter to help manage this biological process. This chapter provides a description of a range of analytical parameters and associated laboratory analyses that, although not critical to permit compliance, may be helpful or necessary for biological and chemical process control that is important to effective nutrient removal.

1.3 Calibration, Validation, and Comparison to Online Instruments

During any given work day, utility operators are busy tending to multiple tasks and customer needs. For this reason, online instrumentation is increasingly being used for process control monitoring. While online instrumentation can provide continuous information, there are several factors to consider, including calibration and validation of the instruments and a comparison of the readings with grab samples that are analyzed using a laboratory method.

Although the required steps to calibrate and maintain specific online analyzers are covered in Chapter 14 and in the analyzer's instruction manual, a pragmatic approach to the procedural requirements for operation and calibration of the online analyzer is necessary for proper operation. To maximize the value of information obtained from online instrumentation, a procedure manual should be developed to guide the operator through the following quality-assurance steps:

- Conduct initial verification of the online analyzer,
- Demonstrate capability of the online analyzer,
- Conduct initial grab sample comparison when the online analyzer is put to use, and
- Conduct routine analyzer checks with grab samples at an appropriate frequency.

1.4 Modeling

Wastewater process modeling currently represents the industry's best understanding of the complex relationships between chemical, physical, and biological processes that provide successful wastewater treatment. Nearly 50 years of research has gone into characterizing the behavior of approximately 60 of the most critical, intricately related wastewater treatment processes. However, the numerical models that have been developed to describe the observed chemical, physical, and biological reactions related to these processes continue to evolve as understanding of these processes improves. Thus, the ability of any model to accurately represent reality depends on the quality of the inputs. Measuring actual conditions at a WRRF is, arguably, the component of modeling that is most susceptible to error. There is uncertainty inherent to sampling and analysis, which suggests that models might not always reflect that which is measured at a facility. Therefore, reliable agreement between model results and facility data is only possible and meaningful if the data are verified to be of the best quality. A description of best practices for developing information for wastewater modeling is included in this chapter.

2.0 BIAS AND VARIATION IN OPERATION, SAMPLING, AND ANALYSIS

Optimizing nutrient removal requires characterization of the various streams into, within, and from a WRRF to make informed and effective process control decisions. Although online instrumentation is being used more often, the traditional wet chemistry approach is still commonly implemented for characterization and instrument calibration. Mass can be calculated by measuring both flow and concentration. Samples for constituent quantity and quality can be collected from influent, primary effluent, final effluent, return activated sludge (RAS), waste activated sludge (WAS), mixed liquor suspended solids (MLSS), sidestreams, and a host of other locations. Samples can be measured for flow, solids, dissolved oxygen, organics, nutrients, metals, pesticides, and compounds of emerging concern. The sampling location, constituent list, sampling mode (grab or composite), and sampling frequency are dependent on the ultimate purpose of the information (i.e., reporting, process control, optimization, design, or modeling) and how critical and sensitive the result is for each use (e.g., a costly fine for an effluent violation or calibration of a process model).

Because sample collection and analyses is time consuming and costly, only the minimum number of samples and replicates that can provide a meaningful result for informed decision making should be collected. The question then becomes, "What is a meaningful result"? What was meaningful a generation ago when NPDES limits were focused only on total suspended solids (TSS) and biochemical oxygen demand (BOD) is far from what may be considered a meaningful result with today's nutrient removal requirements.

Arriving at a reasonably accurate (i.e., minimum difference between the measured and the "true" value) and precise (i.e., minimum difference between replicates of measured values) quantification of constituents is often more complicated and challenging than is widely acknowledged. Accuracy and precision is a function of bias (systemic error) and variation (random error). Bias should be identified so minimization efforts can be taken. Variation is inherent in all material systems because they are heterogeneous, especially wastewater. The concept of heterogeneity is fundamental to sampling theory (Gy, 1992). Variation should be identified to determine how many samples are needed to provide a mean value. This issue is complicated by the fact that both wastewater flow and composition vary disproportionately over the course of a day, which makes variation especially difficult to characterize.

Bias and variation are collectively known as *uncertainty*. The following are three areas where uncertainty is found at a WRRF:

- Analysis, which occurs with analytical methods or equipment in a laboratory; efforts to minimize and quantify laboratory analytical error have been thorough and have become standardized. This is called *analytical uncertainty;*
- Sample collection, either in mode, equipment, method, or location; this is called *sampling uncertainty*; and
- Operation, such as parallel trains or unit processes (bioreactors, clarifiers, or dewatering units), which inevitably lead to unintended differences in hydraulic and/or mass distributions; this is called *operational uncertainty.*

While thorough effort has gone into quantifying and minimizing analytical uncertainty in facility laboratories, comparatively little is typically done to quantify and minimize operational and sampling uncertainty at the facility itself. Some of the same concepts for quantifying and minimizing uncertainty at the laboratory scale can also be applied to sampling and operational scales.

2.1 Analytical Uncertainty

Good laboratory practices that characterize and minimize analytical uncertainty are presented elsewhere, namely Section 1000 of *Standard Methods for the Examination of Water and Wastewater* (APHA et al., 2012). Good laboratory practices involve establishing and following standard operating procedures when performing analytical methods; calibrating measurement equipment such as scales, pipettes, and spectrophotometers on a regular basis; and including reagent blank samples, laboratory-fortified blank samples, laboratory-fortified matrix samples and duplicates, duplicate analyses of the unknown sample, and splitting samples between laboratories, all at prescribed regularity. Good bench-scale practices involve such things as shaking samples brought into the laboratory (pseudo-homogenization) before subsampling an aliquot to be used for actual analysis and preventing cross contamination of samples. Good data processing practices include plotting control charts of results for accuracy and precision; reporting to the correct number of significant figures; calculating standard deviations for both repeatability (i.e., running multiple analysis on the same sample without changing the instrument, analyst, laboratory, or day) and reproducibility (i.e., running the analysis on the same sample, but changing either the instrument, analyst, laboratory, or day); and calculating and reporting method detection limits. All of these practices, if done deliberately and methodically, will allow the analyst to quantify and minimize analytical uncertainty of a sample once it crosses the threshold of the laboratory.

2.2 Sampling Uncertainty

Standard Methods (APHA et al., 2012) states, "Sampling error due to population variability . . . is usually much larger than analytical error components. Unfortunately, sampling error is usually not available and the analyst is usually left with only the published error of the measurement system". This concept is typically forgotten, and the result of a single analysis determined under the auspices of good laboratory practices is then assumed to be the average (mean) concentration of whatever is being sampled or it is assumed that the average measurement will be revealed over time through continued sampling (e.g., assuming the average measurement can be determined by averaging daily values over the course of a month). This is not the case, however, and fundamental statistics must not be overlooked. A single analysis can only be representative of a mean of the population if it is shown to be so, that is, many samples must be taken from a single population (i.e., replicates) to determine the distribution of the population, the standard deviation, and the confidence level of the actual mean. Figure 15.1

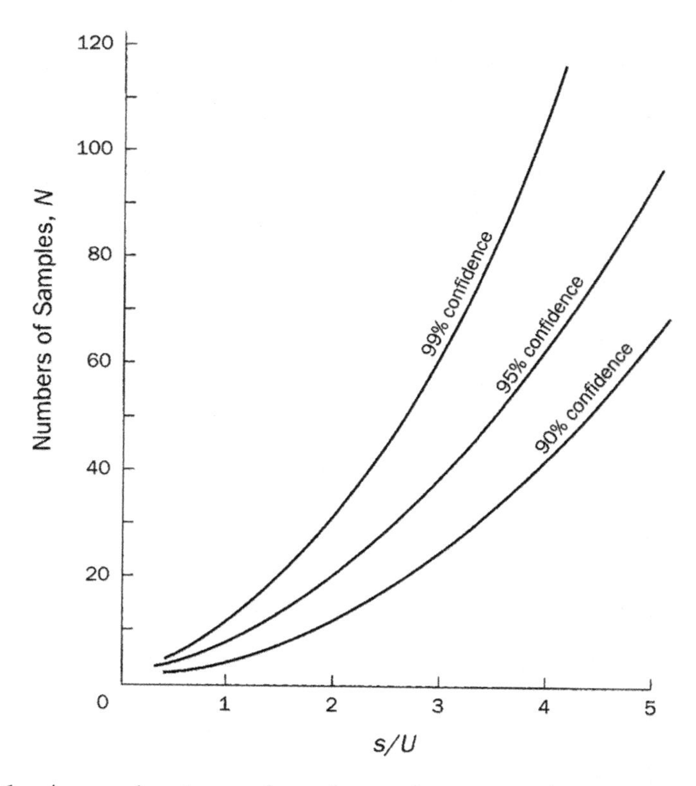

FIGURE 15.1 Approximate number of samples required in estimating a mean concentration. Source: Methods for the Examination of Waters and Associated Materials: General Principles of Sampling and Accuracy of Results. 1980. Her Majesty's Stationery Off., London, England.

quantifies how many samples are typically required to determine the mean value with a given level of confidence, where

N = number of samples required,
s = overall standard deviation, and
U = acceptable level of uncertainty.

For example, for influent BOD concentration, if

s = ±30 mg/L and
U = 20 mg/L, then
N = 20 samples are required for a 95% level of confidence.

This means that 20 random samples are needed to determine the mean for each day with 95% confidence if the standard deviation of the 20 influent BOD analysis results was ±30 mg/L. Unfortunately, characterizing and minimizing the influencing factors of sampling uncertainty and determining a mean value in this way is rarely done. Important sources of uncertainty include the following:

- Unrepresentative sampling location—compositional differences within an influent channel, bioreactor, distribution box, or launder;
- Unrepresentative composite sampling mode—compositional vs flow differences with time and the effect of the selected sampling period, duration, and mode (continuous or discrete, time proportional, flow proportional, or volume proportional);
- Inherent bias in sampling equipment—exclusion of larger particles through sample tubes and preferential settling in composite sample containers;
- Inherent bias in sampling method—grab samples from the surfaces of tanks and the beginning, middle, or end of the solids handling operation; and
- Random error because of the heterogeneity of wastewaters, sludges, and effluents.

2.3 Sampling Mode for Compositing Samples

While wastewater flow and composition are highly variable, both will generally follow repeatable patterns for a given facility under base- or dry-weather conditions (i.e., when storm flows are not contributing to the influent). Time-paced (discrete, uniform aliquot per unit time), discrete flow-paced (discrete, uniform aliquot per given volume), and continuous flow-paced (continuous sample proportional to flowrate) samples will likely provide different results. For many facilities, discrete flow-paced samples are taken with a sampler collecting a volume at given time intervals (e.g., every 15 minutes) and that volume is based on a flow meter. Continuous flow-paced sampling provides a better representation of the wastewater, but is typically impractical because a separate continuous flow line, sampling pump, and adequate sample storage are required. This distinction is important to identify constituents that are present in low volumes or for micropollutants. For example, in a study by Ort and Gujer (2006), continuous flow-paced sampling was needed to identify pharmaceuticals in an influent stream.

Commercially available composite samplers typically operate in either time-paced or discrete flow-paced sampling modes. Unless flowrates are constant, discrete flow-paced composites are most accurate between these two options. For example, take a generic domestic wastewater influent that represents the typical diurnal variability of flow, concentration, and load (Figure 15.2). Hourly uniform volume aliquots result in an average concentration that is 10% less than that of hourly flow-proportional volume aliquots. The actual difference between time-paced vs discrete flow-paced sampling will depend on the variability of the flow and load and the offset between the two at each WRRF. It will also depend on what is being sampled.

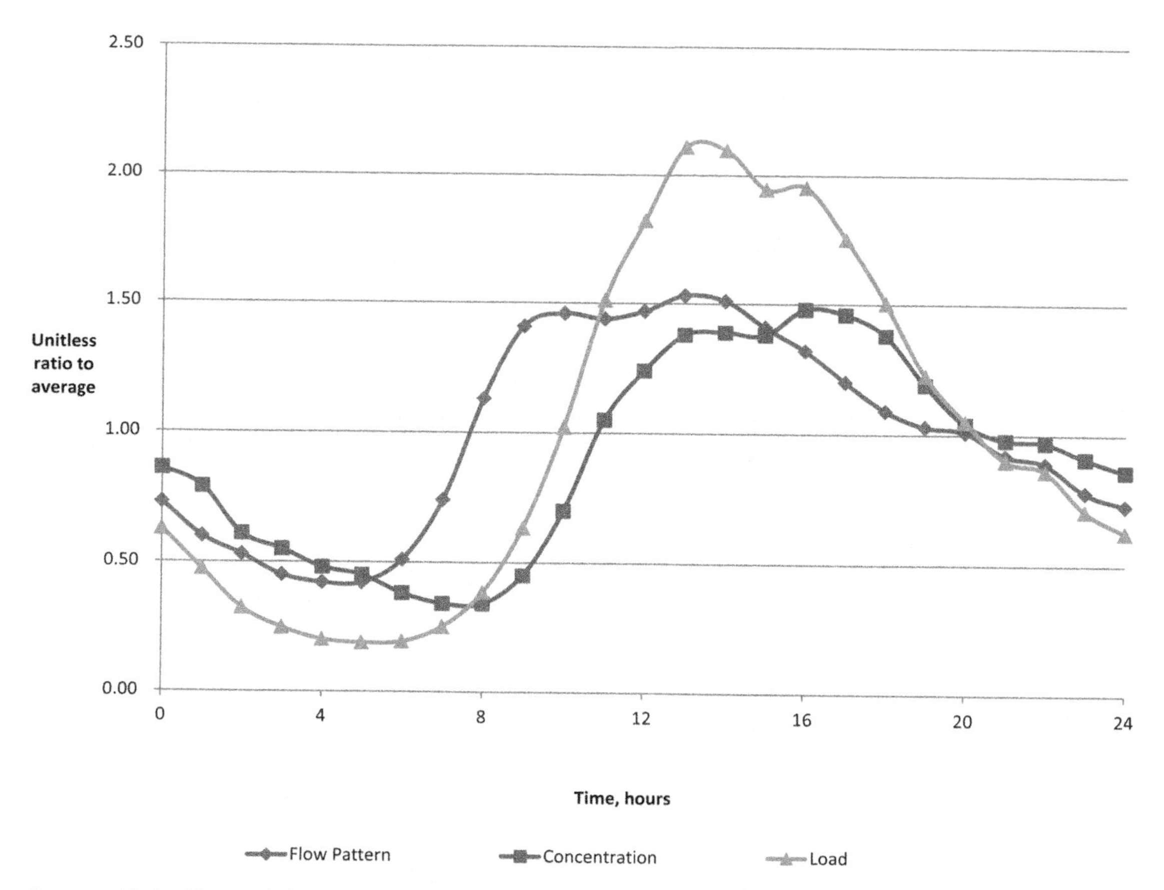

FIGURE 15.2 Typical diurnal relationships for flow, concentration, and load (Grady et al., 1999).

2.4 Operational Uncertainty

Operational uncertainty is simply the uncertainty associated with differences in operation between different process trains. Although the intent might be to operate two parallel trains at a selected split hydraulic or solids loading (50 to 50, 60 to 40, etc.), actual distributions may be different than the intent, hydraulic and solids distributions may be disproportionate, distributions might be variable, or all of these factors. For example, an attentive operator may have the presence of mind to sample MLSS from different trains to determine an average, but it may be difficult to determine if the flow distribution between clarifiers is equal.

3.0 STATISTICS AND RELIABILITY

A common statistical approach for evaluating variability and determining reliability in environmental sampling involves a hierarchical or nested experimental design. The most influential variables that affect sampling and operation uncertainty can be identified and quantified by taking several samples and evaluating the differences statistically.

Nested designs are readily applied to wastewater sampling. They consider the overall effect of one sampled factor (called the *major factor*) on sample variance, the overall effect of a second sampled factor (called the *minor factor*) nested within the major factor on sample variance, and the random error between samples. A type of one-way analysis of variance test can then be performed on nested designs to determine if there is statistical significance associated among major factors, minor factors within major factors, and within minor factors.

A nested design model comprises the following three kinds: fixed effect model, or model I; random effect model, or model II; and mixed effect model, or model III. In model I, all major and minor factors are selected; in model II, a subset of major and minor factors are selected randomly; and, in model III, at least one factor (typically, the major factor) is fixed (Snedecor and Cochran, 1976).

A model I nested design can be represented by the following equation:

$$y_{i,j,k} = \mu + \alpha_i + \beta_{i,j} + \varepsilon_{i,j,k} \tag{15.1}$$

Where

$y_{i,j,k}$ = the *k*th measurement for the *j*th minor factor of the *i*th major factor,
μ = the overall average,
α_i = influence of *i*th level of the major factor,
$\beta_{i,j}$ = influence of *j*th level of the minor factor for the *i*th level of the major factor, and
$\varepsilon_{i,j,k}$ = overall variation among samples, and assumed $N(0, \sigma^2)$.

The variables $\varepsilon_{i,j,k}$ are all assumed to be independent.

A model II nested design can be represented by the following equation:

$$Y_{i,j,k} = \mu + A_i + B_{i,j} + \varepsilon_{i,j,k} \tag{15.2}$$

Where

$Y_{i,j,k}$ = the kth measurement for the jth minor factor of the ith major factor,
μ = the overall average,
A_i = influence of ith level of the random major factor assumed $N(0, \sigma_A^2)$,
$B_{i,j}$ = influence of jth level of the random minor factor for the ith level of the major factor assumed $N(0, \sigma_B^2)$, and
$\varepsilon_{i,j,k}$ = overall variation among samples, and assumed $N(0, \sigma^2)$.

The variables A_i, $B_{i,j}$, and $\varepsilon_{i,j,k}$ are all assumed to be independent.

A model III nested design with the major factor being fixed can be represented by the following equation:

$$Y_{i,j,k} = \mu + \alpha_i + B_{i,j} + \varepsilon_{i,j,k} \tag{15.3}$$

Where

$Y_{i,j,k}$ = the kth measurement for the jth minor factor of the ith major factor,
μ = the overall average,
A_i = influence of ith level of the fixed major factor,
$B_{i,j}$ = influence of jth level of the random minor factor for the ith level of the major factor assumed $N(0, \sigma_B^2)$, and
$\varepsilon_{i,j,k}$ = overall variation among samples, and assumed $N(0, \sigma^2)$.

The variables, $B_{i,j}$, and $\varepsilon_{i,j,k}$, are assumed to be independent. A spreadsheet can be used to perform this analysis for any kind of model because the computational approach is the same.

3.1 Nested Design for Influent Sampling

Sampling of an influent channel can be set up for a nested design. Total suspended solids are likely the most difficult constituent to sample, but the easiest to analyze; therefore, they are appropriate surrogates for all constituents and are ideal for nested design. However, for this example, BOD concentration is considered instead because the results will be used later to demonstrate the effect on the commonly calculated "net yield" parameter.

Suppose there are three composite samplers set up, each for a different influent channel depth, to evaluate the uncertainty with depth; it is often hypothesized that solids stratification may exist depending on site condition (the major factor). Typically, a composite sampler measures 9.5 L (2.5 gal). The entire container is rarely brought to the laboratory; rather, a smaller subsample "split" is typically collected. Multiple splits (the first volume, two middle volumes, and the last volume) are taken of this composite sample to evaluate the uncertainty with the split method. Lastly, each split would be analyzed in duplicate to evaluate the variation among samples. Each of the 24 BOD values and means are shown in Table 15.1. Given the difficulty of getting a representative sample from a large container, an alternative sample collection would be to take the entire sample bottle to the laboratory where it can be properly mixed while aliquots are being removed. This allows clean bottles to be placed in the sampler, thus minimizing the potential for cross contamination. However, this is rarely conducted because of the difficulty in transferring samples. In addition, even in a laboratory or under controlled conditions, proper mixing and subsampling are extremely difficult because of the nature of the sample.

Variances are then calculated between, and among, the different factors by the formulas presented in Table 15.2. Variance is the square of the difference of the value and the mean. For example, to determine the variance between the mean of the first composite sample split taken from mean BOD concentration of the composite sampler collecting samples from the top of the channel (the underlined values in Table 15.1, i.e., 355 and 460 mg/L, respectively), the following equation from Table 15.2 is used (underlined):

$$\sigma^2 = (Y_{\text{bar1}} - Y_{\text{bar1,1}})^2 \quad \text{or } 11025 = (460 - 355)^2. \tag{15.4}$$

This exercise is easily performed in a spreadsheet, taking the data in Table 15.1 and populating Table 15.3.

Individual variances can then be used in the equations and following the example presented in Table 15.4 to determine the degrees of freedom, the sum of squares, and the mean sum of squares. Finally, the overall variance and standard deviation can be determined by using the equations and following the example in Table 15.5.

The result of nested design for BOD concentration determined a BOD concentration of 460 ± 320 mg/L (applying the appropriate number of significant figures at the end of all calculations). The results also indicate that most of the variance comes from the minor factor, which is the method by which the splits were taken. Upon closer inspection of the data, it can be seen that the last split was always much higher than the other values. Visual inspection of the samples confirmed that the last splits were, indeed, more concentrated. It was discovered that the method of shaking and pouring splits resulted in more solids in the last sample because of settling in the composite sample container immediately following shaking and before pouring (i.e., there is much sampling uncertainty). The results also indicate that the difference between depths within the channel and between sample replicates was different, suggesting that a single laboratory analysis of a sample collected at any depth would be sufficiently representative. The minimal difference between laboratory analyses indicates little analytical uncertainty, while the difference between sample splits indicates much sampling

TABLE 15.1 Data for nested design for influent BOD concentration. There are three influent channel depths, four composite sample splits, and two laboratory replicates per split ($3 \times 4 \times 2$).

Major factor α, "Channel depth", a_i; $a = 3$	Top_1				$Middle_2$				$Bottom_3$			
Minor factor β within α $(\beta \mid \alpha)$, "Composite sampler splits", b_j; $b = 4$	$1st_1$ mg/L	$2nd_2$ mg/L	$3rd_3$ mg/L	$Last_4$ mg/L	$1st_1$ mg/L	$2nd_2$ mg/L	$3rd_3$ mg/L	$Last_4$ mg/L	$1st_1$ mg/L	$2nd_2$ mg/L	$3rd_3$ mg/L	$Last_4$ mg/L
ε Random error, replicates, n_k; $n = 2$	350	340	370	800	360	350	370	750	360	400	380	780
	360	360	350	750	330	380	350	730	360	360	420	700
Average each minor factor, $Y_{bar\ i,j}$	355	350	360	775	345	365	360	740	360	380	400	740
Average each major factor, $Y_{bar\ i}$	460				453				470			
Overall average, $Y_{bar\ \ldots}$	461											

TABLE 15.2 Variance equations.

Among major factors	$\sigma^2_{maj\cdot} = (Y_{bar\ i..} - Y_{bar...})^2$
Among minor factors within major factors	$\sigma^2_{min.} = (Y_{bar\ I,j.} - Y_{bar\ i..})^2$
Within minor factors	$\sigma^2_{within} = (Y_{bar\ i,j,k} - Y_{bari,j.})^2$
Total	$\sigma^2_{Total} = (Y_{i,j,k} - Y_{bar...})^2$

uncertainty. Corrective action would be required to remove the bias associated with sample splitting of the composite sample container.

3.2 Nested Design for Waste Activated Sludge Sampling

A nested design for sludge sampling can be applied to other parameters in different ways. A second example is for waste activated sludge (WAS) concentration grab sampling. This was done for a single day's WAS cycle, which occurs over a 4-hour period. The WAS cycle consists of eight 15-minute intervals (15 minutes on, 15 minutes off) for each of the three WAS pumps (same flow for each), which are dedicated to the three online clarifiers. The nested design consisted of taking two grab samples for each WAS pump (major factor) at each 15-minute interval (minor factor) ($3 \times 8 \times 2$). The data are presented in Table 15.6, and the statistical evaluation is presented in Table 15.7. The result of the nested design for WAS concentration determined a WAS concentration of 5800 ± 4800 mg/L after proper application of significant figures. On the surface, this might appear alarming. Upon closer inspection of the data, the variance is split almost equally between the major factor (the different WAS pumps) and the random error between the grab samples at each interval, with the differences of both being statistically significant. The differences between the WAS pumps is an example of much operational uncertainty, while the little difference between sampling intervals indicates little operational uncertainty. The difference between grab samples demonstrates random variability. In this instance, aside from further investigating the operational uncertainty between WAS pumps (e.g., is there a genuine difference in clarifier loading? Or, is it because of the correlation that exists between the distance that the clarifier is from the WAS pump and WAS concentration?), not much can be done to minimize sampling uncertainty. In this instance, an online TSS meter may be beneficial to integrate WAS concentrations with the pump flows from each WAS pump to gain a better understanding of the mass of solids wasted on a given day.

3.3 Nested Design for Mixed Liquor Suspended Solids Concentration

A third example of a nested experimental design is for MLSS concentration. In this last example, MLSS was sampled from three parallel trains of aeration basins

TABLE 15.3 Individual variance for influent BOD concentrations.

Among major factors			**1**				**69**				**84**	
Among minor factors within major factors	11025	12100	10000	99225	11556	7656	8556	82656	12100	8100	4900	72900
Within minor factors	25	100	100	625	225	225	100	100	0	400	400	1600
	25	100	100	625	225	225	100	100	0	400	400	1600
Total	12284	14601	8251	115034	10167	12284	8251	83617	10167	3701	6534	101867
	10167	10167	12284	83617	17117	6534	12284	72451	10167	10167	1667	57201

TABLE 15.4 Degrees of freedom, sum of squares, and mean squares.

	Degree of freedom		Sum of squares (SS)		Mean squares (MS)	
	Formula	Value	Formula	Value	Formula	Value
Among major factors	$a - 1$	2	$SS_{maj\cdot} = b \times n \times \Sigma_i (Y_{bar\ i..} - Y_{bar...})^2$	1233	$MS_{maj} = SS_{maj.}/a - 1$	617
Among minor factors	$a(b - 1)$	9	$SS_{min.} = n \times \Sigma_I \times \Sigma_j (Y_{bar\ I,j.} - Y_{bar\ i..})^2$	681 550	$MS_{min} = SS_{min.}/a(b - 1)$	75 728
Within minor factors	$a \times b(n - 1)$	12	$SS_{within} = \Sigma_I \times \Sigma_j \times \Sigma_k (Y_{bar\ i,j,k} - Y_{bari,j.})^2$	7800	$MS_{within} = SS_{within}/a \times (n - 1)$	650
Total	$a \times b \times n - 1$	23	$SS_{Total} = \Sigma_I \times \Sigma_k (Y_{i,j,k} - Y_{bar...})^2$	690 583		

(major factor), at three different times of the day (minor factor), from three different locations within each basin that were presumably completely mixed (3 × 3 × 3). The data are presented in Table 15.8 and the statistical results are presented in Table 15.7. The result of the nested design for MLSS concentration determined a MLSS concentration of 3700 ± 450 mg/L after proper application of significant figures. Most of the variation was with the major factor (process train) because of the difference, which identified an unexpected difference in flow distribution to the three trains. This is acceptable, however, because understanding that there is a difference between three trains simply means that all three should be sampled to arrive at an average. The results also indicate that the differences over the course of the day and with different locations within each train are statistically insignificant.

Nested designs are time consuming. At the very least, they should be performed once to identify suspected sources of uncertainty for relevant factors to

TABLE 15.5 Overall variances and standard deviations.

	F		σ^2 components Estimate			
	Formula	Value	Formula	Value	% of σ^2 Estimate	σ
Among major factors	MS_{maj}/MS_{min}	0.01	$\sigma^2_\alpha = (MS_{maj} - MS_{min})/n \times b$	9389	20%	97
Among minor factors	MS_{min}/MS_{within}	116.50	$\sigma^2_{\beta\|\alpha} = (MS_{min} - MS_{within})/n$	37 539	79%	194
Within minor factors			$\sigma^2_e = MS_{within}$	650	1%	25
Total			$\sigma^2_{total} = \sigma^2_\alpha + \sigma^2_{\beta\|\alpha} + \sigma^2_\varepsilon$	47 578	100%	316

Table 15.6 Data from nested design for WAS concentration. There are three pumps, eight pump intervals, and two samples per pump interval ($3 \times 8 \times 2$).

Pump	Pump 1								Pump 2								Pump 3							
interval	1st	2nd	3rd	4th	5th	6th	7th	8th	1st	2nd	3rd	4th	5th	6th	7th	8th	1st	2nd	3rd	4th	5th	6th	7th	8th
Units	**mg/L**	**mg/L**	**mg/L**	**mg/L**	**mg/L**	**mg/L**	**mg/L**	**mg/L**	**mg/L**	**mg/L**	**mg/L**	**mg/L**	**mg/L**	**mg/L**	**mg/L**	**mg/L**	**mg/L**	**mg/L**	**mg/L**	**mg/L**	**mg/L**	**mg/L**	**mg/L**	**mg/L**
First sample	8500	7800	7600	7800	8200	7500	8700	9600	16 000	8900	6800	6300	2500	5800	7700	6200	4100	2000	2000	1900	4200	2000	4100	3500
Second sample	10 000	6100	8500	8400	8400	8500	8700	4100	6000	2700	3000	2600	2500	6800	7000	8400	5100	2500	3700	6200	3900	2100	2000	3800
Average of samples	9250	6950	8050	8100	8300	8000	8700	6850	11 000	5800	4900	4450	2500	6300	7350	7300	4600	2250	2850	4050	4050	2050	3050	3650
Average per pump				8025								6200								3319				
Overall average												5848												

TABLE 15.7 Statistics for WAS concentration and MLSS concentration nested designs.

	WAS concentration				MLSS concentration			
	F value	σ^2 value	% of σ^2 value	σ value	F value	σ^2 value	% of σ^2 value	σ value
Among major factors	17.32	5 305 022	51%	2303	9.78	50 576	66%	225
Among minor factors	1.05	124 821	1%	353	4.12	13 086	17%	114
Within minor factors		4 952 708	48%	2225		12 593	17%	112
Total		10 382 552	100%	***4882***		76 255	100%	***452***

give operators, engineers, and regulators a better understanding of the level of uncertainty that exists with facility data. If influent sampling indicates that TSS concentrations are, for example, within 20% at best, this will help frame all other influent results. Results of nested designs could also give operators more comfort as they pursue aggressive optimization efforts. With methodical rigor, nested designs could be performed regularly (perhaps on a quarterly basis) to minimize the uncertainty in sampling and operation.

3.4 Application of Results, Mathematical Rules, and Units

We now have the results of three nested experimental designs that have fleshed out uncertainty for three commonly measured parameters in terms of a mean and a standard deviation. These values are used to calculate net yield for design

TABLE 15.8 Data from nested design for MLSS concentration. There are three trains, three sample times, and three samples per train (3 × 3 × 3).

	Train 1			Train 2			Train 3		
Grab time	a.m.	Noon	p.m.	a.m.	Noon	p.m.	a.m.	Noon	p.m.
Unit	mg/L	mg/L	mg/L	mg/L	mg/L	mg/L	mg/L	mg/L	mg/L
Sample 1	3800	4000	3900	3800	4000	3900	3400	3300	3500
Sample 2	3800	4100	3800	3800	4100	3800	3300	4000	3300
Sample 3	3700	4000	3800	3700	4000	3800	3500	3500	3400
Average of samples	3767	4033	3833	3767	4033	3833	3400	3600	3400
Average per train		3878			3878			3467	
Overall average					3741				

purposes and solids retention time (SRT) for operational control, according to the following equations:

$$Y_{net} = \frac{\text{Mass of TSS produced (WAS flow} \times \text{WAS concentration)}}{\text{Mass of BOD removed (Influent flow} \times \text{Influent BOD concentration)}} \quad (15.5)$$

$$\text{SRT} = \frac{\text{Mass aeration basins (MLSS} \times \text{volume)}}{\text{Mass of TSS wasted (WAS flow} \times \text{WAS concentration)}} \quad (15.6)$$

Assuming that the standard deviation for all flow measurements is 5%, and using the rule for multiplication and division of standard deviation, which is

$$\text{Overall standard deviation} = [\text{result of } n \text{ expressions}] \times \sqrt{\left(\sum_1^n \frac{\sigma_i}{y_{\text{bar},i}}\right)}, \quad (15.7)$$

the net yield is 0.89 ± 0.97 and the SRT is 10 days ± 8.9 days. Although this is not encouraging, there is now a better understanding of how much uncertainty exists with the data collected, which can drive improvements in sampling and operation.

For reference, the rule for addition and subtraction of standard deviation is

$$\text{Overall standard deviation} = [\text{Result of } n \text{ expressions}] \times \sqrt{(\Sigma_1^n \sigma_i^2)}. \quad (15.8)$$

A brief mention of significant figures is warranted. Significant figures are those numbers that have meaning in determining precision. Generally, when performing mathematical operations, the value with the smallest number of significant figures will dictate the number of significant figures that can be expressed in the final result. For more information, refer to the discussion on significant figures in *Standard Methods* (APHA et al., 2012).

Lastly, correct units must be used when reporting a measured value. Common mistakes are the reporting of ammonium, nitrate, or orthophosphate, as these nutrients can be expressed as compounds or as the equivalent amount of elemental nitrogen or phosphorus.

4.0 WASTEWATER CHARACTERISTICS

This section reviews laboratory practices for general wastewater characteristics. The information presented here is mainly based on *Standard Methods* (APHA et al., 2012). Detailed laboratory protocols, sample preservation, and hold times can be found in *Standard Methods* and are listed in this section, as appropriate.

4.1 Temperature

Depending on the location, the mean annual temperature of wastewater influent varies between 3 and 27 °C and is kept between 10 and 30 °C throughout biological treatment processes to support optimal microbial activities. General kinetic models for biological nutrient removal processes are temperature dependent. Biological reaction rates approximately double with every 10 °C increase until inhibitory temperature values are reached. Nitrification reaches a maximum rate at temperatures between 30 and 35 °C, and denitrification occurs efficiently between 5° and 30 °C. Typical temperature ranges can also be found in Chapter 11. Besides the effects on microbial activity, temperature of wastewater also influences the solubility of oxygen and, consequently, affects the oxygen-transfer capacity of the aeration equipment in nitrification.

A calibrated, mercury-filled Celsius thermometer is typically used for temperature measurements (Method 2550; APHA et al. [2012]). At a minimum, the thermometer should have a scale marked for every 0.1 °C, with markings etched on the capillary glass, and should be periodically checked against a precision thermometer.

4.2 pH

pH is one of the most important and frequently tested parameters in water chemistry. It represents the concentration of hydrogen ions in solution and is defined by the following equation:

$$pH = -\log_{10}[H^+] \tag{15.9}$$

where $[H^+]$ is the concentration of hydrogen ions in moles per liter.

The pH of typical wastewater influent varies between 6 and 9. It is important to keep pH within the range of 7 to 8.5 during biological nutrient removal processes for microorganisms to remain sufficiently active. pH is typically reported as unitless or with standard units. Most treatment facilities are able to effectively nitrify with a pH of 6.5 to 7.0, but not below 6.0. Optimum pH values for denitrification are between 7.0 and 8.5. In addition, both nitrification and denitrification will cause a change of pH, and acid-base neutralization is commonly required. Typical pH ranges can also be found in Chapter 11.

The basic principle of the electrometeric pH measurement (Method 4500-H+; APHA et al. [2012]) is based on the potentiometric measurement of hydrogen ion activity within the solution using a standard hydrogen electrode and a reference electrode or a standard pH meter. A circuit is completed through the potentiometer when the electrodes are immersed in the test solution. Many pH meters are capable of reading pH and millivolts with a precision of 0.01 pH (Figure 15.3). pH electrodes require periodic cleaning and calibration to ensure accurate readings.

A pH electrode is relatively free from interference except for a sodium error at pH above 10, which can be reduced by using a special "low-sodium error" electrode. pH measurements are affected by temperature in the following two ways: mechanical effects that are caused by changes in properties of the electrodes and chemical effects caused by equilibrium changes. Standard pH buffers have a specified pH at indicated temperatures.

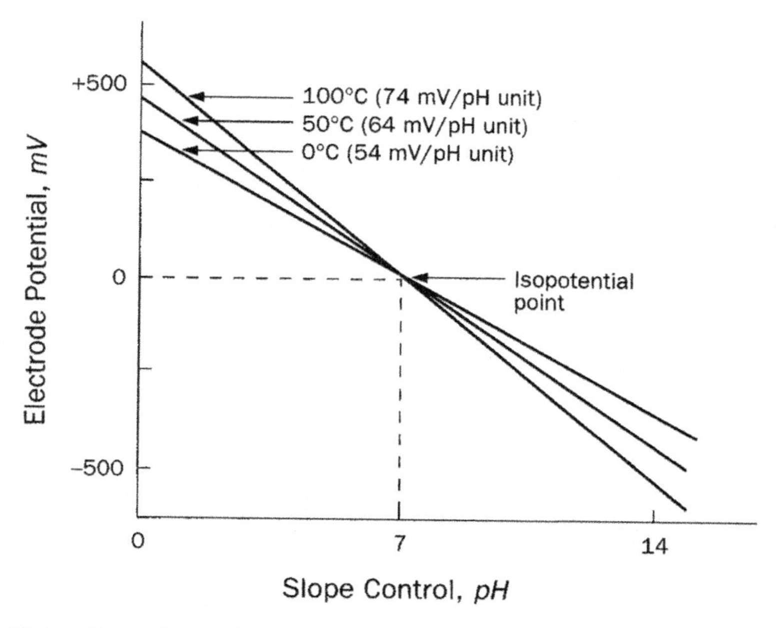

FIGURE 15.3 Typical pH electrode response as a function of temperature.

4.3 Dissolved Oxygen

Dissolved oxygen measures the amount of molecular oxygen present in water or wastewater, typically in milligrams of oxygen per liter. Dissolved oxygen is particularly important to wastewater nutrient removal processes because it fully oxidizes ammonia to nitrate at a typical concentration greater than 1.0 mg/L. Dissolved oxygen also affects the growth of other bacteria in the system, including filamentous organisms that may cause sludge bulking. Therefore, dissolved oxygen concentrations must be maintained in a range that favors desired organisms in different configurations. Dissolved oxygen is a function of temperature, the degree of hardness, and the demand for oxygen in the body of wastewater. The two methods for measuring dissolved oxygen are an electrometric method using a dissolved oxygen electrode and an iodometric method (also called *Winkler titration*). The most common method uses the membrane dissolved oxygen electrode, which compensates for pressure and temperature variations in water, and is the preferred method for measuring dissolved oxygen in polluted and highly colored waters.

The membrane electrode method (Method 4500-O G; APHA et al. [2012]) offers the advantage of analyzing dissolved oxygen in situ and, therefore, is also preferred for in-stream analysis in the field. Oxygen-sensitive electrodes of the polarographic type are composed of two solid metal electrodes in contact with supporting electrolyte separated from the test solution by an oxygen-permeable plastic membrane, which serves as a diffusion barrier against interfering impurities. The diffusion current is linearly proportional to the concentration of molecular oxygen and can be easily converted to concentration units. Because of the effects of temperature changes on membrane permeability, membrane electrodes require temperature compensation by using thermistors in the electrode circuit. There are also other dissolved oxygen measuring technologies such as the luminescent (optical) technology that are popular and are approved for reporting purposes in all states. Additional information on this technology is provided in Chapter 13.

Plastic films in membrane electrode systems are permeable to a variety of gases besides oxygen, such as hydrogen sulfide (H_2S). The interference can be eliminated by frequently changing the membrane and calibrating the electrode.

4.4 Oxidation–Reduction Potential

Certain substances release or lose electrons when dissolved in solution. The intensity or ease of electron loss (oxidation) or electron gain (reduction) is oxidation–reduction potential (ORP). Raw wastewater influent has a typical ORP of -200 mV, and ranges from -400 (strong) to -50 mV (weak). Common ORP values found in nitrification and denitrification systems are $-50\sim +50$ mV and $+100\sim +325$ mV, respectively.

Electrometric measurements (Method 2580; APHA et al. [2012]) are made by potentiometric determination of electron activity using an inert indicator electrode and a suitable reference electrode. The potential difference between the two electrodes equals the ORP of the test solution. Before measuring any samples, ORP electrodes need to be calibrated with a standard solution at the temperature of the sample.

Temperature determines the reference potential of a particular solution and an electrode pair and also affects the reversibility of the redox (reduction–oxidation) reaction, the magnitude of the exchange current, and the stability of the apparent redox potential reading. Oxidation–reduction potential is also sensitive to pH (if hydrogen or hydroxide ions are involved in the redox half cells), and the value increases as pH decreases.

4.5 Alkalinity

Alkalinity is a measure of the ability of the wastewater to neutralize acid. Alkalinity is reported as milligrams of calcium carbonate ($CaCO_3$) per liter. The main sources of alkalinity in wastewater streams are carbonate, bicarbonate, and hydroxide compounds (borates, silicates, and phosphates may also contribute). The full nitrification reaction (conversion of ammonia to nitrate) consumes 7.1 mg/L of alkalinity as $CaCO_3$ for each millgram per liter of ammonia-nitrogen oxidized. Conversely, denitrification produces 3.57 mg of alkalinity as $CaCO_3$/mg of nitrate-nitrogen reduced to nitrogen gas. Therefore, a treatment system that supports both nitrification and denitrification can regain some of the lost alkalinity. In general, a minimum alkalinity of 50 to 100 mg/L is required to ensure adequate buffering during both nitrification and denitrification.

When alkalinity is caused by carbonate or bicarbonate, the pH at the equivalence point of the titration is determined by the concentration of CO_2 at that stage. The end-point pH value varies according to alkalinity concentrations, which are, in turn, determined by the potentiometric titration to the end-point pH, as follows:

$$\text{Alkalinity (mg CaCO}_3\text{/L)} = \frac{A \times N \times 500000}{\text{mL sample}} \tag{15.10}$$

Where

A = mL standard acid used and
N = normality of standard acid.

Soaps, oily matter, suspended solids, or precipitates may coat the glass electrode and cause a sluggish response of the pH electrode.

4.6 Solids

Solids are matters suspended or dissolved in water or wastewater. The suspended and dissolved fractions are further subdivided into volatile (organic) and nonvolatile (fixed or inert) fractions. In raw domestic wastewater, solids are about 50% organic.

4.6.1 *Total Solids*

Total solids in raw wastewater consist of all solids that remain after a sample has been evaporated and subsequently dried in an oven at a defined temperature. Coarse solids, such as rags and debris, are removed before analysis. Total solids affect water clarity. Higher solids decrease the passage of light through water, thereby slowing photosynthesis by aquatic plants. Water will heat up more rapidly and hold more heat; this, in turn, might adversely affect aquatic life that has adapted to a lower-temperature environment.

A well-mixed sample in a weighed dish should be evaporated and dried to a constant weight at 103 to 105 °C in an oven. The increase in weight over that of the empty dish represents the total solids. Highly mineralized water (high concentrations of calcium, magnesium, chloride, and sulfate) may require prolonged drying, proper desiccation, and rapid weighing.

4.6.2 *Total Dissolved Solids*

Total dissolved solids (TDS) are material in the water that will pass through a filter with a 2.0-μm nominal average pore size. Smaller pore sizes have also been used to characterize TDS, such as the Whatman glass fiber filter, with a nominal pore size of 1.58 μm. The TDS is comprised of both colloidal and dissolved solids. The concentration of TDS affects the water balance in the cells of aquatic organisms.

According to Method 2540 C (APHA et al., 2012), the well-mixed sample should be filtered through a standard glass fiber filter and the filtrate should be evaporated to a dryness in a weighed dish and dried to a constant weight at 180 °C. Highly mineralized water containing high concentrations of calcium, magnesium, chloride, and sulfate may require prolonged drying, proper desiccation, and rapid weighing.

4.6.3 *Total Suspended Solids*

The raw wastewater suspended solids concentration can range from 100 to 350 mg/L. Suspended solids are approximately 50% organic solids and 30% inorganic solids. Higher concentrations of suspended solids can serve as carriers of toxics, which readily cling to suspended particles.

According to Method 2540 D (APHA et al., 2012), a well-mixed sample should be filtered through a weighed standard glass fiber filter and the residue retained on the filter should be dried to a constant weight at 103 to 105 °C (the sample size should be limited to yield no more than 200 mg of residue). The increase in weight of the filter represents TSS. The micron rating of the filter should be provided with the result.

Excessive residue on the filter may form a water-entrapping crust; therefore, it may be necessary to exclude large floating particles or submerged agglomerates of nonhomogeneous materials from the sample and also to limit the sample size to that yielding no more than 200 mg of residue.

4.6.4 *Fixed and Volatile Solids*

Fixed solids are the residue of total, suspended, or dissolved solids after being heated to dryness for a specified time at a specified temperature (typically, 550 °C). Volatile solids are calculated by subtracting fixed solids from the total solids value. Determinations of fixed and volatile solids do not distinguish between inorganic and organic matter because the loss upon ignition is not confined to organic matter, but also includes losses because of decomposition or volatilization of some mineral salts. However, the volatile solids fraction is still useful in the controlling WRRF operation because it offers a rough approximation of the amount of organic matter present in the wastewater and activated sludge. Figure 15.4 shows the various solids fractions graphically.

According to Method 2540 E (APHA et al., 2012), the residue should be ignited from total solids, TDS, and TSS tests to a constant weight at 550 °C; the remaining solids represent the fixed fraction of total solids, TDS, and TSS, respectively. Low concentrations of volatile solids in the presence of high fixed solids may result in considerable error. Highly alkaline residues may react with silica in sample or silica-containing crucibles.

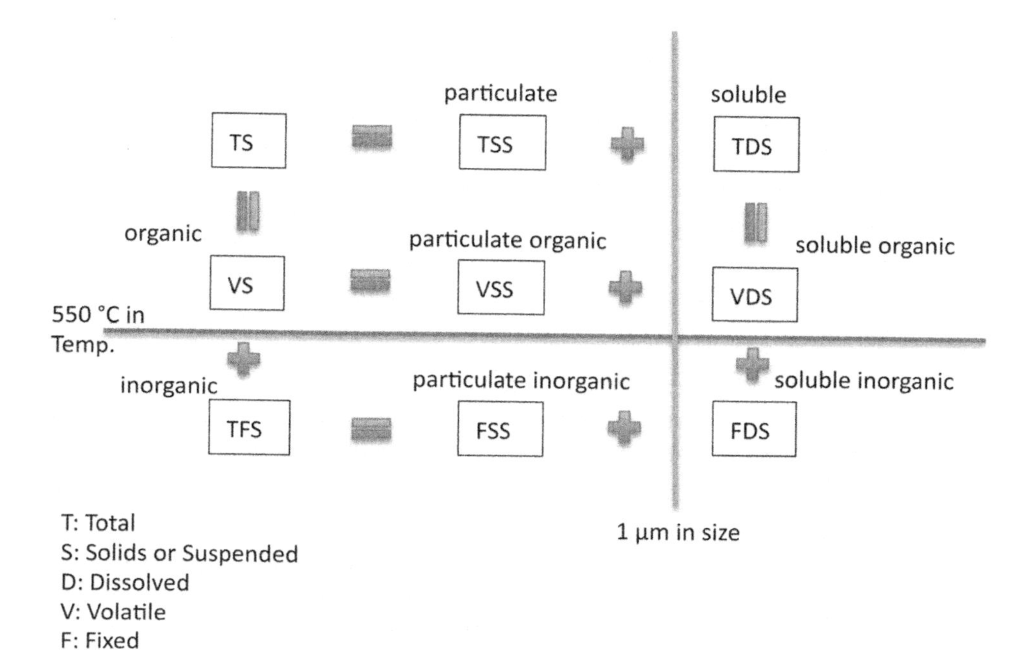

FIGURE 15.4 "Solids matrix" (T = total, S = solids or suspended, D = dissolved, V = volatile, and F = failed).

4.7 Carbonaceous Oxygen Demand

Carbonaceous oxygen demand (COD) is defined as the amount of oxygen required for the chemical oxidation of both biodegradable and non-biodegradable organic substances. The COD test uses dichromate to oxidize the organic material in wastewater.

The principal fractions are particulate and soluble (or filtered) COD. In biological processes, the two are further fractionated to readily biodegradable soluble COD (rbCOD), slowly biodegradable colloidal and particulate COD, non-biodegradable soluble COD, and non-biodegradable colloidal and particulate COD. The rbCOD is often fractionated further into complex COD that can be fermented to volatile fatty acids (VFAs) and short-chain VFAs. Flocculated and filtered COD (ffCOD) is a better measure of truly soluble COD. During ffCOD tests, coagulants (e.g., zinc) are added to enhance precipitation of colloidal solids. The sample is subsequently filtered and the filtrate is subjected to COD analysis, as follows:

$$\text{rbCOD} = \text{ffCOD} - \text{Non-biodegradable COD} \tag{15.11}$$

According to Method 5220 (APHA et al., 2012), when a sample is digested (at 150 °C for 2 hours), the dichromate ion oxidizes COD and changes from the hexavalent state (IV) to the trivalent (III) state. The $Cr_2O_7^{2-}$ absorbs strongly in the 400-nm region, whereas Cr^{3+} absorbs strongly in the 600-nm region. Higher COD (100 to 900 mg/L) is determined based on the increase in Cr^{3+}, and low COD values (<90 mg/L) are determined by following the decrease in $Cr_2O_7^{2-}$ at 420 nm.

Chloride, bromide, or iodide can react with dichromate to produce the elemental form of the halogen and the chromic ion. The interference of chloride can be overcome by complexing with mercuric sulfate before the refluxing procedure. Nitrite exerts a COD of 1.1 mg O_2/mg NO_2-N, and 10 mg of sulfamic acid for each milligram of NO_2-N can be added to eliminate this interference.

4.8 Biochemical Oxygen Demand

Biochemical oxygen demand of wastewater is defined as the amount of oxygen required for the biological decomposition of biodegradable organic matter under aerobic conditions. Theoretically, at 20 °C, an infinite time is required for complete biological oxidation of organic matter; however, for all practical purposes, the reaction may be considered complete in 20 days. Typically, a 20-day period is too long to wait for results in most instances. Experience has shown that a reasonably large percentage (70 to 80%) of total BOD is exerted or oxidized in 5 days.

The rate of biological consumption of organic materials is governed by the following first-order function:

$$\frac{d\text{BOD}r}{dt} = -k_1\text{BOD}r \tag{15.12}$$

$$\text{BOD}r = \text{BOD}u(e^{-k_1 t}) \tag{15.13}$$

Where

$\text{BOD}r$ = amount of waste remaining at time t, in oxygen equivalents, mg/L;

k_1 = first-order reaction rate constant, d^{-1};

$\text{BOD}u$ = ultimate CBOD, mg/L; and

t = time, d.

Therefore, the BOD exerted up to time t (BODt) is given by

$$\text{BOD}t = \text{BOD}u - \text{BOD}r = \text{BOD}u - \text{BOD}u(e^{-k_1 t}) = \text{BOD}u(1 - e^{-k_1 t}) \tag{15.14}$$

Carbonaceous biochemical oxygen demand (CBOD) is a method-defined test measured by the depletion of dissolved oxygen by the biological organisms in a body of water in which the contribution from nitrogenous bacteria has been suppressed by a nitrification inhibitor (the remaining steps are the same as a regular BOD test). It is important to note that the use of a nitrification inhibitor has also been associated with inhibiting CBOD demand. It is important to indicate the use of a nitrification inhibitor and to investigate any potential effects on CBOD removal. On the other hand, the oxygen demand associated with the oxidation of ammonia to nitrite is called *nitrogenous biochemical oxygen demand* (BODn). Figure 15.5 shows the difference between various BOD determinations.

According to Method 5210 (APHA et al., 2012), a well-mixed sample is first suitably diluted with prepared dilution water and then placed in a BOD bottle. This bottle is then filled with dilution water that is saturated with oxygen and contains required nutrients for biological growth. If the testing samples contain low concentrations of microorganisms, seed is needed for the test. After incubation for 5 days at 20 °C, the dissolved oxygen concentration is measured again. The BOD is calculated by

$$\text{BOD, mg/L} = \frac{(D_1 - D_2) - (B_1 - B_2)f}{P} \tag{15.15}$$

Where

D_1 = dissolved oxygen of diluted sample immediately after preparation, mg/L;

D_2 = dissolved oxygen of diluted sample after 5-day incubation at 20 °C, mg/L;

B_1 = dissolved oxygen of seed control before incubation, mg/L;

B_2 = dissolved oxygen of seed control after incubation, mg/L;

F = fraction of seeded dilution water volume in the sample to that in the seed control; and

P = fraction of wastewater in the combined volume.

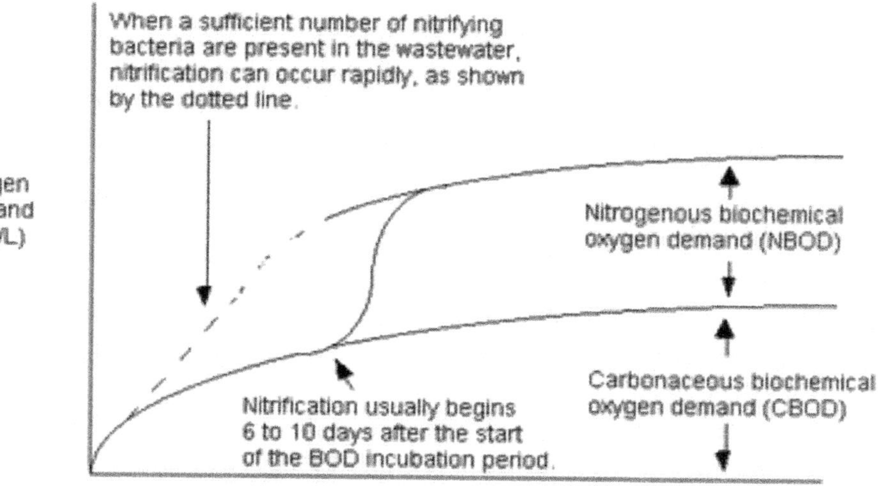

FIGURE 15.5 Components of total BOD.

In raw influent wastewater, 5-day CBOD is approximately 0.85 × BOD_{t5}. However, there is typically variability in BOD tests that can make this ratio not apply to individual samples.

4.9 Volatile Fatty Acids

Low-molecular mass carboxylic acids (C2-C7 monocarboxylic aliphatic acids) are important intermediates and metabolites in biological processes. The most common form of short-chain volatile fatty acids (SCVFAs) in domestic wastewater is acetic acid; however, propionic, butyric, valeric, caproic, and heptanoic acids can also be present in wastewater. The concentration and type of SCVFAs in raw wastewater are particularly important because they have a direct effect on biological phosphorus removal and denitrification performance. In addition, VFAs constitute one of the chemical classes responsible for unpleasant odor generated in wastewater, together with amines and sulfur compounds.

Traditionally, the VFA content in wastewaters and foods has been analyzed by titrimetric or gas chromatographic methods preceded by solvent extraction. The rapid colorimetric determination of organic acids is also in use. Volatile fatty acids contain negative charges when they are ionized (solubilized) in water. Therefore, more recently, some SCVFAs, including acetate, may be measured through ion chromatography without further derivatization and purification. High-pressure liquid chromatography can also be used to determine VFAs in wastewater.

According to Method 5560 (APHA et al., 2012), the gas chromatographic procedure can be used to determine the individual concentrations of acetic, propionic, butyric, isobutyric, valeric, and isovaleric acids. Samples are prepared by acidification, centrifugation, and filtration before injection into the gas chromatogram equipped with a flame ionization detector. Figure 15.6 shows an image of a gas chromatogram. The blank amount of certain fatty acids might be higher than the sample analyte (caused by the buildup of contaminants in the injector and guard column), and needs to be corrected if it exceeds the reporting limits.

5.0 NUTRIENT ANALYSIS

5.1 Nitrogen

The common types of species of nitrogen in wastewater are ammonium ion, un-ionized ammonia, nitrite, nitrate, and organic nitrogen. Fresh samples are always ideal for laboratory analysis. If analysis cannot be done within a short period of time, sample preservation is required to ensure the accuracy of laboratory analysis. This can be done by adding sulfuric acid (H_2SO_4) to drop pH of the sample to 1.5 to 2 and stop any biological reaction (APHA et al., 2012). Neutralization of preserved samples before laboratory analysis by adding sodium and potassium hydroxide (NaOH/KOH) is required. During preservation, the sample should be kept refrigerated at a temperature of 4 °C.

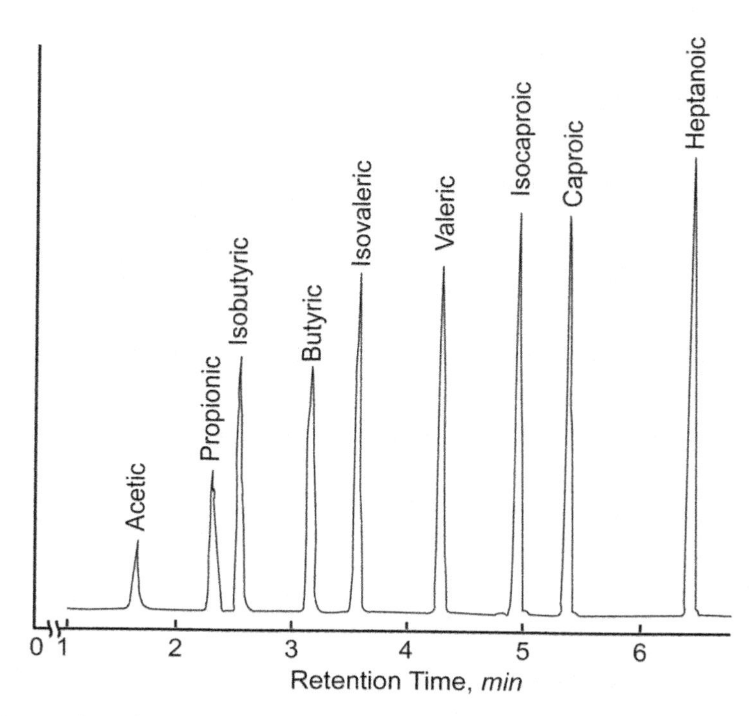

FIGURE 15.6 Gas chromatogram of a fatty acid standard. Column DB-FFAP, 0.53-mm-ID, 30-m, 0.50-μm film thickness, temperature-programmed as described in 5560D.2*d* (APHA et al., 2012).

5.1.1 *Ammonia*

Ammonia-nitrogen is a significant nutrient in raw wastewater and is typically present in the following two forms: ammonium ion and un-ionized ammonia. Measurement of ammonia-nitrogen in the wastewater stream is essential to determine the performance of the biological nutrient removal (BNR) facility. Techniques such as colorimetric method, titrimetric method, ion-selective method, and ion chromatography method can be used to determine ammonia-nitrogen. Measuring the range, difficulty, and presence of interference should be taken into consideration when choosing an appropriate method.

5.1.1.1 *Colorimetric Method*

The Nessler and Phenate methods are the two most common methods for colorimetric determination of ammonia. In each method, chemical reagents with ammonia produce a distinct color and are measured is in a spectrophotometer. The Nessler method (Method 4500-NH3 F; APHA et al. [2012]) typically is used to determine ammonia concentration in treated wastewater effluents with an ammonia-nitrogen concentration of less than 5 mg/L. Turbidity and color interfere with the method and need to be removed by distillation or filtration. If the filtrate is turbid, zinc sulfate should be used to eliminate turbidity (Method 4500-NH3 F;

APHA et al. [2012]). For the detailed procedure and calculation, refer to Method 4500-NH3 F (APHA et al., 2012).

5.1.1.2 *Titrimetric Method*

The titrimetric method for ammonia is used only on samples carried through the primary distillation step described in Method 4500-NH3 C (APHA et al., 2012). Mixed indicators containing methyl red and methylene blue are added to the distilled samples and blanks. The samples are titrated with H_2SO_4 until a pale lavender color is observed (APHA et al., 2012). The ammonia concentration is calculated by taking the difference between the volume of H_2SO_4 titrated for samples and the volume of H_2SO_4 titrated for blanks. For the detailed procedure and calculation, refer to Method 4500-NH3 C (APHA et al., 2012).

5.1.1.3 *Ion-Selective Method*

The ammonia-selective electrode is equipped with a gas-permeable membrane that only allows un-ionized ammonia to diffuse and changes the internal solution pH that is measured by a pH meter with an expanded millivolt scale. Strong base (i.e., NaOH or KOH) needs to be added to the sample to raise the pH to above 11 to convert all ammonium to free ammonia to be detected by the electrode.

The ion-selective method is applicable to analyses of ammonia-nitrogen in wastewaters within the range of 0.03 to 1400 mg/L. High concentrations of other dissolved ions may interfere with the measurements. The method is not affected by high turbidity and color. For the detailed procedure and calculation, refer to Method 4500-NH3 D (APHA et al., 2012).

5.1.1.4 *Ion Chromatography Method*

Ion chromatography is a rapid and simple method to analyze many ions in wastewater. It is accurate, and it eliminates preparation of hazardous reagents. The wastewater sample is injected to a stream of carbonate–bicarbonate and passed through a series of ion exchangers. Each ion has a unique affinity to a separator column and exhibits a different separation time.

If the pH of the wastewater sample is 7.5 or less, pH correction is not necessary. However, pH values of 8.0 to 8.5 require a correction factor of 1.04 and 1.12, respectively. The minimum detection level for ammonia-nitrogen is approximately 0.05 mg/L. Dilution is recommended if the ammonia concentration is higher than 40 mg/L. This method can be directly applied to measure ammonia concentration in raw wastewater, activated sludge, and industrial wastewaters. The following steps must be performed:

- Prepare at least seven standard solutions in the range of low to high (0.2, 1.0, 5.0, 10, 20, 30, and 40 mg/L);
- Use ion-free deionized water during standard solution preparation;

- Take 2 to 3 mL of volume from each standard solution and inject it to the ion chromatograph;
- Prepare a standard curve for ammonia calculations;
- Filter all samples with a 0.45-μm membrane filter;
- Take a 2- to 3-mL sample with a disposable syringe;
- Inject the samples to the ion chromatograph; and
- Calculate the ammonia concentrations based on the standard curve.

5.1.2 *Nitrite and Nitrate*

Nitrite is an intermediate product during nitrification and the final product of nitritation. The nitrite ion is rarely found in a well-operated WRRF. Nitrite formation occurs during the startup period of facility operation and can occur intermittently, but is typically quickly converted to a more stable form of nitrogen.

5.1.2.1 Colorimetric Method

The colorimetric determination of nitrite relies on the formation of a reddish-purple azo dye at pH 2.0 to 2.5 and its photometric measurement at 540 nm. The method is suitable to measure low nitrite concentration in microgram levels of 10 to 1000 $\mu g\ NO_2^-$-N/L.

Suspended solids, free chlorine, and nitrogen trichloride interfere with the measurements. The presence of ions may also interfere with color development. Before analyses, all samples must be filtered through a 0.45-μm membrane filter. Analysis of fresh samples is recommended for accurate results. If sample storage is necessary, filtering and freezing of samples is sufficient to preserve the samples for a couple of days. Acid preservation should never be used for nitrite analysis (APHA et al., 2012). For the complete procedure and calculation, refer to Method 4500-NO_2^- B (APHA et al., 2012).

There is also a proprietary spectrophotometer that is commercially available for nitrite measurements. However, the details of any proprietary system should be reviewed for U.S. Environmental Protection Agency (U.S. EPA) approval.

5.1.2.2 Ion Chromatography Method

Ion chromatography is a rapid, easy, and accurate method to determine nitrite-nitrogen and nitrate-nitrogen of wastewaters. It eliminates the use of hazardous chemicals. The anions are separated on the basis of their relative affinities for a strongly basic anion exchanger (separator column). The separated anions are directed onto a strongly acidic cation exchanger (suppressor column), where they are converted to their acidic form. The anions, in their acidic form, are measured by conductivity. The quantification is performed based on the peak area for each ion (APHA et al., 2012). The common anions in wastewater treatment are chloride (Cl^-), nitrite (NO_2^-), phosphate (PO_4^{3-}), nitrate (NO_3^-), and sulfate (SO_4^{2-}). Some organic acids have been reported to interfere with measurements (APHA et al., 2012). The following procedure is recommended for nitrite and nitrate determination by ion chromatography:

- Prepare a series of nitrite and nitrate standards by weighing sodium nitrite and sodium nitrate (deionized water should be used during standard preparation);
- Take 2 to 3 mL from the combined standard solutions and inject them to the ion chromatographer;
- Repeat standard injection at least twice; produce a standard curve based on area and concentration relationship;
- Filter samples with a 0.45-μm pore diameter filter (if prompt determination is not possible, freeze the filtered samples at −10 °C or below);
- Make dilution, if necessary (use deionized water during dilution);
- Inject samples into the ion chromatographer; and
- Calculate nitrite and nitrate calculations from the standard curve.

Nitrate is the highest oxidation state of nitrogen. Nitrate nitrogen is part of total nitrogen and total inorganic nitrogen.

5.1.2.3 Colorimetric Method (Cadmium Reduction Method)

A proprietary spectrophotometer is commercially available for nitrate measurements. More than 16 substances can cause interference effects at certain concentration levels. The test can only be used if COD is less than 500 mg/L. The details of any proprietary system should be reviewed for U.S. EPA approval.

5.1.2.4 Nitrate Electrodes Screening Method

This method uses a selective sensor for the nitrate ion that develops a potential across a thin, inert membrane (Method 4500-NO_3^- D; APHA et al. [2012]). The electrode responds only to the nitrate ion in the range of 0.14 to 1400 mg/L (APHA et al., 2012). The electrode responds to nitrate activity rather than nitrate concentration. Interference includes chloride, bicarbonate, and nitrite ions. Chloride and bicarbonate concentrations exceeding the nitrate concentration by 5 and 10 times, respectively, may interfere with the nitrate. Using a buffer solution containing silver sulfate (Ag_2SO_4) may minimize such interferences. Reducing the pH to 3 eliminates bicarbonate ions. Adding sulfamic acid will minimize the nitrite interference. For the detailed procedure and calculation, refer to Method 4500-NO_3^- D (APHA et al., 2012).

5.1.3 *Organic Nitrogen*

Organic nitrogen in wastewater does not represent all the nitrogen found in organic form. Rather, it is the portion related to protein, urea, and amino compounds.

Organic nitrogen decomposition releases organic nitrogen into the medium. In a well-operated BNR facility, effluent organic nitrogen concentration is typically 0.5 to 2.0 mg/L.

To determine the organic nitrogen concentration, the Kjeldahl method is performed after removing ammonia-nitrogen in the medium. Otherwise, the organic nitrogen concentration is estimated by taking the difference between total Kjeldahl nitrogen (TKN) and ammonia-nitrogen.

5.1.4 *Total Kjeldahl Nitrogen*

In wastewater, TKN is the sum of unionized ammonia, ammonium ion, and organic nitrogen. For total nitrogen, the nitrite and nitrate concentration is added to TKN. The macro-Kjeldahl method and semi-micro-Kjeldahl method are available to determine Kjeldahl nitrogen. The most reliable results are obtained from fresh samples. If prompt analysis is not possible, acid preservation should be used followed by 4 °C storage. Mercury chloride addition is not recommended because it reduces the ammonia in the sample.

5.1.4.1 Macro-Kjeldahl Method

The macro-Kjeldahl method is applicable for samples containing a broad range of organic nitrogen. It requires a relatively large volume for analysis and is recommended if enough sample volume can be obtained (APHA et al., 2012). The three main steps of the process include digestion, distillation, and ammonia determination. For the complete procedure and calculation, refer to Method 4500-N_{org} B (APHA et al., 2012).

5.1.4.2 Semi-Micro-Kjeldahl Method

The semi-micro-Kjeldahl method is applicable to samples with high concentrations of organic nitrogen. For the detailed procedure and calculation, refer to Method 4500-N_{org} C (APHA et al., 2012). Proprietary test kits are also available for quick TKN analysis; however, it is important to note that the kits are not based on *Standard Methods*. Carbonaceous oxygen demand and chloride will interfere at certain concentration levels.

5.2 Phosphorus

Phosphorus is essential to the growth of biological organisms. However, excessive phosphorus content in surface water causes issues such as accelerated plant growth, algal bloom, low dissolved oxygen, and the death of certain invertebrates and fish. Typical forms of phosphorous in wastewater include orthophosphates, condensed phosphates, and organically bound phosphate. Orthophosphates are reactive phosphates in wastewater that respond to the colorimetric test or ion chromatography without preliminary hydrolysis or digestion, such as PO_4^{3-}, HPO_3^-, $H_2PO_4^-$, and H_3PO_4. Organically bound phosphates are phosphate fractions that are converted to orthophosphates by digestion of organic matter (APHA et al., 2012).

Orthophosphate and total phosphorus are of particular interest when it comes to phosphorus analysis. If the soluble form of phosphate is of interest, the sample should be filtered promptly after collection and stored at −10 °C. If longer storage is desired, 40 mg/L of mercury chloride should be added. If total phosphorus is of interest, then 1 mL of concentrated HCl should be added to the unfiltered samples. If the analysis is performed within 48 hours, then freezing without acid addition will be sufficient enough to preserve the samples (APHA et al., 2012). Use of phosphate-containing detergents to clean phosphorus glassware should be avoided (APHA et al., 2012).

5.2.1 *Total Phosphorus*

Total phosphorus includes all orthophosphates and condensed phosphates, both soluble and particulate, and organic and inorganic fractions. It is determined by using digestion methods and orthophosphate measurement. There are three digestion methods: persulfate digestion, which is the most common, perchloric acid digestion, and sulfuric–nitric acid digestion (Method 4500-P B; APHA et al. [2012]). Unfiltered activated sludge samples require 1 to 50 to 1 to 200 times dilution before the aforementioned digestion step. Following digestion, all complex phosphorus is converted to orthophosphate form and is determined by colorimetric, spectrophotometric, or chromatographic methods.

Although proprietary test kits are available for a quick total phosphorus analysis, it is important to note that the kits are not based on *Standard Methods*. The details of any proprietary system should be reviewed for U.S. EPA approval.

5.2.2 *Dissolved Phosphorus*

The dissolved phosphorus test measures that fraction of total phosphorus that is in solution in the water (as opposed to being attached to suspended particles). It is determined by first filtering the sample through a 0.45-μm pore diameter filter, then analyzing the filtered sample for total phosphorus.

5.2.3 *Orthophosphate*

Orthophosphate can be measured by directly adding a complex agent such as ammonium molybdate, which will form a colored complex with phosphate, followed by performing a colorimetric test using a spectrophotometer. Although proprietary test kits are also available for quick orthophosphate analysis, it is important to note that the kits are not based on *Standard Methods.* The details of any proprietary system should be reviewed for U.S. EPA approval.

5.2.3.1 *Vanadomolydophosphoric Acid Colorimetric Method*

Orthophosphate reacts with ammonium molybdate under acidic conditions to form molybdophosphoric acid. Vanadium is added to promote vanadomolybdophosphate, which produces a yellow color. The color intensity is proportional to phosphorus in the sample. The color is spectrometrically detected at a wavelength of 470 nm (recommended) for measurements (APHA et al., 2012). A standard curve is prepared in a phosphorus range of 0 (blank) to 20 mg/L. The detection limit of the method is 200 μg P/L. If the digestion step is included, the measured phosphorus values will reflect the total phosphorus. For the complete procedure and calculation, refer to Method 450-P C (APHA et al., 2012).

5.2.3.2 *Ascorbic Acid Method*

Similar to the method described in Section 5.2.3.1, in the ascorbic acid method (Method 450-P E; APHA et al. [2012]), orthophosphates react with ammonium molybdate and potassium antimonyl tartrate in a highly acidic medium to form intensely colored molybdenum blue by ascorbic acid. The developed color should be measured within 30 minutes following the color development at 880 nm. An arsenate concentration as low as 0.1 mg As/L interferes with the color development. Hexavalent chromium and nitrite may also interfere with the results. The minimum detectable concentration of phosphorus is approximately 10 μg/L. For the complete procedure and calculation, refer to Method 450-P E (APHA et al., 2012).

5.2.4 *Refractory Phosphorus*

When faced with low total phosphorus limits (<0.2 mg/L total phosphorus), it may be helpful to identify the soluble component that is not responsive to biological treatment (conventional or enhanced biological phosphorus removal), advanced phosphorus removal methods (tertiary processes that include adsorption onto metal hydroxide solids, flocs, or coated silica sand), and high solids capture. The soluble phosphorus that passes through both biological and advanced treatment is considered refractory or recalcitrant phosphorus. For most domestic wastewater, the background concentration of recalcitrant phosphorus is between 0.005 and 0.02 mg/L. However, with certain industrial contributions, concentrations may be higher. The magnitude of recalcitrant phosphorus is only significant when it might be a dominant portion of the effluent total phosphorus because it makes meeting the effluent objective difficult.

If a facility has a low total phosphorus limit and is able or planning to achieve high solids capture (<5 mg/L TSS) and near-complete adsorption of orthophosphate (nondetect, which can be as low as 0.005 mg/L phosphorus, depending on the method) through some tertiary process, recalcitrant phosphorus can be determined by a test. This is done by taking a secondary effluent sample and treating it in a way that simulates a tertiary phosphorus removal process by adding sufficient metal salt, mixing it to adsorb the orthophosphate, passing it through a 0.45-μm filter, and analyzing the filtrate for total phosphorus and orthophosphate. The difference will be the sum of the dissolved acid hydrolyzable and dissolved organic phosphorus, which will be what remains following biological treatment and advanced phosphorus removal. This is considered recalcitrant phosphorus. The analytical method that provides the lowest detection limit should be used, which is best done with long cell path cuvettes and a spectrophotometer. The test kits do not provide a sufficiently low detection limit.

Excessive metal dose can cause a matrix interference with the low detection method. In addition, if the metal salt used is a strong acid, or even a weak acid, but too much is used, the pH becomes so low that solubility limitations are exceeded. A metal salt that is a weaker acid, such as sodium aluminate, is recommended for use, or a buffer should be added to ensure that the pH is not too low. The total phosphorus analysis goes through persulfate digestion followed by colorimetry. The orthophosphate analysis, which is sometimes considered "reactive", is simply direct colorimetry.

6.0 MICROBIOLOGY

Just as microorganisms play a significant role in biological processes, the same is true for BNR in WRRFs. Microorganisms enable the removal of most nitrogenous and phosphoric compounds found in wastewater. However, like many other treatment processes, microorganisms have long been viewed as the "black boxes" of a facility (Oehmen et al., 2007; Seviour et al., 2003). The knowledge of engineering and operational features of BNR processes predates an understanding of its microbiology.

Today, after much advancement in the field of wastewater microbiology, engineers and operators agree that a thorough knowledge of the structure and function of these complex communities is essential to successful operation and process optimization of BNR processes. Understanding how these microorganisms function, monitoring their growth and abundance in activated sludge, and proactively predicting their behavior can help minimize operational upsets. Thus, microbiological examinations of activated sludge samples are important and provide an opportunity for engineers and operators to decide on ways to maintain healthy activated sludge conditions and to prevent and correct unhealthy conditions (Gerardi, 2008).

Bacteria are the group of microbes most directly involved in BNR processes. They dominate the microbial composition of activated sludge, both in numbers and biomass. However, their presence also allows for the growth and existence of several eukaryotic predator organisms such as protozoa and metazoa (Curds, 1982). Although protozoa and metazoa are not directly responsible for the primary degradation of pollutants in wastewater, they contribute to the elimination of particulate matter and act as indicators of environmental conditions in the treatment process. For example, the presence of stalked ciliates, rotifers, and nematodes indicate long SRTs or low organic loading. A predominance of metazoa is often described as "old sludge" and a predominance of flagellates and amoeboids is described as "young sludge". Because ammonia removal is associated with sensitive nitirifying bacteria, microscopic examination can help determine SRT, which can be modified, if needed, to provide nitrification. This section provides laboratory methods for microbiological examinations of activated sludge processes focusing on BNR.

6.1 Microscope Use

All microbiological examinations require the use of microscopes. The two basic types of microscopes used for analyzing activated sludge samples are light, or optical, microscopes and electron microscopes.

Light, or optical, microscopes require light waves to provide illumination and are the most commonly used devices for all general wastewater laboratory work. Light microscopes can be bright field, dark field, phase contrast, or fluorescence, with lenses that can accommodate magnifications ranging from 10 to 1000×, depending on specific laboratory analytical needs. The most commonly used laboratory microscopes in WRRFs are the bright-field and phase-contrast microscopes. The fluorescence microscope uses UV light as its light source, and is used mainly in advanced molecular analytical methods such as fluorescent in situ hybridization (FISH) and fluorescence staining techniques.

Electron microscopes use electrons to provide the illumination. Because the beam has an exceptionally short wavelength, it strikes most objects in its path and increases the resolution of the microscope, enabling extremely small objects such as subcellular components or viruses to be viewed. Electron microscopes can be either transmission or scanning and are much more expensive than light microscopes. These microscopes are not commonly used in WRRF laboratories; rather, they are often used for research purposes. For more details on the aforementioned microscopes, types of lenses, filters, microscopic measurements, and how to use them, refer to literature by Csuros and Csuros (1999), Gerardi (2008), and Jenkins et al. (2004).

6.2 Protozoa and Metazoa

Biological wastewater treatment relies on the activity of a mixed community of bacterial culture for the removal of organic substances and nutrients. However, their presence also allows for the growth of predator organisms such as protozoa and metazoa. While these organisms are of little value to the process of BNR directly, microscopic examinations of protozoa and metazoa are a common and widespread practice as indicator organisms. In addition, they help with the removal of nonflocculated and loosely flocculated bacteria from wastewater producing more clarified effluent and may contribute to biomass flocculation through the production of fecal pellets and mucus (Jenkins et al., 2004). Protozoa and metazoa constitute approximately 5%, by weight, of the activated sludge biomass and are represented by about 200 species (Curds, 1975).

Protozoa are a large collection of organisms with considerable morphological and physiological diversity. They are all eukaryotic and are considered to be unicellular. Common protozoa found in activated sludge include amoeba, flagellates, and ciliates. A useful guide to identifying protozoa in wastewater treatment systems can be found in literature by Patterson (1998).

Metazoa are multicellular eukaryotic organisms that include worms such as rotifers and nematodes and arthropods such as crustacea and insects. Most of them are aerobic, but some organisms can survive under low dissolved oxygen conditions. The main metazoa present in activated sludge are rotifers and nematodes (Richard, 1989).

To observe these organisms under a microscope, Jenkins et al. (2004) recommend placing one drop of activated sludge on a microscopic slide, adding a cover slip on it, and examining it at 100× magnification using phase-contrast illumination. All protozoa and other higher life forms should be counted by scanning the entire cover slip area using the mechanical stage of the microscope; an average of the results for four to five separate preparations should be used for purposes of enumeration. For further details, refer to literature by Jenkins et al. (2004).

6.2.1 *Flagellates*

Flagellets are typically small, oval, or elongated forms that possess a whip-like structure or flagella that provide locomotion for the organism. Their presence can indicate significant soluble BOD levels. Many of these occur at low dissolved oxygen concentrations accompanied by high organic loads. Jenkins et al. (2004) and Patterson (1998) provide illustrations of common flagellates found in activated sludge.

6.2.2 *Amoebae*

Amoebae, which represent the simplest life forms, vary in shapes and sizes. They are slow-moving organisms and have a cytoplasm or gut content that flows against a flexible cell membrane, permitting locomotion. Two types of amoebae are known to exist in activated sludge: naked and testate. Amoebae grow well on particulate organic matters and are able to tolerate low dissolved oxygen conditions. Jenkins et al. (2004) and Patterson (1998) provide illustrations of amoeba found in activated sludge.

6.2.3 *Ciliates*

Ciliates are free-swimming, crawling, or stalked organisms. Ciliates are typically found under good floc-forming conditions and are generally indicators of satisfactory activated sludge operations. They are sensitive organisms and their presence or absence may indicate toxicity. Jenkins et al. (2004) and Patterson (1998) provide illustrations of the different kinds of ciliates found in activated sludge.

6.2.4 *Rotifers*

These metazoa have a variety of shapes and are much larger and more complex in structure than protozoa. Most of them are motile and are able to attach themselves to activated sludge flocs with contractile feet. They occur over a wide range of SRTs and some species are indicative of high SRT. Jenkins et al. (2004) provides illustrations of common rotifers found in activated sludge.

6.2.5 *Higher Invertebrates*

Higher forms of organisms found in activated sludge include nematodes, tardigrades, gasterotrichs, and annelids. They are generally observed at higher SRTs. Tardigrades, gasterotrichs, and annelids appear to occur only in nitrifying systems. Jenkins et al. (2004) provide illustrations of common higher invertebrates found in activated sludge.

6.3 Bacteria

Biological nutrient removal processes involve the metabolic activities of various bacteria to remove nitrogenous and phosphoric compounds from wastewater. Common, favorable ones include polyphosphate-accumulating organisms (PAOs), ammonia oxidizers (ammonia-oxidizing bacteria), nitrite, and nitrate oxidizers (nitrate-oxidizing bacteria), denitrifiers, denitrifying PAOs, and anaerobic ammonia oxidizers. Common, unfavorable ones include filamentous bacteria. Several species of bacteria have been identified to fall into each of these categories. Moreover, the contributions of interdisciplinary research in wastewater microbiology have led to the development of numerous advanced molecular methods, such as FISH, cloning of 16S rDNA, denaturant gradient gel electrophoresis, and real-time polymerase chain reactions, to characterize and understand the structure and function of these bacteria. However, only a handful of WRRF laboratories are equipped to run these advanced methods. They are more commonly used for research and not for monitoring the health of the treatment process, although they are highly capable of accomplishing that purpose.

6.3.1 *Filamentous Bacteria*

The presence of filamentous bacteria in activated sludge is known to cause sludge bulking and foaming. These problems are commonly reported in BNR systems and several literature citations provide evidence that BNR favors the growth of undesirable and excessive filamentous bacteria (Eikelboom, 2000; Lee et al., 1996; Metcalf and Eddy, 2003). The causes for the growth of filamentous bacteria include several operational conditions, such as low oxygen concentration, low food-to-microorganism ratio (F/M), septicity, nutrient deficiency, low pH, and high grease and oil. Different types of filamentous bacteria are known to thrive under different operating conditions. Microscopic examination of filamentous bacteria can provide a wealth of information on the nature and causes of sludge foaming and bulking and the ability to correct them.

Determining the number and type of filamentous microorganisms is the first step in controlling this problem. Such examination requires a phase contrast microscope with the capability of 100 to 1000× magnification. Details on sample collection, transport, storage, and sampling frequency are provided in Jenkins et al. (2004).

Common microscopic methods of enumerating and identifying filamentous bacteria involve counting and staining methods. Counting methods include

- Simplified filament counting technique,
- Total extended filament length measurement method, and
- Nocardioform organism filament counting technique.

Staining methods include

- Gram stain,
- Neisser stain,
- Sulfur stain,
- India ink reverse stain,
- Polyhydroxyalkanoate (PHA) stain, and
- Crystal violet sheath stain.

For more details on the counting methods, refer to Jenkins et al. (2004).

6.3.2 *Polyphosphate-Accumulating Organisms*

The fluorescent stain, 4′, 6-diamidino-2-phenylindol (DAPI), binds strongly to DNA in cells and fluoresces blue when excited with light at 340 nm using a microscope. However, DAPI also stains polyphosphate granules accumulated within any bacteria to emit yellow fluorescence. This staining technique is commonly used to differentiate between PAOs and non-PAOs. For direct polyphosphate staining using DAPI, mixed liquor samples collected from the aeration tank should be washed twice with 1X phosphate-buffered saline (PBS) (a mixture of 8 g NaCl, 0.2 g KCl, 1.44 g Na_2HPO_4, and 0.24 g KH_2PO_4 in 1 L of water; pH 7.4 and filter sterilized) and purged at least 30 times using a 26-gauge needle to disrupt flocs. The samples should then be fixed with 4% paraformaldehyde solution in 1X PBS for 45 minutes at room temperature. The fixed cells will be filtered through a 0.22-μm polycarbonate filter paper and the filter paper containing the fixed cells will be transferred onto gelatin-coated sterile glass slides. The cells on the slide will then be stained with 5 μg/mL of DAPI in the dark for 30 minutes and the polyphosphate positive cells can be viewed using an epifluorescence microscope.

6.4 Microbiological Activity

Monitoring microbiological activity can be useful in optimizing the operation of a bioreactor, particularly for systems operating in nutrient removal mode. It can aid in detecting influent toxicity or inhibition or a potential slug loading into the facility. In addition, having the ability to quickly and easily measure active biomass concentration creates many opportunities for optimizing bioreactors, including management of biomass inventory, which, consequently, aids in controlling the F/M to improve settleability and the optimization of aeration. This allows operators to adjust oxygen delivery to the active biomass, leading to more efficient operation. There are available methods for monitoring at least two compounds that are indicators of microbiological activity. These include nicotinamide-adenine dinucleotide (NADH) and adenosine triphosphate (ATP).

6.4.1 *Nicotinamide-Adenine Dinucleotide*

Microbiological activity can be monitored using the concentration of a coenzyme NADH, which is found in all living cells. In cellular metabolism, nicotinamide-adenine dinucleotide (NAD+) is involved in redox reactions, carrying electrons from one reaction to another. The coenzyme is, therefore, found in the following two forms: NAD+ (oxidizing agent that accepts electrons) and NADH (reducing agent to donate electrons). These electron transfer reactions are the main function of NAD+, although it is involved in other cellular processes. This metabolic intermediate can be characterized by fluorescence at particular wavelengths. For a given biomass, the NADH fluorescence signal reflects the ratio between NADH and NAD+, allowing determination of the redox state of the biomass.

Monitoring NADH involves the use of a sensor (probe) to monitor the strength of a fluorescence signal reflecting from the biomass; this, in theory, allows for more reliable monitoring and control of the biological state or activity in the aeration tanks at low dissolved oxygen concentrations than with a dissolved oxygen probe. The fluorescence signal is dependent on the concentration of NADH in the biomass in the aeration tanks. The fluorescence signal reflected from the biomass originates from the probe, which emits a UV light with a wavelength of 340 nm; NADH has a unique property in that it fluoresces at 460 nm when excited with UV irradiance at a wavelength of 340 nm. There is a proprietary control system available on the market that has been tested at full-scale treatment facilities; the SymBio® process by Eimco Water Technologies uses online monitoring of NADH to determine changes in biological demands. Using the online NADH signal, airflow to the basin is controlled to promote simultaneous nitrification–denitrification of wastewater. The NADH monitor uses a fluorescence sensor to detect changes in NADH, which, in turn, provides information on the status of biological wastewater treatment processes. Weerapperuma and de Silva (2004) reported that the NADH sensor requires minimal maintenance and can provide real-time information for process control. The manufacturer claims a 25 to 30%

energy savings compared to nitrifying facilities without this control technology; however, no independent data from full-scale facilities has been published to verify these claims (U.S. EPA, 2010).

6.4.2 *Adenosine Triphosphate*

Adenosine triphosphate is the keystone of metabolic activity in living cells (Lehninger, 1982). Most of the energy within microorganisms is stored and transmitted via ATP. Adenosine triphosphate is produced as microbial food and is consumed and subsequently used for cell maintenance and the synthesis of new cells and biochemical reactions. As such, ATP can be used for monitoring biological processes. The value of monitoring ATP in biological wastewater treatment was recognized 35 years ago (Patterson et al., 1970). More recently, Archibald et al. (2001), in a study using a suite of respirometric tests on mixed liquor from paper mill activated sludge processes, concluded that ATP measurements provided a useful monitor of the proportion of viable cells and a toxicity indicator in an activated sludge process. These and other studies have provided direct measurement of ATP as a monitoring tool in biological treatment processes; however, most studies have not used methods that distinguish between extracellular ATP and ATP contained only within microorganisms. In efforts to distinguish between these forms, a related stress index that is based on the ratio of dissolved ATP to total ATP, referred to as the *biomass stress index*, has been developed as part of a proprietary monitoring tool. This commercially available tool has been demonstrated in laboratory and full-scale reactors; as stresses such as suboptimal pH, anoxia, toxicity, and nutritional deficiencies were applied to the microbial populations, the stress index increased (Cairns et al., 2005).

7.0 REFERENCES

American Public Health Association; American Water Works Association; Water Environment Federation (2012) *Standard Methods for the Examination of Water and Wastewater,* 22nd ed.; American Public Health Association: Washington, D.C.

Archibald, F.; Me´thot, M.; Young, F.; Paice, M. G. (2001) A Simple System to Rapidly Monitor Activated Sludge Health And Performance. *Water Res.,* **35** (10), 2543–2553.

Cairns, J. E.; Whalen, P. A.; Whalen P. J.; Tracey, D. R.; Palo, R. E. (2005) Dissolved ATP—A New Process Control Parameter for Biological Wastewater Treatment. *Proceedings of the 78th Annual Water Environment Federation Technical Exhibition and Conference* [CD-ROM], Washington, D.C., Oct 29–Nov 2; Water Environment Federation: Alexandria, Virginia.

Csuros, C.; Csuros, M. (1999) *Microbiological Examination of Water and Wastewater;* CRC Press: Boca Raton, Florida.

Curds, C. R. (1975) Protozoa. In *Ecological Aspects of Used Water Treatment: The Organisms and their Ecology*; Curds, C.R., Hawkes, H.A., Eds.; Academic Press: New York.

Eikelboom, D. H. (2000) *Process Control of Activated Sludge Plants by Microscopic Investigation*; IWA Publishing: London.

Gerardi, M. H. (2008) Microscopic Examination of the Activated Sludge Process; Wiley & Sons: Hoboken, New Jersey.

Gy, P. M. (1992) *Sampling of Heterogeneous and Dynamic Material Systems—Theories of Heterogeneity, Sampling, and Homogenizing*; Elsevier Science: Amsterdam, Netherlands.

Jenkins, D.; Richard, M. C.; Daigger, G. T. (2004) *Manual on the Causes and Control of Activated Sludge Bulking and Foaming*, 3rd ed.; Lewis Publishers: Boca Raton, Florida.

Lee, N. M.; Carlsson, H.; Aspegren, H.; Welander, T.; Andersson, B. (1996) Stability and Variation in Sludge Properties in Two Parallel Systems for Enhanced Biological Phosphorus Removal Operated with and without Nitrogen Removal. *Water Sci. Technol.*, **34** (1–2), 101–109.

Metcalf and Eddy, Inc. (2003) *Wastewater Engineering, Treatment and Reuse*, 4th ed.; Tchobanoglous, G., Burton, F., Eds. McGraw-Hill: New York.

Oehmen, A.; Lemos, P. C.; Carvalho, G.; Yuan, Z.; Keller, J.; Blackall, L. L.; Reis, M. A. M. (2007) Advances in Enhanced Biological Phosphorus Removal: From Micro to Macro Scale. *Water Res.*, **41**, 2271–2300.

Ort, C.; Gujer, W. (2006) Sampling for Representative Micropollutant Loads in Sewer Systems. *Water Sci. Technol.*, **54** (6–7), 169–176.

Patterson, D. J. (1998) *Free-Living Freshwater Protozoa: A Colour Guide*; Manson Publishing: London.

Patterson, J. W.; Brezonik, P. L.; Putnam, H. D. (1970) Measurement and Significance Of Adenosine Triphosphate in Activated Sludge. *Environ. Sci. Technol.*, **4** (7), 569–575.

Richard, M. (1989) *Activated Sludge Microbiology*; The Water Pollution Control Federation: Alexandria, Virginia.

Seviour, R. J.; Mino, T.; Onuki, M. (2003) The Microbiology of Biological Phosphorus Removal in Activated Sludge Systems. *FEMS Microbiol. Rev.*, **27**, 99–127.

Snedecor, G. W.; Cochran, W. G. (1976) *Statistical Methods*, 6th ed.; The Iowa State University Press: Ames, Iowa.

U.S. Environmental Protection Agency (2004) *EPA Guidelines for Water Reuse*; U.S. Environmental Protection Agency: Washington, D.C.

U.S. Environmental Protection Agency (2010) *Evaluation of Energy Conservation Measures for Wastewater Treatment Facilities*; U.S. Environmental Protection Agency: Washington, D.C.

Water Environment Federation (2012) *Basic Laboratory Procedures for the Operator-Analyst*, 5th ed.; Water Environment Federation: Alexandria, Virginia.

8.0 SUGGESTED READINGS

Curds, C. R. (1982) The Ecology and Role of Protozoa in Aerobic Sewage Treatment Processes. *Annu. Rev. Microbiol.*, **36**, 27–46.

Dold, P.; Bye, C.; Chapman, K.; Brischke, K.; White, C.; Shaw, A.; Barnard, J.; Latimer, R.; Pitt, P.; Vale, P.; Brian, K. (2010) What Do We Model, How Should We Model? *Proceedings of the Wastewater Treatment Modeling 2010 Seminar (WWTmod 2010)*; March 28–30, Monte St. Anne, Quebec, Canada; pp 133–150.

Grady, Jr., C. P. L.; Daigger, G. T.; Lim, H. C. (1999) *Biological Wastewater Treatment*, 2nd ed.; Marcel Dekker: New York.

Chapter 16

Case Studies—Nitrification and Denitrification

Jason Beck, P.E.; Vincent Apa, P.E., BCEE; and William C. McConnell, P.E., BCEE

1.0 INTRODUCTION

This chapter documents eight case studies of facilities that operate nitrification and denitrification processes. These case studies cover full-scale municipal treatment facilities that are specifically designed and operated for nitrogen removal. All facilities used in the case studies in this chapter have been in operation for a minimum of 2 years so the reader can gain insight to the challenges encountered by facility operators during the startup and commissioning phases of the project and the design and operational changes that were required following facility startup.

2.0 CASE STUDY 1 (MODIFIED LUDZACK–ETTINGER)—ALBUQUERQUE, NEW MEXICO, SOUTHSIDE WATER RECLAMATION PLANT

2.1 Facility History

The Albuquerque, New Mexico, Southside Water Reclamation Plant (SWRP) Nitrogen Removal Project was mandated by the U.S. Environmental Protection Agency. The project included conversion of the entire 288-ML/d (76-mgd) SWRP to a modified Ludzack–Ettinger (MLE) treatment system for biological nutrient removal (BNR). The system consists of a two-stage, single-sludge, suspended growth biological system capable of attaining high ammonia-nitrogen (NH_3-N) removal through nitrification and partial denitrification of nitrate (NO_3).

The design included demolition of older trickling filter and secondary clarifier facilities that were previously operated in parallel with the six aeration basins to provide space for six of the eight new basins. Air for fine-bubble aeration equipment in the new MLE basins is provided by four new 335-kW (450-hp) blowers housed in a new blower and chemical feed building. Modifications were made to store and feed soda ash and acetic acid, which can also be used to feed methanol as an alternate chemical. Other significant work included design of four new final clarifiers that were each 41 m (135 ft) in diameter. Improvements were completed in 1998, and SWRP is currently operating well within regulatory requirements.

2.2 Process Description

The wastewater system consists of small-diameter collector sewers, wastewater lift stations, and large-diameter interceptor sewers conveying wastewater flows to the SWRP. The facility is designed to treat average and peak influent flows of 288 ML/d (76 mgd) and 454 ML/d (120 mgd), respectively, and serves more than 563,000 people. Before being discharged to the Rio Grande River, the treatment facility provides preliminary screening; aerated grit removal; primary clarification and sludge removal; and advanced secondary treatment including nitrogen removal, final clarification, and UV disinfection.

The MLE process was designed at a 9-day aerobic solids retention time (SRT) and a minimum water temperature of 15 °C. There are 13 MLE process basins with a total anoxic volume of 40.5 ML (10.7 mil. gal) and a total oxic volume of 72 ML (19.1 mil. gal). There is also a swing zone (oxic or anoxic) that has a total volume of 30 ML (8 mil. gal). The facility typically runs the swing zone as anoxic. There are 12 41-m (135-ft) diameter final clarifiers. Disinfected effluent is sent to the rotary disc filters for filtration before discharge. Additionally, a 32-km (2-mi)-long distribution pipeline delivers up to 19 ML/d (5 mgd) of reclaimed wastewater for irrigating turf areas at 26 parks, fields, and other areas.

Waste activated sludge (WAS) is sent to dissolved air flotation thickeners. Thickened WAS (TWAS) is blended with primary sludge, anaerobically digested, and dewatered with centrifuges. The SWRP has used clean energy for more than 25 years, currently supplying about 30% of its total power needs with renewable biogas-fueled combined heat and power (CHP) and up to the remaining 70% with natural-gas-fueled CHP. The facility's average power requirements are 4.5 MW. The digesters supply enough biogas for 1.6 MW, leaving the rest to be generated by natural gas or purchased from the utility, Public Service Company of New Mexico. (Refer to Figure 16.1 for a process flow diagram of the overall facility, Table 16.1 for a list of the facility's wastewater characteristics, and Table 16.2 for a list of the facility's process design criteria.)

The SWRP has summer and winter daily maximum ammonia (NH_3) limits of 1.5 and 4.5 mg/L, respectively. Additionally, the total inorganic nitrogen (TIN) 30-day average maximum concentration is 9.7 mg/L and the daily maximum concentration is 14.6 mg/L. Total suspended solids (TSS) and 5-day carbonaceous biochemical oxygen demand ($CBOD_5$) monthly limits are 30 and 15 mg/L, respectively.

2.3 Design and Operation Modifications

There have been no process modifications to the original design and both swing zones are run in the anoxic mode.

2.4 Performance

The SWRP has been treating an average of 230 ML/d (60 mgd), which is below the design flow. The influent biochemical oxygen demand (BOD) concentration is lower than the design value, and TSS and total Kjeldahl nitrogen (TKN) are higher than the design value. Effluent ammonia-nitrogen and TIN from 2011 and 2012 are shown in Figure 16.2. Effluent TIN exceeded the facility's effluent limit on a few occasions during 2011 due to blower malfunctions. Effluent ammonia-nitrogen averaged 0.5 mg/L and effluent BOD, TSS, and total phosphorus averaged 5, 6, and 7 mg/L, respectively.

3.0 CASE STUDY 2 (SEQUENCING BATCH REACTOR)—KINGSTON, MASSACHUSETTS, WASTEWATER TREATMENT PLANT

3.1 Facility History

The Kingston Wastewater Treatment Plant (WWTP) is located in the Town of Kingston, Massachusetts. Facility construction was completed in 2002 as a BNR treatment process to provide treatment for an average daily flow of 1.42 ML/d (0.375 mgd) and a peak-hour flow of 5.67 ML/d (1.50 mgd). The facility discharges

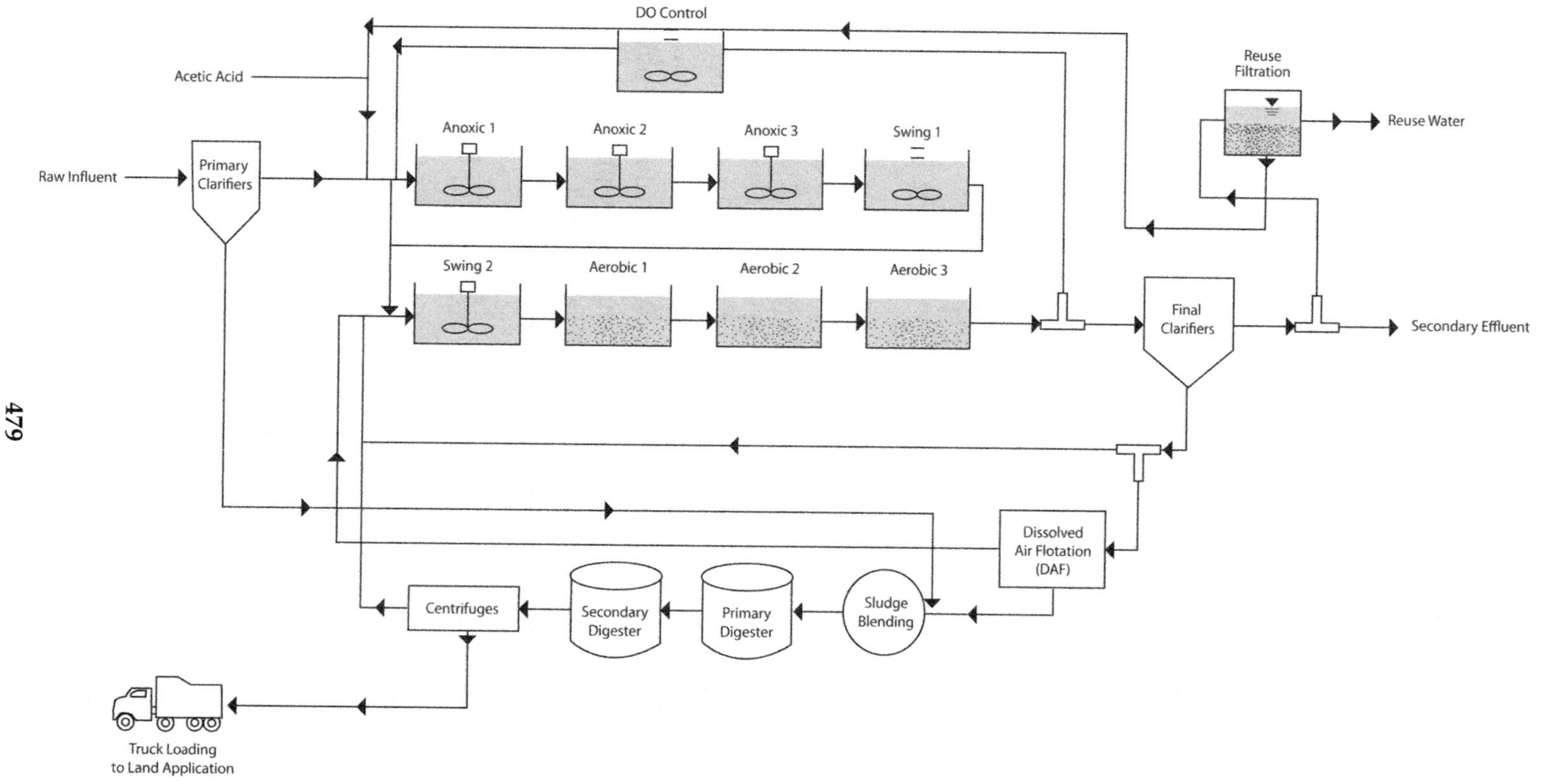

FIGURE 16.1 Albuquerque, New Mexico, SWRP flow schematic.

TABLE 16.1 Albuquerque, New Mexico, SWRP wastewater characteristics.

Influent flow	288 ML/d (76 mgd)
Total suspended solids	250 mg/L
Biochemical oxygen demand	270 mg/L
Total Kjeldahl nitrogen	40 mg/L
Total phosphorus	10 mg/L

its effluent in accordance with a permitted groundwater discharge permit, or, the effluent may be pumped to a holding pond for possible use as supplemental irrigation water at a nearby golf course.

3.2 Process Description

The preliminary treatment process provides for measurement of the facility influent flow and grit and screenings removal. The equipment used in the preliminary treatment process includes an aerated inlet tank, mechanically cleaned rotary bar screen, and an aerated grit tank.

Raw wastewater enters the influent box at the head end of two inlet tanks through two 200-mm (8-in.) force mains. Septage and facility recycles are pumped from a nearby holding tank and also discharged at this point. The raw wastewater flows through two sections of the inlet tank, each of which is aerated to keep solids in suspension and to strip off odors. Coarse-bubble aeration is provided.

Flow is then directed to a 6-mm (0.25-in.) spaced, mechanically cleaned, influent screen. If the mechanically cleaned influent screen is taken out of service, the flow can be directed to the bypass channel, which is equipped with a manually cleaned bar rack. The wastewater then enters an aerated grit tank. An airlift pump removes the settled grit and pumps it to a grit classifier. The grit tank can be bypassed, if necessary.

TABLE 16.2 Albuquerque, New Mexico, SWRP process design criteria.

Effluent nitrate	8 mg/L (no phosphorus limit)
Aerobic SRT	9 days
Minimum water temperature	15 oC
MLE process basins	13
Total anoxic volume	40.5 ML (10.7 mil. gal)
Total aerobic volume	72.3 ML (19.1 mil. gal)
Total swing zone (aerobic or anoxic) volume	30.3 ML (8 mil. gal)
Final clarifiers (41-m [135-ft] diameter)	12

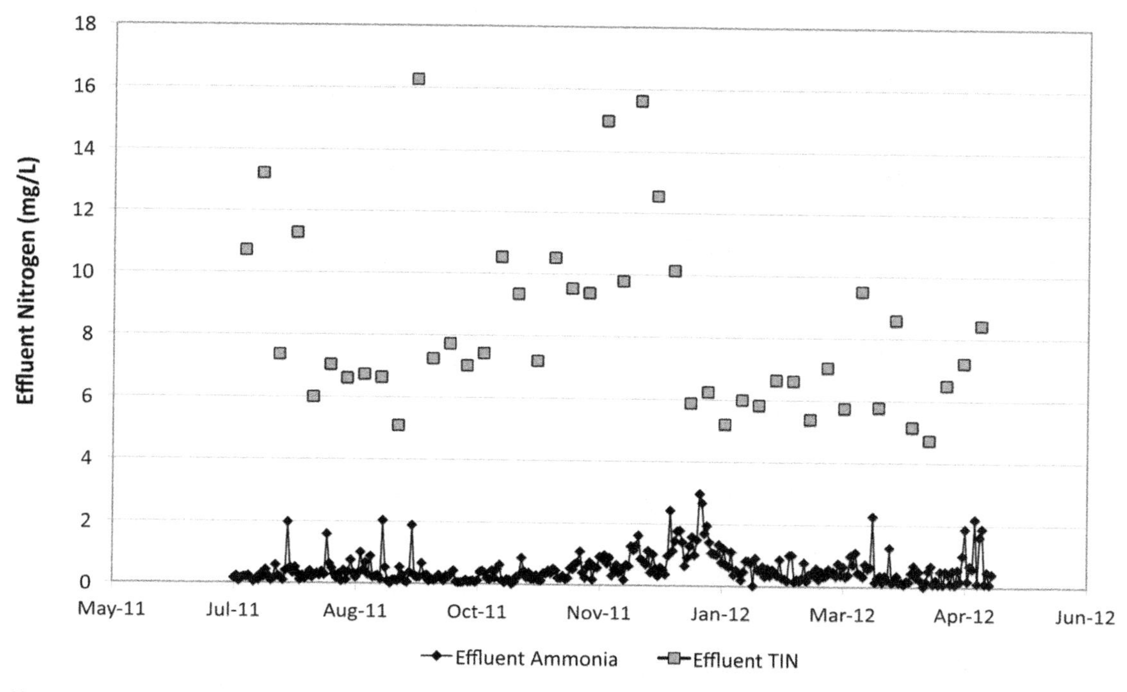

FIGURE 16.2 Albuquerque, New Mexico, SWRP performance (effluent ammonia).

Effluent from the grit tank flows by gravity to one of two sequencing batch reactors (SBRs). Before entering an SBR, soda ash may be introduced to the flow stream to increase alkalinity.

The facility is designed to receive a significant quantity of septage considering the overall capacity of the facility, which represents about 25% of the total average design influent BOD load and about 20% of the total average design influent TKN load. The existing septage-receiving facility consists of two holding tanks with approximately 91 000 L (24 000 gal) of total storage with diffused aeration. The septage-receiving facility consists of two offloading stations with 44-mm (1.75-in.) manual bar racks. The tanks are used to store septage, filtrate from the gravity belt thickener (GBT), and other facility drainage.

The Kingston facility uses a two-SBR tank system to provide continuous treatment. One SBR is filled at a time, and, while the one tank is in midcycle, the second tank in filling and vice versa. Each SBR is 18.3 m (60 ft) by 18.3 m (60 ft), with a low water level (LWL) of 5.4 m (17.6 ft) and a high water level of 5.6 m (21 ft), providing a LWL-treatment volume of 1.79 ML (0.474 mil. gal) per SBR.

Each SBR typically runs through five cycles per day, although, occasionally, a six-cycle-per-day operation is used. When running at five cycles per day, each cycle is set to take 4.8 hours and is broken into five basic stages during normal operation to achieve the desired degree of nitrification and denitrification. These stages are fill (including an anoxic mixing period and an aerobic mixing period), react (aerobic treatment), settle, decant, and idle. Each treatment cycle operates at the stage times shown in Table 16.3. Each SBR is equipped with a fixed fine-bubble diffuser system and three 37-kW (50-hp) rotary-displacement blowers are installed to supply the required process air.

The Kingston facility has a two-disk effluent filter. This system is used to remove suspended solids and to provide polishing of the SBR effluent before UV disinfection and discharge. The filter unit consists of two disks; each disk is comprised of six segments. The disk filter has an average-daily design hydraulic capacity of 1.77 ML/d (0.467 mgd) and a design peak hydraulic capacity of 3.53 ML/d (0.933 mgd).

The UV disinfection system uses two UV reactors. Each reactor has two banks that are contained within a single flow channel; each bank contains four modules and each module accommodates two lamps.

The facility uses one GBT to thicken WAS from the two SBRs. Polymer is injected to the sludge feed through a four-port tangential injection ring by a polymer metering pump and blended by a static mixer. A process flow diagram is shown in Figure 16.3.

The Kingston facility has National Pollutant Discharge Elimination System (NPDES) maximum monthly average limits of 30 mg/L on 5-day BOD (BOD_5) and TSS. A maximum monthly nutrient limit on total nitrogen requires an effluent total nitrogen of 10 mg/L. Additionally, more stringent discharge limits must be met when discharging to the holding point for effluent reuse as golf course irrigation water. Effluent daily average BOD_5 and total nitrogen must not exceed 10 mg/L and effluent daily TSS must not exceed 5 mg/L. Table 16.4 summarizes the Kingston facility's influent design loading for the maximum monthly design condition and Table 16.5 summarizes the Kingston facility's significant unit process design criteria.

3.3 Performance

The Kingston facility has been in compliance with its NPDES permit since startup and the BNR process has performed well (Figure 16.4). Monthly average total

TABLE 16.3 Stage times for treatment cycles at the Kingston, Massachusetts, facility.

Stage	Time (min.)
Fill static and mix (anoxic)	60
Fill (aerobic)	84
React (aerobic)	60
Settle	60
Decant	24
Total	288

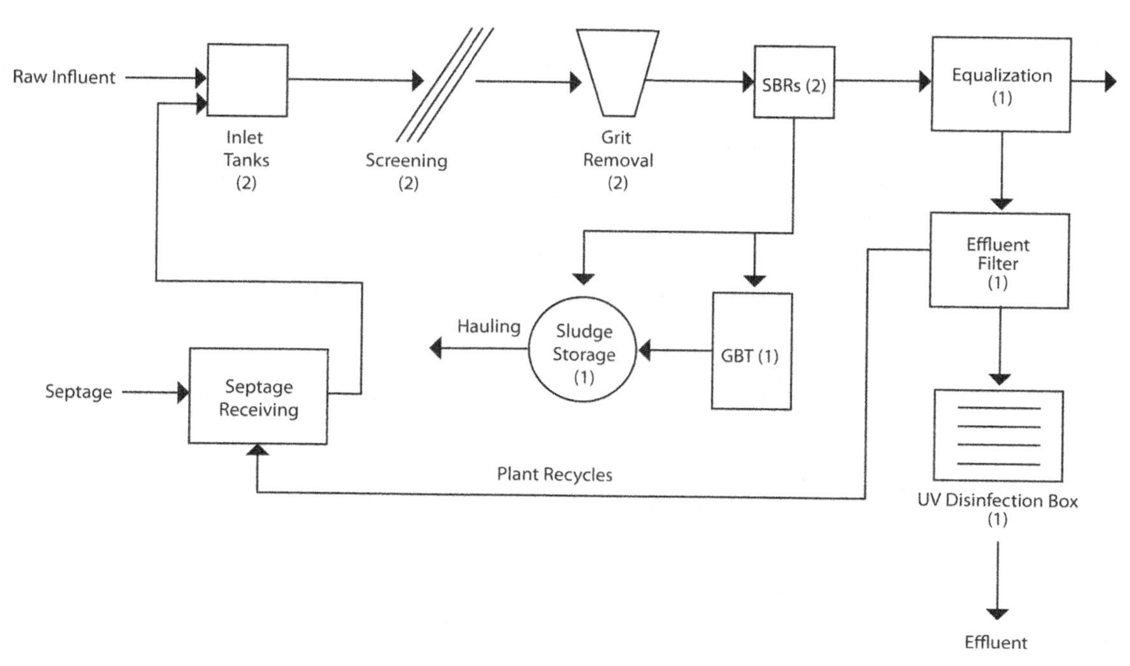

FIGURE 16.3 Kingston, Massachusetts, facility process flow diagram.

nitrogen, ammonia-nitrogen, nitrite-nitrogen, and nitrate-nitrogen are plotted. The facility's average monthly effluent total nitrogen concentration has ranged from 3 to 8 mg/L.

4.0 CASE STUDY 3 (4-STAGE BARDENPHO)—H.L. MOONEY ADVANCED WATER RECLAMATION FACILITY

4.1 Facility History

The H.L. Mooney Advanced Water Reclamation Facility is operated by the Prince William County Service Authority and serves the eastern portion of Prince William County, Virginia. The facility was brought online in 1981 and, until recently,

TABLE 16.4 Kingston facility influent design maximum monthly loads.

Contaminant	Load (kg/d)[(lb/d)]
Biochemical oxygen demand	766 (1685)
Total suspended solids	1350 (2970)
Total Kjeldahl nitrogen	152 (335)

TABLE 16.5 Kingston facility design criteria.

Description	Value
Preliminary treatment	1
• Number of inlet tanks	8.1 × 2.1 × 2.4 m (26.5 × 7 × 7.75 ft) sidewater depth
– Dimensions	
• Influent screen	
– Number of screens	1
– Bar spacing	6 mm (0.25 in.)
• Number of grit chambers	1
– Dimensions	2.4 × 2.4 × 3.7 m (8 × 8 × 12 ft) sidewater depth
SBRs	2
• Number of units	18.3 × 18.3 × 5.6 m high water level (HWL) (60 × 60 × 21 ft)
• Dimensions (each)	
• Aerobic SRT at maximum month load	12 days
• MLSS (at low water level)	4700 mg/L
Effluent filtration	1
• Number of filters	10.0 m^2 (108 sq ft)
• Filter surface area	7.3/14.6 m/h (3/6 gpm/sf)
• Hydraulic loading rate (average and maximum)	
Disinfection	Ultraviolet medium pressure
• Type	2
• Number of reactors	16
• Total number of lamps	
Sludge processing	470/770 kg/d
• Waste sludge load (average and maximum monthly)	(1030/1685 lb/d) GBT
• Type of thickening	1
• Number of GBTs	4%
• Thickened sludge concentration	

was rated to provide a treatment capacity of 68 ML/d (18 mgd). Because of anticipated increases in flow and load to the facility, a significant treatment facility upgrade was undertaken and completed in 2011. The purpose of the upgrade was to increase the facility's design flow to 91 ML/d (24 mgd) while complying with its effluent nutrient limits.

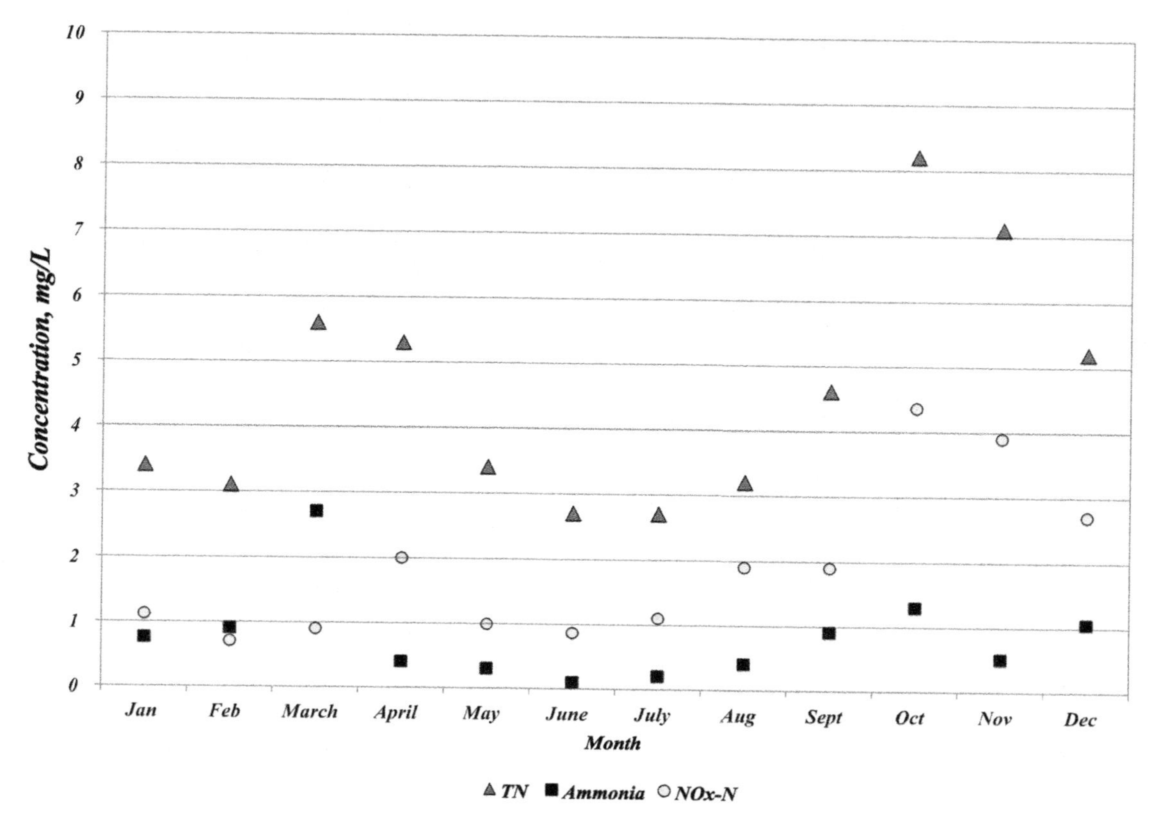

FIGURE 16.4 Kingston facility BNR process performance results.

4.2 Process Description

The H.L. Mooney facility provides preliminary and chemically enhanced primary, secondary, and tertiary treatment, achieving enhanced nutrient removal and disinfection. Of particular interest are BNR processes that consist of a suspended growth BNR process followed by denitrification filters. The recently upgraded suspended growth BNR process basins are configured as plug flow reactors with the flexibility to operate as either an MLE system for nitrification and denitrification or as a four-stage Bardenpho system (refer to Figure 16.5 for the bioreactor configuration).

The inlet end of each basin is configured as a three-stage anoxic reactor, with the first two stages sized to promote selection of preferred microorganisms and to "deselect" problematic microorganisms. The third anoxic selector is sized to provide increased volume for nitrate removal. Following the anoxic selector is a swing zone that can run in either anoxic or aerobic mode.

Finally, an aerobic reactor provides nitrification year-round with sufficient aerobic SRT. The last stage of the aerobic volume includes both diffusers and

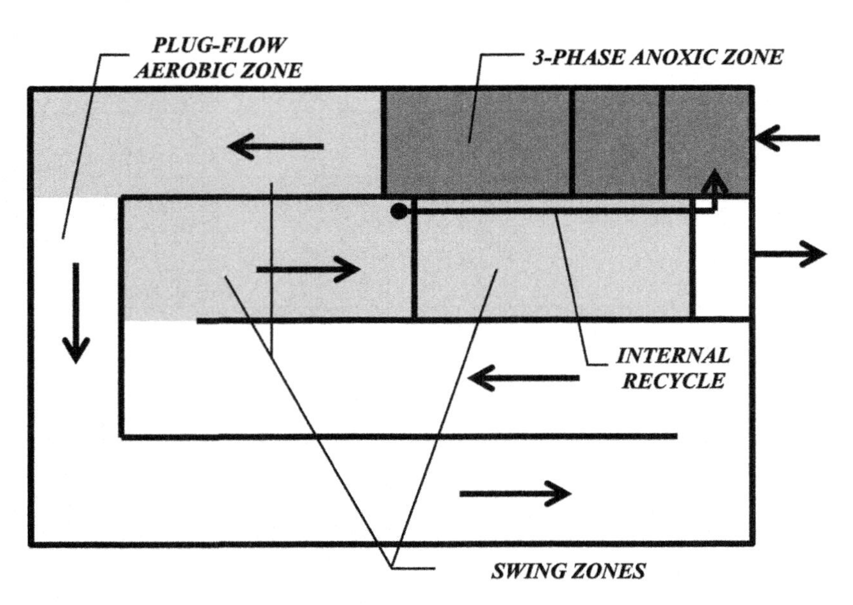

Figure 16.5 H.L. Mooney AWRF process flow diagram.

mixers so that the system is not mixing-limited and excessive dissolved oxygen concentration is not a problem.

Downstream of the aerobic volume are two swing-zone stages that can be operated either anoxically or aerobically. During periods when the wastewater is coldest, operators have the flexibility to run these zones aerobically to maintain nitrification through worst-case conditions. The swing zones can then be converted to anoxic zones during warmer conditions, enabling the process to provide improved denitrification. The facility has the ability to add methanol as a supplemental carbon source to both the initial anoxic zones and the post-swing zones.

Downstream of the final clarifiers that serve the suspended growth system, tertiary denitrification filters further polish the effluent and allow the facility to achieve compliance with its annual average effluent total nitrogen limit. Methanol is added upstream of the filters to provide a carbon source. The filters also achieve a low effluent TSS concentration, which, in turn, enables the facility to comply with its monthly average effluent total phosphorus limit. A UV-light disinfection facility is used to disinfect the treated effluent. Before discharge to Neabsco Creek, a tributary of the Potomac River, the effluent is reaerated through a cascade. Solids processing includes gravity thickening, centrifugation, and incineration.

The H.L. Mooney facility has effluent monthly average permit limits on total nitrogen and total phosphorus of 3 and 0.18 mg/L, respectively. The facility also has a monthly average $CBOD_5$ limit of 5 mg/L and a TSS limit of 6 mg/L.

TABLE 16.6 H.L. Mooney facility influent design maximum monthly loads.

Parameter	Load (kg/d) [(lb/d)]
Five-day BOD	22 400 (49 300)
Total suspended solids	26 500 (58 400)
Total Kjeldahl nitrogen	3600 (7930)
Total phosphorus	418 (921)

Tables 16.6 and 16.7 summarize the H.L. Mooney facility's influent and primary effluent design loading for the maximum monthly design condition and Table 16.8 summarizes the facility's significant unit process design criteria.

4.3 Performance

The H.L. Mooney facility has been in compliance with its NPDES permit since startup and the suspended growth BNR process has performed well. Monthly average total nitrogen, ammonia-nitrogen, nitrite-nitrogen, and nitrate-nitrogen (NO_x-N) and the quantity of methanol added to the suspended growth and denitrification filter processes are shown in Figure 16.6. A significant improvement in effluent total nitrogen occurred in the June–July timeframe, at the same time that methanol addition was reduced. Methanol addition was reduced because swing zones were modified to operate in anoxic mode, thus maximizing use of available carbon in the suspended phase. Effluent monthly average total nitrogen from the suspended growth system was below 3 mg/L after this modification was made.

Denitrification filters provide additional total nitrogen removal with biological treatment from biomass that is attached to the filter media. During 2011, influent total nitrogen to the denitrification filters averaged 4.85 mg/L, which represented an 88% reduction of total nitrogen in the suspended growth system. An additional 2.58 mg/L of total nitrogen removal occurred in the denitrification filters, producing final effluent total nitrogen of 2.27 mg/L. Although this is minor compared to total nitrogen removal that occurs in suspended growth

TABLE 16.7 H.L. Mooney facility NPDES primary effluent design maximum monthly loads.

Parameter	Load (kg/d) (lb/d)
Five-day BOD	16 000 (35 100)
Total suspended solids	9500 (21 000)
Total Kjeldahl nitrogen	3730 (8210)
Total phosphorus	310 (690)

Table 16.8 H.L. Mooney facility process design criteria.

Description	Value
Preliminary treatment	2
• Number of screens	1.2-m (4-ft) wide × 1.8-m (6-ft) deep
– Screen size	6 mm (0.25 in.)
– Perforations	2
• Number of grit chambers	Vortex
– Type	3.7 m (12 ft)
– Diameter	
Primary clarifiers	5
• Number of units	29 m (95 ft)
• Diameter	1.3/2.7 m/h
• Surface overflow rate (average and peak)	(790/1580 gpd/sq ft)
Bioreactors	5
• Number of basins	4
• Passes per basin	9.1 ML (2.4 mil. gal)
• Volume per basin	1.2 ML (0.31 mil. gal)
– Anoxic volume per basin	7.0 ML (1.86 mil. gal)
– Aerobic volume per basin	0.87 ML (0.23 mil. gal)
– Swing volume per basin	9 days
• Aerobic SRT at maximum monthly load	2500 mg/L
• MLSS (average)	6 × average daily flow (ADF)
• Internal recycle pumping capacity	
Secondary clarifiers	9
• Number of units	6 at 29 m (95 ft); 3 at 39 m (125 ft)
• Diameter	0.5/1.0 m/h
• Surface overflow rate (average and peak)	(300/600 gpd/sq ft)
Filtration	24
• Number of filters	890 m^2 (9576 sq ft)
• Filter surface area	1.8 m (6 ft)
• Media bed depth	4.6/8.8 m/h
• Hydraulic loading rate (average and maximum)	(1.9/3.6 gpm/sq ft)
Disinfection	Ultraviolet medium pressure
• Type	2
• Number of reactors	16
• Total number of lamps	
Sludge processing	470/770 kg/d (1030/1685 lb/d)
• Waste sludge load (average and maximum monthly)	Gravity thickeners
• Type of thickening	4
• Number of gravity thickeners	3 to 5%
• Thickened sludge concentration	Centrifuges
• Type of dewatering	3
• Number of centrifuges	Fluid bed incinerator
• Disposal	

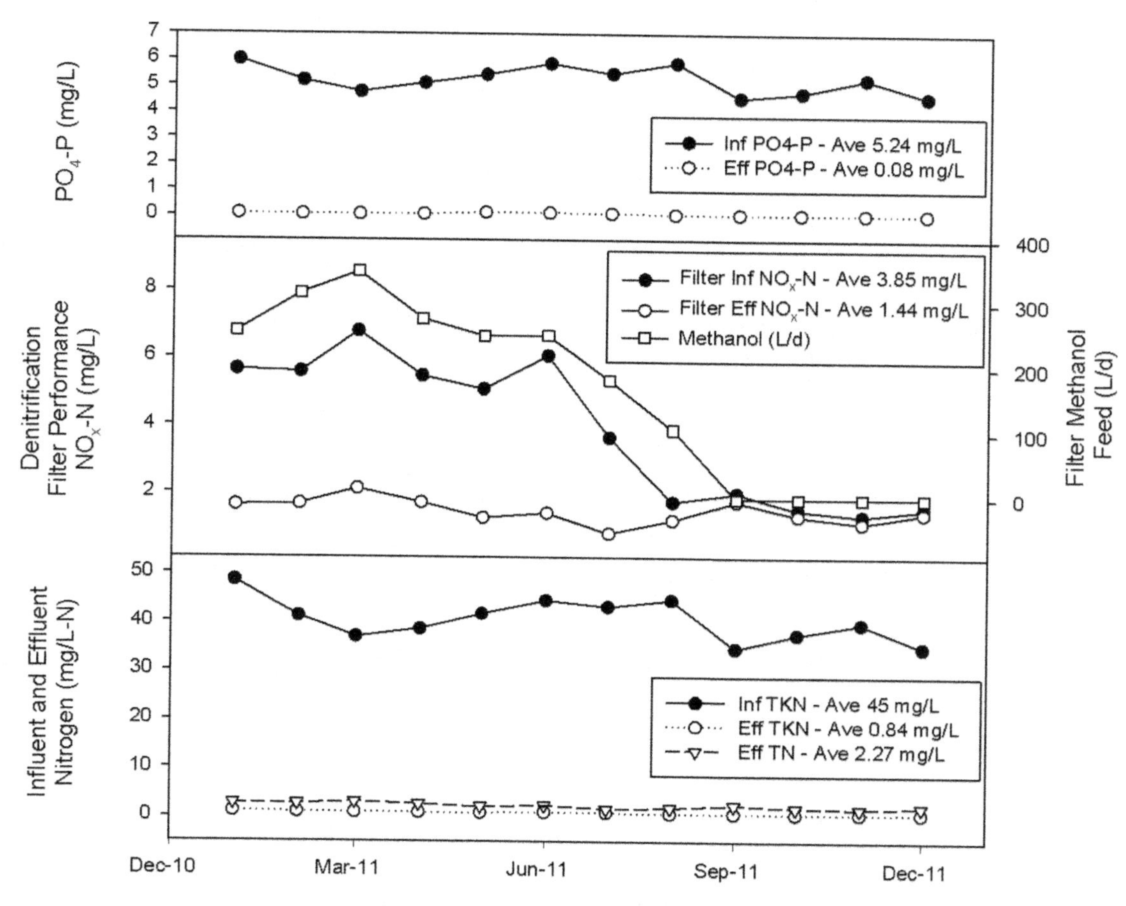

FIGURE 16.6 H.L Mooney facility influent and effluent performance.

systems, denitrification filters are a critical process for the H.L. Mooney facility because of the stringent nutrient limits that must be met on a monthly basis. The denitrification filters provide mitigation for process upsets that may occur in the suspended growth system and polish off residual nitrate-nitrogen that is present in the secondary effluent.

5.0 CASE STUDY 4 (OXIDATION DITCH)—GILDER CREEK WASTEWATER TREATMENT PLANT

5.1 Facility History

The Gilder Creek Wastewater Treatment Plant is owned and operated by Renewable Water Resources (ReWa), and is responsible for providing secondary wastewater treatment for a portion of the Gilder Creek subbasin in Greenville County, South Carolina, including part of the City of Mauldin. The treatment facility is constructed on a 95-ac site located on the eastern edge of Greenville County near the confluence of the Enoree River and Gilder Creek and adjacent to East Georgia Road.

The treatment facility discharges to the Enoree River, upstream of the confluence to Gilder Creek. Operation of the Gilder Creek facility began in 1987, with an average daily maximum monthly flow capacity of 15.1 ML/d (4 mgd). The treatment facility operated from 1987 until mid 1993 as an operation and maintenance (O&M) privatization project. However, on August 1, 1993, ReWa assumed responsibility for O&M of the facility. In 2000, Phase I construction began with the addition of flow equalization facilities, increasing the facility treatment capacity to 18.9 ML/d (5 mgd). The Phase II upgrade and expansion was completed in 2005, increasing the facility's treatment capacity to 30.3-ML/d (8-mgd) average flow and 75.7-ML/d (20-mgd) peak flow.

In 2007, daily facility influent data from May 2005 through June 2007 was analyzed; influent concentrations for BOD, TSS, TKN, and total phosphorus were found to be less than the design concentrations used for the previous facility expansion up to 30.3 ML/d (8 mgd). Therefore, desktop evaluations and full-scale field testing at the facility were completed to determine the capacity of each process at the lower influent concentrations. The evaluation indicated that the limiting treatment process within the secondary process was the aeration system at 42.8 ML/d (11.3 mgd). Consequently, in 2009, ReWa applied for, and received, an NPDES permit modification with a rerated capacity of 42.8 ML/d (11.3 mgd).

5.2 Process Description

Raw influent enters the facility by gravity from a 1.07-m (42-in.) Gilder Creek trunk sewer. Wastewater is screened with mechanical bar screens and then pumped to grit removal or an offline flow equalization basin. After influent pumping, the following processes provide preliminary, primary, and secondary treatment:

- Primary clarifiers (3),
- Oxidation basins (4),
- Secondary clarifiers (4),
- Deep bed filters (5 cells), and
- Ultraviolet disinfection.

Solids handling facilities and processes include a primary solids and WAS sludge storage tank, primary solids and WAS thickening rotary drum thickeners (2), anaerobic digesters (2), digested sludge rotary drum thickeners (2), thickened digested sludge storage (1), and land application. Figure 16.7 shows a process flow diagram of the facility.

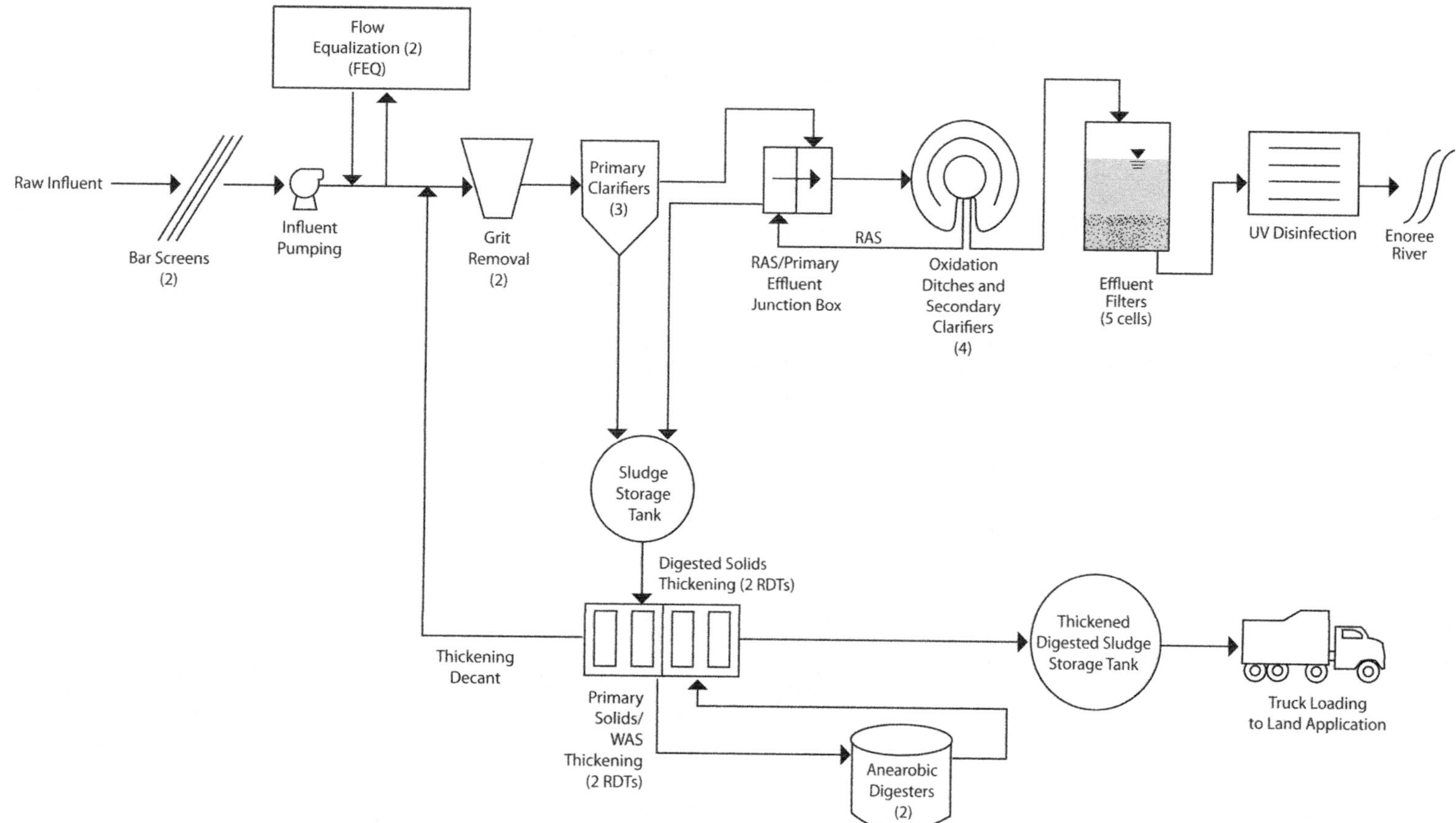

FIGURE 16.7 Gilder Creek facility process flow diagram.

TABLE 16.9 Gilder Creek facility design hydraulic and organic loading parameters.

Parameter	Design value
Average daily flow	43 ML/d (11.3 mgd)
Five-day BOD, mg/L	303
Total suspended solids, mg/L	190
Total Kjeldahl nitrogen, mg/L	44

The four oxidation basins are carrousels that were constructed with a secondary clarifier in the middle. The basins have a combination of mechanical surface aerators and coarse-bubble diffusers for BOD removal and nitrification.

Permit requirements for the Gilder Creek facility include effluent limits on BOD and TSS of 85% removal, or 22.5 and 30 mg/L, respectively. Summer ammonia-nitrogen limits are 3.6 and 5.4 mg/L on a monthly and weekly average basis and winter ammonia-nitrogen limits are 9.4 and 14.1 mg/L on a monthly and weekly average basis, respectively.

The Gilder Creek facility facilities were designed to treat the influent wastewater characteristics and loadings given in Table 16.9. Process design criteria for the secondary treatment process are provided in Table 16.10.

5.3 Design and Operation Modifications

Aeration for odor control is supplied through paddle mixers and coarse-bubble diffusers that are located in the middle of the basins. Facility operators can modify the paddle-mixer depth and speed to vary the degree of aeration. However, modifications to the paddle mixers have not been made because effluent limits

TABLE 16.10 Gilder Creek facility design criteria.

Parameter	Design value
Carrousel system design	
Aeration volume per basin, ML (mil. gal)	5.37 (1.42)
Number of carrousel aeration basins	4
Number of mechanical surface aerators per basin	2
Horsepower per aerator	125
Number of dissolved oxygen meters per basin	1
Process design	
Solids retention time for nitrification, days	15
Mixed liquor suspended solids, mg/L	2000

have not required increased denitrification and total nitrogen removal that may occur with reduced aeration.

5.4 Performance

The Gilder Creek facility has averaged 0.2 and 3.6 mg/L of effluent ammonia-nitrogen and BOD, respectively, at an average effluent flow of 14.6 ML/d (3.9 mgd). The historical performance for the facility is shown in Figure 16.8. The benefit of an oxidation ditch is the ability to achieve nitrogen removal without the cost of internal recycle pumping. The Gilder Creek facility has performed well under their permit limits while achieving some total nitrogen removal (approximately 30%) and alkalinity recovery.

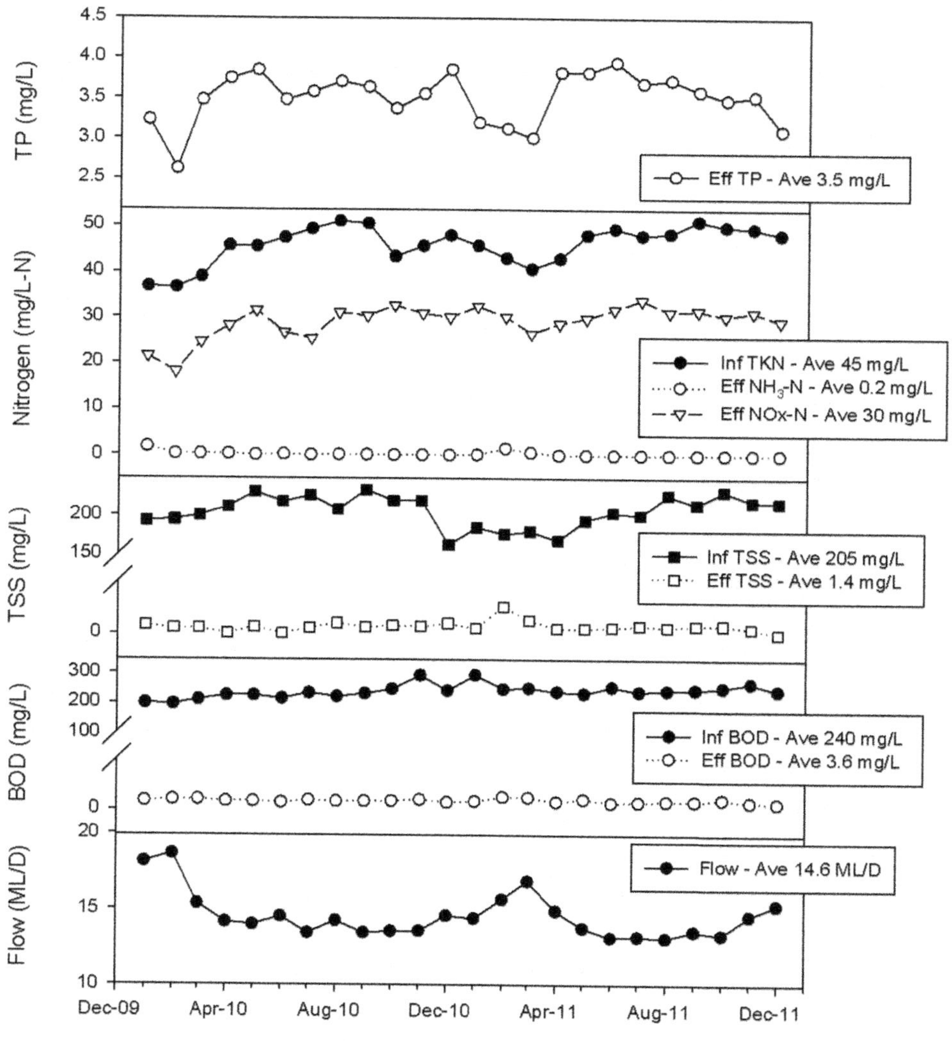

FIGURE 16.8 Gilder Creek facility influent and effluent performance.

6.0 CASE STUDY 5 (INTEGRATED FIXED-FILM ACTIVATED SLUDGE)—CITY AND COUNTY OF BROOMFIELD, COLORADO, WASTEWATER TREATMENT PLANT

6.1 Facility History

The City and County of Broomfield Wastewater Treatment Plant is located in Broomfield, Colorado, and was originally constructed in 1954 with a single treatment train, including primary clarification, trickling filter, and secondary clarification. Following an expansion in 1988, the secondary treatment train included an activated sludge process downstream of the biotower, which provided "roughing" treatment of the primary effluent to meet discharge limits for CBOD and TSS of 25 and 30 mg/L, respectively (Rutt et al., 2006). In 1999, a wastewater treatment master plan was completed that identified improvements necessary to increase the capacity of the facility up to 45 ML/d (12 mgd) by the year 2020 in two expansion phases.

One of the alternatives considered for expansion was the use of plastic media carriers known as *integrated fixed-film activated sludge* (IFAS). At the time, no installations of IFAS existed in North America. As such, pilot testing was completed and produced favorable results that led to retrofitting the existing activated sludge process and the installation of IFAS media in 2003. With the installation, the City and County of Broomfield facility became the first facility to use IFAS in North America. The Phase 1 expansion increased the treatment capacity of the facility from 20 ML/d (5.4 mgd) to 30 ML/d (8 mgd) within the existing tank volume. The Phase I IFAS installation was operated and monitored for 4 years and produced such positive results that it was selected for use in the Phase II expansion in 2010 that increased the facility's capacity to 45 ML/d (12 mgd).

6.2 Process Description

The City and County of Broomfield facility has three stages of BNR and operates as a combined nitrification/denitrification and biological phosphorus removal facility. Primary effluent and return activated sludge (RAS) enter a preanoxic zone to denitrify nitrates before the anaerobic zone. Following the anaerobic zone, mixed liquor recycle is added in the anoxic zone. Finally, an aerobic zone provides BOD removal and nitrification (Figure 16.9). The facility provides the following unit processes:

- Preliminary treatment including screening, influent pumping, and grit removal;
- Flow equalization;
- Primary clarifiers;
- Three trains of secondary treatment (anaerobic, anoxic, and aerobic);

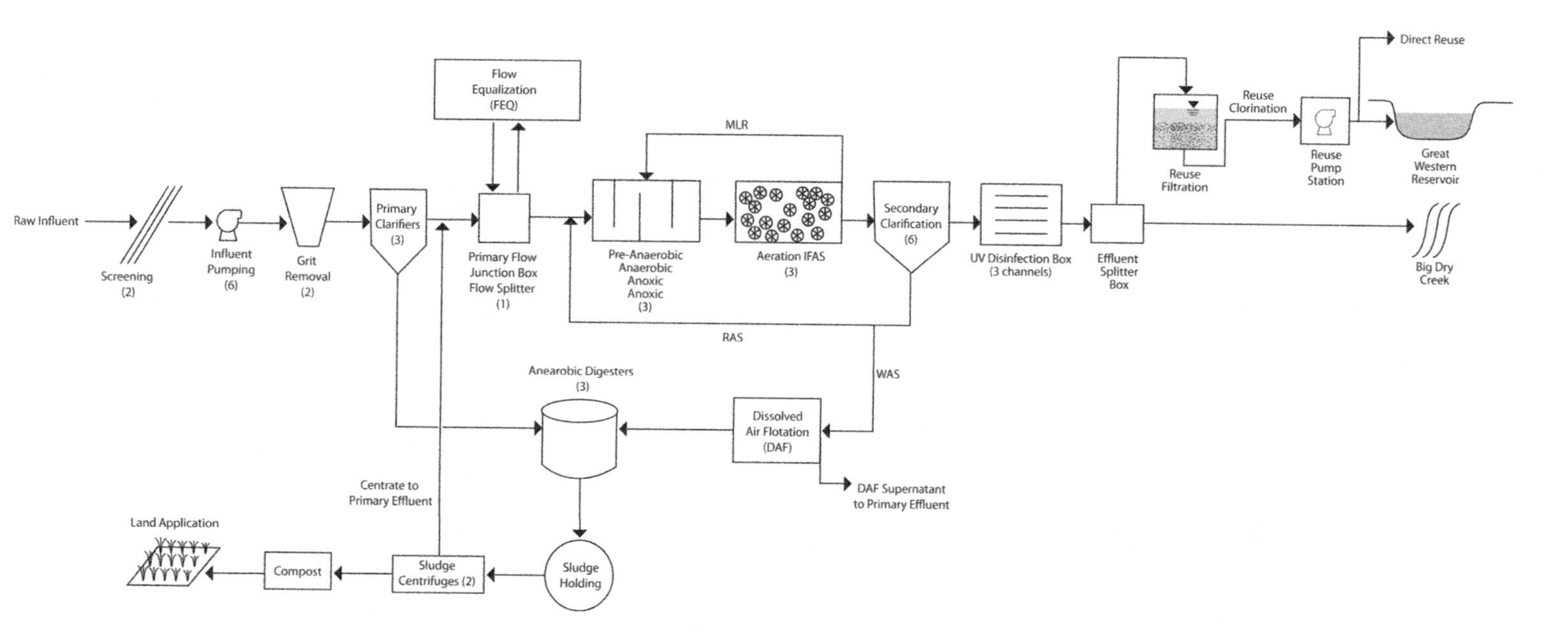

FIGURE 16.9 City and County of Broomfield facility process flow diagram.

- Secondary clarifiers; and
- Ultraviolet disinfection.

As part of the Phase I expansion, facilities were added downstream of the secondary treatment process to provide 23 ML/d (6 mgd) of re-use capacity and included filtration and chlorine disinfection. Solids handling facilities include

- Dissolved air flotation thickening,
- Anaerobic digestion,
- Centrifuge dewatering, and
- Composting followed by land application.

The IFAS media are installed in the aerobic cells of the three secondary treatment trains that include the following components that were installed to accommodate the addition of IFAS:

- Media retention screens, including cylindrical through the wall and wall top screens, and
- Coarse-bubble diffused aeration.

The City and County of Broomfield facility has effluent permit limits on CBOD, TSS, and ammonia-nitrogen of 25, 30, and 3.5 mg/L, respectively, and locally imposed re-use limits on total phosphorus of 0.1 mg/L and nitrate of 10 to 15 mg/L. The influent wastewater characteristics of the City and County of Broomfield facility, which are in the typical range expected for a municipal treatment facility, are provided in Table 16.11 (Rutt et al., 2006).

6.3 Process Design Criteria

The design for the Phase I expansion included the addition of IFAS media at a 48% fill fraction. However, the fill fraction was lowered to 30% based on the field conditions and operational experience at the flows and organic loading observed during design (completed in 2004). Process data collected from the treatment process between 2004 and 2008 confirmed that the fill fraction could be reduced

TABLE 16.11 City and County of Broomfield, Colorado, facility NPDES influent wastewater characteristics (2011–current).

Parameter	Average value
Average daily maximum monthly flow	22 ML/d (6.0 mgd)
Carbonaceous BOD_5	250 mg/L
Total suspended solids	360 mg/L
Ammonia-nitrogen	33 mg/L

from 48 to 30% in the Phase II expansion. Design criteria of the treatment process are provided in Table 16.12.

6.4 Design and Operation Modifications

Since the installation of IFAS media in late 2003, operators have refined the treatment process and made minor modifications based on the original design. The performance of the IFAS media exceeded expectations and operators were able to reduce mixed liquor suspended solids (MLSS) from the design value of 3500 mg/L to approximately 2000 mg/L and to reduce the IFAS media fill fraction as mentioned previously (Phillips et al., 2008). During the Phase II expansion, the city elected to reduce the number of blowers from what was installed during Phase I based on a substantial amount of historical data (Phillips et al., 2010).

6.5 Performance

The City and County of Broomfield facility has been successfully operating in an IFAS configuration since 2004; the IFAS media have performed better than expected, as indicated by the reduction in media fill fraction for the Phase II expansion. From 2011 to 2012, the facility effluent ammonia and total phosphorus averaged 0.3 and 0.1 mg/L, respectively, at an annual average flow of 22.5 ML/d (6.0 mgd). The IFAS system has been able to achieve treatment objectives while maintaining a mixed liquor concentration of 2000 mg/L that is less than the design value of 3500 mg/L (Phillips et al., 2008). The IFAS systems are often considered to be "self regulating" because of the two sources of active biomass: suspended and attached (retained). As temperatures increase and decrease during the year, biomass on the media decreases and increases, respectively, in response. One of the unique features of IFAS is that each reactor cell can be treated independently of other cells because of the media being retained and the conditions that occur in each cell. During 2006, the biomass on the media in the first and second cells averaged 13.7 and 7.4 g/m^2, respectively (Phillips et al., 2008). The higher biomass

TABLE 16.12 City and County of Broomfield, Colorado, facility process design criteria (Phillips et al., 2008).

Parameter	Value
Integrated fixed-film activated sludge fill fraction in aerobic zones	30%
Total SRT (including biomass attached to IFAS media)	10 days
Suspended growth SRT (not including attached biomass)	4.7 days
Mixed liquor suspended solids	3500 mg/L
Mixed liquor recycle rate	350% of influent flow
Wastewater temperature	13 to 25 °C
Residual dissolved oxygen	4 mg/L

that occurred in the first cell is attributed to the high organic loading compared to the second cell. Effluent performance of the facility is shown in Figure 16.10.

7.0 CASE STUDY 6 (CYCLIC AERATION)—TOWN OF CARY, NORTH CAROLINA, WATER RECLAMATION FACILITY

7.1 Facility History

The wastewater collection system and water reclamation facilities of Cary, North Carolina, have evolved over the years. As a small town, until 1970 the primary means

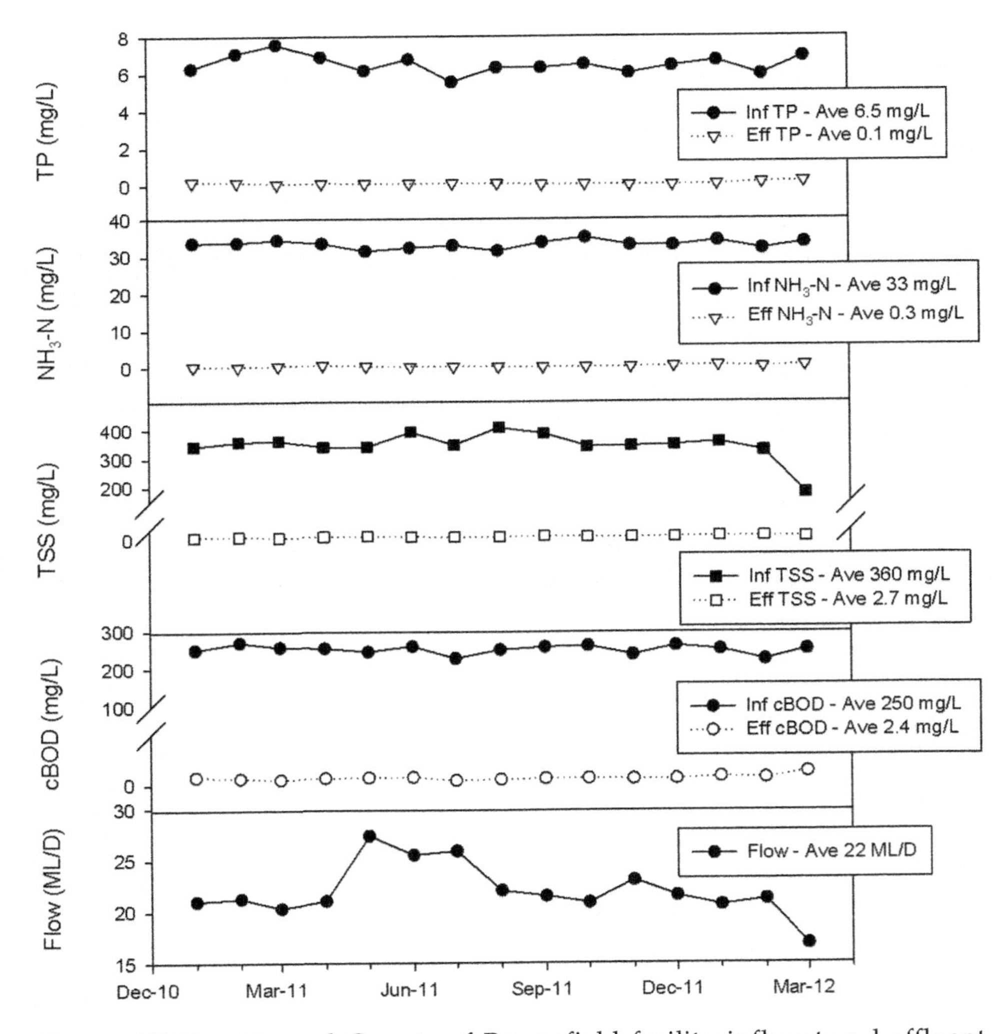

FIGURE 16.10 City and County of Broomfield facility influent and effluent performance.

of wastewater disposal was via septic tank. As the town grew, wastewater pumping stations began collecting sewer influent in several areas and then conveying the wastewater to a neighboring city for treatment. During the 1980s, small sections of the town were served by small treatment facilities, until 1984, when Cary completed construction and put into operation the North Cary Water Reclamation Facility.

7.2 Process Description

The 45-ML/d (12.0-mgd) North Cary facility uses the Biodenipho BNR process that consists of three trains of paired oxidation ditches where the active process volume is continuously alternated between aerobic and anoxic conditions to achieve BNR. The current influent flowrate is approximately 25 ML/d (6.5 mgd), or 55% of design flow. Therefore, one train remains offline. Each Biodenipho process train is followed by multistage postdenitrification reactors and a postaeration zone.

Each biological process train at the North Cary facility consists of an anaerobic selector for biological phosphorous removal, a pair of oxidation ditches, and a secondary anoxic zone for nitrate trimming. Each of the oxidation ditches is equipped with four surface brush aerators (45 kW, or 60 hp), one submersible aerator (30 kW, or 40 hp), and two submersible mixers. In addition, each oxidation ditch is equipped with a dissolved oxygen probe for automatic dissolved oxygen control. Supplemental carbon addition to the secondary anoxic zone is currently not required to meet total nitrogen removal goals.

The Town of Cary facility has an effluent permit limit on total nitrogen of 64 975 kg/yr (143 246 lb/yr), which is equivalent to 3.92 mg/L at the facility's permitted capacity. The facility also has an effluent quarterly average limit on total phosphorus of 2 mg/L.

7.3 Process Design

Current treatment processes at the North Cary facility consist of the following: coarse bar screens; two influent pump stations; fine bar screens; grit removal facilities; two separate biological treatment trains, each consisting of a four-stage anaerobic selector, two oxidation ditches, a three-stage secondary anoxic zone and a reaeration zone; two secondary clarifiers; effluent filters; UV disinfection facilities; and a cascade aerator for reaeration before discharge.

Waste activated sludge at the North Cary facility is pumped to two GBTs for thickening before being pumped to three aerobic digesters for solids stabilization and storage. Digested biosolids from the digesters are hauled as a liquid and land applied to privately owned agricultural land. Table 16.13 includes a summary of process design parameters.

7.4 Design and Operation Modifications

The isolation ditch is operated in a 240-minute cycle. For the first 30 minutes, both ditches operate in an aerobic mode. For the next 90 minutes, ditch 1 operates in

TABLE 16.13 Town of North Cary, North Carolina, facility process design parameter summary.

Tank/process	Number	Size (each)	Size (total)
Anaerobic selectors	4 each train	0.35 ML (0.093 mil. gal)	1.41 ML (0.372 mil. gal)
Oxidation ditch	2 each train	5.7 ML (1.5 mil. gal)	11.4 ML (3 mil. gal)
Secondary anoxic zone	3 each train	0.42 ML (0.111 mil. gal)	1.26 ML (0.333 mil. gal)
Reaeration zone	1 each train	0.42 ML (0.111 mil. gal)	0.42 ML (0.111 mil. gal)
Clarifiers	2 each	39.6-m (130-ft) diameter	N/A

an anoxic mode while ditch 2 continues in an aerobic zone. A second 30-minute cycle follows in which both ditches operate aerobically. During the last 90 minutes, ditch 1 continues to operate in an aerobic mode, while ditch 2 operates in an anoxic mode. This alternating strategy allows nitrification to occur, followed by denitrification while in the anoxic mode.

The low total nitrogen concentrations are achieved with a final anoxic tank similar to a five-stage Bardenpho process. The wastewater is then reaerated before the secondary clarifiers to prevent denitrification that could cause rising sludge. The RAS rate is 55% of the average flow. The RAS enters a holding tank with a 30-minute detention time to minimize the amount of nitrates that enter the anaerobic tank. The sludge age is approximately 16 to 18 days. The secondary effluent passes through a deep-bed sand filter. Methanol can be fed to aid in denitrification, if needed; however, methanol has never been fed. Figure 16.11 shows the North Cary facility nitrogen removal performance.

8.0 CASE STUDY 7 (MEMBRANE BIOREACTOR)—CITY OF NORTH LAS VEGAS, NEVADA, WATER RECLAMATION FACILITY

8.1 Facility History

The City of North Las Vegas Water Reclamation Facility is an advanced tertiary level wastewater treatment facility designed to remove conventional pollutants, such as nitrogen and phosphorous, through either enhanced BNR or chemical addition. On a 40-ac site, the facility has average daily and peak-hour flow capacities of 95 ML/d (25 mgd) and 190 ML/d (50 mgd), respectively. The facility uses a membrane bioreactor (MBR) and provides reclaimed water that is disinfected before being discharged. The MBR process removes all substances from the wastewater that are larger than 3 μm. Coupled with disinfection, the treatment process provides a pathogen-safe reclaimed water product that can be used for irrigating areas such as public golf courses or that can be discharged back to Lake Mead.

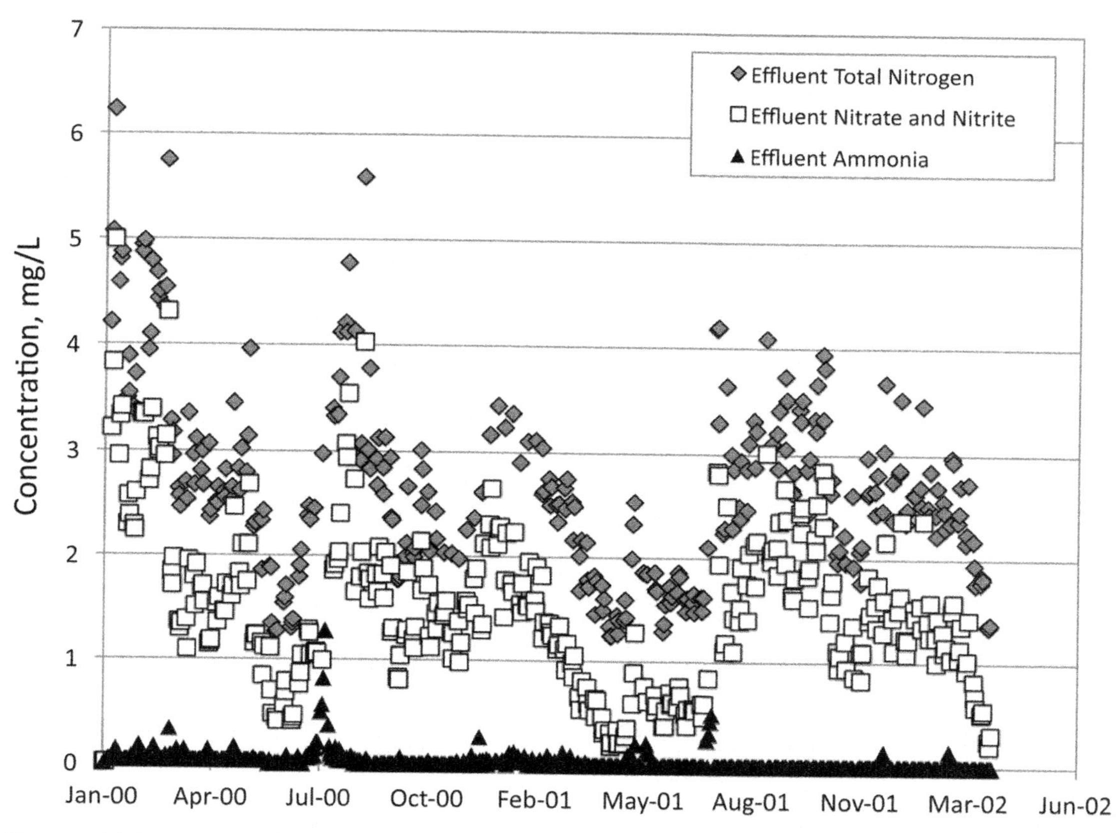

FIGURE 16.11 Town of North Cary, North Carolina, facility effluent nitrogen data.

In the past, all of the wastewater generated within the city limits of North Las Vegas was conveyed to the City of Las Vegas and the Clark County Water Reclamation District for treatment. Once conveyed to these agencies, the potential for water reuse by the City of North Las Vegas was gone, and the water commodity was owned by those agencies. In addition, almost 68% of the wastewater fund's operating budget paid for treatment costs to these third-party agencies.

In January 2004, the City of North Las Vegas city council authorized an in-depth analysis of wastewater treatment options for the city. In October 2004, the city council directed staff to pursue construction of a wastewater reclamation facility and authorized the city manager to pursue a site for the facility. After obtaining all necessary approvals and before construction, the proposed wastewater reclamation facility and all associated infrastructure was estimated to cost approximately $321.3 million and was to be funded with bond proceeds and capital reserves.

Construction of this facility began in November 2008, with the facility being substantially completed in August 2011. The City of North Las Vegas facility has been in operation since June 2011 and influent flows currently average 64 ML/d (17 mgd). This allows independence from the city of Las Vegas wastewater treatment rates and provides North Las Vegas with the ability to maximize efficiency of operations and water resources.

When it was placed online, the City of North Las Vegas facility was the largest MBR facility treating wastewater in North America and one of the largest MBR facilities in the world. Advanced nutrient removal is used to remove phosphorous and nitrogen before discharging the highly treated wastewater to the Sloan Channel, where it is conveyed to Lake Mead. The City of North Las Vegas facility uses either enhanced biological phosphorus removal (EBPR) or chemical phosphorus precipitation for phosphorus removal. A modified three-stage A2/O process is used for simultaneous nitrification and denitrification. The City of North Las Vegas facility has a staff of 21 highly trained personnel who operate and maintain the facility 24 hours a day, 365 days a year, through an extensive computerized supervisory control and data acquisition operational system.

8.2 Process Description

The incoming wastewater first enters the facility through a 10.65-m-deep influent pump station and then through coarse screens. After grit removal, the flow proceeds through fine screens and then 12 MBRs for treatment. The treated water is disinfected with sodium hypochlorite at the chlorine contact building and then dechlorinated. The City of North Las Vegas facility uses either EBPR or chemical phosphorus precipitation (Figure 16.12).

8.2.1 *Liquid Treatment Processes*

The liquid treatment processes include preliminary treatment, secondary treatment, and disinfection. The following subsections discuss each process in detail.

8.2.1.1 Preliminary Treatment (Headworks Facility)

The headworks facility (HWF) contains all the preliminary treatment for the facility. Preliminary treatment includes coarse screening, influent pumping, grit removal, and fine screening. Processing units for the collected screenings and grit are also located in the HWF.

8.2.1.2 Secondary Treatment

Biological treatment is performed in six parallel bioreactor basins. Each basin is divided into a series of zones that are isolated from each other by submerged baffle weir walls. The configuration and sizing of these zones is based on the basic principles of BNR using a configuration adapted for use in the MBRs that optimizes the enhanced biological removal of nitrogen and phosphorus while recognizing the unique differences in the quality and flow of RAS in MBR systems. Membrane bioreactor air scour is controlled by an airflow setpoint.

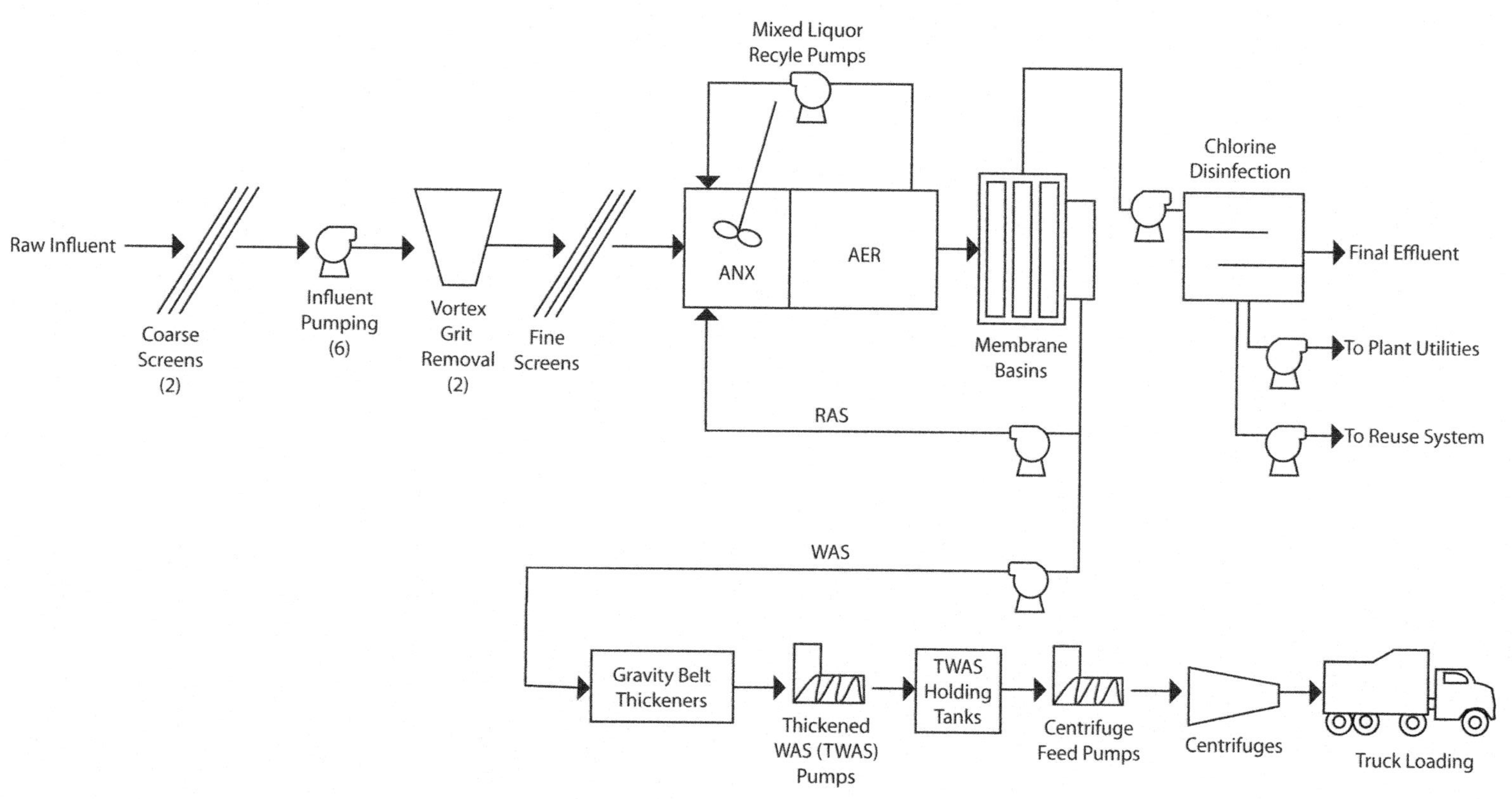

Figure 16.12 City of North Las Vegas facility process flow diagram.

8.2.1.3 Disinfection

Facility effluent can be discharged as a surface water discharge to the Las Vegas Wash ("Wash"), pumped to re-use applications and uses, or pumped to the facility water system. Chlorine solution is injected to the 1.22-m (48-in.) chlorine contact basin (CCB) influent pipe upstream of a jet mixing system to ensure even distribution of chlorine throughout the CCB influent. Membrane permeate enters the chlorine contact basin via one of two 0.91-m (36-in.) pipes from the MBR system. Chlorine residual is monitored at both the upstream end of the CCB (for the influent) and the downstream end (for the effluent). Any residual chlorine in the effluent discharged to the Wash is dechlorinated using sodium bisulfite. Effluent for facility water is withdrawn after dechlorination.

8.2.2 *Solids Treatment Processes*

Thickening and dewatering equipment is designed to remove excess water from the solids to reduce the volume and weight of waste transported to a nearby landfill. The solids thickening equipment consists of GBTs. Waste activated sludge is pumped from the membrane bioreactor to the solids handling facility. The WAS is injected with diluted polymer before being discharged onto the GBT's rotating belt. The TWAS is then pumped to horizontal bowl centrifuges. The TWAS is injected with diluted polymer before entering the centrifuge. The separated water is collected and transported back to the facility's headworks for treatment. The cake is stored in the fully enclosed, odor-controlled hoppers before being loaded and hauled to a landfill for disposal.

The City of North Las Vegas facility has effluent permit limits for BOD and TSS of 30 mg/L. Additionally, the facility must meet a 30-day average effluent ammonia-nitrogen concentration of 200 kg/d (91 lb/d), which is equivalent to 0.43 mg/L at the facility's design capacity of 94.6 ML/d (25 mgd). The facility has a total nitrogen limit of 10 mg/L and a total phosphorus limit of 77 kg/d (35 lb/d), which is equivalent to 0.17 mg/L at the design capacity.

8.3 Process Design Criteria

The following are process design criteria for the City of North Las Vegas facility: GE/Zenon Model ZeeWeed (ZW 500D);

- Design average daily flow of 94.6 ML/d (25.0 mgd);
- Design peak hour flow of 189.3 ML/d (50.0 mgd);
- Membrance bioractor sludge recycle flow (4*average day maximum month (ADMM) of 454.3 ML/d (120 mgd);
- Twelve trains;
- Basin side water depth of 6.35 m (20.83 ft) W × 17.4 m (57 ft) L × 3.7 m (12 ft);

- A working volume for each train of 404 000 L (106 700 gal);
- Total working volume of 4 841 070 L (1 278 875 gal);
- Sixteen cassette slots per train, 192 total cassettes slots (plus four extra future cassette spaces per train);
- Forty-eight modules per cassette, 9216 total modules (10 cassettes each train with 48 modules, two cassettes each train with 32 modules, and 6528 total modules);
- Membrane area per module of 31.6 m^2 (340 sq ft);
- Total membrane area of 206 200 m^2 (2 219 520 sq ft);
- Design maximum total membrane pressure (TMP) of 82.7 kPa (12 psig);
- Operating TMP of 6.9 to 69.0 kPa (1 to 10 psig);
- Two 1305-kW (1750-hp) process blowers and two 932-kW (1250-hp) process blowers; and
- Membrane scour (two) at 634 kW (850 hp).

Table 16.14 summarizes the influent wastewater characteristics.

8.4 Design and Operation Modifications

The facility-operating SRT has been reduced from approximately 20 days at startup to a target SRT of 8 days. This was done to increase the amount of phosphorus being removed from the facility by WAS.

The facility was started up with a RAS rate of 61 ML/d (16 mgd) per bioreactor (four bioreactors were used at startup and a RAS flow of 4 times the influent flow, totaling 242 ML/d [64 mgd]) and an internal mixed liquor rate (MLR) of 17 ML/d (4.5 mgd) per bioreactor. Because of the low MLR rate, both the anoxic and swing zones were anaerobic, which caused additional phosphorus release in the anoxic zone and continued into the swing zone, downstream of

TABLE 16.14 City of North Las Vegas facility wastewater characteristics.

Parameter	Effluent criteria	Maximum
Five-day BOD	300 mg/L	325 mg/L
Total suspended solids	330 mg/L	357 mg/L
Total Kjeldahl nitrogen	43 mg/L nitrogen	47 mg/L nitrogen
Ammonia–nitrogen	27 mg/L nitrogen	28 mg/L nitrogen
Total phosphorus	7 mg/L phosphorus	7.6 mg/L nitrogen
Minimum temperature	18 °C	18 °C
Average temperature	24 °C	24 °C
Maximum temperature	30 °C	30 °C

the anoxic zone. The relatively high RAS rate maximized the amount of nitrate being recycled into the anaerobic zone, reducing the quantity of volatile fatty acids available to drive EBPR.

The facility diurnal flow variability has been more severe than anticipated. The facility can experience influent swings greater than 106 ML/d (28 mgd) during late morning to early afternoon flows to flow of less than 23 ML/d (6 mgd) during the early morning hours. This variation causes the organic-to-nutrient ratios in the facility influent to show their highest levels in the early afternoon through the evening and then rapidly drop overnight. These wide variations cause wide swings in bioreactor dissolved oxygen levels, which exasperates the high nitrate recycled to the anaerobic zone.

8.5 Performance

The City of North Las Vegas facility has had excellent performance since startup and has not exceeded any conventional effluent quality parameters except total phosphorus. Biochemical oxygen demand, TSS, fecal coliforms, ammonia, and nitrate–nitrogen results have been at or below method-detection limits since startup, with TIN results averaging below 5 mg/L (see Figures 16.13 and 16.14).

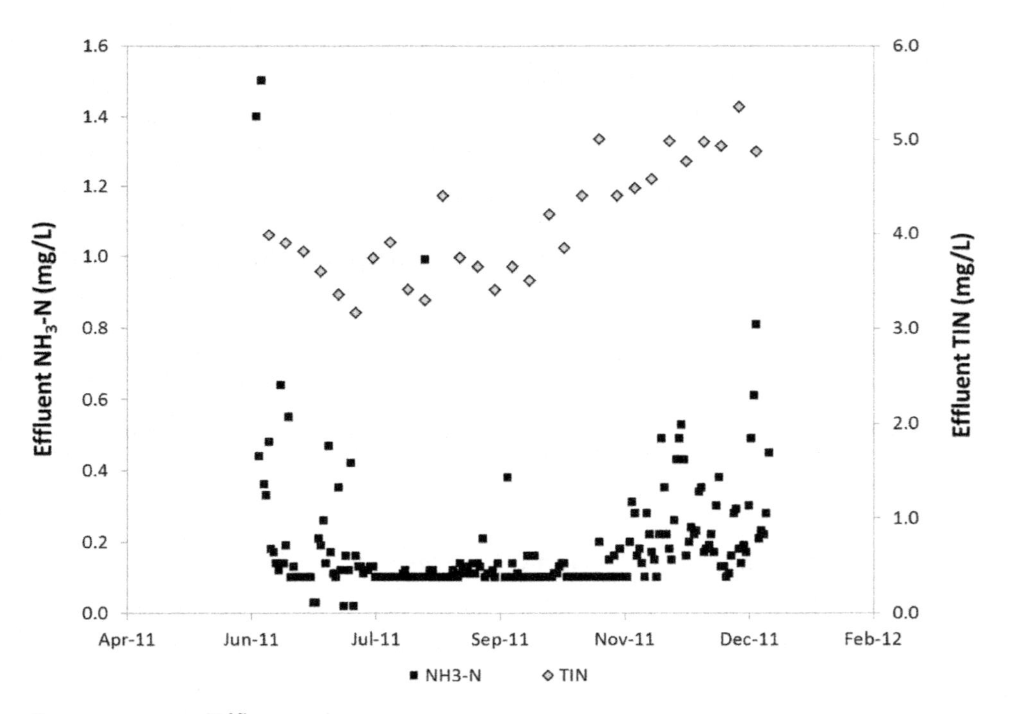

FIGURE 16.13 Effluent nitrogen.

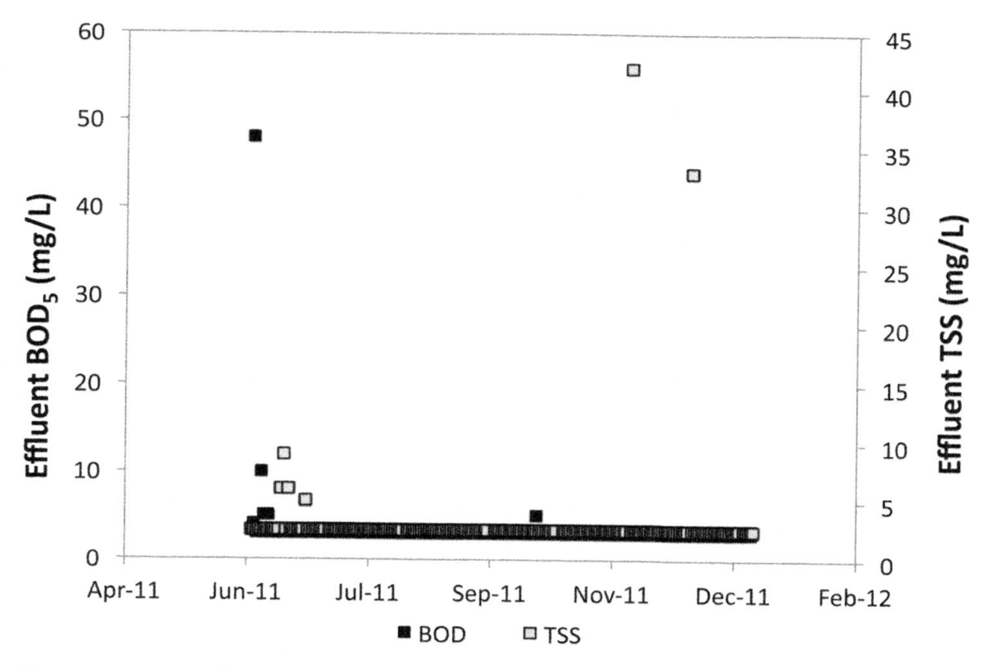

FIGURE 16.14 Effluent BOD5 and TSS.

The only operations issue that has affected the City of North Las Vegas facility has been the facility's ability to remove total phosphorus using EBPR. The facility has experienced excellent phosphorus removal using ferric chloride during the months of November 2011 and March 2012 (Figure 16.15). During the first 2.5 months of operation, the facility experienced excellent total phosphorus removal using EBPR; however, from the middle of August and September 2011, the facility experienced excursions in total phosphorus that were not stabilized until ferric addition was implemented in the middle of October 2011. Ferric chloride feed was stopped in December 2011 to reestablish EBPR. The facility was almost stabilized by February 2012 when a fifth bioreactor was placed in service, which destabilized the EBPR process, at which point ferric addition had to be reinstated.

Chemicals are used to clean the membranes and to keep them within the operating TMP range. Sodium hypochlorite (10.3%) is used for

- Maintenance cleans twice per week per train (12 trains)—132 000 L/yr (35 000 gal/yr) and
- Recovery cleans semiannually per train—52 000 L/yr (13 600) gal/yr.
- Citric acid (50%) is used for

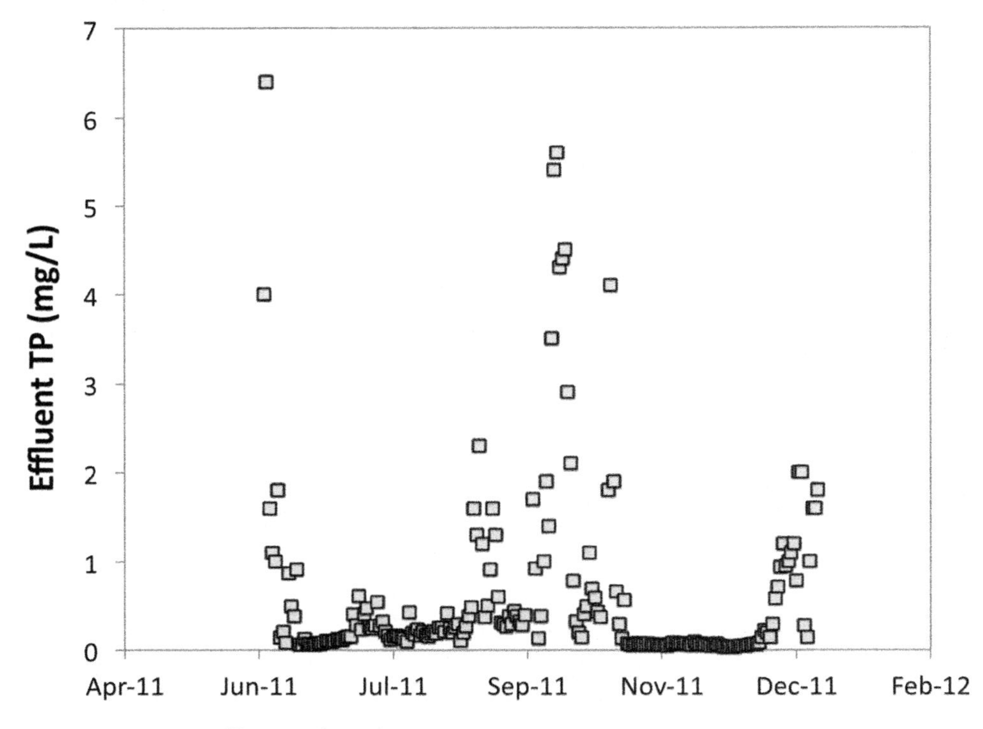

FIGURE 16.15 Effluent phosphorus.

- One maintenance clean once per week per train (12 trains)—33 000 L/yr (8750 gal/yr) and
- One recovery clean annual per train—10 000 L/yr (2650 gal/yr).

9.0 CASE STUDY 8 (STEP-FEED)—PISCATAWAY, MARYLAND, WASTEWATER TREATMENT PLANT

9.1 Facility History

The Piscataway Wastewater Treatment Plant, located in Accokeek, Maryland, was first put into service in the 1960s. During the 1960s and 1970s, the facility was expanded to its current design capacity of 116-ML/d (30-mgd) average daily flow and 320-ML/d (85-mgd) peak hydraulic flow. In the early 2000s, the facility was upgraded to a step-feed activated sludge process, which allows the facility to achieve seasonal BNR-level treatment. The step-feed configuration also allows the facility to be operated with varying aerobic and anoxic volumes. Swing zones allow the facility to vary its configuration, as needed, to achieve its effluent standards. The target effluent total nitrogen concentration is 5.6 mg/L year round based on the allowed annual allocation at the 115-ML/d (30-mgd) average design capacity. The Piscataway facility serves about 193,000 customers in Prince George's County and discharges into the Potomac River. The State of Maryland has imposed higher levels of nutrient treatment for water resource recovery facilities (WRRFs) discharging into the Chesapeake Bay, its rivers, or tributaries. As part of this initiative, the Piscataway facility is currently under agreement to meet enhanced nutrient removal (ENR)-level treatment goals of 3.0 mg/L total nitrogen by 2013. The facility's current effluent total phosphorus limit of 0.18 mg/L will not change with the ENR permit.

9.2 Process Description

The processes at the Piscataway facility include

- Preliminary treatment including aerated grit chambers and fine screens (6.3-mm [0.25-in.] opening),
- Primary clarification,
- Step-feed BNR reactors (two independent trains),
- Chemical feed for phosphorous removal (alum) and secondary clarification,
- Tertiary filters (dual media),
- Ultraviolet disinfection, and
- Gravity cothickening of primary sludge and WAS followed by lime stabilization to belt press dewatering and land application (Class B biosolids).

The Piscataway facility has two step-feed activated sludge treatment trains to remove total nitrogen. The first train consists of five sets of anoxic and aerobic basins in series, with the capability to feed primary effluent to the first four anoxic zones. There are four sets of anoxic and aerobic zones in series. Primary effluent can be fed to the first three anoxic zones. Return activated sludge enters the first anoxic zone in both trains. The annual average total nitrogen concentration in the effluent from the Piscataway facility is 2.58 mg/L. Methanol or an alternative carbon source is not needed because primary effluent can be fed to each anoxic zone, with the exception of the last zone in each train. In addition, internal recycles are not necessary because of the multiple anoxic and aerobic zones in series. Sodium hydroxide is added upstream of the step-feed activated sludge process to increase the alkalinity of the wastewater. This is necessary because alkalinity is consumed as part of the nitrification process, although only a portion is returned following denitrification.

Current effluent limits include total nitrogen of less than 3 mg/L and total phosphorus of 0.18 mg/L. The wastewater characteristics for the Piscataway facility are shown in Table 16.15, process design criteria for the Piscataway facility are shown in Table 16.16, and the effluent performance of the Piscataway facility is shown in Table 16.17.

TABLE 16.15 Piscataway, Maryland, facility wastewater characteristics.

Parameter	Design value
Design average daily flow	114 ML/d (30.0 mgd)
Actual average daily flow	83.3 ML/d (22.0 mgd)
Influent BOD_5	110 mg/L
Primary effluent BOD_5	80 to 100 mg/L
Primary effluent TKN	20 mg/L

10.0 SUMMARY

Biological nitrogen removal occurs in a variety of configurations and has evolved over the last couple of decades. The configuration of the treatment basins, internal recycle locations, and flows, swing zones, tertiary filtration, chemical addition, and solids handling processes discussed herein are unique to each treatment facility and have various advantages and disadvantages. However, the core reactions of biological nitrogen removal, autotrophic aerobic nitrification, and heterotrophic anoxic denitrification fueled by a carbon source have remained virtually unchanged since its first use in WRRFs. Conventional BNR using the highest level of treatment has been shown to produce effluent total nitrogen concentrations of less than 3 mg/L with external carbon addition. However, the level of total nitrogen removal is highly dependent on site-specific conditions (e.g., temperature) and nitrogen characteristics in the raw wastewater. The level of treatment that is appropriate for each facility is dependent on several cost and noncost factors. Current and anticipated future permit limits on nitrogen species, physical constraints such as land availability and topography, local energy and operating costs, and ease of operation must all be considered to select the most efficient and cost-effective treatment process for BNR.

TABLE 16.16 Piscataway, Maryland, facility process design criteria.

Number of trains	5
Passes per train in train 1	5
Passes per train in train 2	4
Total volume train 1	10 ML (2.65 mil. gal)
Total volume train 2	10.4 ML (2.76 mil. gal)
Percent anoxic volume in train 1	28%
Percent anoxic volume in train 2	34%
Return rate	100% of influent flow

TABLE 16.17 Piscataway, Maryland, 2010 performance.

Flow	83.3 ML/d (22.1 mgd)
Effluent BOD5	2 mg/L
Effluent TSS	1 mg/L
Effluent total nitrogen	3.4 mg/L
Effluent total phosphorus	0.08 mg/L
	(Influent = 2.54 mg/L, 97% total phosphorus removal)
Alum usage	800 tons/year

11.0 REFERENCES

Phillips, H. M.; Maxwell, M.; Johnson, T.; Barnard, J.; Rutt, K.; Seda, J.; Corning, B.; Grebenc, J. M.; Love, N.; Ellis, S. (2008) Optimizing IFAS and MBBR Designs Using Full-Scale Data. *Proceedings of the Water Environment Federation Technical Exhibition and Conference* [CD-ROM]; Chicago, Illinois, Oct 18–22; Water Environment Federation: Alexandria, Virginia.

Phillips, H. M.; Steichen, M. T.; Johnson, T. L. (2010) The Second Generation of IFAS and MBBR: Lessons to Apply. *Proceedings from the Water Environment Federation Biofilm Reactor Technology Conference;* Portland, Oregon, August 15–18; Water Environment Federation: Alexandria, Virginia.

Rutt, K.; Mayo, D.; Dalsoglio, D. (2008) Balancing Growth and Water: Broomfield, Colorado's First "Model" City, Follows Fundamental Vision to Create Community. *Colorado Public Works J.*, **4** (1), 9–19.

Rutt, K.; Seda, J.; Johnson, C. (2006) Two Year Case Study of Integrated Fixed Film Activated Sludge (IFAS) at Broomfield, CO, WWTP. *Proceedings of the 79th Annual Water Environment Federation Technical Exhibition and Conference* [CD-ROM]; Dallas, Texas, Oct 21–25; Water Environment Federation: Alexandria, Virginia.

Chapter 17

Case Studies—Enhanced Biological Phosphorus Removal

Barry Rabinowitz, Ph.D., P.Eng., BCEE; Samuel S. Jeyanayagam, Ph.D., P.E., BCEE; and M. Kim Fries, M.A.Sc., P.Eng.

1.0 INTRODUCTION

This chapter presents eight case studies of full-scale facilities that are specifically designed and operated for optimal enhanced biological phosphorus removal (EBPR) from municipal wastewaters to meet effluent total phosphorus limits between 0.05 and 1.5 mg/L. It should be noted that most EBPR facilities in North America are also designed to simultaneously achieve some degree of nitrogen removal in a single sludge system, that is, to either meet an effluent ammonia-nitrogen (NH_3-N) standard though nitrification or an effluent total nitrogen standard through nitrification and denitrification. All facilities used as case studies in this chapter have been in operation for a minimum of 2 years to provide the reader with insight to the challenges encountered by facility operators during the startup and commissioning phases of the project and the design and operational changes that were required following commissioning to ensure reliable and effective EBPR.

2.0 CASE STUDIES

2.1 Case Study 1—Pine Creek Wastewater Treatment Plant, Calgary, Alberta, Canada

2.1.1 *Facility History*

The 100-ML/d Pine Creek Wastewater Treatment Plant (Figure 17.1) is a state-of-the-art biological nutrient removal (BNR) facility that serves the rapidly growing southern areas of the City of Calgary, Alberta, Canada. Final effluent is discharged through an outfall to the environmentally sensitive Bow River, which is an important recreational river and source of potable water for several downstream communities. The facility has a unique process design that incorporates both conventional BNR and step-feed BNR operating capabilities. Primary sludge is fermented in two single-stage fermenters and thickeners, and the volatile fatty acid (VFA)-rich fermenter supernatant is discharged directly to the anaerobic zone of the BNR bioreactor.

FIGURE 17.1 Stage 1 Pine Creek facility and Bow River, Calgary, Alberta (courtesy of The City of Calgary).

2.1.2 *Process Description*

The Stage 1 facility is intended to handle wastewater from 250,000 people, although the facility is expandable in stages to eventually handle up to 1.75 million people. The facility was constructed in a series of contracts between 2005 and 2009. The monthly average effluent criteria for the Pine Creek facility are as follows:

- Five-day biochemical oxygen demand (BOD_5), 15 mg/L;
- Total suspended solids (TSS), 15 mg/L;
- Total phosphorus, 0.5 mg/L;
- Ammonia-nitrogen, 5 and 10 mg/L (during summer and winter, respectively);
- Total nitrogen, 15 mg/L; and
- Fecal coliforms, 200 per 100 mL.

Although the effluent total phosphorus limit set by the regulator is 0.5 mg/L, the Pine Creek facility has to meet an effluent concentration of about 0.3 mg/L to comply with the city's total mass loading limit for phosphorus. Figure 17.2 presents a process flow diagram showing key unit processes. The liquid treatment stream consists of a headworks (fine screening and grit removal), four rectangular primary clarifiers, two multicell bioreactors based on the Westbank process configuration with step-feed capabilities for cold and wet weather operation, four secondary clarifiers, 12 tertiary fabric filters, and two UV disinfection channels. The four-pass bioreactor configuration, shown in Figure 17.3, includes the following novel design features:

- Multiple primary effluent feed points to the preanoxic, anaerobic, anoxic, and "swing" zones to protect the preanoxic and anaerobic zones from hydraulic variations;

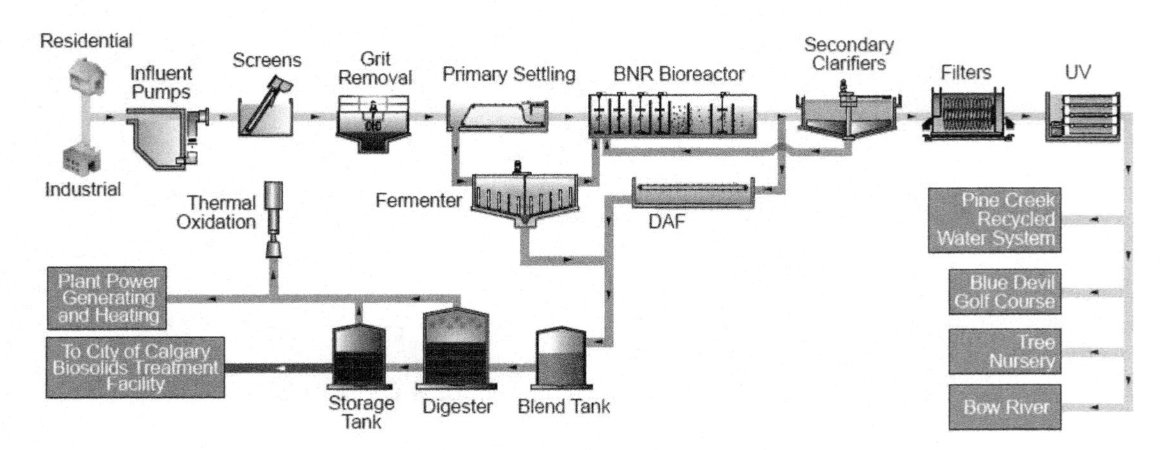

FIGURE 17.2 Pine Creek facility process flow diagram (courtesy of The City of Calgary).

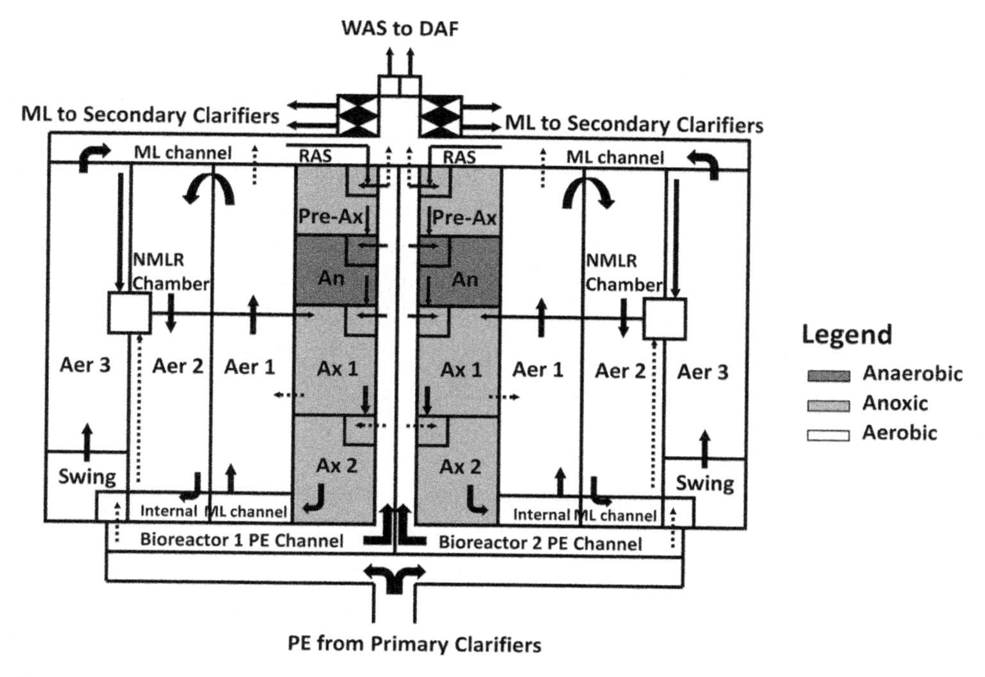

FIGURE 17.3 Pine Creek facility bioreactor layout (courtesy of The City of Calgary).

- Multiple feed points for the VFA-rich fermenter supernatant to enable optimal use for EBPR and denitrification;
- The ability to step-feed excess primary effluent to a small swing cell at the head end of the fourth pass to minimize flux loading to the secondary clarifiers during wet weather flows;
- The ability to the maximize denitrification through the use of swing cells to optimize the anoxic mass fraction;
- The flexibility to take any two passes of either bioreactor out of service for maintenance while keeping the remaining two passes of that bioreactor in service; and
- A unique intercell baffle design combined with surface mixed liquor wastage that eliminates the accumulation of biological foam and scum on the bioreactor surface.

The solids treatment stream consists of two single-stage primary sludge fermenters, two dissolved air flotation (DAF) thickeners, a sludge blend tank, two mesophillic anaerobic digesters, and a digested sludge storage tank. The purpose of the primary sludge fermenters is to generate the short-chain VFAs required to promote reliable EBPR and to serve as a source of readily biodegradable carbonaceous oxygen demand (COD) for denitrification in the bioreactor. The fermenters also gravity thicken the primary sludge to a concentration of 4 to 6% (as dry solids) before anaerobic digestion. The primary sludge pumping rate is typically set at 2 to 3% of the facility flow, and the fermenters are operated at a solids retention time (SRT) between 3 and 6 days. A steady stream of nitrate-rich final effluent or DAF subnatant is added to the fermenter inlet as a means of controlling methane and hydrogen sulfide (H_2S) formation in the units, and to act as a carrier to convey the VFAs produced to the BNR process. Thickened fermented primary sludge and thickened waste activated sludge (WAS) are combined in the sludge blend tank and pumped to anaerobic digestion for stabilization. The digested sludge is pumped to a remote site for lagoon dewatering, and the dewatered biosolids are applied to agricultural land as a soil amendment.

2.1.3 *Facility Commissioning*

The liquid treatment unit processes were built during the first phase of construction and commissioned in October 2008, with the effluent total phosphorus limit being met through in-facility alum addition. Before introducing wastewater, approximately 1200 m^3 of mixed liquor was trucked in from a nearby BNR facility, where it was discharged into one of the facility's bioreactors. When this seeding exercise was completed, the bioreactor mixed liquor suspended solids (MLSS) concentration was approximately 1000 mg/L and raw wastewater was diverted to the trunk sewer feeding the facility. The flows were gradually increased as the facility proved it could handle the flow and develop a stable BNR biomass in its bioreactors. The facility started producing a high quality effluent almost immediately upon startup because of the seeding of the bioreactors. Sludge settleability was excellent and effluent turbidity was minimal through the initial operating period.

The solids stream unit processes were commissioned in October 2009. The anaerobic digesters were seeded with about 3000 m^3 of digested sludge from a nearby facility. This material was deposited in one of the digesters and used to establish digestion in short order after beginning the transfer of sludge from the liquid stream process to the solids stream process. The primary sludge fermenters were started by transferring primary sludge to those units. Volatile fatty acid generation commenced quickly and, by the beginning of 2010, EBPR had been established to the point that alum addition was virtually eliminated.

2.1.4 *Process Monitoring and Control*

Key parameters that are monitored and used to control the EBPR process include the following:

- Primary sludge fermenters
  - Primary sludge pumping rate and concentration;
  - Elutriation water flowrate;

- ○ Mechanism rotational speed and torque;
- ○ Sludge blanket height;
- ○ Thickened sludge pumping rate and concentration;
- ○ Fermenter SRT; and
- ○ Supernatant flowrate and characteristics (VFA, COD, soluble BOD, etc.); and

- Biological nutrient removal process
  - ○ Primary effluent flow and characteristics (VFA, COD, total phosphorus, total Kjeldahl nitrogen [TKN], etc.);
  - ○ Anaerobic, anoxic, and aerobic zone supernatant characteristics (orthophosphate [PO_4], ammonia [NH_4], oxidized nitrogen [NO_x], etc.);
  - ○ Aerobic zone dissolved oxygen and MLSS concentrations;
  - ○ Bioreactor SRT; and
  - ○ Secondary effluent characteristics (PO_4, NH_3, NO_x, etc.).

2.1.5 *Operating Data*

Figure 17.4 shows the monthly average effluent total phosphorus concentrations for 2009, 2010, and 2011. During this period, the monthly average effluent total nitrogen concentrations ranged from 7 to 12 mg/L. With the exception of

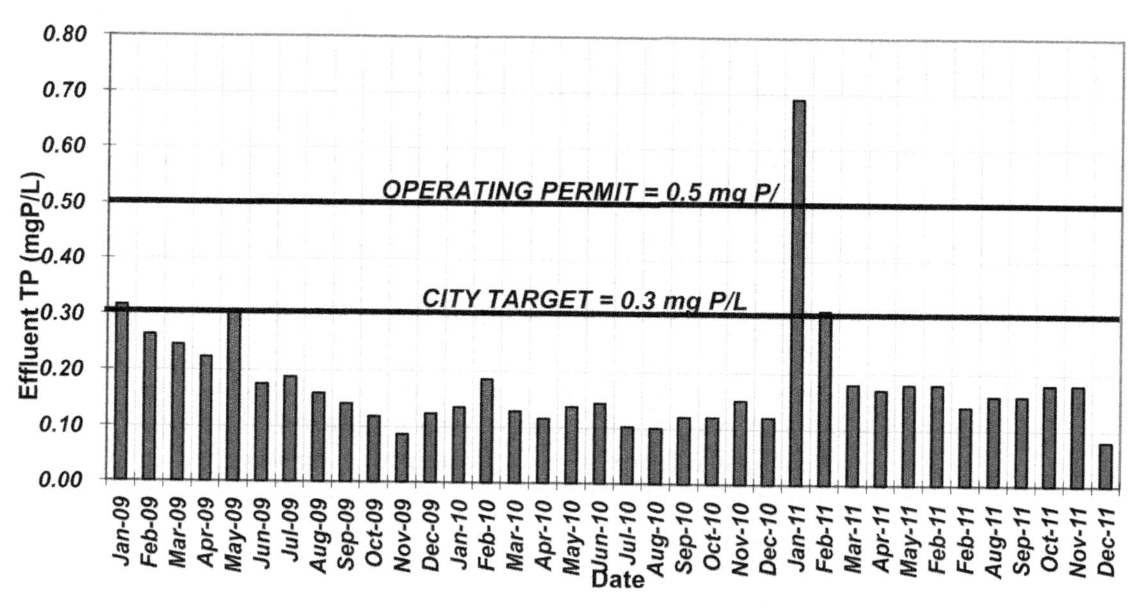

FIGURE 17.4 Pine Creek facility effluent total phosphorus concentrations from 2009 to 2011 (courtesy of The City of Calgary).

January 2011, when the effluent total phosphorus limit was exceeded, the effluent total phosphorus and total nitrogen concentrations have been consistently well below effluent limits. The unstable process performance during January 2011 was attributed to low return activated sludge (RAS) pumping rates, which caused excessive solids to flow over the secondary clarifier weirs, poor SRT control, and a severe biological foaming event. The delay in construction of the primary sludge fermenters allowed a comparison of the performance of the BNR process with and without fermentation over an extended period of time. Data from the first 2 years of facility operation indicated that, with primary sludge fermentation, the facility was capable of reliably meeting its effluent total phosphorus target of less than 0.3 mg/L. Without fermentation, the facility required the in-facility addition of approximately 50 mg/L of alum to meet the same effluent total phosphorus quality standard (Rabinowitz and Fries, 2010).

2.1.6 *Operating Experience*

While the Pine Creek facility has consistently met or exceeded its effluent quality requirements, there were numerous issues that arose during commissioning that needed to be addressed so that the facility could operate effectively. These issues varied from small malfunctions to fairly significant equipment problems. The following summaries are examples of the more serious of these operating problems:

- Fermenter mechanisms—during the early months of fermenter operation, there was an excessive buildup of scum on the fermenter surface, which filled up the scum boxes and had to be manually removed. The problem was eventually solved by modifying the scum removal mechanisms such that the mechanisms were fitted with torque sensors that shut down the drives under high-torque conditions. However, there were persistent inequalities between the torque sensors and the drive power draw;
- Nitrification and denitrification optimization—nitrification was quickly established in the secondary treatment process because of the seeding of bioreactors with sludge from a nearby BNR facility. However, during the early operating period, it was difficult to obtain the turndown necessary on the blowers, recycle pumps, and aeration system so that truly oxygen-free conditions could be achieved in the anoxic zones of the bioreactor. To reduce the oxygen transferred to the bioreactor, mixed liquor recycles were modulated (one of two duty pumps was operated); all swing zones that could operate as either anoxic or aerobic zones were operated as anoxic zones (mixing only); and aeration to the last pass of the bioreactor was severely constricted to limit the dissolved oxygen concentration. Fortunately, the blowers had sufficient turndown to handle this situation, although the provision in the design for the blow-off of excess air was frequently used to reduce the air supply to the bioreactor. Nitrogen removal

was achieved within a relatively short period of time and, as the load to the facility increased, these operating concerns abated. There is now sufficient organic load to achieve oxygen-free conditions in the anoxic zones so that denitrification effectively lowers the effluent total nitrogen concentration to below 10 mg/L; and

- Tertiary filtration—the facility has one of the largest fabric filter installations for effluent filtration in the world. The vendor's standard design was modified to reduce the number of appurtenances associated with operation of this system. Furthermore, a central backwash system was used rather than two pumps per filter and backwashing of four disks at one time was used rather than the standard two discs, substantially reducing the number of backwash control valves. The vendor for the system provided a packaged control system for the filters, with one control panel (including programmable logic controllers [PLCs]) for each pair of filters. These controls interface with the facility's central digital control system (DCS) through a fieldbus interconnection. Communication between the vendor-supplied PLCs and the DCS tended to cause problems with the PLCs, which, for no apparent reason, would stop operating. When the PLCs were rebooted, the failure history was lost and, therefore, it was impossible to diagnose the root cause of the problem. Ultimately, communication devices were updated and improved so that the two systems could operate without one interfering with the other.

2.2 Case Study 2—Nine Springs Wastewater Treatment Plant, Madison, Wisconsin

2.2.1 *Facility History*

The 160-ML/d (42-mgd) Nine Springs Wastewater Treatment Plant serves the City of Madison, Wisconsin, and several surrounding communities in the Dane County area. The Madison metropolitan area surrounds a chain of four ancient glacial lakes, and, in 1959, the district began pumping treated effluent 5 miles to Badfish Creek to avoid discharge to the chain of lakes. The facility was initially constructed on the current site in 1929. The first facility was a trickling filter facility with Imhoff tanks for sludge treatment; within 5 years, an activated sludge train and anaerobic digestion were added. Several facility upgrades have been constructed in response to population growth and changes in effluent discharge limits. There have been 10 significant upgrades to the facility since 1929, and the eleventh upgrade is currently under construction.

2.2.2 *Process Description*

In the early 1980s, the activated sludge system was modified for single-stage nitrification and the trickling filter was removed. The last significant change to address revised discharge limits occurred in 1997, when the facility was modified for EBPR without the construction of additional tankage. Because the facility did not have an effluent total nitrogen standard, the selected process configuration was the University of Cape Town process without nitrified mixed liquor recycle. Figure 17.5 presents a process flow diagram of the current facility operation. The effluent criteria for the Nine Springs facility are as follows:

- Five-day BOD, 7 and 16 mg/L (during summer and winter, respectively);
- Total suspended solids, 10 and 16 mg/L (during summer and winter, respectively);
- Total phosphorus, 1.5 mg/L;
- Ammonia-nitrogen, 1.3 and 3.8 mg/L (during summer and winter, respectively); and
- Fecal coliforms, 400 per 100 mL (during summer).

2.2.3 *Process Monitoring and Control*

The facility uses dissolved oxygen control for biological process control and energy savings. There are 10 EBPR trains with two dissolved oxygen probes in each train. This is the main operational control used for the process. The laboratory performs single daily colorimetric tests for effluent orthophosphorus Monday through Friday for a quick check on process performance, but batches samples for auto-analyzer analyses for regulatory reporting. Phosphorus removal has been consistent and this test has proved to be adequate to ensure stable

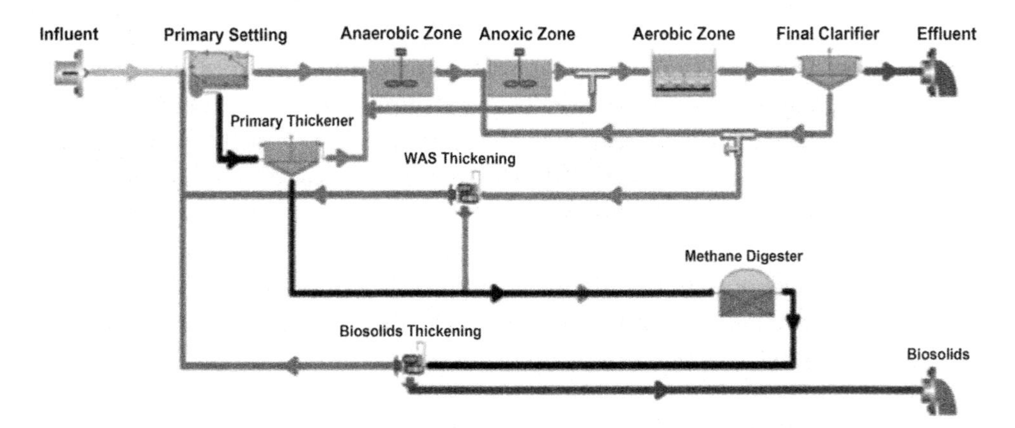

FIGURE 17.5 Nine Springs facility process flow diagram (courtesy of the Madison Metropolitan Sewerage District).

EBPR operation. The daily samples for compliance monitoring are batched and analyzed on a once- or twice-weekly basis.

Key parameters that are monitored and used to control the EBPR process include the following:

- Primary effluent characteristics (total phosphorus, NH_3, BOD_5, and TSS) (routinely monitored);
- Anaerobic, anoxic, and aerobic zone supernatant characteristics (PO_4, NH_3, NO_x, etc.) (checked occasionally in conjunction with research projects or other studies);
- Aerobic zone dissolved oxygen and MLSS concentrations;
- Bioreactor SRT; and
- Secondary effluent characteristics (PO_4, NH_3, NO_x, etc.) (routinely monitored).

A portion of RAS is wasted to maintain bioreactor SRT. Samples of mixed liquor and RAS are analyzed for MLSS concentration three times a week to verify that the EBPR process is being maintained at the target SRT.

2.2.4 *Operating Data*

The facility produces an effluent with a far better quality than that required by the discharge permit. The effluent total phosphorus limit is 1.5 mg/L, but the facility has averaged 0.39 mg/L since 1997 without chemical addition or tertiary filtration. The effluent ammonia-nitrogen and nitrate-nitrogen (NO_3-N) concentrations have averaged 0.16 and 14.8 mg N/L, respectively, while the effluent TSS concentration has averaged 4.8 mg/L. The average monthly effluent total phosphorus and ammonia-nitrogen concentrations are shown in Figures 17.6 and 17.7, respectively.

2.2.5 *Operating Experience*

The successful application of EBPR is attributable, in part, to a conservative secondary clarifier design surface overflow rate of 12.2 $m^3/m^2{\cdot}d$ (300 gpd/sq ft) at an average annual flow (AAF) of 190 ML/d (50 mgd). Before upgrading to EBPR, the facility was frequently subject to filamentous bulking problems, with sludge volume indexes (SVIs) in the 200- to 300-mL/g range. The highest monthly average SVI value since the facility's conversion to EBPR has been 140 mL/g, with average values being around 100 mL/g. Effluent total phosphorus concentrations have been decreasing with time, most likely because of increasing wastewater ratios of BOD_5 to total phosphorus. Operating experience at the Nine Springs facility since the upgrade to EBPR can be summarized as follows:

- Waste activated sludge production and handling—after the upgrade to EBPR, WAS production increased by approximately 9% at the same SRT. This was attributable to increased uptake of phosphorus and other

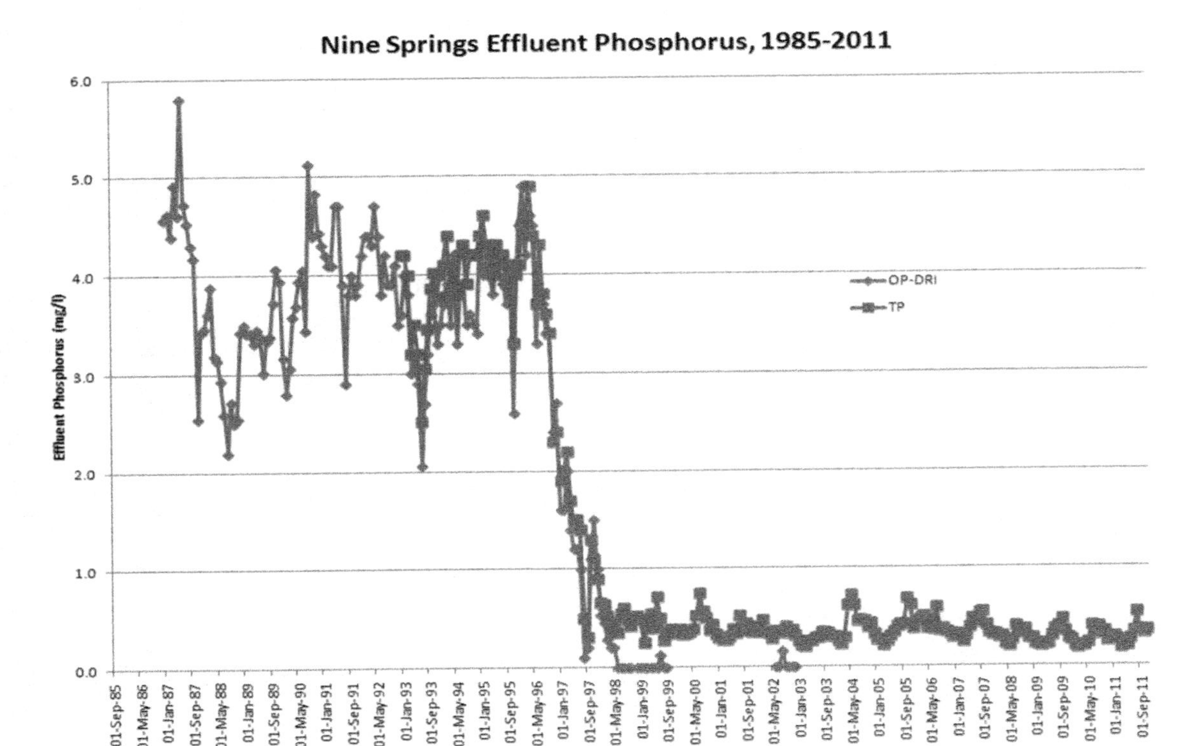

FIGURE 17.6 Nine Springs facility effluent orthophosphorus and total phosphorus concentrations from 1985 to 2011 (courtesy of the Madison Metropolitan Sewerage District).

ions into the activated sludge cell mass and improved TSS removal. The increase in WAS production was offset, to some degree, because the sludge thickened better at lower SVIs. The DAF thickened sludge concentration increased from an average of 3.7% solids to 4.2% solids without the use of polymer;

- Ultraviolet disinfection—maintenance requirements for the UV disinfection system decreased significantly with EBPR. Before EBPR, quartz tube scaling required cleaning with citric or phosphoric acid approximately every 3 weeks. After the upgrade, the quartz tubes no longer required cleaning during the 6-month disinfection season. The reason was thought to be lower effluent TSS and the luxury uptake of phosphorus, magnesium, and potassium in the EBPR process;
- Aeration system—offgas testing of the aeration system demonstrated that, in spite of the uptake of soluble organics in the anaerobic and anoxic zones, the alpha values of the fine-bubble aeration system were not significantly greater than at the head end of the process before incorporation of the anaerobic and anoxic zones. Alpha values were 0.3 to 0.4 at the head end

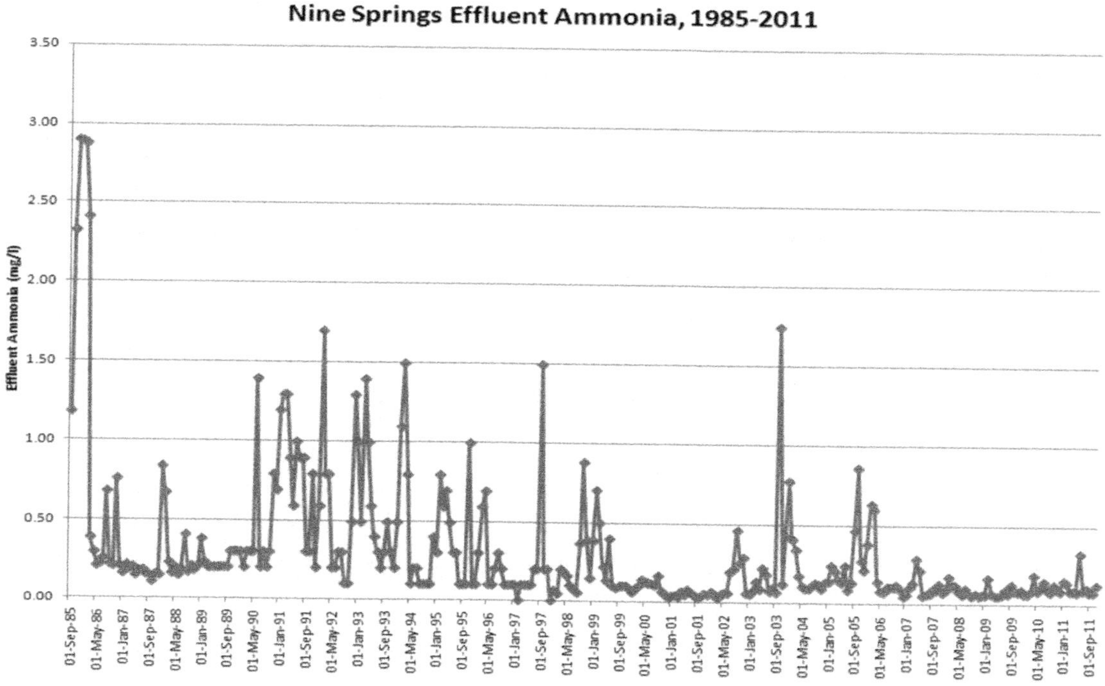

FIGURE 17.7 Nine Springs facility effluent ammonia-nitrogen concentrations from 1985 to 2011 (courtesy of the Madison Metropolitan Sewerage District).

of the aerobic zone, and increased to ~0.7 as aeration proceeded. This was similar to the offgas testing results obtained before the EBPR upgrade. Low dissolved oxygen operation was successfully tried in one facility in 2003. Maintaining low dissolved oxygen concentrations throughout the aerobic zone did not affect EBPR, had a small effect on ammonia-nitrogen removal, and increased the total nitrogen removal. Aeration savings were significant compared to a control facility. There is the possibility this operation resulted in increased *microthrix* predominance; however, the SVIs did not exceed 140 mL/g during the low dissolved oxygen testing; and

- Recycle rate pacing—the mixed liquor recycle from the anoxic zone to the anaerobic zone is paced with the influent flowrate. The recycle rate typically does not affect process performance significantly except under high wet weather flow conditions. It was discovered early during operation of the EBPR process that unless the recycle flow was increased proportionately to influent flow during high-flow events, the solids mass fraction in the anaerobic zone can drop significantly and EBPR can suffer. However, pacing the recycle rate with the influent flowrate has resolved this problem.

2.2.6 *Future Operation*

The success of EBPR has been followed by challenges in the anaerobic digestion system. The area uses groundwater with high hardness for drinking water. The magnesium concentration in raw wastewater entering the facility is approximately 45 mg/L. Although struvite problems have always been present in the digestion system, they became more prevalent after conversion to EBPR. Ferric chloride has been added to the digester feed and filtrate return to mitigate struvite problems and to control the phosphorus return to the secondary treatment process. The struvite problem became even more severe when the anaerobic digestion system was converted from mesophilic to thermophilic operation, and ferric chloride addition could no longer control these problems. The digestion system was returned to mesophilic operation and the current facility upgrade is being constructed to address these problems. Steam heating of raw sludge, acid-phase digestion, phosphorus release from WAS before digestion, thermophilic methane-phase operation, and struvite harvesting using the Ostara Pearl process are being included in the current upgrade. Figure 17.8 presents a process flow diagram of the upgraded facility. A small recycle stream from the acid-phase digester will be mixed with WAS in the phosphorus release tank and held for 6 to 8 hours before WAS thickening. The combination of WAS thickening filtrate and digested sludge thickening filtrate will be pumped to the Ostara Pearl process for struvite harvesting.

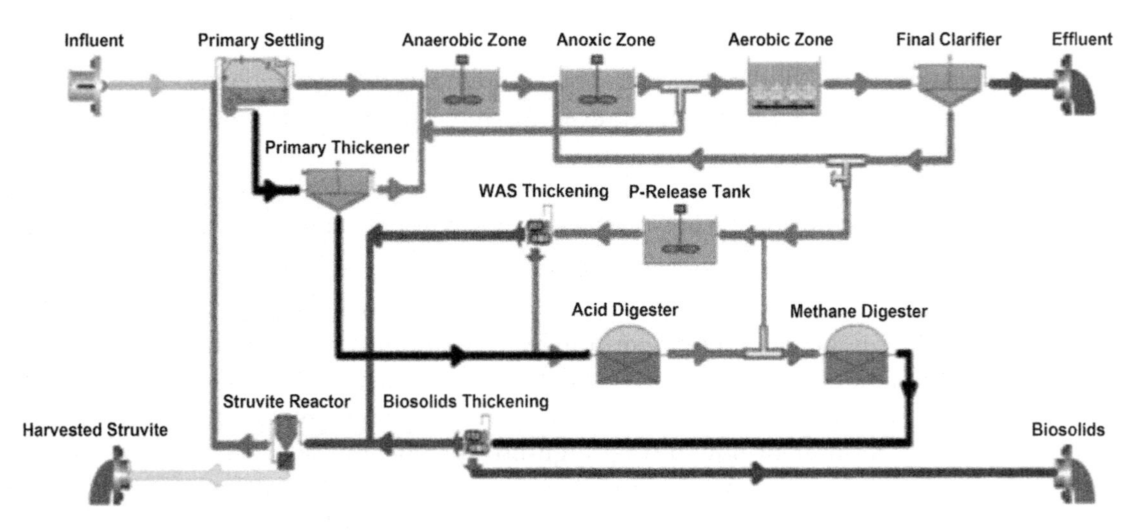

FIGURE 17.8 Nine Springs facility process schematic with EBPR and future phosphorus recovery (courtesy of the Madison Metropolitan Sewerage District).

2.3 Case Study 3—F. Wayne Hill Water Resources Center, Gwinnett County, Georgia

2.3.1 *Facility History*

The F. Wayne Hill Water Resources Center (FWHWRC), located in Buford, Georgia, began operation in March of 2001 as a 75-ML/d (20-mgd) advanced water resource recovery facility (WRRF). An upgrade and expansion to 225 ML/d (60 mgd) was completed in 2006. The FWHWRC is Gwinnett County's largest and most advanced WRRF.

2.3.2 *Process Description*

Existing unit processes include screening and grit removal, primary clarifiers, three- and five-stage BNR activated sludge basins, and secondary clarifiers. After secondary treatment, the effluent is split into two treatment trains. The first treatment train is rated for 75 ML/d (20 mgd) and includes tertiary chemical clarification and granular media filtration. The second treatment train is rated for 150 ML/d (40 mgd) and includes chemical flocculation and clarification and membrane filtration. The effluent is combined and treated through preozonation, granular activated carbon filtration, and ozone disinfection.

The FWHWRC includes 10 activated sludge basins. Bioreactors 1 to 4 are five-stage processes with anaerobic, primary anoxic, aerobic, secondary anoxic and swing (currently aerated), and reaeration zones. The newer basins, Bioreactors 5 to 10, have three stages: anaerobic, anoxic, and aerobic zones. Figures 17.9a and 17.9b show the process flow diagrams for the facility. Solids handling includes mechanical WAS thickening, anaerobic digestion, chemical solids gravity thickeners, and dewatering centrifuges. Waste activated sludge is mechanically thickened on centrifuges and pumped to the digesters, while primary sludge is pumped directly to the digesters. After digestion, the sludge is stored in a holding tank before being dewatered.

The final effluent is discharged at two locations, the Chattahoochee River and Lake Lanier. The facility has the flexibility to pump 75 ML/d (20 mgd) of effluent approximately 20 miles, combine it with effluent from the Crooked Creek Water Reclamation Facility, and discharge it to the Chattahoochee River. The remaining 150 ML/d (40 mgd) is pumped to Lake Lanier. The river discharge must meet the "metro limits" developed for Chattahoochee River discharge in the metropolitan Atlanta area. The Lake Lanier discharge must meet the following, more stringent National Pollutant Discharge Elimination System (NPDES) monthly effluent limits:

- Carbonaceous oxygen demand, 18 mg/L;
- Total suspended solids, 3 mg/L;
- Ammonia-nitrogen, 0.4 mg/L;

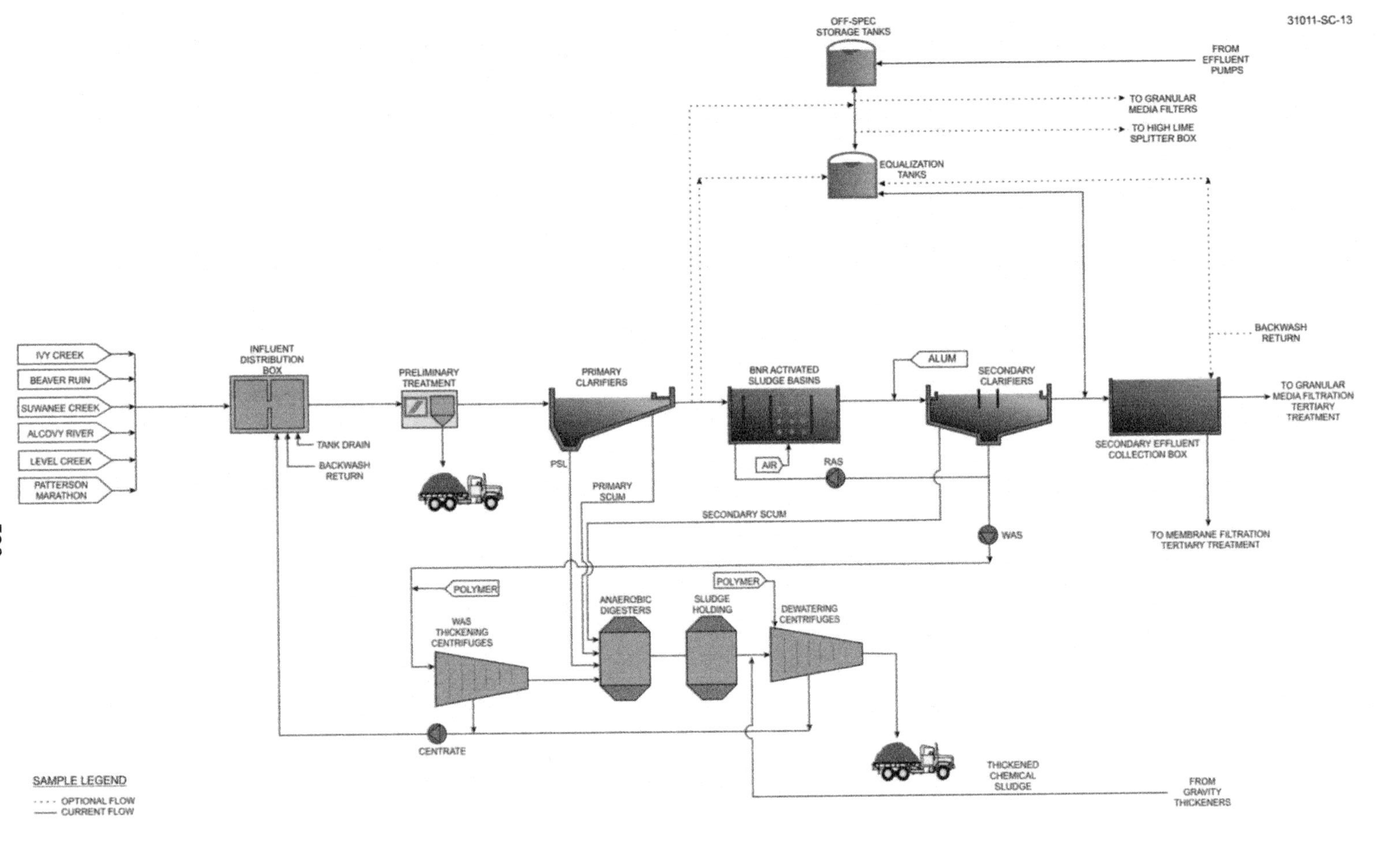

FIGURE 17.9A F. Wayne Hill Water Resources Center BNR process flow diagram (courtesy of Gwinnette County).

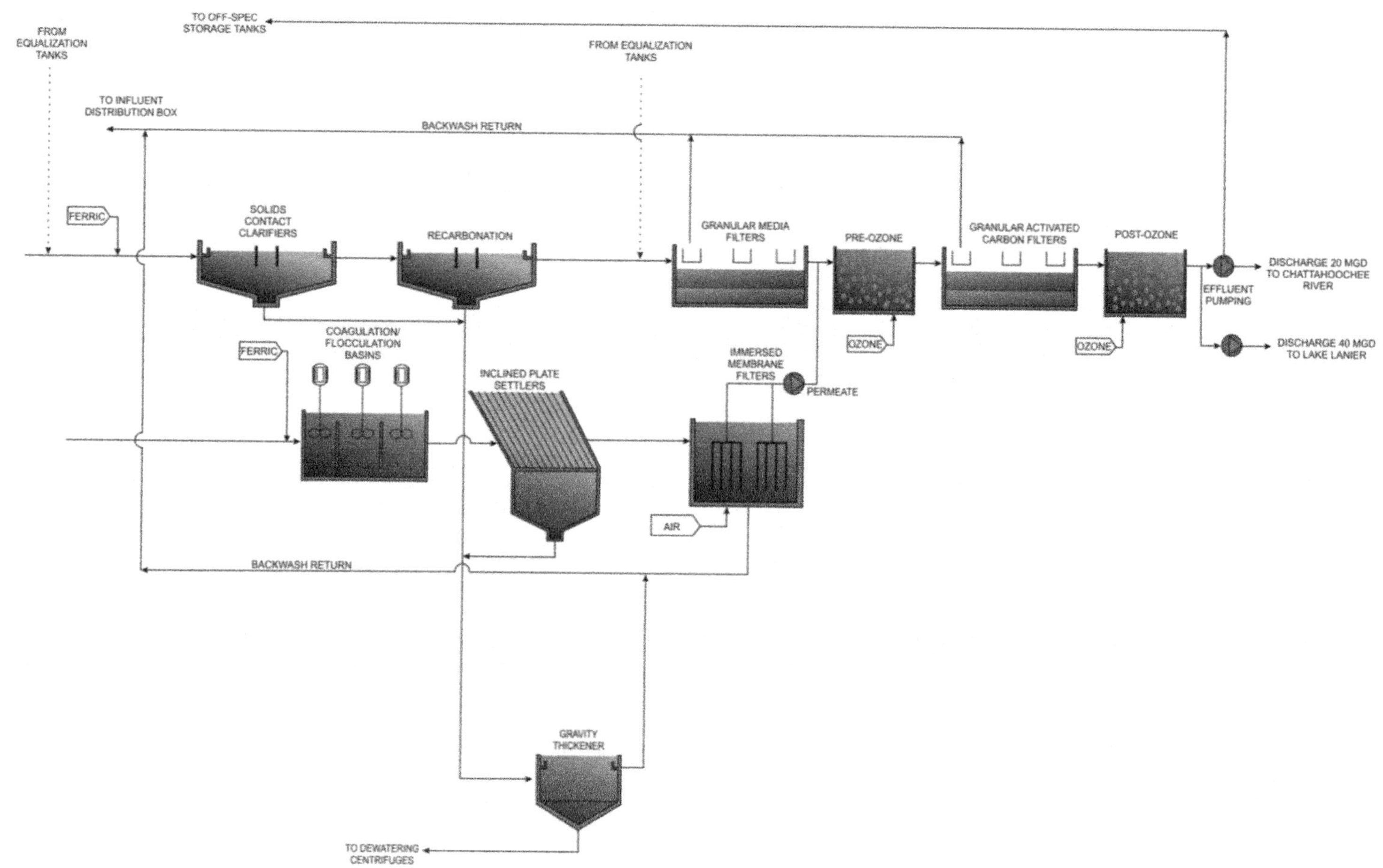

FIGURE 17.9B F. Wayne Hill Water Resources Center tertiary treatment process flow diagram (courtesy of Gwinnette County).

- Total phosphorus, 0.08 mg/L;
- Fecal coliforms, two per 100 mL;
- Turbidity, 0.5 NTU;
- Dissolved oxygen greater than 7.0 mg/L;
- pH of 6.0 to 9.0; and
- Temperature change of 2 °F.

2.3.3 *Operating Data*

The average daily wastewater flow has steadily increased since the FWHWRC came online in March 2001 (Figure 17.10). The increase in flow is largely attributed to the closing of smaller WRRFs and offloading the Yellow River Water Reclamation Facility for construction and reallocating its flow to FWHWRC.

Phosphorus removal at FWHWRC has been consistently good, averaging 0.05 mg/L in the final effluent (Figure 17.11). The average required alum dosage of 20 mg/L [as $Al_2(SO_4)_3 \times 14H_2O$] is relatively low for this effluent quality because of highly efficient EBPR. The alum is added to the mixed liquor just upstream of the secondary clarifiers. Ferric chloride dosing to the tertiary clarifiers is also low at 4 mg/L. The multiple filtration steps in the tertiary process lead

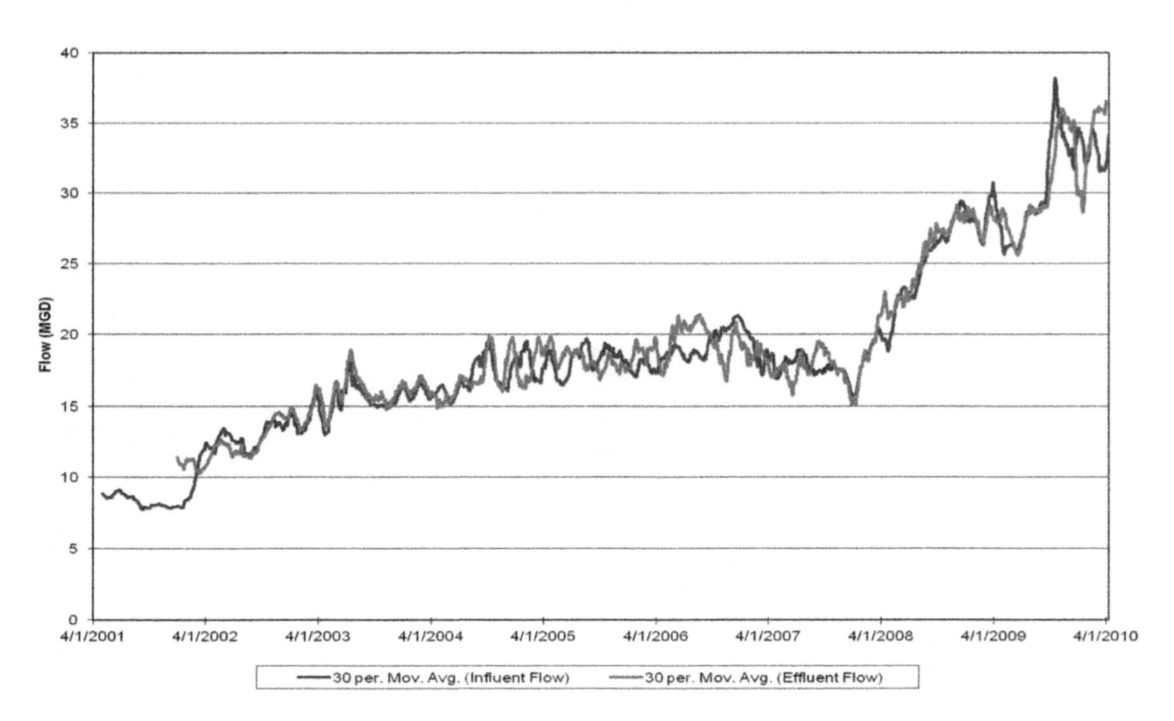

FIGURE 17.10 F. Wayne Hill Water Resources Center average influent and effluent flowrates (courtesy of Gwinnette County).

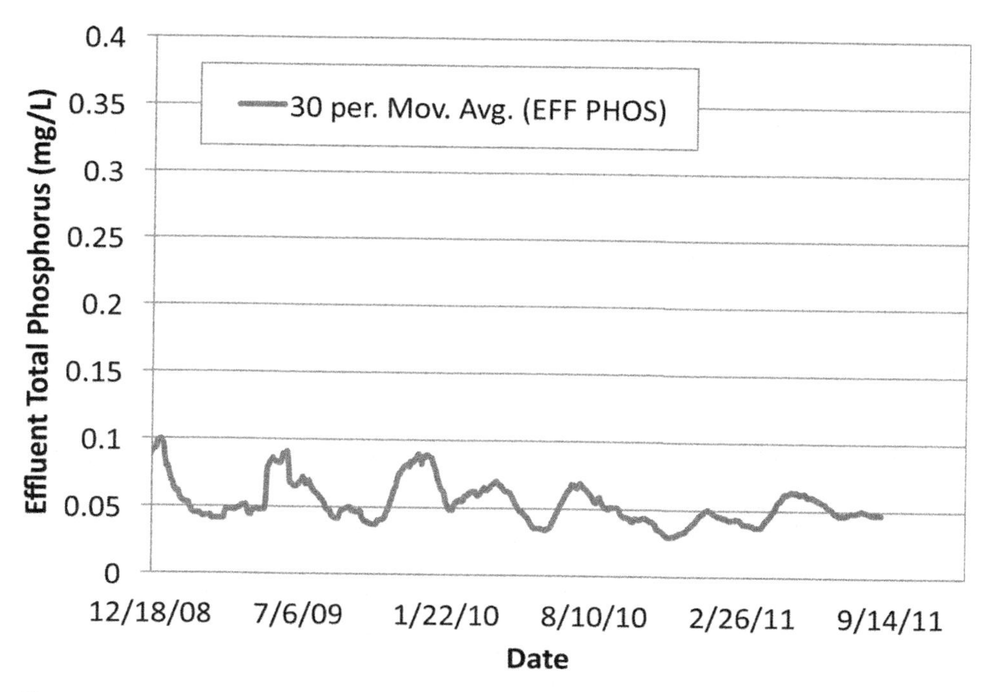

FIGURE 17.11 F. Wayne Hill Water Resources Center tertiary effluent total phosphorus concentrations (courtesy of Gwinnette County).

to an essentially TSS-free effluent, which further optimizes system phosphorus removal. The FWHWRC nitrification performance has been consistently good, with an average effluent ammonia-nitrogen concentration of 0.15 mg/L; in addition, only 2% of the data points exceed 1 mg/L and only a few data points exceed the monthly permit limit of 0.5 mg/L.

2.3.4 *Operating Experience*

While the FWHWRC has consistently met its stringent effluent quality requirements, the following issues must be addressed in facility design and operation:

- Bioreactor configuration—operators at FWHWRC have optimized bioreactor operation in terms of swing zone flexibility and internal recycle location to maximize EBPR. The mixed liquor leaving the bioreactors has a relatively high nitrate concentration (between 8 and 15 mg/L as nitrogen). Facility staff have set relatively low RAS pumping rates at between 30 and 40% of AAF. Sludge blankets in the secondary clarifiers are maintained in the 0.9- to 1.2-m (3- to 4-ft) range, which results in full denitrification of RAS and phosphorus release in the secondary clarifiers and RAS. Facility staff have attempted to balance denitrification and phosphorus release in RAS. Periodic nutrient profiles are conducted to verify BNR performance (Figure 17.12);

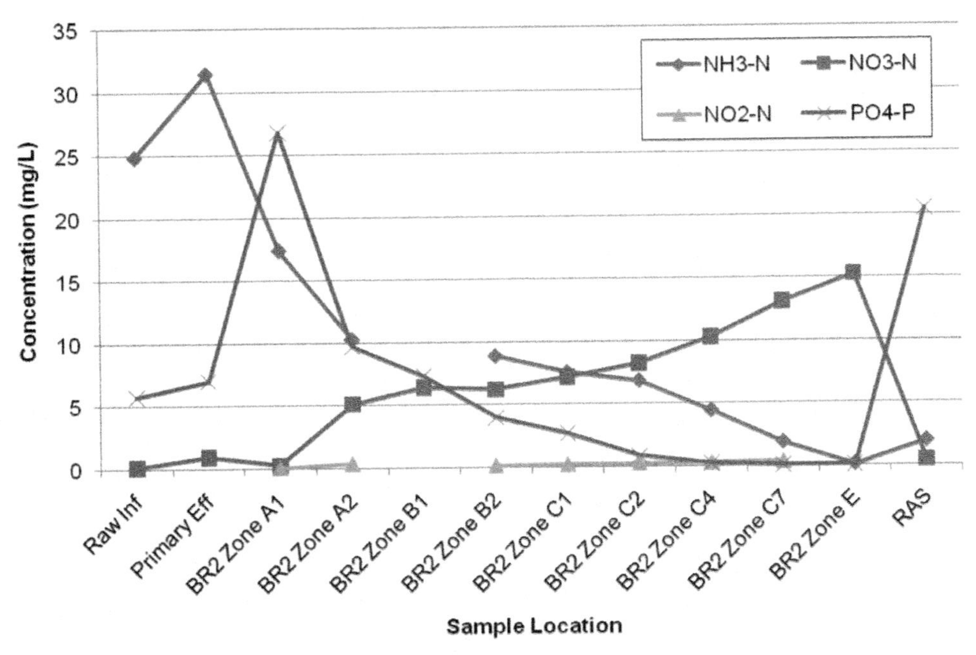

FIGURE 17.12 F. Wayne Hill Water Resources Center typical bioreactor phosphorus and nitrogen profile (courtesy of Gwinnette County).

- Struvite formation—the phosphorus concentration in centrate return from the anaerobic digestion process is low at FWHWRC compared to other EBPR facilities. This is because significant amounts of struvite in the sludge cake serve as a sink for the phosphorus released in anaerobic digesters. This actually provides a significant benefit to the FWHWRC through phosphorus recycle load reduction, but results in significant operational headaches with clogged centrate pipes and so on. Gwinnett County is planning to install a nutrient recovery process combined with intentional phosphorus release from the WAS using the WASSTRIP process to precipitate struvite in a controlled manner, reducing the nuisance struvite formation and maintaining a low phosphorus concentration in the centrate return stream;
- Sludge thickening capacity—the FWHWRC is in the process of replacing the thickening centrifuges with rotary drum thickeners (RDTs) for co-thickening of primary sludge and WAS to increase thickening and digestion capacity; and
- Fats, oil, and grease (FOG) and high-strength waste (HSW) receiving facility—the FOG/HSW truck receiving station allows the injection of high strength wastes directly into the digesters to increase biogas production.

2.4 Case Study 4—Durham Advanced Wastewater Treatment Plant, Tigard, Oregon

2.4.1 *Facility History*

Clean Water Services (CWS) in Hillsboro, Oregon, uses an effective combination of advanced wastewater treatment and innovative watershed management to ensure that the water quality of the nearby Tualatin River is protected in a sustainable, cost-effective manner. Recently, CWS has faced several challenges including rapid growth that has resulted in a significant increase in flows and loads within the service area of its Durham Advanced Wastewater Treatment Plant located in Tigard, an uncertain regulatory environment in which more restrictive effluent limits may be implemented, and increased pressure to provide more economical and sustainable wastewater treatment.

In 1988, CWS received the first phosphorus total maximum daily load (TMDL) permit in the United States. The agency's 100-ML/d (25-mgd) Durham facility has an effluent total phosphorus permit limit of 0.11 mg/L during the 6-month summer nutrient removal season (May 1 to October 31). Clean Water Services also received the first watershed-based NPDES permit in the United States in 2004. This watershed-based permit integrates the four treatment facilities in the CWS district into one permit mechanism. Only two of the four treatment facilities operate during the summer nutrient removal season and both facilities must meet phosphorus discharge limits. The Durham facility was constructed in a series of contracts starting in 1976. The effluent criteria for the facility are as follows:

- Five-day BOD, 5 and 10 mg/L (during summer and winter, respectively);
- Total suspended solids, 5 and 10 mg/L (during summer and winter, respectively);
- Total phosphorus, 0.11 mg/L (summer only);
- Ammonia-nitrogen, 1.0 mg/L (summer only); and
- *Escherichia coli,* 126 per 100 mL.

The effluent ammonia permit limit depends on river flow. The allowable ammonia loading becomes more stringent as the flow in the Tualatin River decreases through the summer, and the facility is designed to fully nitrify, when necessary.

2.4.2 *Process Description*

Figure 17.13 presents a process flow diagram of the Durham facility. The facility has four secondary treatment trains, each with an independent A_2/O process. The secondary effluent is treated in three tertiary treatment trains, each consisting of a mixing and coagulation tank and a tertiary clarifier. The tertiary effluent

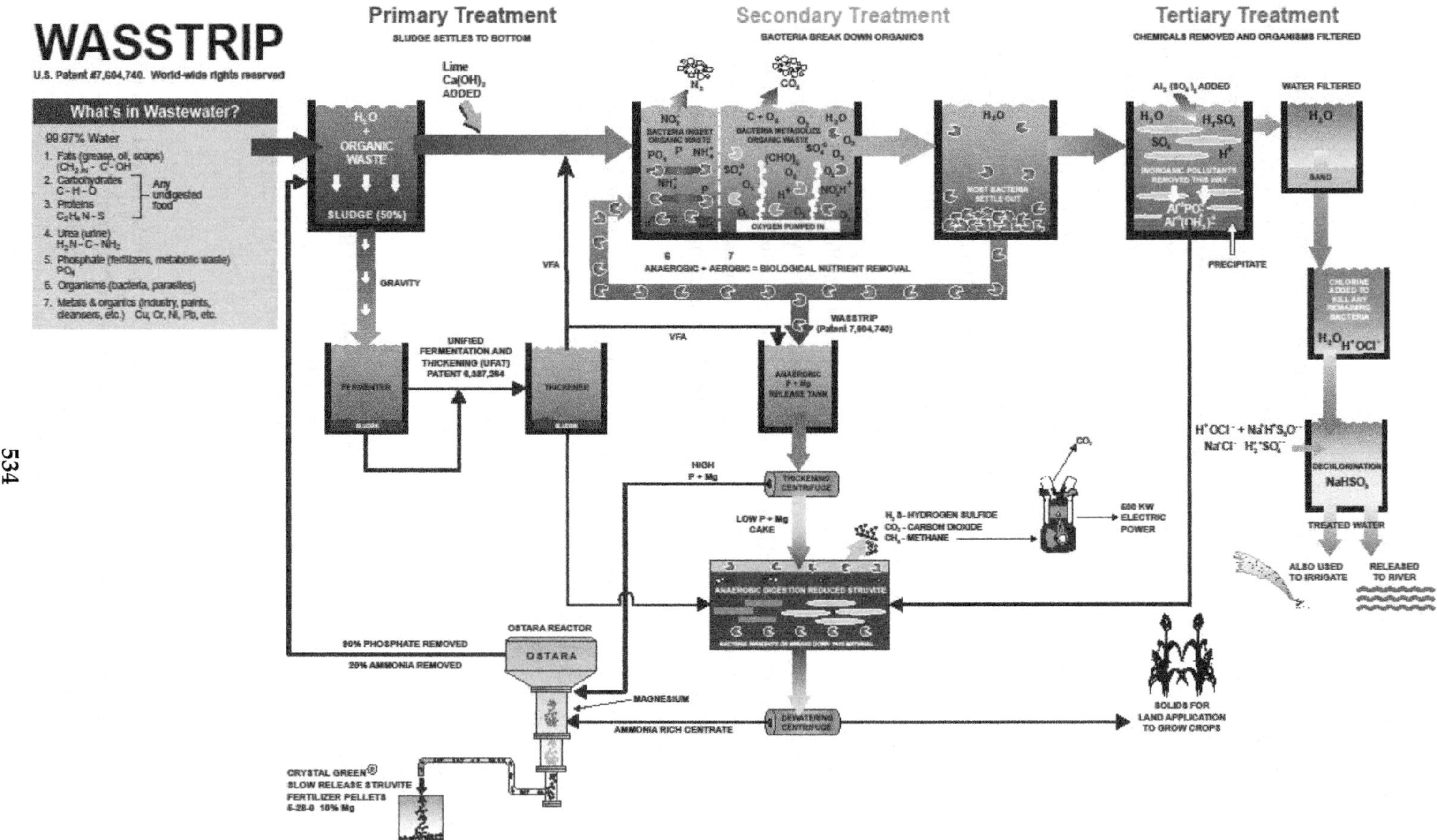

FIGURE 17.13 Process flow diagram of the Durham facility (courtesy of Clean Water Services).

is chlorinated and filtered on dual media filters. The final effluent is either discharged to the Tualatin River or sent to re-use customers.

To reduce both chemical and chemical sludge disposal costs, the facility has moved toward "greener" technologies by transitioning to EBPR. The facility has undergone several expansions and upgrades to keep pace with growth in the service area and advances in treatment technology and to meet stricter discharge standards. Alum can be added at the primary, secondary, and tertiary treatment stages. To meet its stringent effluent orthophosphorus limit, the facility initially used a 30-mg/L alum dose to the primary clarifiers and a second 30-mg/L alum dose to the tertiary clarifiers. However, when EBPR is effective, the tertiary alum dose is maintained between 10 and 20 mg/L and no alum is added to the primary clarifiers or secondary process provided that the EBPR process meets its 0.5-mg/L orthophosphorus target in the secondary effluent.

Lime is added to the primary effluent for pH and alkalinity control. The lime dosage varies and is adjusted to maintain an effluent alkalinity above 75 mg/L (as $CaCO_3$). Primary sludge is pumped to the patented Unified Fermentation and Thickening (UFAT) process (CWS, 2002), while WAS and tertiary sludge are blended in an aerated sludge storage tank before centrifuge thickening. The three sludges are blended in a digester feed before being fed to the anaerobic digesters.

2.4.3 *Operating Data*

Since the initial startup, effluent total phosphorus and ammonia-nitrogen concentrations have consistently been well below the effluent limits. Figure 17.14 shows the monthly average effluent orthophosphorus concentrations from January 2007 to December 2011. The data indicate that the facility generally meets its effluent orthophosphorus limit of less than 0.1 mg/L during the summer nutrient removal season. Data records also indicate that the facility readily meets its effluent ammonia-nitrogen limit during the summer nutrient removal season.

2.4.4 *Operating Experience*

Since it was originally commissioned in 1976, the Durham facility has been modified and upgraded to meet an increasingly stringent effluent total phosphorus standard while decreasing its reliance on chemical phosphorus removal, thus necessitating the optimization of EBPR. Many of these process upgrades were driven by facility operating staff, the most notable of which are as follows:

- Primary sludge fermentation—by converting the facility's three primary sludge gravity thickeners to the two-stage UFAT process, the facility was able to achieve EBPR in the secondary treatment process and eliminate primary and secondary alum addition. The UFAT process consists of a static fermenter and a gravity thickener in series. The static fermenter underflow and supernatant are remixed upstream of the gravity thickener. This

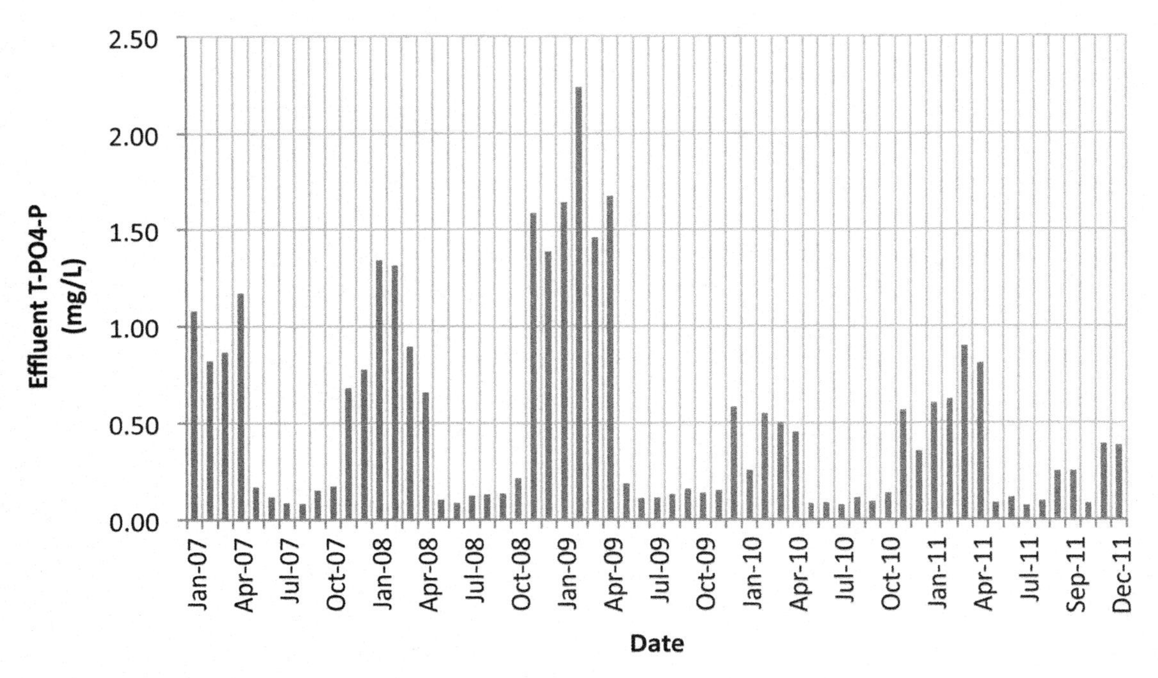

FIGURE 17.14 Durham facility effluent orthophosphorus concentrations from 2007 to 2011 (courtesy of Clean Water Services).

remixing accomplishes two tasks: it elutriates the generated VFA from the fermenter sludge blanket and strips out trapped gases. This provides a significantly higher VFA net yield compared to operating static fermenter to thickeners in parallel and delivers far superior thickening results. During summer months, the UFAT process increases the primary effluent VFA concentration by between 15 and 20 mg/L, thus maintaining a ratio of VFA to phosphorus of more than 4.0 and stabilizing EBPR. More UFAT VFA is needed to supplement the low-influent VFA during the colder months of the nutrient removal season (Benisch et al., 2009).

- Centrate treatment for phosphorus recovery—the Ostara Pearl process removes phosphorus from the dewatering centrate to reduce the phosphorus load to the EBPR process. The process uses struvite crystallization to remove approximately 85% of the orthophosphorus and 14% of the ammonia from the dewatering centrate. In 2010, struvite recovery from the centrate reduced alum use at the facility by almost 40% during the summer months. During the non-nutrient removal season, EBPR is maintained and the struvite production brings in increased revenue through the sale of Crystal Green as a slow-release fertilizer product. The Ostara process has several benefits resulting from a reduction in the phosphorus loading to the facility. These include decreased alum and lime costs, reduced chemical sludge handling costs, and increased revenue from struvite production (Baur et al., 2011).

- Waste activated sludge phosphorus stripping—the WASSTRIP process (CWS, 2009) takes the Ostara process one step further. The EBPR bacteria take up and release phosphorus along with magnesium as counter ions in the EBPR process. In the presence of VFAs, whether in the anaerobic zone of the aerators or in the digesters, polyphosphate-accumulating organisms (PAOs) release both phosphorus and magnesium. Struvite is formed in the digester in the presence of ammonia. One mode of the WASSTRIP process adds VFAs to WAS in an anaerobic reactor. Both phosphorus and the counter ion magnesium are released from bacterial cells. The WAS is thickened and the phosphorus and magnesium-rich centrate is sent to the Ostara reactor, where it is combined with ammonia-rich dewatering centrate to make struvite. The thickened WAS, stripped of releasable phosphorus and magnesium, is sent to the digester. Full-scale operation of the WASSTRIP process started in June 2011. Initial results showed 33 kg (72 lb) of magnesium removed from WAS for a reduction of 330 kg (720 lb) of struvite in the digester, and the same additional quantities available for capture in the Ostara reactors as a revenue source (Baur et al., 2011).

2.5 Case Study 5—Kurt R. Segler Water Reclamation Facility, Henderson, Nevada

2.5.1 *Facility History*

The 120-ML/d (32-mgd) Kurt R. Segler Water Reclamation Facility (KRSWRF) is a state-of-the-art BNR facility that serves the City of Henderson, Nevada. Final effluent is discharged through an outfall to the environmentally sensitive Las Vegas Wash ("Wash"), which drains into Lake Mead. The lake is an important recreational lake and source of potable and irrigation water for several communities in Arizona, California, Nevada, and Mexico. A portion of the effluent is also reclaimed and used to irrigate local golf courses, a cemetery, and highway green areas.

2.5.2 *Process Description*

The KRSWRF was constructed in a series of contracts from 1994 to 2008. The effluent criteria for the facility are as follows:

- Five-day BOD, 30 mg/L;
- Total suspended solids, 30 mg/L;
- Total phosphorus, <18 kg/d (40 lb/d);
- Ammonia-nitrogen, <53 kg/d (116 lb/d); and
- Fecal coliforms, 200 per 100 mL.

Effluent total phosphorus limits are applied from March 1 to October 31, while effluent ammonia-nitrogen limits are applied from April 1 to September 30. The KRSWRF limits are part of a total limit set for the Wash, which receives effluent from North Las Vegas, Las Vegas, and the Clark County Water Reclamation District (CCWRD). In 2001, there was an algae bloom in Lake Mead and the dischargers volunteered to meet effluent total phosphorus and ammonia-nitrogen limits during each month. The Nevada Department of Environmental Protection accepted this offer and kept the seasonal limits in the NPDES permits.

Figure 17.15 presents a process flow diagram showing the key unit processes. The liquid treatment stream includes the headworks comprised of fine screening, grit removal, and influent pumping. Three flow equalization basins are used to shave off the peak flows and allow a steady flow to each complex. The flow is then divided into the East and West Complexes. The facility does not have primary treatment. The East Complex is a single-train, 30-ML/d (8-mgd), completely

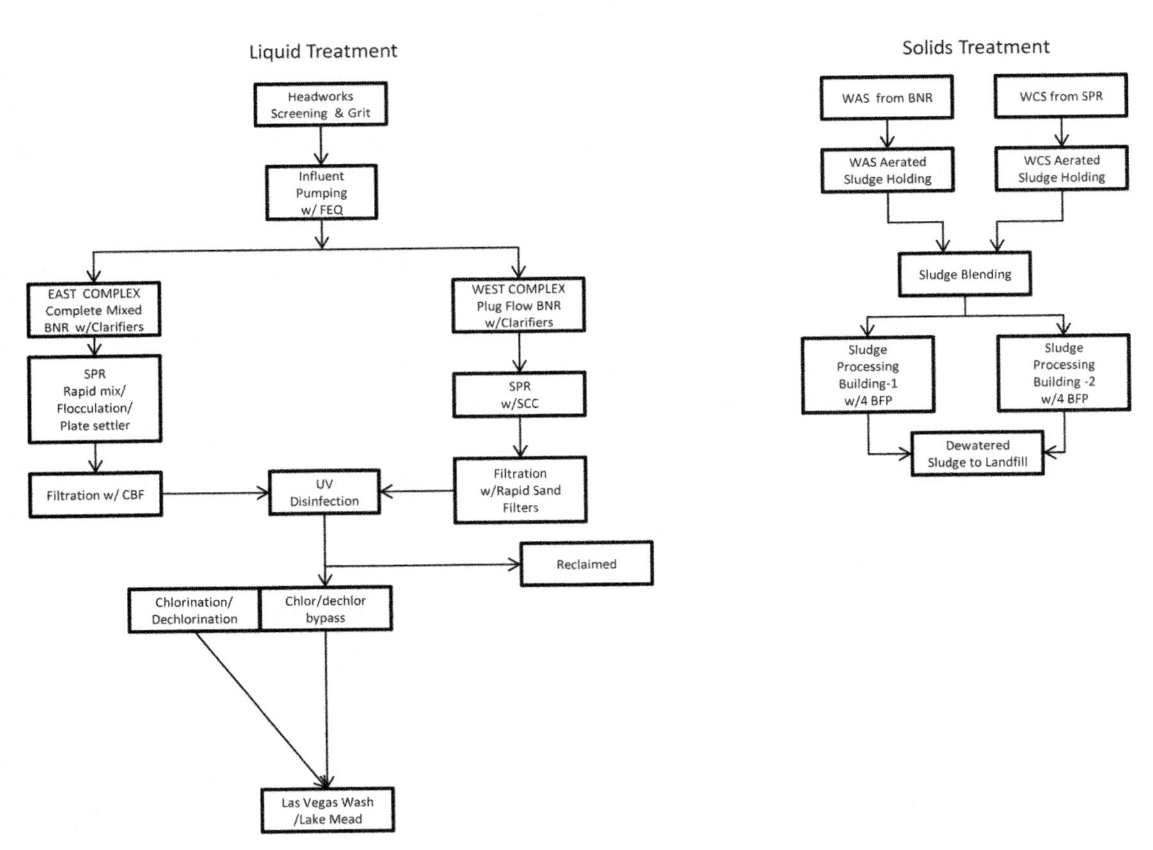

FIGURE 17.15 Kurt R. Segler facility process flow diagram (courtesy of City of Henderson).

mixed process and the West Complex is a two-train, 90-ML/d (24-mgd), plug flow process. Each complex includes supplementary phosphorus removal (SPR) using alum and polymer in separate coagulation processes followed by effluent filtration. Supplementary phosphorus removal is needed to meet the low total phosphorus TMDL limits of less than 18 kg/d (40 lb/d). Ultraviolet irradiation is used for effluent disinfection, and chlorine is available as a backup to UV. The WAS and chemical sludge are dewatered on belt filter presses (BFPs) and hauled to a landfill.

The bioreactor design is based on the modified Johannesburg (JHB) process (Figure 17.16). Both complexes use preanoxic, anaerobic, anoxic, and aerobic zones for deoxygenation, phosphorus release, denitrification, and nitrification and phosphorus uptake, respectively. The East Complex has one preanoxic zone that receives RAS for deoxygenation. The anaerobic zone that receives the influent wastewater can be comprised of two or three cells, and the anoxic zone can be comprised of three or four cells. Nitrified mixed liquor is returned to the anoxic zone for denitrification. The aerobic zone is a completely mixed basin equipped with diffused air aeration, and is followed by two secondary clarifiers. Return activated sludge is pumped back to the preanoxic zone. Alum and polymer are added to the secondary effluent in a two-cell, rapid-mix chamber. The effluent can flow to a three-cell flocculation chamber followed by a plate settler and four continuously backwashed filters (CBF) or it can be directly filtered by CBFs.

The West Complex has two 45-ML/d (12-mgd) trains. Each train has one preanoxic zone that deoxygenates RAS. The anaerobic zone can be comprised of three

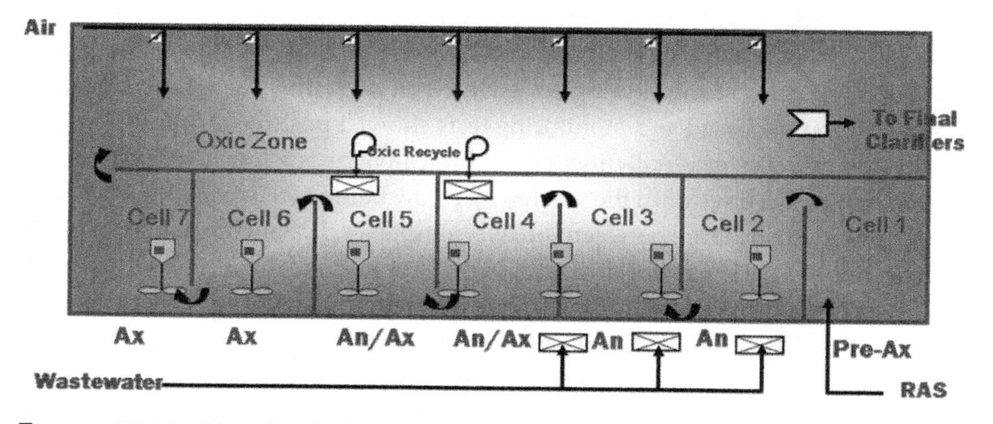

FIGURE 17.16 Kurt R. Segler facility bioreactor layout (courtesy of City of Henderson).

or four cells and the anoxic zone can be comprised of three or four cells. Nitrified mixed liquor is returned to the anoxic zone for denitrification. The aerobic zone is a plug flow process in an oxidation-ditch configuration. Oxygen is added to the aerobic zone using mechanical aerators and diffused-air aeration. Mixed liquor from the two aerobic zones is combined and flows to four secondary clarifiers. The combined RAS is divided between the two trains. Alum and polymer are added to the secondary effluent, and the water is pumped to four solids contact clarifiers followed by 16 mono-media rapid sand filters. Filter backwash is returned to the headworks. Filtered effluents from the East and West Complexes are combined and flow through two UV disinfection channels with two banks of lamps in each channel. Following UV disinfection, the water can flow via gravity or be pumped to reclaimed water pumping stations for distribution. Water that is not reclaimed flows to the Wash.

The modified JHB process configuration has several unique design features, including multiple wastewater feed points to the preanoxic and anaerobic zones. The mixer in Anaerobic Zone 2 is operated intermittently to allow mixed liquor settlement and fermentation to occur in this cell (Barnard et al., 2010). The WAS and waste chemical sludge (WCS) are pumped to aerated sludge holding basins (ASHB). The WAS and WCS are held in separate ASHBs. The two sludges flow into a blend tank and the blended sludge is pumped to two solids process buildings. Four BFPs in each building are used to dewater the sludge. Polymer is added to the sludge before the BFPs. The dewatered sludge drops into trailers and is hauled to a landfill for disposal.

2.5.3 *Facility Commissioning*

The BNR process for the East and West Complexes was commissioned in phases because of modifications related to the construction project. Activated sludge was transferred from ASHBs or from complex to complex to seed the phase being commissioned. Initially, half of the expected MLSS concentration was added to the activated sludge process. Wastewater was introduced at 50% of the train's capacity and process monitoring tests were run each day. As the mixed liquor concentration increased and monitoring tests showed positive results, the wastewater flowrate was gradually increased. Typically, the wastewater flowrate was increased every 2 days until the wastewater flow being received by the KRSWRF was properly distributed between the two complexes. Enhanced biological phosphorus removal was not initially achieved.

Before commissioning, it was discovered that the influent VFA concentration ranged from 10 to 15 mg/L. During commissioning of the East Complex, the mixer in the second anaerobic cell was shut off and operated for 15 minutes each day. This adaptation produced enough VFAs to allow phosphorus to be released in the anaerobic zone and phosphorus uptake to occur in the aerobic zone. During the first commissioning exercise, the initial secondary effluent dissolved phosphorus concentration ranged between 0.15 and 0.3 mg/L. After both complexes were fully commissioned, the alum dosage was decreased to approximately 10% of the original value. Biochemical oxygen demand, TSS, and ammonia-nitrogen limits were met without any problems during commissioning; in addition, all effluent limits are currently being met.

2.5.4 *Process Monitoring and Control*

Key parameters that are monitored and used to control the BNR process include the following:

- Wastewater flow and characteristics (BOD, TSS, VFA, NH_3, TKN, total phosphorus);
- Preanoxic, anaerobic, anoxic, aerobic zone, and supernatant characteristics (PO_4, NH_3, and NO_x);
- Secondary effluent, SPR, and filter effluent characteristics (PO_4, NH_3, NO_x, and pH);
- Aerobic zone dissolved oxygen, MLSS, and settleability;
- Activated sludge SRT; and
- Chemical dose (alum and polymer).

2.5.5 *Operating Data*

Since the initial startup of the facility, effluent total phosphorus and ammonia-nitrogen concentrations have consistently been well below effluent limits. Figure 17.17 shows monthly average effluent total phosphorus concentrations for 2009, 2010, and 2012.

2.5.6 *Operating Experience*

While KRSWRF has consistently met or exceeded its effluent quality requirements since it was commissioned, several operating problems have emerged that relate to the EBPR performance. These include

- Low influent VFA concentration—during the design phase of the BNR upgrade, the wastewater VFA concentration was tested and proved to be sufficient to support stable EBPR. The ratio of VFA to total phosphorus was greater than 5 to 1. Before commissioning, the ratio was found to be less than 3 to 1. As noted previously, the mixer in the second anaerobic zone was operated 15 minutes a day and enough VFA was produced to allow the EBPR to stabilize (Barnard et al., 2010). The mixer was started and stopped manually. Eventually, timers were added to the mixer controls to allow operators to select the time of day and the duration of the intermittent mixer operation;

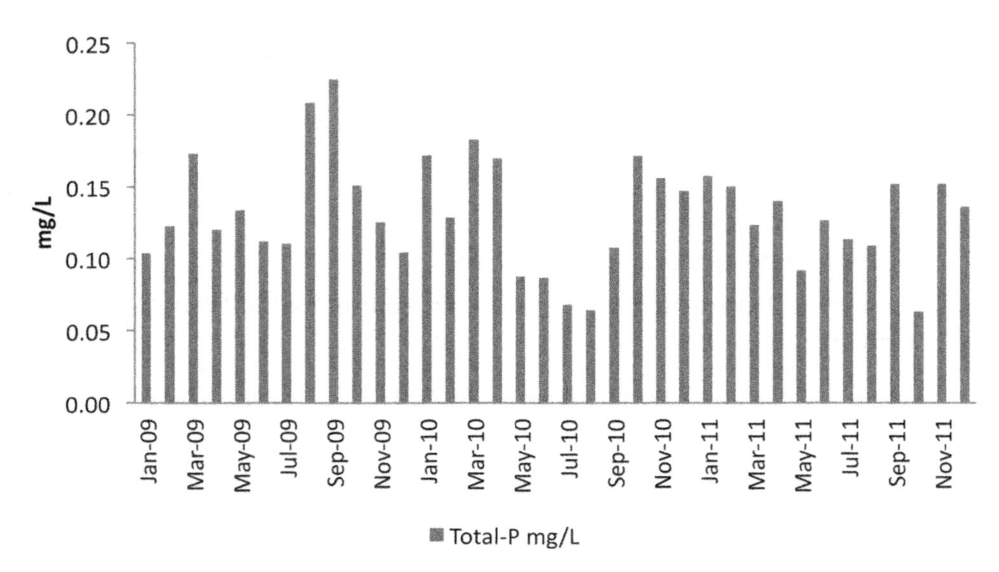

FIGURE 17.17 Kurt R. Segler facility effluent total phosphorus concentrations from 2009 to 2011 (courtesy of the City of Henderson).

- Enhanced biological phosphorus removal optimization—the submersible mixers in the preanoxic, anaerobic, and anoxic zones (23 in all) began to fail. Working with CCWRD, facility staff were able to source a reliable mixer, and found that the motor horsepower could be reduced from 9 to 4.5 kW (12 to 6 hp). The reduced power requirements of the new mixers resulted in considerable power savings for the facility. The secondary effluent orthophosphorus concentration is now consistently below 0.1 mg/L and, at times, averages 0.05 mg/L. The lower secondary effluent orthophosphorus concentrations have resulted in lower alum doses, which are, at times, less than 10 mg/L. Before making the change to the lower horsepower mixers, alum doses ranged from 25 to 35 mg/L. The reduced mixing power and alum dosage requirements resulted in operating cost savings of more than $175,000 in 2010;
- Heat-related EBPR problems—Henderson is in a desert climate. During the summer months, the ambient temperature can reach 46 °C (115 °F) and higher. Starting in June and lasting through August or September, ambient temperatures typically range between 38 and 43 °C (100 and 110 °F). When the temperature exceeds 38 °C (100 °F), the mixed liquor temperature will climb to greater than 28 °C and, at times, be 30 °C. When the mixed liquor temperature exceeds 28 °C, the competition between the PAO and glycogen-accumulating organism (GAO) populations is greatly increased. During the first year of operation, the GAOs outcompeted the PAOs, the effluent dissolved phosphorus concentration hovered around

1 mg/L, and the MLSS concentration was approximately 2000 mg/L for about 6 weeks. The next year, the MLSS concentration was lowered to 1800 mg/L and the problem lasted about 3 weeks. The following year, the MLSS was lowered to 1500 mg/L and the problem lasted about 1 week. This same phenomenon occurred again in 2011. When operating at an MLSS concentration of 1500 mg/L, the bioreactor SRT is around 3.5 days. To date, operating in this range has not overloaded solids processing and the operating values have been on target. The primary issue at the lower MLSS is overwasting from the bioreactors. The facility is manned for 10 hours a day, 7 days a week. If the WAS control system fails, there could be a problem with too much sludge being wasted. When facility staff see an increase in the effluent orthophosphorus concentration from around the 0.05- to 0.1-mg/L range to the 0.2- to 0.3-mg/L range in a 4-day period, it becomes apparent that GAOs are taking over. During these times, the alum dosage is increased so that the facility meets its effluent total phosphorus limit.

2.6 Case Study 6—Broad Run Water Reclamation Facility, Ashburn, Virginia

2.6.1 *Facility History*

The Broad Run Water Reclamation Facility is a 42-ML/d (11-mgd) facility in Ashburn, Virginia, that is owned and operated by Loudoun Water. The facility is subject to a specific state regulation, the Dulles Watershed Regulation, which requires stringent discharge standards because it is upstream of a drinking water supply in the Potomac River. The facility was commissioned in 2008. The effluent criteria for the Broad Run facility are as follows:

- Carbonaceous oxygen demand, 10 mg/L;
- Total phosphorus, 0.1 mg/L;
- Total nitrogen, 4 mg/L;
- Turbidity, 0.5 NTU; and
- *E. coli.*, 200 per 100 mL.

2.6.2 *Process Description*

The process design incorporates preliminary screening and grit removal, primary clarification, fine screening (2-mm openings), flow equalization, a membrane bioreactor (MBR), activated carbon, and effluent UV disinfection. The WAS is centrifuge thickened, combined with primary sludge, stabilized by anaerobic digestion, and centrifuge dewatered before land application. The MBR is operated as a five-stage Bardenpho process that is modified to save aeration energy by recycling the highly oxygenated RAS to the first aerobic stage. The bioreactor is one of the application points of alum addition, but alum and polymer can also be added to the primary clarifiers for phosphorus removal and enhanced TSS removal. Methanol can be added to the second anoxic zone to meet the low effluent total nitrogen limit.

The facility has a design flow of 21 ML/d (5.5 mgd). It has three bioreactors, each with a volume of 7190 m^3, and 12 membrane tanks equipped with GE/Zenon ZW500d membrane cassettes. Figure 17.18 presents a schematic of the bioreactor configuration. The large number of membrane tanks allows for in situ recovery cleaning of the membranes without significantly diminishing treatment capacity. The bioreactor volume distribution is 13% anaerobic, 13% first anoxic, 51% aerobic, 9% swing, and 14% second anoxic. The facility is located in the Chesapeake Bay watershed and is designed to accommodate a minimum wastewater temperature of 12 °C. The facility also incorporates a fully automated supervisory control and data acquisition system to optimize treatment and energy and chemical consumption.

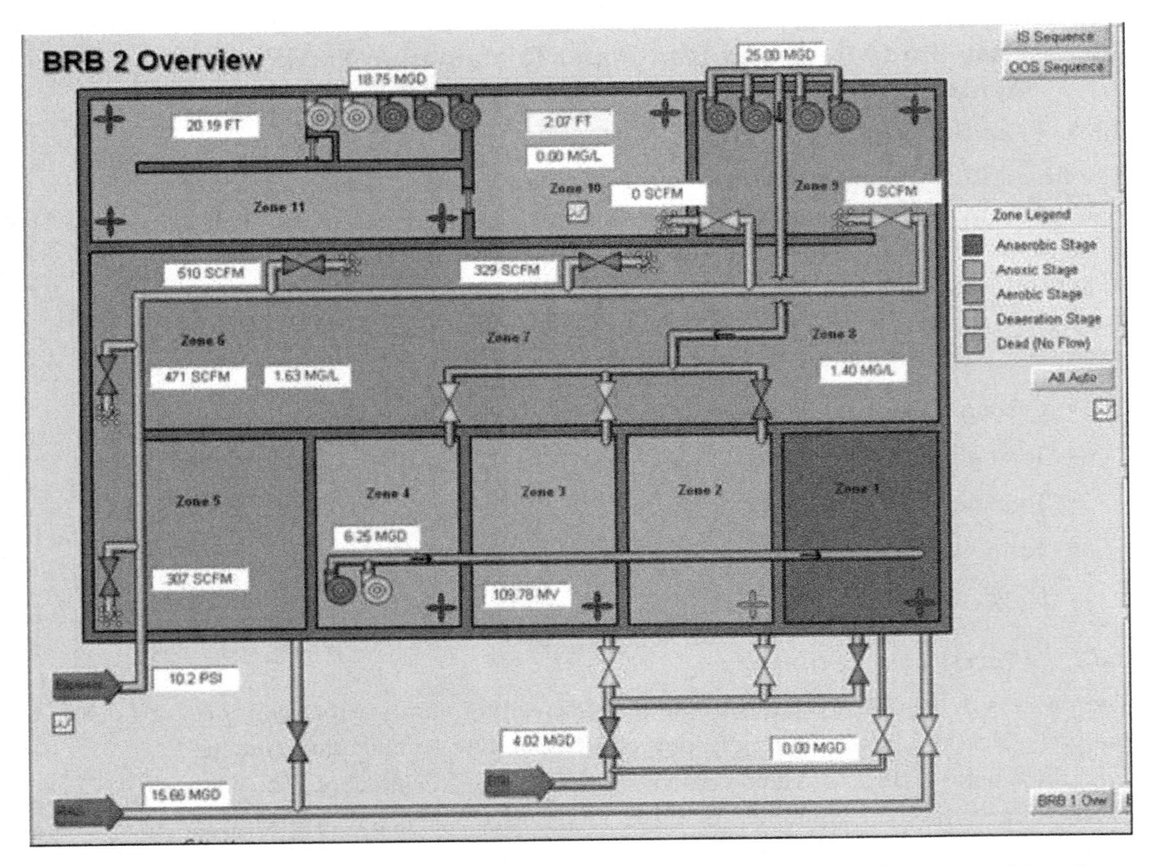

FIGURE 17.18 Overview of Broad Run facility biological reactor layout (courtesy of Loudoun Water).

2.6.3 *Facility Commissioning*

During its first year of operation, the facility treated approximately 35% of its average design flow capacity of 38 ML/d (10 mgd). The wastewater characteristics were typical for a service area with wastewater of primarily domestic origin. The average concentrations of total COD and soluble COD in the primary effluent were 214 and 110 mg/L, respectively. With an average raw wastewater total ratio of COD to TKN of less than 7 to 1 and a soluble ratio of TKN to BOD_5 of less than 3 to 1, its amenability for BEPR and biological denitrification was considered to be weak to moderate without the addition of supplemental carbon. The primary clarifiers removed more than 75% of the incoming TSS and approximately 20% of the incoming total phosphorus. The average primary effluent total phosphorus and orthophosphorus concentrations were 5.9 and 4.4 mg/L, respectively. The average primary effluent TKN and ammonia-nitrogen concentrations were 31 and 25 mg/L, respectively. Neither the thickening nor the dewatering centrates were recycled to the facility headworks during this first year of operation because the solids processing facilities were started up at a later date in 2008.

2.6.4 *Operating Data*

Figure 17.19 presents the final effluent total phosphorus concentrations and alum dosages for 2010. The data clearly show that, while EBPR in the MBR contributes

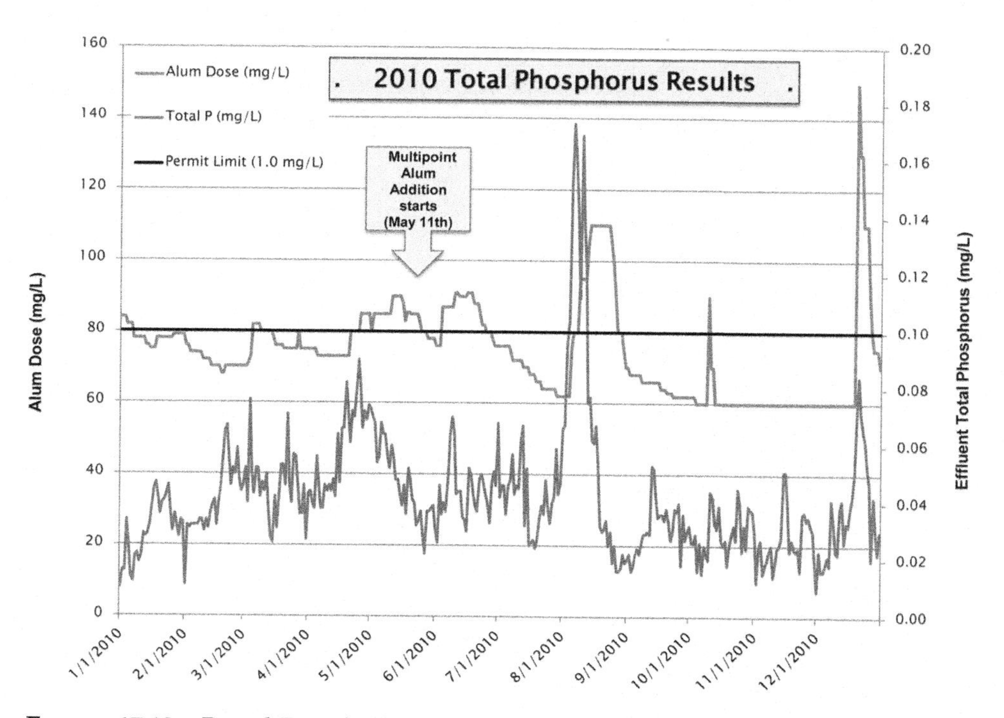

FIGURE 17.19 Broad Run facility effluent total phosphorus concentrations and alum dose for 2010 (courtesy of Loudoun Water).

to phosphorus removal, chemical addition was the primary means to meet the 0.1-mg/L effluent total phosphorus limit. Figure 17.20 presents the final effluent total nitrogen concentrations and methanol dosages for 2010. Methanol addition was stopped in June 2010, and the data clearly show that the facility is capable of meeting its effluent total nitrogen limit of 4.0 mg/L without methanol addition most of the time.

2.6.5 *Operating Experience*

During the Broad Run facility's first year of operation, numerous process modifications were made in an effort to optimize the total phosphorus and total nitrogen removal characteristics. These are described as follows:

- Flow equalization—the primary modification that improved both total phosphorus and total nitrogen removal was the equalization of flow into the MBR. This was accomplished using two 19-ML (5-mil. gal) equalization tanks to deliver a constant flow to the membrane bioreactors. Flows greater than the average daily flow were diverted to these equalization tanks after the fine screens and released from the equalization tanks to the influent pumping station during low-flow periods. Although the facility is designed to accommodate a peak flow of 71 ML/d (18.8 mgd) through

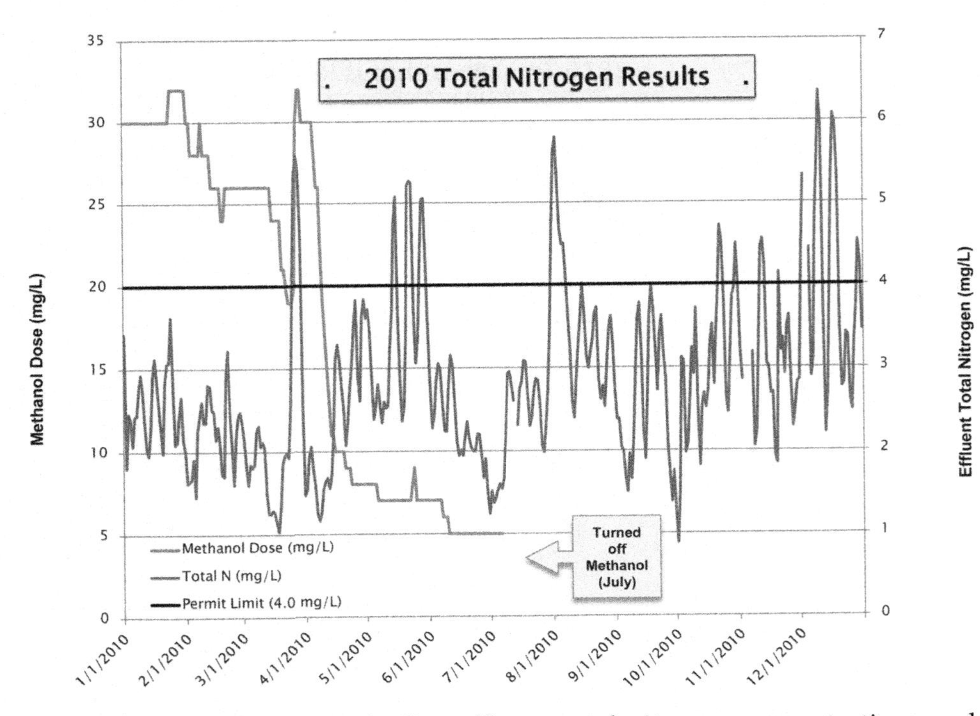

FIGURE 17.20 Broad Run facility effluent total nitrogen concentrations and methanol dose for 2010 (courtesy of Loudoun Water).

the membrane bioreactors and downstream processes, facility staff decided to equalize both the diurnal flow variations and the higher wet weather flows to equalize the ratio of influent COD to TKN that typically fluctuates under these conditions;

- Optimization for phosphorus removal—although the facility is designed to allow multipoint addition of alum, a single addition point was used throughout the first year of operation. The alum was fed into the chamber receiving flow of the mixed liquor pumps that vertically lift the mixed liquor from Zone 9 of the bioreactors approximately 3.4 m (11 ft) to the membrane tanks. During the first 3 months of operation, the orthophosphorus concentration in the mixed liquor upstream of the alum addition point was monitored and used to control alum dosage based on a target molar ratio of aluminum to phosphorus between 2 and 4 to 1. However, inconsistent analytical results of orthophosphorus concentrations in the mixed liquor, potentially related to the concentration of solids and foam in the mixed liquor sample, caused erratic changes in the dosage rate. From then on, the facility began controlling the alum dose based solely on the membrane permeate orthophosphorus concentration. By August 2008, the high alum dose had stabilized total phosphorus removal, but had also increased the MLSS concentration from 5000 to 9000 mg/L because of the increased alum solids. While low alum dosages to the mixed liquor are known to improve membrane filterability of mixed liquor, excessive mixed liquor alum solids reduces the permeability of the membranes and increases the frequency and complexity of membrane recovery cleanings. To control membrane filtration performance, facility staff took steps to reduce the inert solids concentration in MLSS by increasing the wasting rate and step-wise lowering the alum dose. Facility staff carefully monitored this reduction in alum dosage to ensure that the effluent total phosphorus limit was met. By early October, the alum dose was less than 70 mg/L (a molar ratio of aluminum to phosphorus of approximately 1.3 to 1) and the concentration of total phosphorus in the membrane permeate were consistently less than 0.05 mg/L. The alkalinity was controlled to be greater than 50 mg/L (as $CaCO_3$). The contribution of EBPR to the system total phosphorus removal was suggested by anecdotal evidence as the increase in the orthophosphorus concentration in the membrane permeate that occurred twice during the fall of 2008 after dissolved oxygen control problems caused an increase in the nitrate concentration in the first anoxic zone and in the membrane permeate. The relatively low average molar ratio of aluminum to phosphorus at 99% total phosphorus removal with single-point alum addition further suggested that EBPR was contributing to the system phosphorus removal in a significant manner. After optimization, the molar ratio of aluminum to phosphorus was further reduced to less than 1.3 to 1 until early

November, while warm waters favored biological activity and the SRT had not yet been increased in preparation for winter. In December, the SRT was increased from 14 days to 21 days to stabilize nitrification during the cold weather. The increase in SRT was expected to cause a reduction in EPBR and, indeed, an increase in the molar ratio of aluminum to phosphorus to approximately 1.8 to 1 was required to meet the 0.1-mg/L effluent total phosphorus limit;

- Optimization for nitrogen removal—complete nitrification is particularly important because the final effluent must meet the TKN limit of 1.0 mg/L. To optimize nitrification, the dissolved oxygen concentration in the bioreactor aerobic zones is controlled at a minimum of 1.0 mg/L, and a minimum alkalinity requirement of 50 mg/L (as $CaCO_3$) was established. Sodium hydroxide is added to the bioreactor influent flow to maintain this minimum level. Facility staff closely monitor alkalinity throughout the year because denitrification rates vary with temperature and alum addition rates vary depending on the dosage needed to meet the effluent total phosphorus limit; and
- Aeration control—an automatic dissolved oxygen control system is used to maintain the dissolved oxygen concentration between 1.5 and 3.0 mg/L in the aerobic zones. Control system effectiveness was increased by regular cleaning of the dissolved oxygen probes by facility staff, by fine tuning PLC-based logic using a most-open-valve protocol, and, finally, by shutting down the aeration diffusers in the last aerobic zone (Zone 9).

2.7 Case Study 7—Metro Wastewater Treatment Plant, St. Paul, Minnesota

2.7.1 *Facility History*

Metropolitan Council Environmental Services (MCES) treats wastewater from the Minneapolis–St. Paul Area. The Metro facility, which is the largest of MCES's seven treatment facilities, has a rated annual average of 946 ML/d (250 mgd) and discharges effluent to the Mississippi River. Its current NPDES permit includes an effluent total phosphorus limit of 1.0 mg/L (annual average) and seasonal nitrification. In a series of projects between 1997 and 2004, the facility was upgraded from step-feed nitrification to an EBPR process. The change to EBPR involved several years of engineering studies and full-scale demonstration testing of the proposed process changes (Block et al., 2008).

2.7.2 *Process Design*

Figure 17.21 presents a process flow diagram of the Metro facility. The 16 existing four-pass aeration tanks were converted to highly flexible bioreactors that can be operated in a number of EBPR process configurations, including RAS reaeration

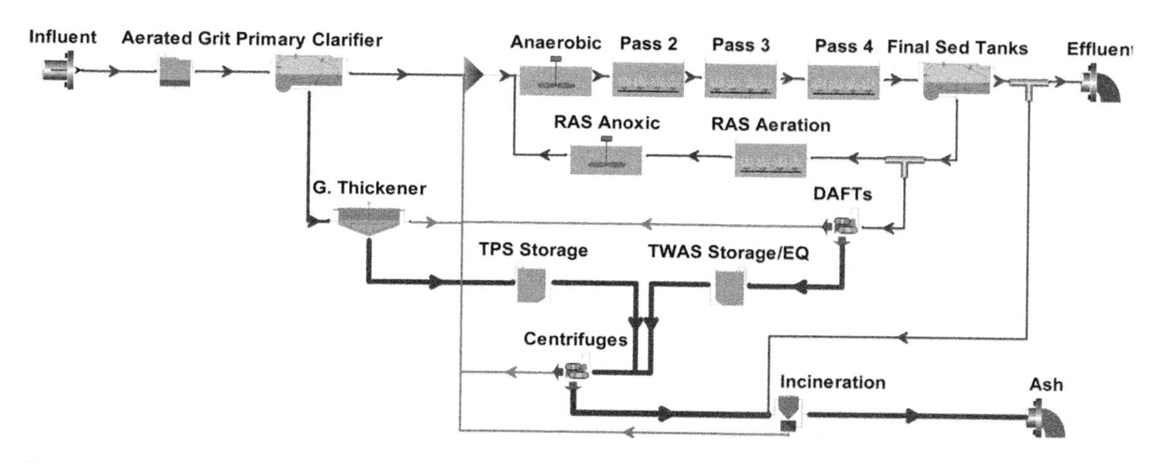

FIGURE 17.21 Metro facility process flow diagram (courtesy of Metropolitan Council Environmental Services).

and RAS denitrification. To facilitate the variable-mode selector, structural baffle walls were used to partition the first two passes of each bioreactor into five zones, some of which are swing zones equipped with both fine-bubble diffusers and mechanical mixers (Figure 17.22). The process design also provides multiple

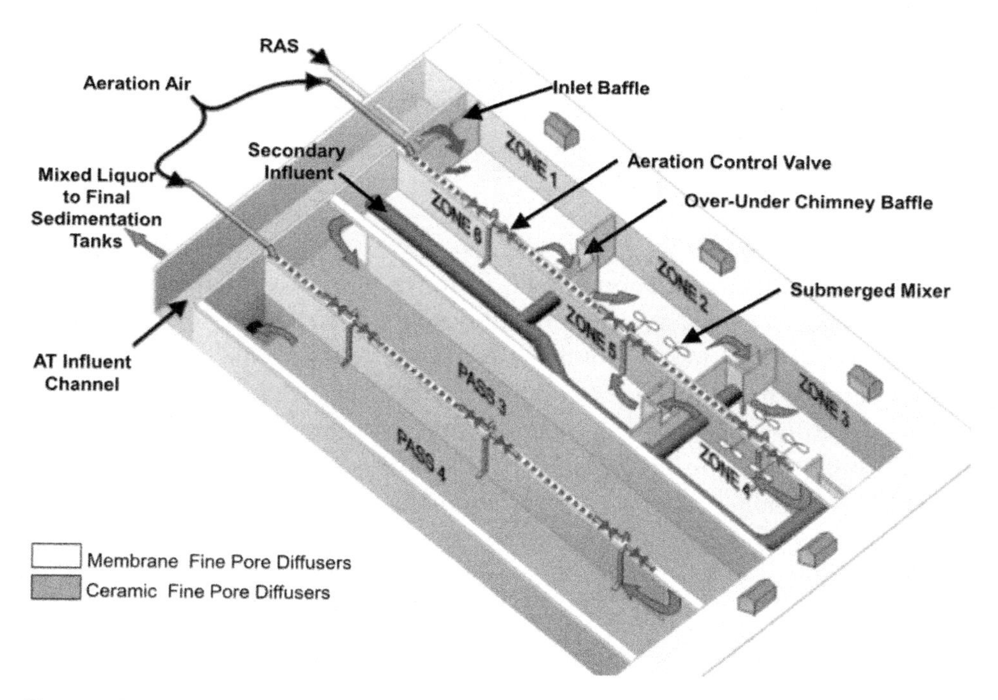

FIGURE 17.22 Metro facility bioreactor layout (courtesy of Metropolitan Council Environmental Services).

influent feed points to accommodate the various operating modes. The effluent criteria for the Metro facility are as follows:

- Carbonaceous BOD_5, 14 mg/L (June);
- Carbonaceous BOD_5, 10 mg/L (July to September);
- Carbonaceous BOD_5, 24 mg/L (October to May);
- Total suspended solids, 30 mg/L;
- Total phosphorus, 1.0 mg/L (annual average);
- Ammonia-nitrogen, 13 and 8 mg/L (May and June, respectively);
- Ammonia-nitrogen, 5 mg/L (July to September);
- Ammonia-nitrogen, 9 and 21 mg/L (October and November); and
- Fecal coliforms, 200 per 100 mL.

2.7.3 *Process Monitoring and Control*

Key parameters that are monitored and used to control the EBPR process include the following:

- Secondary influent flow and characteristics (VFA, COD, total phosphorus, TKN, NH_3, PO_4, temperature, etc.);
- Aerobic zone dissolved oxygen and MLSS concentrations;
- Bioreactor SRT and SVI;
- Final sedimentation tank RAS flow, effluent TSS, and solids loading rate; and
- Secondary effluent characteristics (TSS, carbonaceous BOD_5, COD, total phosphorus, PO_4, NH_3, etc.).

2.7.4 *Operating Data*

The Metro facility treats annual average flows of approximately 680 ML/d (180 mgd) and average wet weather monthly flows of 795 ML/d (210 mgd). These flows are between 60 and 70% of the rated design capacity. The facility has four EBPR treatment trains that are monitored and controlled separately. The facility reliably produces a combined effluent with a total phosphorus concentration below 0.5 mg/L (Figure 17.23). Similarly, the facility reliably produces an effluent with an ammonia-nitrogen concentration of less than 5 mg/L during winter months and less than 2 mg/L during the remaining months of the year.

2.7.5 *Operating Experience*

The operating experience of the facility since its conversion to EBPR is summarized as follows:

- Process configuration—Metro staff varied the selector operating modes during the first year of EBPR operation. During this period, a difference

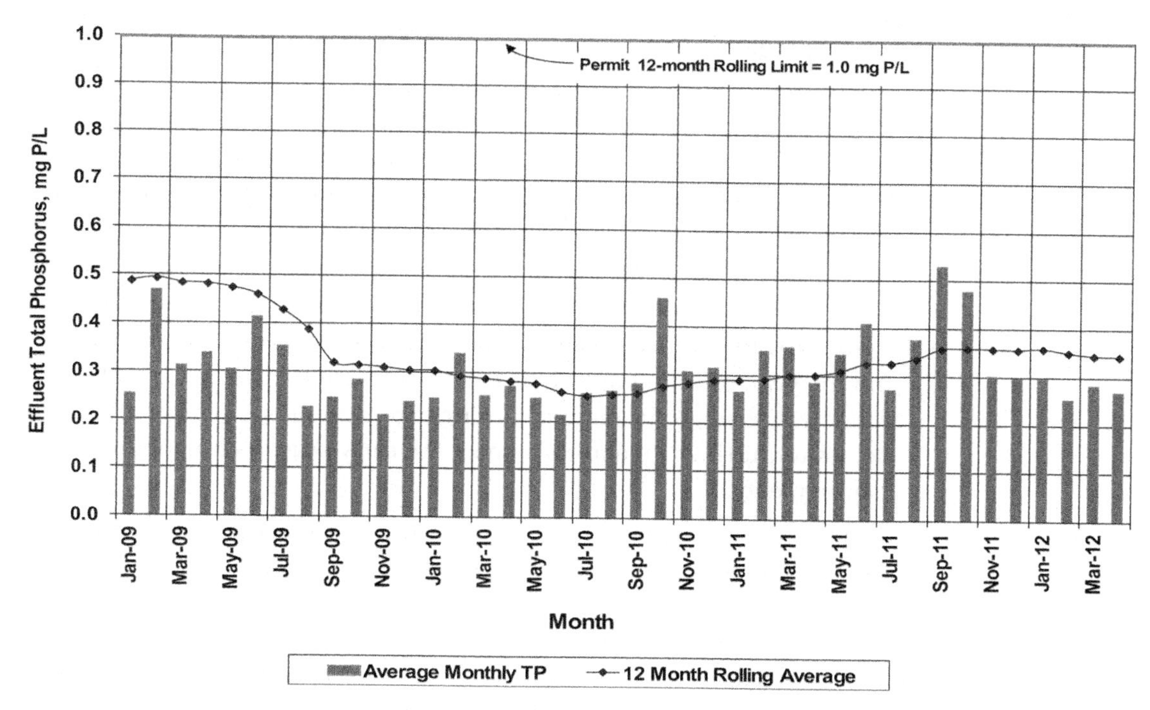

FIGURE 17.23 Metro facility effluent total phosphorus concentrations from 2009 to April 2012 (courtesy of Metropolitan Council Environmental Services).

was noted between EBPR performance in the summer and winter operating modes. The winter operating mode eliminated the RAS denitrification zone to increase the RAS aerobic mass fraction, thereby enhancing nitrification. However, operating without an endogenous RAS denitrification zone can hinder EBPR if the secondary influent contains insufficient VFAs to promote EBPR and to denitrify the RAS in the anaerobic selector. The summer mode uses the RAS aeration and denitrification zones, two anaerobic zones to promote EBPR, and the aerobic zones for nitrification and phosphorus uptake. Today, operators use only the summer mode, and phosphorus removal and nitrification remain relatively consistent throughout the year;

- Availability of VFAs—the most prominent sources of VFAs are influent sewers and gravity thickener overflow. Because no dedicated fermenter or other VFA sources exist, VFA concentrations entering the EBPR process can vary dramatically, particularly during wet weather events or periods of low wastewater temperature. Volatile fatty acid concentrations in the secondary influent range from approximately 25 mg/L (as COD) in the summer to 18 mg/L in the winter. Despite the VFA variability, the Metro facility reliably achieves total phosphorus discharges that are less than

0.5 mg/L during both summer and winter periods without an external carbon source or metal salt addition;

- Cold weather operation—to sustain a base inventory of nitrifiers during the winter, the project team developed a process-configuration matrix to vary the SRT, anoxic and anaerobic zones, and influent feed point in response to wastewater temperatures and effluent ammonia requirements. As noted previously, the facility found that the most reliable effluent total phosphorus was achieved by operating in the summer mode. At current loadings, the Metro facility operates at a total SRT of approximately 10 days throughout the year, resulting in full nitrification during non-winter months and roughly 90% ammonia removal during winter months, which provides the necessary nitrifier inventory to meet the early season effluent ammonia requirements;
- Final sedimentation tank (FST) modifications—the Metro facility has 24 Gould Type II rectangular FSTs with center sludge draw-off. Based on empirical observations and feedback from facility operators, it was determined that the tanks suffered from extensive hydraulic short-circuiting and poor solids collection. The inlet configuration of the tanks did not promote even distribution of flow across the tank width or depth and the sludge withdrawal from one side of the tank resulted in asymmetric circulatory currents. To improve the clarifiers' influent hydraulic characteristics, the team installed a staggered configuration of vertical baffling at each tank inlet. The baffles provided a small flocculation zone and promoted more uniform flow across the tank. To improve solids collection, the team installed a fixed suction header in each tank's RAS withdrawal trough. During full-scale testing, it was found that the concentration of solids withdrawn from the modified tanks was more than double that from the unmodified tanks. The improvement in sludge withdrawal also allowed the tanks to be loaded up to their full theoretical solids loading limit so the "last" half of the tank could be used for thickening and clarification. These improvements optimized clarifier performance and increased capacity, thus eliminating the need to construct new clarifiers.

2.8 Case Study 8—Kelowna Pollution Control Centre, Kelowna, British Columbia, Canada

2.8.1 Facility History

The Kelowna Pollution Control Centre (KPCC) is located in the semi-arid climate of south-central British Columbia, Canada. Effluent from KPCC is discharged to Okanagan Lake. This 145-km (90-mile) long lake has an hydraulic retention time (HRT) of approximately 75 years and represents an important drinking water source and active recreational area. In 1978, the city elected to build North

America's first full-scale BNR process to minimize eutrophication of the lake. The process design, which originated in South Africa, comprises a highly flexible and conservative bioreactor configuration to facilitate facility optimization. Initially, KPCC demonstrated the ability to remove orthophosphorus to less than 0.05 mg/L; however, facility performance was inconsistent. Consequently, facility staff embarked on 20 years of optimization and experimentation, during which time the understanding of EBPR evolved and ultimately enabled consistently low effluent phosphorus and nitrogen concentrations.

The regulator initially applied a total phosphorus discharge limit of 2.0 mg/L and encouraged process experimentation to evaluate the capabilities of the process. By 1987, total phosphorus removal had stabilized and the regulator revised the required effluent discharge requirements, as follows:

- Carbonaceous BOD_5, 7 mg/L;
- Total suspended solids, 6 mg/L;
- Total phosphorus, 0.2 mg/L (2012 annual average);
- Total nitrogen, 6 mg/L (daily limit); and
- Fecal coliforms, 200 per 100 mL.

2.8.2 *Process Description*

Figure 17.24 presents the original process flow diagram showing the key unit processes. The liquid treatment stream consisted of a headworks (comminutor and grit removal), three rectangular primary clarifiers, two multicell bioreactors based on the five-stage Bardenpho process configuration with step-feed capabilities for

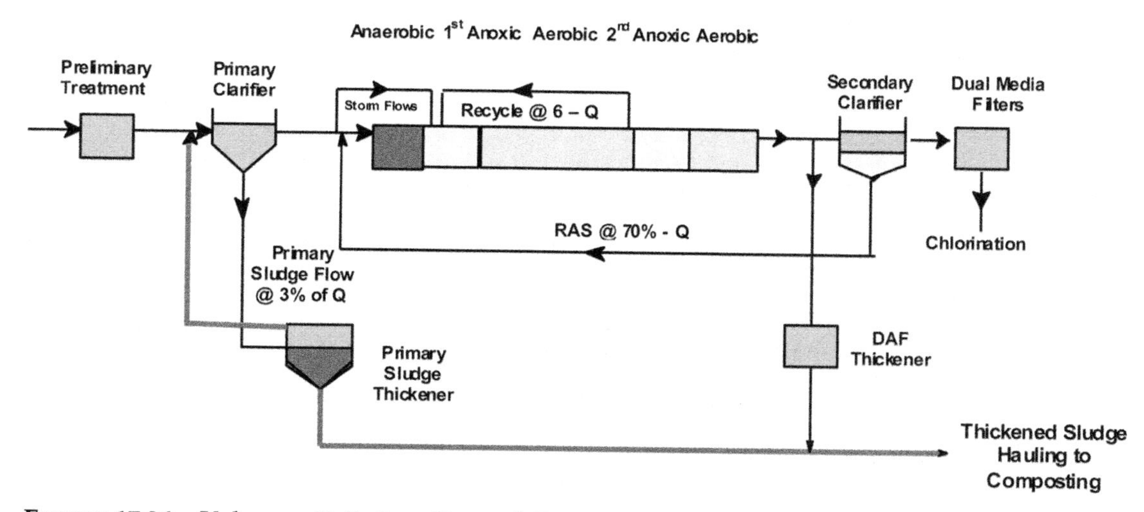

FIGURE 17.24 Kelowna Pollution Control Centre original process flow diagram (courtesy of the City of Kelowna).

storm flow conditions, three secondary clarifiers, four dual-media effluent filters, and a chlorine contact tank. The solids train consisted of DAF thickeners for removal of WAS directly from the bioreactor. Primary sludge was pumped to a gravity thickener at 3% of facility flow to prevent excessive anaerobic conditions in the thickener. Overflows from the gravity thickener and DAF were returned to the anaerobic zone of the bioreactor together with the primary effluent. Thickened WAS and primary sludge were trucked to an off-site aerobic composting process. The 21-cell bioreactor design enabled a flexible process configuration for control of anoxic and aerobic conditions. Each train was equipped with

- Three anaerobic cells;
- Four anoxic cells with fixed-speed mixers;
- Fourteen multipurpose cells, with two-speed mixers (high-speed aerobic or low-speed anoxic), submerged turbine aerators with dissolved oxygen control, and air control valves to turn off the air supply under anoxic conditions; and
- Six times AAF denitrification recycle flow between the first aerobic and first anoxic zones.

2.8.3 *Facility Commissioning*

The new KPCC was commissioned in June 1982. Nitrification and denitrification were achieved within 3 weeks and EBPR started soon thereafter. Initially, one bioreactor train was commissioned and phosphorus and nitrogen removal met permit requirements within 2 months. However, with the commissioning of the second bioreactor and by increasing the nominal HRT to more than 30 hours, nutrient removal decreased. This provided early evidence that excessive retention times in EBPR processes are not beneficial.

2.8.4 *Process Monitoring and Control*

Key parameters that are monitored and used to control the EBPR process include the following:

- Primary sludge fermenters
  - Primary sludge pumping rate;
  - Sludge blanket height/process SRT;
  - Thickened sludge pumping rate and concentration; and
  - Supernatant flowrate and characteristics (VFA, COD, and BOD); and
- Biological nutrient removal process
  - Primary effluent flow and grab and composite samples (VFA, COD, total phosphorus, and TKN);

- Cell-by-cell measurement of nutrients (orthophosphorus, NH_3, and NO_3);
- Aerobic zone dissolved oxygen controls;
- Bioreactor SRT and MLSS concentrations;
- Return activated sludge (RAS, PO_4, NH_4, and NO_3);
- Secondary effluent characteristics (TSS, PO_4, NH_4, and NO_3); and
- Effluent filter discharge and backwash grab and composite samples (TSS, PO_4, NH_4, NO_3, and total and fecal coliforms).

2.8.5 *Operating Data*

Figure 17.25 presents effluent total phosphorus and orthophosphorus concentrations for a 7-year period, which included the conversion of the Kelowna KPCC from the five-stage Bardenpho process to the three-stage Westbank process. The more stable effluent phosphorus concentrations in the later years shows that the three-stage Westbank Process demonstrated more stable operation at the shorter HRT.

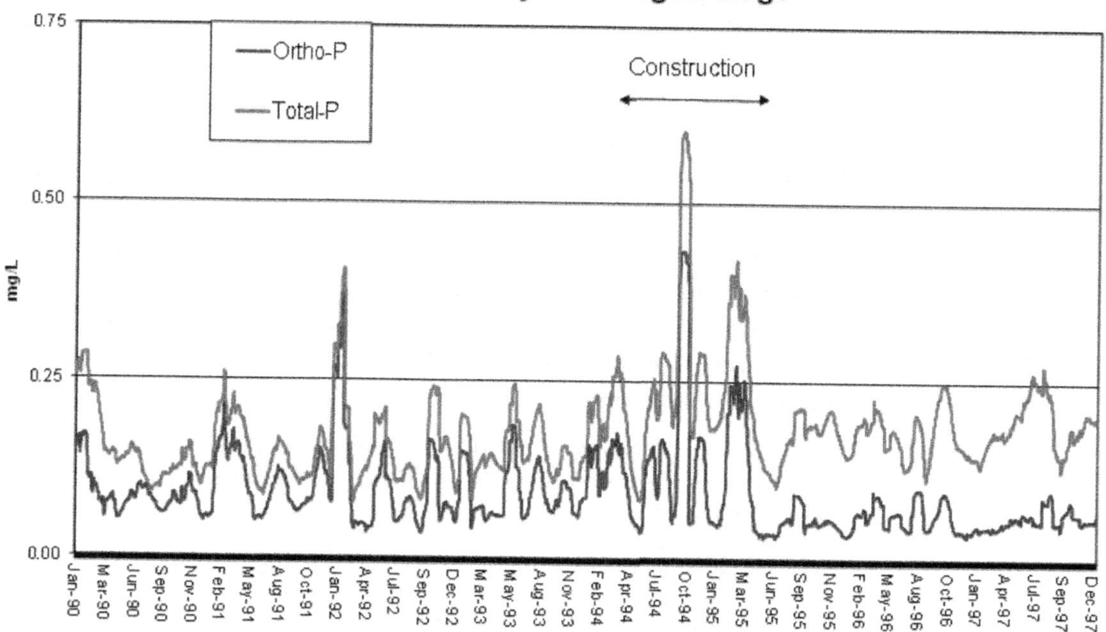

FIGURE 17.25 Kelowna Pollution Control Centre effluent total phosphorus and orthophosphorus concentrations from 1990 to 1997 (courtesy of the City of Kelowna).

2.8.6 *Operating Experience*

Process research was underway on a global scale in the early 1980s as biological and chemical phosphorus removal mechanisms were actively investigated. Research at the University of British Columbia (Vancouver, British Columbia), the University of Cape Town (Cape Town, South Africa), and the city of Johannesburg (South Africa) produced significant data to demonstrate that the EBPR removal mechanism required phosphorus release and uptake under sequential anaerobic and aerobic conditions. The key contributions of KPCC to the understanding of the EBPR mechanism are summarized as follows:

- Role of VFAs in EBPR—the KPCC was uniquely positioned to test this theory because it had a primary sludge gravity thickener in the process configuration. Approximately 250 mg/L of VFAs were being produced at a primary sludge flowrate equivalent to 3% of influent flow. Overflow from the thickeners flowed by gravity back to the primary clarifiers and entered the bioreactor anaerobic zone with the primary effluent. A simple experiment to direct the thickener overflow to only one of the bioreactors resulted in an effluent orthophosphorus concentration of 0.03 mg/L in the bioreactor receiving the thickener overflow and 3.0 mg/L in the effluent from the bioreactor that received no thickener overflow. After a 1-month period, the thickener overflow was redirected to the other bioreactor and the orthophosphorus removals quickly reversed. This provided conclusive evidence that the thickener overflow contained the VFAs responsible for EBPR (Oldham and Stevens, 1984);
- Reduction in anaerobic zone HRT—following the understanding that VFAs entering the anaerobic zone were responsible for initiating the EBPR mechanism, experiments completed at the University of British Columbia concluded that VFA uptake was essentially complete in less than 45 minutes (Comeau et al., 1986; Rabinowitz, 1985). Again, KPCC was uniquely positioned to test a step reduction in anaerobic zone retention time and to measure orthophosphorus release and uptake before effluent discharge. Experiments were conducted with 3, 2, and 1 anaerobic cells in service with the following results:
    - A 3-hour anaerobic zone retention time released approximately 25 mg/L of orthophosphorus and produced effluent orthophosphorus concentrations of 0.9 mg/L (80% removal),
    - A 2-hour anaerobic zone retention time released approximately 18 mg/L of orthophosphorus and produced effluent orthophosphorus concentrations of 0.68 mg/L (85% removal), and
    - A 1-hour anaerobic zone retention time released approximately 10 to 12 mg/L of orthophosphorus and produced effluent orthophosphorus concentrations of 0.45 mg/L (90% removal).

- Clearly, more orthophosphorus was released at longer anaerobic retention times. However, all of the orthophosphorus released could not be taken up in the subsequent aerobic zone. Therefore, it was postulated that orthophosphorus release in the presence of surplus VFAs resulted in more effective EBPR. The continued release of orthophosphorus in subsequent anaerobic cells with no external VFA source proved to be a less effective in promoting the biological phosphorus release and uptake mechanism. These two types of releases were defined as being primary and secondary phosphorus release (Barnard, 1984);
- Separation of the primary effluent and VFA source—spiral secondary clarifier mechanisms were installed in the KPCC in the 1980 expansion and this type of mechanism effectively removed all NO_x from the RAS, facilitating the denitrification of 3 to 6 mg/L of nitrogen in the clarifier without any associated rising sludge problems. In this instance, little primary effluent COD is required to protect the anaerobic zone from RAS nitrates. With RAS nitrates near zero and little COD required for RAS denitrification, experiments were begun to use the original storm flow bypass for continuous primary effluent bypass to the first anoxic zone. The diversion of a large fraction of the primary effluent COD to the anoxic zone resulted in smaller anoxic zone sizing for complete denitrification and a larger allocation of tankage to the aerobic zone for nitrification; and
- Conversion from the five-stage Bardenpho process to the three-stage Westbank process—In 1994 and 1995, the KPCC bioreactor was converted from two modules having the five-stage Bardenpho process configuration to four modules having the three-stage Westbank process configuration with step-feed capabilities (Figure 17.26). This resulted in more effective use of the bioreactor volume and more stable EBPR performance.

3.0 SUMMARY

Since the introduction of full-scale EBPR processes to North America in the early 1980s, researchers, designers, and operators have worked hand-in-hand to optimize the performance of these processes. The important role of VFAs in the EBPR mechanism cannot be overstated, and operators have endeavored to manage the available readily biodegradable carbon to stabilize system phosphorus removal. To this end, carbon management for optimum biological phosphorus and nitrogen removal in single sludge systems has become important because the PAOs and denitrifying heterotrophs compete for the same carbon sources. Furthermore, EBPR bioreactors often have the flexibility to vary the relative sizes of the anaerobic, anoxic, and aerobic zones so that phosphorus and nitrogen removal can be optimized under changing operating conditions. Use of primary sludge fermentation remains popular in temperate climates, where VFAs naturally present

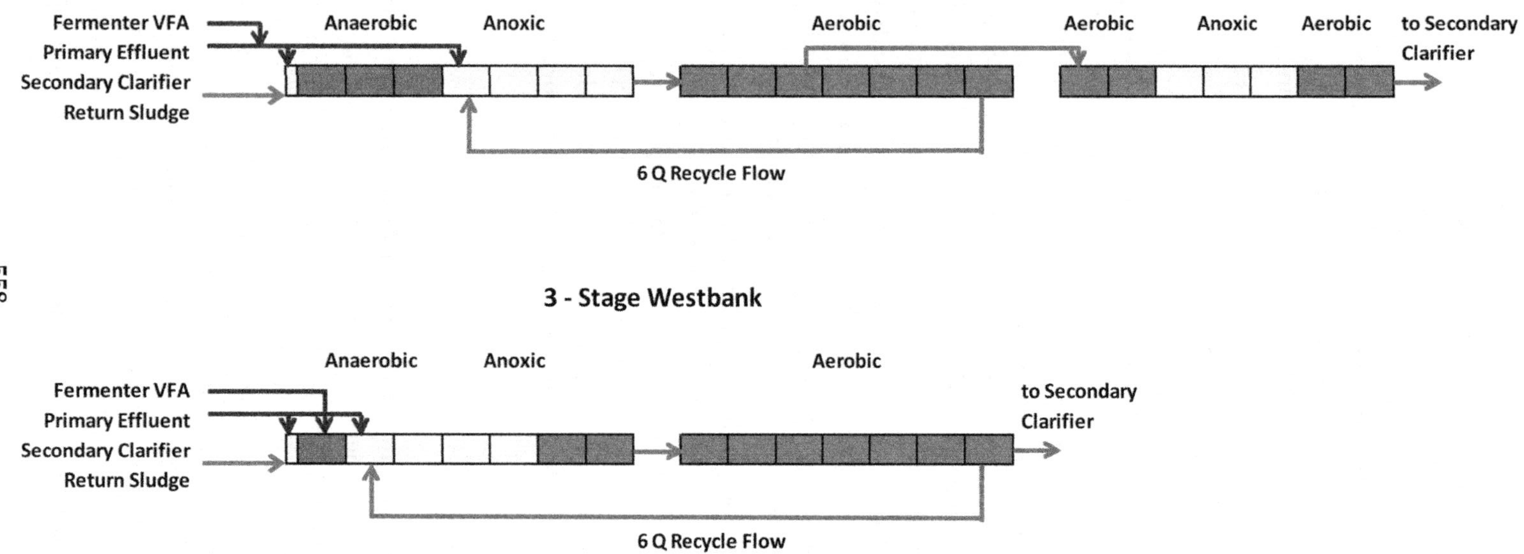

FIGURE 17.26 Kelowna Pollution Control Centre original five-Stage Bardenpho and retrofitted three-Stage Westbank process configurations (courtesy of the City of Kelowna).

in influent wastewater are insufficient to ensure reliable EBPR in the facility. Treatment of sludge handling return streams for phosphorus removal through intentional struvite formation and harvesting has served to reduce the recycled phosphorus loading to EBPR facilities, thus improving their performance and potentially generating an operating revenue stream through the production of a slow-release fertilizer product.

Facilities designed for EBPR now reliably achieve effluent total phosphorus concentrations below 0.1 mg/L. However, achieving these low concentrations is somewhat dependent on tertiary filtration, metal salt addition, or a combination of the two. Although MBR processes designed for EBPR have become increasingly popular to meet stringent effluent quality parameters, these systems still rely on in-facility alum addition to achieve stringent effluent total phosphorus limits. Full-scale operating experience has shown that incorporating phosphorus removal has facility-wide effects. For example, implementing EBPR increases both the recycle loads and the potential for struvite formation in the sludge handling system. Inclusion of chemical phosphorus removal can increase sludge production and affect UV disinfection, particularly if iron salt is used. However, thoughtful facility design and operation can minimize these effects.

4.0 REFERENCES

Barnard, J. L. (1984) Activated Primary Tanks for Phosphate Removal. *Water SA*, **10**, 3, 121–126.

Barnard, J. L.; Houweling, D.; Analla, H.; Steichen, M. (2010) Saving Phosphorus Removal at the Henderson NV Plant. *Proceedings of the 83rd Annual Water Environment Federation Technical Exhibition and Conference* [CD-ROM]; New Orleans, Louisiana, Oct 2–6; Water Environment Federation: Alexandria, Virginia.

Baur, R.; Britton, A.; Prasad, R. (2010) Full Scale Nutrient Recovery by Struvite Crystallization at Clean Water Services Durham AWWTP: Operational Data Analysis. *Proceedings of the 83rd Annual Water Environment Federation Technical Exhibition and Conference* [CD-ROM]; New Orleans, Louisiana, Oct 2–6; Water Environment Federation: Alexandria, Virginia.

Baur, R.; Cullen, N.; Laney, B. (2011) Nutrient Recovery: One Million Pounds Recovered—With Benefits. *Proceedings of the 84th Annual Water Environment Federation Technical Exhibition and Conference* [CD-ROM]; Los Angeles, California, Oct 15–19; Water Environment Federation: Alexandria, Virginia.

Benisch, M.; Baur, R.; Neethling, J. B.; Zaklikowski, A. (2009) Results from a Full Scale UFAT VFA Generation Capacity Study. *Proceedings of the 82nd Annual Water Environment Federation Technical Exhibition and Conference* [CD-ROM]; Orlando, Florida, Oct 10–14; Water Environment Federation: Alexandria, Virginia.

Block, T. J.; Rogacki, L.; Voigt, C.; Esping, D. G.; Parker, D. S.; Bratby, J. R.; Gruman, J. A. (2008) No Chemicals Required. *Water Environ. Technol.,* , **20,** 1, 42–47.

Clean Water Services (2002) Unified Fermentation and Thickening Process. U.S. Patent 6,387,264.

Clean Water Services (2009) Waste Activated Sludge Stripping to Remove Internal Phosphorus. U.S. Patent 7,604,740.

Comeau, Y.; Hall, K. H.; Hancock, R. E. W.; Oldham, W. K. (1986) Biochemical Model for Enhanced Biological Phosphate Removal. *Water Res.,* **20** (12), 1511–1521.

Oldham, W. K.; Stevens, G. M. (1984) Initial Operating Experience of a Nutrient Removal Process (Modified Bardenpho) at Kelowna, British Columbia. *Can. J. Civil Eng.,* **11** (3), 474–479.

Rabinowitz, B.; Fries, M. K. (2010) Primary Sludge Fermenters in BNR Plants: Are They Cost-Effective for Meeting Effluent Phosphorus Limits? *Proceedings of the 83rd Annual Water Environment Federation Technical Exhibition and Conference* [CD-ROM]; New Orleans, Louisiana, Oct 2–6; Water Environment Federation: Alexandria, Virginia.

Rabinowitz, B. (1985) The Role of Specific Substrates in Excess Biological Phosphorus Removal. Ph.D. Thesis, Department of Civil Engineering, University of British Columbia, Vancouver, British Columbia.

Appendix A

Optimization and Troubleshooting Guides

William C. McConnell, P.E., BCEE

Nutrient removal processes require continuous monitoring and adjustment to maintain optimal treatment efficiency. Unusual or upset conditions can occur, and are identified when indicators show changes from normal operating conditions. The upset condition may lead to effluent quality that does not meet permit limits for one or more of the following parameters: 5-day biochemical oxygen demand (BOD_5) (or 5-day carbonaceous biochemical oxygen demand [$CBOD_5$]), total suspended solids (TSS), ammonia-nitrogen (NH_3-N), total nitrogen, total inorganic nitrogen, and total phosphorus. The first step in resolving an upset condition is to find the problem causing the upset, whether it is loading, aeration, biomass inventory, clarifier operation, internal recycle, pH and alkalinity, toxicity or a combination of factors. Similarly, if the process is performing at less than desired efficiency or reliability, the same procedures can be used to optimize the process.

When the problem is identified, the next step is to go through an appropriate troubleshooting exercise to resolve the issue. The optimization and troubleshooting guides included in this appendix are intended to assist in identifying the problem and in determining the appropriate response.

1.0 OPTIMIZATION AND TROUBLESHOOTING GUIDE FORMAT

Tables A.1 through A.11 each address a different operational aspect of nutrient removal processes, as follows:

Table A.1—nitrification,

Table A.2—denitrification,

Table A.3—enhanced biological phosphorus removal (EBPR),

Table A.4—chemical phosphorus removal,

Table A.5—aeration and mixing systems (diffused aeration),

Table A.6—aeration and mixing systems (mechanical aeration),

Table A.7—biomass inventory,

Table A.8—clarifier operation,

Table A.9—internal recycle,

Table A.10—pH and alkalinity, and

Table A.11—toxicity.

Tables A.1 through A.11 are arranged in columns with the following headers:

- "Indicator and observations"—this shows the condition(s) observed or reported by the operator;
- "Probable cause"—this shows possible causes of the upset or inefficiency;
- "Check or monitor"—the operator should check the system for specific information to document the current condition and aid in diagnosing the problem. After the solution has been implemented, the operator should perform the listed monitoring step to verify process performance improvement until the process is optimized or recovered from upset; and
- "References"—the numbers listed in this column show where additional information is located in either another table or in one of the chapters of this manual.

2.0 Optimization and Troubleshooting Guides

Tables A.1 through A.11 present the optimization and troubleshooting guides. Reading across the page of each table, follow the numbers and letters. For example, solution 1a in the "Solutions" column refers to the corresponding item 1a in the "Probable cause" and "Check or monitor" columns. Information from a decision-tree-type troubleshooting guide specific to EBPR, developed by Benisch et al. (2004), was incorporated to these tables.

3.0 REFERENCE

Benisch, M.; Baur, R.; Neethling, J. B. (2004) Decision Tree for Troubleshooting Biological Phosphorus Removal. *Proceedings of the 77th Annual Water Environment Federation Technical Exhibition and Conference* [CD-ROM]; New Orleans, Louisiana, Oct 2–6; Water Environment Federation: Alexandria, Virginia.

TABLE A.1 Process upset—nitrification.

Indicator or observations	Probable cause	Check or monitor	Solutions	Reference
(1) Higher effluent NH_3-N than normal, accompanied by raw wastewater odor and/or dark color in the aerobic zone.	(1a) Excessive loading of BOD and/or TKN.	(1a) Check NH_3-N at the end of the process before clarification for nitrification performance.		Chapter 3 Chapter 4
		(1a1) Check BOD and TKN concentrations in influent to the facility.	(1a1) Check for and discourage discharges to the collection system that are causing unusually high strength in the influent wastewater.	
		(1a2) Check BOD and TKN concentrations in the influent to the BNR process.	(1a2) See if sidestreams from solids handling are causing periodically high loads and adjust operations to even out BOD and TKN loadings to the process. If the facility has primary clarifiers, check removal efficiency and consider adjustments.	
		(1a3) If activated sludge, check dissolved oxygen in the aeration zone; if rotating biological contactor, check dissolved oxygen in the first or second compartment; if trickling filter, check dissolved oxygen in the trickling filter effluent. If moving bed biofilm reactor (MBBR) or IFAS, check bulk liquid dissolved oxygen concentration. Dissolved oxygen checks should consider diurnal variation as daily or once per shift dissolved oxygen readings may miss low-dissolved oxygen periods.	(1a3) If activated sludge with diffused aeration and mixing, see Table A.5; if activated sludge with mechanical aeration and mixing, see Table A.6. If trickling filter, try to increase aeration as a short-term solution; if MBBR or IFAS, increase the aeration rate to increase dissolved oxygen concentration, if possible. Also, see 1a2.	

(continued)

TABLE A.1 Process upset—nitrification (*Continued*).

Indicator or observations	Probable cause	Check or monitor	Solutions	Reference
	(1b) Effective capacity of process has been reduced.	(1b1) Determine the ratio of volatile suspended solids to TSS of the mixed liquor.	(1b1) If there is a lower ratio of volatile suspended solids to TSS than is historically seen, check the performance of upstream grit removal and chemical addition systems. Operate based on mixed liquor volatile suspended solids (i.e., volatile suspended solids) on SRT until the ratio of volatile suspended solids to TSS returns to normal.	Chapter 4
		(1b2) Observe mixing and aeration patterns.	(1b2) If activated sludge with diffused aeration and mixing, see Table A.2. If activated sludge with mechanical aeration and mixing, see Table A.3.	
		(1b3) If a biofilm system, look for excessive growth or debris that hinders aeration or airflow and causes short-circuiting.	(1b3) Increase fluid velocity to decrease biofilm thickness by flooding, increase recirculation rate, flush with hose, and increase aeration rate.	
(2) Higher effluent NH_3-N than normal, accompanied by appearance of uneven loading (odors, mixed liquor suspended solids color, and dissolved oxygen); MLSS concentrations or dissolved oxygen different in different trains.	(2a) Unequal flow distribution.	(2a) Check flow of process influent to each train. Check NH_3-N and NO_2 + NO_3-N at the end of each process train before clarification.	(2a) Adjust flow-splitting devices to provide equal flow to same-sized trains or to provide equal loadings to different-sized trains.	
	(2b) Unequal RAS distribution.	(2b) If activated sludge and RAS enters separately, check RAS flow to each train.	(2b) Adjust flow-splitting devices on RAS to provide equal flow to same-sized trains or to provide equal loadings to different-sized trains.	
	(2c) Unequal MLSS distribution.	(2c) If process influent and RAS mix and then are split, check for good mixing and check flow to each train.	(2c) Increase mixing if poor mixing is observed before flow distribution. Adjust mixed liquor suspended solids flow-splitting devices to provide equal flow to same-sized trains or to provide equal loadings to different-sized trains.	

(3) Higher effluent NH_3-N than normal, accompanied by dark diluted color of mixed liquor suspended solids and sometimes white, sudsy foam in aerobic zone.	(3) Mixed liquor suspended solids (and solids retention time) is lower than desired because of excessive wasting.	(3) Check mixed liquor suspended solids and dissolved oxygen and check NH_3-N at end of process before clarification.	(3) If mixed liquor suspended solids is low, dissolved oxygen in aeration zones is high, and NH_3-N is high, decrease wasting to allow mixed liquor suspended solids to increase for a solids retention time adequate for nitrification to occur. See Chapter 3 and Table A.7.	Chapter 3
(4) Higher effluent NH_3-N than normal, accompanied by low influent pH or low alkalinity.	(4a) Low alkalinity in potable water combined with decomposition of wastewater in collection system during warm weather.	(4a) Check influent pH and alkalinity.	(4a1) Decrease solids retention time by reducing mixed liquor suspended solids through increased wasting if nitrification is occurring adequately or if nitrification is not desired. (4a2) Add alkalinity through chemical addition. See Chapters 3 and 6.	Chapter 3 Chapter 6
	(4b) Acidic discharge(s) to collection system.	(4b) Check influent pH, alkalinity, and other parameters that could identify the source of acidic discharge(s).	(4b) Identify and eliminate source of acidic discharge(s). For short-term solution, see (4a2) above	
(5) Higher effluent NH_3-N concentration, accompanied by high dissolved oxygen.	(5) Inhibition of nitrification.	(5) Check pH profile through process, dissolved oxygen in aeration zones, and toxicity.	(5) Check solids retention time and dissolved oxygen in process; adjust higher, if possible. Refer to Tables A.10 and A.11.	

Table A.2 Process upset—denitrification.

Indicator or observations	Probable cause	Check or monitor	Solutions	Reference
(1) Higher effluent NO_2 + NO_3 concentration than normal.	(1a) Excessive loading of total Kjeldahl nitrogen.	(1a) Check NO_2 + NO_3 at end of process before clarification for nitrification performance.		
		(1a1) Check total Kjeldahl concentrations in influent to the facility.	(1a1) Check for and discourage discharges to collection system that are causing unusually high strength in the influent wastewater.	
		(1a2) Check total Kjeldahl concentrations in influent to the biological nutrient removal process.	(1a2) See if sidestreams from solids handling are causing periodically high loads and adjust operations to even out total Kjeldahl nitrogen loadings to process.	
	(1b) Effective capacity of process has been reduced.	(1b1) Check for grit deposits at bottom of bioreactors	(1b1) If there is a lower ratio of volatile suspended solids to TSS than historically seen, check the performance of upstream grit removal and chemical addition systems. Operate based on mixed liquor volatile suspended solids (i.e., volatile suspended solids) on solids retention time until the ratio of volatile suspended solids to TSS returns to normal.	Chapter 6
		(1b2) Observe mixing patterns.	(1b2) If available, increase mixing intensity. Install and adjust baffling to improve mixing. Refer to Chapter 6.	

	(1c) The ratio of BOD to total Kjeldahl nitrogen has changed.	(1c1) Check BOD and TKN concentrations in influent to BNR process.	(1c1) Look at facility operations to see if sidestreams from solids handling are causing periodically high loads and adjust operations to reduce or even out BOD and total Kjeldahl nitrogen loadings to process. See Chapter 12.	Chapter 5 Chapter 6 Chapter 12
		(1c2) Check BOD and total Kjeldahl nitrogen concentrations in influent to facility.	(1c2) If BOD to total Kjeldahl nitrogen is low, there may be insufficient BOD for denitrification in anoxic zones.	
		(1c2a) High total Kjeldahl nitrogen in influent.	(1c2a) Check for and discourage discharges to collection system that are causing unusually high total Kjeldahl nitrogen.	
		(1c2b) Low BOD in influent.	(1c2b) Add carbon source to increase the ratio of BOD to total Kjeldahl nitrogen and improve denitrification. See Chapters 5 and 6.	
	(1d) Inhibition of denitrification	(1d) Check $NO_2 + NO_3$, dissolved oxygen, and/or oxidation–reduction potential profiles through the process.	(1d) Optimize the anoxic conditions in anoxic zones; see Tables A.5 and A.6 for controlling aeration and Table A.9 for controlling internal recycle. Verify supplemental carbon addition flow.	
(2) Odors in anoxic zones	(2) Under-loading of process is causing excessive detention time.	(2) Check detention time and oxidation–reduction potential in anoxic zones.	(2a) If possible, decrease volume of anoxic zone.	
			(2b) Add some dissolved oxygen to the beginning of anoxic zone.	
			(2c) Increase internal recycle to return more NO_2 + NO_3-N and dissolved oxygen.	

TABLE A.3 Process upset—EBPR.

Indicator or observations	Probable cause	Check or monitor	Solutions	Reference
(1) Higher effluent total phosphorus concentration than normal.	(1a) Ratio of BOD to total phosphorus has changed.	(1a1) Check BOD, soluble BOD, total phosphorus, and orthophosphorus in influent to the biological nutrient removal process.	(1a1) Review facility operations to see if sidestreams from solids handling are causing periodically high loads and adjust operations to reduce or even out total phosphorus loadings to process.	Chapter 7 Chapter 8
		(1a2) Check BOD, soluble BOD, total phosphorus, and orthophosphorus in influent to facility.	(1a2) If BOD to total phosphorus or soluble BOD to orthophosphorus is low, there may be insufficient VFAs for biophosphorus removal.	
		(1a2a) High total phosphorus in influent.	(1a2a) Check for and discourage discharges to collection system that are causing unusually high total phosphorus.	
		(1a2b) Low soluble BOD (low volatile fatty acids) in anaerobic zone.	(1a2b) Chemicals added to the collection system for odor control may inhibit fermentation and, thereby, formation of volatile fatty acids. Consider using different chemicals if this is the case. If possible, raise the sludge blanket in the primary clarifiers 0.31 to 0.61 m (1 to 2 ft). This will turn the primaries into a first-stage fermenter and produce some VFAs that will help recover the process.	
			(1a2c) Add VFAs, such as acetic acid, to anaerobic zone to increase biophosphorus removal. See Chapter 7.	
		(1a3) Insufficient soluble BOD (volatile fatty acids) and additional volatile fatty acids unavailable.	(1a3) Use chemical phosphorus removal on short-term basis or, if this is continuing situation, use chemicals and recalculate solids retention time and wasting based on higher inorganic content of mixed liquor suspended solids resulting from chemical solids production. See Chapter 8.	
			(1b) Optimize the anaerobic conditions in the anaerobic zone and ensure that there is no short circuiting; see Tables A.5 and A.6 for controlling aeration. Increase anaerobic zone volume, if possible.	

	(1b) Inhibition of phosphorus release in anaerobic zone.	(1b) Check orthophosphorus, NO_2 + NO_3, NH_3-N, dissolved oxygen, and/or oxidation–reduction potential profiles through process. Observe mixing in anaerobic zone.		
	(1c) Inhibition of biological phosphorus uptake.	(1c1) Check orthophosphorus, dissolved oxygen profiles through process. Observe mixing in aerobic zone(s). Check aerobic hydraulic retention time.	(1c1) Ensure that there is sufficient hydraulic retention timeand no short circuiting; see Tables A.5 and A.6 for controlling aeration. Install baffling, if needed, to block strong currents from inlet to outlet of aeration zones.	
		(1c2) Check influent volatile fatty acid and volatile fatty acid orthophosphorus profiles through process.	(1c2) If shock load of influent volatile fatty acid occurs, excessive phosphorus release in the anaerobic zone may exceed biological phosphorus uptake in the aerobic zone(s). Eliminate excess volatile fatty acid discharge or equalize volatile fatty acid load.	
		(1c3) Check pH, temperature, and microorganism population for glycogen-accumulating organisms.	(1c3) If other causes of low biological phosphorus uptake have been eliminated from consideration, the treatment facility conditions (low pH, high temperature, etc.) may favor glycogen-accumulating organisms over phosphorus-accumulating organisms. Consider pH adjustment or chemical phosphorus removal.	
		(1c4) Check the PO4-P/TP (ratio of orthophosphorus to total phosphorus) to determine if increase is from particulate phosphorus in TSS or soluble phosphorus (PO_4-P). If chemical addition has been on an increasing trend, a microscopic analysis should be performed to check if chemical addition has cut off polyphosphate-accumulating organisms.	(1c4) Adjust chemical coagulant dosage if orthophosphorus has increased. Improve effluent suspended solids quality if particulate. Refer to Table A.8.	
(2) Odors in anaerobic zone	(2) Underloading of process is causing excessive detention time.	(2) Check detention time and oxidation–reduction potential in anaerobic zone.	(2a) If possible, decrease volume of anaerobic zone (e.g., decrease number of basins in service or lower operating water level). (2b) Increase return activated sludge flowrate to return more NO_2 + NO_3-N.	Chapter 7 Chapter 9

TABLE A.4 Process upset—chemical phosphorus removal.

Indicator or observations	Probable cause	Check or monitor	Solutions	Reference
(1) Sudden increase in effluent total phosphorus concentration.	(1a1) Chemical feed pump failure.	(1a1) Verify that the pump has power.	(1a1) Provide power.	
		(1a2) Check the manufacturer's troubleshooting information.	(1a2) Perform system checks and conduct maintenance as directed by the manufacturer.	
	(1b) Chemical precipitates have formed in the piping, plugging chemical feed.	(1b) Verify that chemical is reaching the application point and potentially break the piping in search of the suspected restriction.	(1b) If the chemical flow is restricted, attempt to remove the restriction or restricted piping. Muriatic acid has been successful in dissolving chemical buildup. Contact the chemical distributor for additional advice. Consider discontinuing carrier water that may be promoting precipitate formation within the pipe.	
	(1c) High slug loading from an unknown source.	(1c1) Monitor key industrial contributors.	(1c1) Work with industries to control loads.	
		(1c2) Check sidestream contributions.	(1c2) Sidestreams should be treated carefully. Contingencies may include adding chemical to the sidestreams, equalizing the flow, and recycling sidestreams during periods of low loadings.	
	(1d) Solids carryover and breakthrough from solids separation process.	(1d) Monitor effluent suspended solids.	(1d) Improve effluent suspended solids quality. Refer to Table A.8.	

	(1e) Analytical error.	(1e) Laboratory quality assurance and quality control.	(1e) If laboratory data are in error, the data should be excluded from the data set for process control purposes and should be noted, as required, if used for regulatory reporting.
	(1f) Change in ratio of orthophosphorus to total phosphorus.	(1f) Check ratio of orthophosphorus to total phosphorus to determine if increase in total phosphorus is attributable to phosphorus in soluble or particulate form.	(1f) Adjust chemical coagulant dosage if orthophosphorus has increased. Improve effluent suspended solids quality if particulate. Refer to Table A.8.
(2) Gradual increase in effluent total phosphorus concentration.	(2a) Additional process loading because of industrial contributions.	(2a1) Monitor key industrial contributors.	(2a1) Work with industries to control loads.
		(2a2) Check sidestream loads.	(2a2) Sidestreams should be treated carefully. Contingencies may include adding chemical to the sidestreams, equalizing the flow, and recycling sidestreams during periods of low loadings.
	(2b) Inconsistent chemical strength.	(2b) Monitor chemical strength.	(2b1) Increase the dose rate when using weaker chemicals; decrease the dose rate when using stronger chemicals.
			(2b2) Require more consistent product from supplier.
	(2c) pH has shifted, resulting in less efficient phosphorus removal.	(2c) Check pH	(2c) Adjust pH with appropriate chemical addition.

TABLE A.5 Aeration and mixing systems—diffused aeration.

Indicator or observations	Probable cause	Check or monitor	Solutions	Reference
(1) Low dissolved oxygen in aeration zones.	(1a) Poor oxygen transfer in aeration zones.	(1a) Check for diffuser problems.	(1a) Repair diffusers if needed; resolve fouling if noted.	
	(1b) Insufficient aeration.	(1b) Check airflow vs calculated airflow for loadings. Check dissolved oxygen profile through process.	(1b) Increase airflow to aeration zones until dissolved oxygen is approximately 2.0 mg/L. If airflow per diffuser is too high and additional aeration zones are available, put additional aeration zones in service.	
	(1c) Changes in influent characteristics, decreasing field oxygen transfer efficiency	(1c) Check airflow vs calculated airflow for loadings. Check dissolved oxygen profile through the process.	(1c) Increase airflow to aeration zones if possible.	
(2) High dissolved oxygen in aeration zones.	(2) Flow, BOD, and/or total Kjeldahl nitrogen are lower than design.	(2a) Check airflow vs calculated airflow for loadings. Check dissolved oxygen profile through process.	(2a) Decrease airflow to aeration zones until target dissolved oxygen is reached. If airflow per diffuser is too low, take aeration zone(s) out of service to maintain at least the minimum recommended airflow per diffuser.	
		(2b) Compare actual flow, BOD, and total Kjeldahl nitrogen to design values. Calculate required airflow.	(2b) If diffusers allow, reduce the amount of air added by periodically reducing or shutting off air (not more than 1.5 to 2 hours off to prevent odors). Use a combination of on time and either reduced airflow time or off time throughout the day to better match aeration to the oxygen demand.	

			(2b) If multiple trains are available, reduce number of trains in service.
(3) Significant dissolved oxygen into anoxic zones.	(3a) Too much dissolved oxygen at internal recycle suction.	(3a) Check dissolved oxygen in aeration zone where internal recycle originates.	(3a) For activated sludge process with internal recycle, taper dissolved oxygen profile for low dissolved oxygen at internal recycle suction. Low oxygen uptake at end of basin may require supplemental mixing to allow reduced airflow below that required for mixing to reduce dissolved oxygen.
	(3b) Too much turbulence at influent.	(3b) Observe turbulence or splashing at influent.	(3b) Adjust basin levels, use baffles, and/or modify inlet ports to minimize turbulence and introduction of dissolved oxygen to anoxic zones.
(4) Significant dissolved oxygen and/or NO_2 + NO_3-N into anaerobic zone.	(4a) Too much dissolved oxygen returning with return activated sludge.	(4a) Check dissolved oxygen in return activated sludge. Check return activated sludge concentration and flow.	(4a) Reduce dissolved oxygen in aeration zone immediately upstream of clarifiers. If clarifier operation allows, reduce RAS flow
	(4b) Too much NO_2 + NO_3-N returning with return activated sludge.	(4b) Check NO_2 + NO_3-N concentration in return activated sludge. Check return activated sludge concentration and flow.	(4b) If nitrification is not needed, reduce solids retention time to reduce NO_2 + NO_3- N concentration going to clarifiers. If clarifier operation allows, reduce return activated sludge flow.
	(4c) Too much turbulence at influent.	(4c) Observe turbulence or splashing at influent.	(4c) Adjust basin levels, use baffles, and/or modify inlet ports to minimize turbulence and introduction of dissolved oxygen to anaerobic zones.

(continued)

Table A.5 Aeration and mixing systems—diffused aeration (*Continued*).

Indicator or observations	Probable cause	Check or monitor	Solutions	Reference
(5) Poor mixing pattern.	(5a) Diffusers need cleaning or repair.	(5a) Visual observation of mixing pattern.	(5a) Bump or chemically clean in-place diffusers; take basin out of service and manually clean diffusers, replace broken diffusers, or periodically replace all diffusers if beyond expected service life.	
	(5b) If blow-offs are present, air is releasing through blow-offs.	(5b) Check for excessive diffuser headloss or blow-off malfunction.	(5b) Clean or replace diffusers as in 5a and/or repair blow-off system.	
	(5c) Diffusers not installed at same elevation.	(5c) Check that diffusers are installed level and at the same elevation.	(5c) Take basin out of service and adjust diffuser support systems and/or laterals so that they are level and that diffusers are at the same elevation.	
(6) No or too little turbulence.	(6a) Too few blowers in operation.	(6a) Check number of blowers in operation and control system.	(6a) Increase number of blowers in operation or adjust controls to bring on more blowers if automatically controlled.	
	(6b) Blower malfunction.	(6b) Compare airflow at each blower to its performance curve(s).	(6b) Diagnose and repair and replace malfunctioning parts. Check setting of inlet air vanes and valves.	
	(6c) Dirty inlet filter.	(6c) Check inlet air pressure between filter and blower.	(6c) Replace inlet filters; clean filters if washable.	
	(6d) Valves need adjustment.	(6d) Check airflow to basin.	(6d) Adjust manual valves to better distribute air.	

	(6e) Aeration control system needs adjustment.	(6e) Compare control parameter values to set point(s).	(6e) If insufficient air is provided under automatic operation, adjust controls to increase airflow when setpoints are maintained.
	(6f) Air rate too low for proper operation of diffusers.	(6f) Compare airflow divided by number of diffusers to acceptable low value.	(6f) Increase airflow to provide at least minimum recommended airflow per diffuser.
(7) Excessive turbulence over entire basin.	(7a) Too many blowers in operation.	(7a) Check dissolved oxygen in aeration zones and number of blowers in operation.	(7a) If dissolved oxygen is above desired levels, decrease airflow. Stay within range of recommended airflow per diffuser as in 7c.
	(7b) Aeration control system needs adjustment.	(7b) Compare control parameter values to setpoint(s).	(7b) If excess air is provided under automatic operation, adjust controls to decrease airflow when setpoints are maintained.
	(7c) Air rate too high for proper operation of diffusers.	(7c) Compare airflow divided by number of diffusers to acceptable high value.	(7c) Decrease airflow to provide, at most, the maximum recommended airflow per diffuser.
	(7d) Leak(s) in air distribution piping system.	(7d) Check for leaks in air-distribution piping system.	(7d) Repair leaks.

TABLE A.6 Aeration and mixing systems—mechanical aeration.

Indicator or observations	Probable cause	Check or monitor	Solutions	Reference
(1) Low dissolved oxygen in aeration zones.	(1a) Poor oxygen transfer in aeration zones.	(1a) Check for aerator problems.	(1a) See solutions 5, 6a, and 7.	
	(1b) Insufficient aerators in service.	(1b) Check total horsepower in service vs calculated horsepower based on design oxygen transfer. Check dissolved oxygen profile through process.	(1b) Increase number of aerators in service until dissolved oxygen is approximately 2.0 mg/L. If additional aeration zones are available, put them in service.	
(2) High DO in aeration zones.	(2) Flow, BOD, and/or TKN are lower than design.	(2a) Check total horsepower in service vs calculated horsepower. Check dissolved oxygen profile through process.	(2a) Decrease level in basin to lower impeller submergence until dissolved oxygen is approximately 2.0 mg/L. If dissolved oxygen is still too high, take aerator out of service, use low speed if two-speed aerators, or take an aeration zone out of service to maintain minimum impeller submergence.	
		(2b) Compare actual flow, BOD, and total Kjeldahl nitrogen to design values. Calculate required aeration horsepower.	(2b) If aerators allow, periodically reduce the speed of the aerator to reduce oxygen transfer but maintain mixing. If aerator is able to resuspend solids, consider shutting off aerator(s) (not more than 1.5 to 2 hours off to prevent odors). Use a combination of on time and either reduced speed time or off time throughout the day to better match aeration to the oxygen demand. For large aerators, on–off operation may be too hard on gear reducers; consider soft-start of motors if on–off operation is used.	

(3) Significant dissolved oxygen into anoxic zones.	(3a) Too much dissolved oxygen at internal recycle suction.	(3a) Check dissolved oxygen in aeration zone where internal recycle suction originates.	(3a) For activated sludge process with internal recycle, locate internal recycle suction where dissolved oxygen is lowest and adjust submergence, as in 2a. If NO_2 + NO_3-N concentration in effluent is low, reduce internal recycle rate.
	(3b) Too much turbulence at influent.	(3b) Observe turbulence or splashing at influent.	(3b) Adjust basin levels, use baffles, and/or modify inlet ports to minimize turbulence and introduction of dissolved oxygen to anoxic zones.
(4) Significant dissolved oxygen and/or NO_2 + NO_3-N into anaerobic zone.	(4a) Too much dissolved oxygen returning with return activated sludge.	(4a) Check dissolved oxygen in return activated sludge. Check return activated sludge concentration and flow.	(4a) Reduce dissolved oxygen in aeration zone immediately upstream of clarifiers. If clarifier operation allows, reduce return activated sludge flow.
	(4b) Too much NO_3-N and NO_2-N returning with return activated sludge	(4b) Check NO_3-N and NO_2-N returning with return activated sludge. Check return activated sludge concentration and flow.	(4b) If nitrification is not needed, reduce SRT to reduce NO_2 + NO_3-N concentration going to clarifiers. If clarifier operation allows, reduce return activated sludge flow.
	(4c) Too much turbulence at influent.	(4c) Observe turbulence or splashing at influent.	(4c) Adjust basin levels, use baffles, and/or modify inlet ports to minimize turbulence and introduction of dissolved oxygen to anaerobic zones.
(5) Surging noise and waves in aeration zone.	(5a) Water level too low for mechanical aerator impeller.	(5a) Visual observation of waves in basin and surging noise; check impeller submergence level.	(5a) Increase level in basin to provide recommended minimum impeller submergence; repair leaks in weirs or gates that allow water level to drop below desired level at low flows.

(*continued*)

Table A.6 Aeration and mixing systems—mechanical aeration (*Continued*).

Indicator or observations	Probable cause	Check or monitor	Solutions	Reference
	(5b) Basin design prone to wave formation at aerator.	(5b) Visual observation of waves reflecting off surfaces and creating standing wave.	(5b) Adjust aerator speed or position, if possible, or modify basin using baffles to reduce the standing wave effect.	
(6) Motor overload.	(6a) Water level too low for mechanical aerator impeller.	(6a) Visual observation of waves in basin, surging noise; check impeller submergence level.	(6a) See 5a in this guide.	
	(6b) Water level too high for mechanical aerator impeller.	(6b) Check impeller submergence level, especially at peak flow.	(6b) Decrease level in basin to limit impeller submergence to recommended maximum at peak flow.	
(7) Vibration, reduced splashing, low dissolved oxygen	(7a) Impeller fouled with debris.	(7) Check for vibration, visual observation of impeller, and visual observation of splash pattern.	(7a) Take aerator out of service and remove debris from impeller.	
	(7b) Impeller fouled with ice.		(7b) Take aerator out of service and remove ice from impeller, install manufacturer-approved shields that keep splash confined to basin, and minimize ice formation.	

TABLE A.7 Biomass inventory.

Indicator or observations	Probable cause	Check or monitor	Solutions	Reference
(1) Wastewater temperature is low, dissolved oxygen is higher; higher effluent total Kjeldahl nitrogen than normal.	(1) Cold weather is slowing down biomass activity.	(1) Check wastewater temperature in process.	(1) Increase solids retention time to continue nitrification, if required. If nitrification is not required during cold weather, then adjust process to discourage nitrification (lower solids retention time, less air, and lower mixed liquor suspended solids and return activated sludge flowrate).	
(2) Mixed liquor suspended solids is low, dark diluted color of mixed liquor suspended solids, accompanied by				
(2a) Higher effluent total Kjeldahl nitrogen than normal.	(2a1) Mixed liquor suspended solids is lower than desired because of excessive wasting.	(2a1) Check mixed liquor suspended solids and dissolved oxygen and check NH_3-N and NO_2 + NO_3-N at end of process before clarification.	(2a1) Decrease wasting to allow mixed liquor suspended solids to increase for a solids retention time adequate for nitrification to occur. Select a better time to obtain more consistent mixed liquor suspended solids concentration and/or use moving average of 7 or more days on which to base wasting. Institute operational limits on the amount of wasting that can occur in a single day.	Table A.8
	(2a2) High effluent TSS.	(2a2) Check effluent TSS.	(2a2) Refer to Table A.8.	
(2b) If using an enhanced biophosphorus removal process without nitrification, higher effluent total phosphorus than normal.	(2b) Mixed liquor suspended solids is low to inhibit nitrification, but is too low for the temperature to enhanced biophosphorus removal population.	(2b) Check mixed liquor suspended solids and dissolved oxygen and check orthophosphorus at end of process before clarification.	(2b) Decrease wasting as in 2a1. If nitrification occurs and there is insufficient aeration capacity to maintain stable operation, operate at low mixed liquor suspended solids and implement chemical phosphorus removal until temperature increases to where biophosphorus population is restored; see Chapter 8.	Chapter 8

(continued)

Table A.7 Biomass inventory (*Continued*).

Indicator or observations	Probable cause	Check or monitor	Solutions	Reference
(3) Mixed liquor suspended solids is high, pin floc in secondary clarifier effluent, high SVI, sometimes dark tan foam on aeration basin. Higher effluent total Kjeldahl nitrogen or total phosphorus than normal.	(3a) High mixed liquor suspended solids resulting from reduced wasting or seasonally higher temperatures.	(3a1) If multiple trains are in service, calculate loadings if fewer trains are in service.	(3a1) If calculated loadings are within design values with fewer trains in service, decrease number of trains in service.	
		(3a2) Check mixed liquor suspended solids and waste activated sludge solids concentrations and waste activated sludge flowrate and calculate solids retention time.	(3a2) Lower mixed liquor suspended solids concentration to lower solids retention time and consider lowering return activated sludge rate.	
(4) Secondary release of phosphorus in the second anoxic basin yielding high total phosphorus concentration in the effluent.	(4a) Detention time in the second anoxic is too high.	(4a) Monitor return activated sludge recycle rate.	(4a) Increase return activated sludge recycle rate.	
	(4b) Lack of carbon in the second anoxic basin preventing denitrification.	(4b) Biochemical oxygen demand or carbonaceous oxygen demand into the second anoxic zone.	(4b) Add organic carbon (methanol, etc.) to the second anoxic basin.	
	(4c) First anoxic basin completing denitrification.	(4c1) Monitor the mixed liquor suspended solids recycle rate to the first anoxic basin.	(4c1) Decrease the mixed liquor suspended solids recycle to the first anoxic basin.	
		(4c2) Monitor the oxygen supply to the aerobic zone to allow carryover to the second anoxic zone.	(4c2) Increase the oxygen supply to the last stage before the second anoxic zone.	
(5) High chemical demand for phosphorus removal affecting the nitrification performance of the system.	(5) Too much chemical sludge produced by metal salts used for phosphorus precipitation.	(5) Verify where the metals salts are being fed. Verify the ratio of mixed liquor volatile suspended solids to mixed liquor suspended solids in the system.	(5) Implement staged chemical addition to reduce the dosage fed to the secondary clarifiers and trim at a tertiary treatment process.	

TABLE A.8 Clarifier operation.

Indicator or observations	Probable cause	Check or monitor	Solutions	Reference
(1) Floating solids at surface, sometimes thick solids layer, possibly bulking solids. Higher effluent TSS or total phosphorus than normal.	(1a) Denitrification in clarifier causing rising sludge.	(1a) Check NO_2 + NO_3-N and dissolved oxygen concentrations entering and leaving clarifiers.	(1a) Optimize denitrification process (increase internal recycle, lower dissolved oxygen in internal recycle, add supplemental carbon, etc.) to achieve <10 mg/L nitrate in final effluent. Reduce sludge blanket in clarifier. Increase dissolved oxygen in clarifier feed. Decrease number of clarifiers in service if loading allows.	
	(1b) Return activated sludge not being removed quickly enough.	(1b) Monitor blanket height and check return activated sludge rate and detention time in clarifiers. Check orthophosphorus in return activated sludge.	(1b) Increase return activated sludge return rate and/or increase dissolved oxygen in clarifier influent. Verify that return activated sludge is flowing normally from each clarifier. Reduce solids detention time in clarifiers (lower blanket height, increase return activated sludge rate, and reduce number of clarifiers in service).	
	(1c) Excessive turbulence causing air bubbles in floc.	(1c) Observe turbulence and presence of bubbles on floc from mixed liquor effluent.	(1c) Reduce turbulence in aeration tank immediately upstream of clarifiers and reduce turbulence between aeration tank and clarifier inlet.	
	(1d) Filamentous organisms in mixed liquor causing bulking sludge.	(1d) Check settleability and sludge volume index.	(1d) Minimize formation of filamentous organisms by adjusting conditions in process basin.	

(*continued*)

Table A.8 Clarifier operation (*Continued*).

Indicator or observations	Probable cause	Check or monitor	Solutions	Reference
(2) Solids overflowing clarifier weir. Higher effluent TSS and BOD than normal.	(2a) Return activated sludge not being removed quickly enough.	(2a) Monitor blanket height and check return activated sludge rate.	(2a) See Solution 1b.	
	(2b) Unequal flow distribution.	(2b) Check flow to each clarifier.	(2b) Adjust flow-splitting devices to provide equal flow to same-sized clarifiers or provide equal loadings to different-sized clarifiers.	
	(2c) Hydraulic and/or solids overloading	(2c) Check surface overflow rates at peak flow. Check solids loading rate.	(2c) If available, place another clarifier in service. If equipped, switch to step-feed operation to reduce solids loading. Reduce surges to the clarifiers by controlling pumping stations in collection system and recycle streams from solids handling to equalize flows. Decrease solids inventory if process allows. Alter mode of operation to improve settleability.	
	(2d) Turbulence from collection rake.	(2d) Observe solids carryover when rake passes by.	(2d) If possible, reduce rake travel speed.	
(3) Phosphorus in floc. Higher effluent TSS than normal.	(3a) Excessive turbulence upstream of clarifiers.	(3a) Observe turbulence upstream and check dissolved oxygen in last aeration zone.	(3a) Reduce aeration in aeration zone upstream of clarifiers. Adjust flow-splitting devices or water levels to reduce turbulence upstream of clarifiers. Add chemical flocculant (polymer) on an interim basis.	Chapter 12

	(3b) Solids retention time too long.	(3b) Calculate solids retention time (check mixed liquor suspended solids and waste activated sludge solids concentration and waste activated sludge flowrate).	(3b) Reduce solids retention time by increasing sludge wasting. Limit wasting increase to 10% higher for 2 times SRT to minimize unstable operation.
	(3c) Short-circuiting in clarifier.	(3c) Look for areas where floc carryover is heaviest. Observe water level along effluent weir.	(3c) Level weirs. If possible, add or adjust baffling to reduce excessive velocities and density currents.
	(3d) Upset resulting from loading or colloidal solids in sidestreams from solids handling	(3d) Measure TSS and turbidity in sidestreams from solids handling. Check dissolved oxygen profile through process	(3d) Look at facility operations to see if sidestreams from solids handling are causing periodically high loads and adjust operations to reduce colloidal solids and even out BOD and total Kjeldahl nitrogen loadings to the process. See Chapter 12.
	(3e) Toxicity	(3e) Refer to Table A.11.	(3e) Refer to Table A.11.
(4) Thin return activated sludge. Higher effluent total phosphorus than normal.	(4a) Plugging of sludge withdrawal.	(4a) Check return activated sludge concentration, observe sludge withdrawal equipment.	(4a) Backflush return activated sludge collection system.
	(4b) Return activated sludge return rate too high.	(4b) Check return activated sludge concentration.	(4b) Reduce return activated sludge return rate.
	(4c) Poor sludge settling.	(4c) Check sludge volume index and compare to historical values.	(4c) Minimize formation of filamentous organisms by adjusting conditions in process basin.
(5) Turbid effluent. Higher effluent TSS or total Kjeldahl nitrogen than normal.	(5) Toxic or acid constituents in wastewater.	(5) See Table A.11.	(5) See Table A.11.

TABLE A.9 Internal recycle.

Indicator or observations	Probable cause	Check or monitor	Solutions	Reference
(1) If removing nitrogen through internal recycle, high NO_2 + NO_3-N in effluent upstream of clarifiers. Higher effluent NO_2 + NO_3-N than normal.	(1a) Internal recycle flowrate is insufficient.	(1a1) Verify that internal recycle pumps are operating. (1a2) Compare internal recycle flowrate vs design rate. Look for clog in valve, line, or pump.	(1a1) If pumps are not operating, fix the problem and restore internal recycle flow. (1a2) If pumps are operating and flowrate is lower than normal, unclog internal recycle system and restore internal recycle flow.	
	(1b) Internal recycle is returning too much.	(1b) Check dissolved oxygen concentration in internal recycle.	(1b) Reduce dissolved oxygen concentration in internal recycle. See Tables A.5 and A.6.	
	(1c) Insufficient carbonaceous BOD to BOD for denitrification in anoxic zones.	(1c) See Table A.2.	(1c) See Table A.2.	
(2) DO in anoxic zone. Higher Effluent NO_2 + NO_3-N than normal.	(2) Internal recycle rate is too high for the loading to the process.	(2) Calculate internal recycle flowrate needed to achieve nitrogen removal and compare to actual rate.	(2) Lower internal recycle flowrate.	
(3) Low dissolved oxygen in aeration zone with aeration system at maximum; average or high NH_3-N and low NO_2 + NO_3-N in effluent upstream of clarifiers.	(3) Oxygen-limited condition in aeration zone is causing denitrification to occur simultaneously in aeration zone.	(3) Check profiles of dissolved oxygen, NH_3-N, and NO_2 + NO_3-N through the process.	(3) For short-term solution, consider discontinuing internal recycle flow if aeration zone is large enough to accomplish simultaneous nitrification and denitrification. Increase aeration in aeration zone to reduce effluent NH_3-N, raise dissolved oxygen in aeration zone, and allow resumption of internal recycle.	

TABLE A.10 pH and alkalinity.

Indicator or observations	Probable cause	Check or monitor	Solutions	Reference
(1) pH is average in influent, but there is low pH in process effluent. Higher effluent total Kjeldahl nitrogen than normal.	(1a) Solids retention time is too high.	(1a) Check pH profile through process and effluent NH_3-N.	(1a) Decrease solids retention time by reducing mixed liquor suspended solids through increased wasting if nitrification is occurring adequately or if nitrification is not desired.	Chapter 3 Chapter 6
	(1b) Insufficient alkalinity to maintain pH.	(1b1) Check mixed liquor suspended solids, waste activated sludge flowrate, and waste activated sludge concentration and calculate solids retention time.	(1b1) Add alkalinity through chemical addition. See Chapters 3 and 6.	
		(1b2) Check effluent alkalinity.	(1b2) Consider implementing anoxic zone if none exists to recover alkalinity through denitrification. The anoxic zone can be created by shutting off aeration in the zone and ensuring good mixing through other means. Internal recycle may be used to maximize alkalinity recovery. See Chapter 6.	
(2) Low pH or alkalinity in influent to biological nutrient removal process.	(2a) Low pH or alkalinity entering the facility.	(2a) Check influent pH and alkalinity.	(2a) See Table A.1.	Chapter 3 Chapter 6 Chapter 12
	(2b) Sidestreams in facility or acidic discharge in facility.	(2b) Check pH and alkalinity in sidestreams.	(2b) Look at facility operations to see if sidestreams are causing low pH or alkalinity and adjust operations to even out pH and alkalinity entering the process. If acid is used in the facility, consider neutralizing before returning to facility. See Chapter 12.	
	(2c) Insufficient alkalinity.	(2c) Check pH and alkalinity in process influent and effluent.	(2c) Add alkalinity through chemical addition. See Chapters 3 and 6.	

TABLE A.11 Toxicity.

Indicator or observations	Probable cause	Check or monitor	Solutions	Reference
(1) Light color and/or high dissolved oxygen in aeration zones. Higher effluent TSS, total Kjeldahl nitrogen, or total phosphorus than normal.	(1) Toxic load inhibited nitrification, leading to overaeration.	(1a) Check pH profile through process, dissolved oxygen in aeration zones, NH_3-N, and NO_2 + NO_3-N through process.	(1a) Maintain as much biomass in areas unaffected by the toxic load through rerouting of RAS, reduction in aeration or mixing, or isolation of basins to preserve biomass for reseeding. Consider an increase in return sludge flowrate for more dilution to the influent that has toxic substance.	
		(1b) Check raw wastewater samples and return streams from solids handling for toxic components.	(1b) Identify and eliminate the source of toxic load.	
(2) Raw or chemical odor in process, pin floc in clarifier effluent or cloudy effluent. Higher effluent TSS or total Kjeldahl nitrogen than normal.	(2) Toxic load has affected the biomass.	(2a) Take sample and look for dead (inactive) microorganisms under the microscope to confirm.	(2a) See solution 1a.	
		(2b) See 1b above.	(2b) See solution 1b.	
(3) For attached growth process, excessive sloughing with little or no biomass on media. Higher effluent TSS or total Kjeldahl nitrogen than normal.	(3) Toxic load has affected the biomass.	(3) See 1b and 2a.	(3) See Solution 1b. If recirculation is being used, turn off recirculation until toxic load has passed; restart recirculation to dilute any toxic residue after the load has passed.	

Index

CPSIA information can be obtained
at www.ICGtesting.com
Printed in the USA
BVOW07s2122130617
486440BV00036B/86/P